Men's Health and Wellbeing

Sanchia S. Goonewardene
Oliver Brunckhorst • David Albala
Kamran Ahmed

Editors

Men's Health and Wellbeing

 Springer

Editors
Sanchia S. Goonewardene
The Princess Alexandra Hospital
Harlow, UK

David Albala
Associated Medical Professionals
Syracuse, NY, USA

Oliver Brunckhorst
King's College London
Guys' Hospital Campus
MRC Centre for Transplantation
London, UK

Kamran Ahmed
King's College London
London, UK

King's College Hospital
London, UK

Sheikh Khalifa Medical City
Abu Dhabi, UAE

Khalifa University
Abu Dhabi, UAE

ISBN 978-3-030-84754-8   ISBN 978-3-030-84752-4   (eBook)
https://doi.org/10.1007/978-3-030-84752-4

This Springer imprint is published by the registered company Springer Nature Switzerland AG
The registered company address is: Gewerbestrasse 11, 6330 Cham, Switzerland

Foreword

Welcome to *Trends in Men's Health*. On behalf of my team and I, it has been a pleasure putting this together for you. This is an important topic for both patients and clinicians alike and was specifically written for you.

Years ago, when I was a fellow at Guy's and St Thomas Hospitals, I came across this topic as a young registrar. Clearly, it is a broad area, encroaching on many areas in medicine. Two years ago, Kamran Ahmed and I realised this book needed to be written. By chance, we ended up on a flight back from America together and the rest, as they say, is history.

We are lucky to have experts from around the world contributing to this book. There are many difficult topics we deal with, from STIs to recreational drug use, medical and surgical topics, all of which can impact on *Trends in Men's Health*. The key to understanding why we do this is to always put the care of your patient first— do this and you will always succeed.

Harlow, UK
London, UK
Syracuse, NY, USA
London, UK

Sanchia S. Goonewardene
Oliver Brunckhorst
David Albala
Kamran Ahmed

Acknowledgements

For my family and friends—always supporting me in what I do.

For my Guardian Angel, one of a kind, for all the times you have saved me, without me knowing, and gone against the rest of the world to do it-good deeds are always rewarded.

For my team at Springer Nature, for always giving me a chance to get published.

Karen Ventii, my most amazing, Harvard-based copyeditor—a really great job done.

For all the amazing clinicians who contributed to this book.

Oliver Brunckhorst—an amazing co-editor.

Prof Albala—for your ongoing help and support.

Kamran Ahmed—first Registrars together at Guy's Hospital, now Consultant and Senior Lecturer!! Truly amazing.

Contents

About the Editors

Sanchia S. Goonewardene MBChB (Hons. Clin. Sc), BMedSc, PGCGC, Dip. SSc, MRCS, MPhil Urology Registrar, Princess Alexandra Hospital HarlowSanchia S. Goonewardene qualified from Birmingham Medical School with Honours in Clinical Science and a BMedSc Degree in Medical Genetics and Molecular Medicine. She has a specific interest in academia during her spare time, with over 727 publications to her credit with 2 papers as the number 1 most cited in fields (Biomedical Library), and has significantly contributed to the Urological Academic World—she has since added a section to the European Association of Urology Congress on Prostate Cancer Survivorship and Supportive Care and is an associate member of an EAU Guidelines Panel on Chronic Pelvic Pain. She has been the UK lead in an EAU-led study on salvage prostatectomy. She has also contributed to the BURST IDENTIFY study as a collaborator.Her background with research entails an MPhil, the work from which went on to be drawn up as a document for PCUK then, endorsed by NICE. She gained funding from the Wellcome Trust for her Research Elective. She is also an Alumni of the Urology Foundation, who sponsored a trip to USANZ trainee week. She also has six books published—*Core Surgical Procedures for Urology Trainees* (ranked third in Book Authorities' 100 Greatest Urology Books of All Time), *Prostate Cancer Survivorship*, *Basic Urological Management*, *Management of Non Muscle Invasive Bladder Cancer*, *Salvage Therapy in Prostate Cancer* and *Muscle Invasive Bladder Cancer*.She has supervised her first thesis with King's College London and Guy's Hospital (BMedSci Degree gained first class, students' thesis score 95%). Recently, she has gained her first Associate Editor position with the *Journal of Robotic Surgery*, and is responsible as Urology Section Editor. She is an editorial board member of the *World Journal of Urology* and was invited to be Guest Editor for a Special Issue on Salvage Therapy in Prostate Cancer.She is also a review board member for BMJ case reports. Additionally, she is on the International Continence Society Panel on Pelvic Floor Dysfunction and Good Urodynamic Practice Panel, is an ICS abstract reviewer and has been EPoster Chair at ICS.More recently, she has been ICS Ambassador and is ICS Mentor. She has also chaired semi-live surgery at YAU-ERUS and presented her work as part of the Young Academic Urology Section at ERUS.

Oliver Brunckhorst MBBS, BSc (Hons), MRCS (Eng)Urology PhD Fellow and South London Urology TraineeMr Oliver Brunckhorst undertook his medical training at Imperial College London where he graduated with distinctions in medical science, clinical science and clinical practice and received the Imperial College Faculty of Medicine Prize for his overall performance in finals. He undertook an intercalated BSc at King's College London in Anatomy, Developmental and Human Biology graduating with first class honours. He was then appointed as a surgical Academic Foundation Doctor at Imperial College London and North West Thames Deanery and subsequently as an NIHR Urology Academic Clinical Fellow at King's College London and South Thames Deanery. Currently, Oliver is taking some time out of clinical training to undertake a PhD at King's College London on Mental Wellbeing and Quality of Life in Prostate Cancer and Survivorship.His interest in academic urology and surgery includes mental wellbeing and quality of life in men's health including malignant and benign pathology, surgical education and curriculum development, simulation-based training, non-technical skills in operating theatres and surgical innovation. He has published numerous original articles and systematic reviews/meta-analyses on both clinical and non-clinical topics in high impact journals. His work has been presented in several national and international urological and surgical conferences, having won several best poster prizes for his work. Additionally, he has authored six book chapters in clinical, education and anatomical subjects, including a chapter in the highly regarded *Gray's Surgical Anatomy* textbook. Oliver is also a keen educator regularly teaching medical students in preparation for their clinical finals, as a clinical skills tutor and lecturer on the Surgical Sciences intercalated BSc programme at King's College London, and as a faculty member in numerous international urology courses.

David Albala MD, Dr. David Albala graduated with a geology degree from Lafayette College in Easton, Pennsylvania. He completed his medical school training at Michigan State University and went on to complete his surgical residency at the Dartmouth-Hitchcock Medical Center. Following this, Dr. Albala was an endourology fellow at Washington University Medical Center under the direction of Ralph V. Clayman. He practised at Loyola University Medical Center in Chicago and rose from the ranks of Instructor to Full Professor in Urology and Radiology in 8 years. After 10 years, he became a tenured professor at Duke University Medical Center in North Carolina. At Duke, he was Co-Director of the Endourology fellowship and Director for the Center of Minimally Invasive and Robotic Urological Surgery. He has over 180 publications in peer-reviewed journals and has authored five books in endourology and general urology. He is the Editor-in-Chief of the *Journal of Robotic Surgery*. He serves on the editorial board for *Medical Reviews in Urology*, *Current Opinions in Urology* and *Urology Index and Reviews*. He serves as a reviewer for eight surgical journals.Presently, he is Chief of Urology at Crouse Hospital in Syracuse, New York, and a physician at Associated Medical Professionals (a group of 29 urologists). He is considered a national and international authority in laparoscopic and robotic urological surgery and has been an active teacher in this area for over 20 years. His research and clinical interests have focused on robotic urological surgery. In addition, other clinical interests include minimally invasive treatment of benign prostatic hypertrophy (BPH) and the use of fibrin sealants in surgery. He has been a visiting professor at numerous institutions across the USA as well as overseas in countries such as India, China, Iceland, Germany, France, Japan, Brazil, Australia, and Singapore. In addition, he has done operative demonstrations in over 32 countries and 23 states. He has trained 16 fellows in endourology and advanced robotic surgery.In addition, Dr. Albala is a past White House Fellow who acted as a special assistant to Federico Pena, Secretary of Transportation, on classified and unclassified public health-related issues.

Kamran Ahmed MBBS, MRCS, PhD, FRCS Urol

(1) Consultant Urological Surgeon & Andrologist at King's College Hospital and Senior Clinical Lecturer (Associate Professor) at King's College London. (2) Consultant Urologist at Sheikh Khalifa Medical City, Abu Dhabi, UAE.

Kamran's areas of interests include core and specialist urology. He is a specialist in andrology (men's health, male sexual/erectile dysfunction, prosthetics, Peyronie's disease), male fertility management, vasectomy reversal and urethral stricture surgery (endoscopic and reconstructive, i.e. urethroplasty). He completed his specialist urology training in London and undertook clinical fellowship training at University College London Hospital in andrology, men's reconstructive surgery and infertility management.

He holds a PhD degree in Surgical Education from Imperial College London and is also an NHS Clinical Entrepreneur. His research interests include surgical education (curriculum and assessment tools development and validation), men's health and survivorship in benign and malignant urological conditions, success of sperm retrieval techniques and management of infertility in non-obstructive azoospermia. He has received a number of awards, and clinical and educational grants, from various organisations, including the European Association of Urology, the British Association of Urological Surgeons, the Royal College of Surgeons, the National Institute for Health Research, Coptcoat Charity and Pelican Group. Mr Ahmed has published widely, with more than 260 peer-reviewed publications and around 300 national and international conference presentations. He is the editor and author of three urological textbooks and has written numerous book chapters. He has served on the Editorial Board of BJUI. He has been the lead for Surgical Sciences iBSc modules at KCL. He is education lead for EULIS section of European Association of Urology (EAU), board member of Junior ERUS-EAU and tutor on the European Urology Resident Education Programme (EUREP). He has been invited as visiting professor or lecturer to institutions in the USA, Europe and the Middle East.

Abbreviations

AAOT	Alcohol assertive outreach team
ABD	Androgen binding protein
ABV	Alcohol by volume
ACTs	Alcohol care teams
ADH	Alcohol dehydrogenase
ADT	Androgen deprivation therapy
AFP	Alpha-fetoprotein
AGE	Advanced glycation end-product
AH	Alcoholic hepatitis, acute alcoholic hepatitis
AIDS	Autoimmune immunodeficiency disease
AII	Angiotensin
A1AT	Alpha 1 antitrypsin
AIN	Anal intraepithelial neoplasia
ALT	Alanine aminotransferase
ALP	Alkaline phosphatase
AJCC	American Joint Committee on Cancer
AS	Androgen suppression
AST	Aspartate aminotransferase
AMA	Antimitochondrial antibodies
AMPK	5′-Adenosine monophosphate-activated protein kinase pathway
5ARI	5-Alpha reductase inhibitors
AR	Androgen receptor
ART	Assisted reproduction techniques
AQP1	Aquaporins
ArLD	Alcohol-related liver disease
ATP	Adenosine triphosphate
AUDIT	Alcohol Use Disorders Inventory Test
AUR	Acute urinary retention
AWS	Alcohol withdrawal syndrome
AZF	Azoospermia factor

5AR	5-Alpha-reductase
b-HCG	Beta subunit of human chorionic gonadotropin
BDI	Beck's Depression Inventory
BEP	Bleomycin, etoposide and platinum
BIS	Body Image Scale
BMD	Bone mineral density
BMI	Body mass index
BOO	Bladder outflow obstruction
BPE	Benign prostatic enlargement
BPH	Benign prostatic hyperplasia
bPFS	Biochemical progression free survival
CABI	Concern about body image
CAP	Controlled attenuation parameter scores
CAYA-T	Cancer Assessment for Young Adults-Testicular
CBAVD	Congenital bilateral absence of the vas deferens
CBT	Cognitive behavioural therapy
CCs	Circadian clocks
CES-D	Center for Epidemiologic Studies Depression Scale
CFTR	Cystic fibrosis transmembrane receptor
CGRP	Calcitonin gene-related peptide
cGMP	Cyclic guanosine monophosphate
CH+PE+TM	Clinical history, plus physical examination, plus tumour markers
CHD	Coronary heart disease
CHT	Chemohormonal therapy
CI	Confidence interval
CIWA-Ar	Clinical Institute Withdrawal Assessment for Alcohol-Revised
CNS	Central nervous system
CMNI	Conformity to Masculine Norms Inventory
CON	Control condition
COPD	Chronic obstructive pulmonary disease
CT	Computerised tomography
CRS	Cytoreductive surgery
CRPM	Colorectal peritoneal metastases
CS	Clinical stage
CP	Child–Pugh score
CRP	C reactive protein
DALYs	Disability adjusted life years
DAMPs	Damage-associated molecular patterns
DBD	DNA binding domain
DEXA	Dual-energy x-ray absorptiometry
DFA	Direct fluorescent antibody
DHEA	Dehydroepiandrosterone
DHT/5a-DHT/5β-DHT	Dihydrotestosterone

DNA	Deoxyribonucleic acid
DRE	Digital rectal examination
DSM-5	Diagnostic and Statistical Manual of Mental Disorders
E2	17β-estradiol
EAU	European Urological Association
EBRT	Brachytherapy
EBV	Epstein–Barr virus
EC	Endothelial cells
ED	Erectile dysfunction
EDs	Endocrine disruptors
EDHF	Endothelium-derived hyperpolarising factor
ELF	Enhanced liver fibrosis
ENI	Pelvic elective nodal irradiation
EORTC-QOL-C30	European Organisation for Research and Treatment of Cancer Quality of Life of Cancer Patients
ERCP	Endoscopic retrograde cholangiopancreatography
ERK	Extracellular signal-regulated kinase
ET	Endothelin
EUS	Endoscopic ultrasound
EIA	Enzyme immunoassay
FACT-G	Functional Assessment of Cancer Therapy
FAP	Familial adenomatous polyposis
FertiQOL	Fertility Quality of Life Questionnaire
FFA	Free fatty acids
FBC	Full blood count
FDG-PET	Fluorodeoxyglucose-positron emission tomography
FRAX	Fracture Risk Assessment Tool
FSE	Frozen section examination
FSH	Follicle stimulating hormone
fT	Free testosterone
GAD-7	Generalised anxiety disorder-7
GAHS	Glasgow alcoholic hepatitis score
GCNIS	Germ cell neoplasia in situ
GCT	Germ cell tumours
GEM	Genetic evaluation of men
GETUG	Groupe d'Etude des Tumeurs Uro-Genitales
GERD	Gastroesophageal reflux disease
GI	Gastrointestinal tract
GIST	Gastrointestinal stromal tumour
GGT	Gamma glutamyl transferase
GnRH	Gonadotropin-releasing hormone
GLUT	Glucose transporter
GORD	Gastro-oesophageal reflux disease
GU	Genitourinary cancers
HADS-D/HADS	Hospital Anxiety and Depression Scale

HB	Haemoglobin
HCPs	Healthcare professionals
HCG	Human chorionic gonadotropin
HCV	Hepatitis C virus
HDCT	High-dose chemotherapy
HDL	High-density lipoproteins
HCC	Hepatocellular carcinoma
HIPEC	Hyperthermic intraperitoneal chemotherapy
HIV	Human immunodeficiency virus
HOLEP	Holmium laser enucleation of the prostate
Holmium YAG	Holmium-yttrium-aluminium-garnet
HPG	Hypothalamic–pituitary–gonadal
HSV/HSV2/HPV	Herpes simplex virus/Herpes virus
HVPG	Hepatic venous pressure gradient
ICD-10	International Classification of Diseases
ICSI	Intracytoplasmic sperm injection
INR	International normalised ratio
IBD	Inflammatory bowel disease
IG	Immunoglobulins
IL	Interleukin
IPSS	The International Prostate Symptom Score
IP3	Inositol triphosphate
IR	Insulin resistance
IVF	In vitro fertilisation
IU	International units
KTP	Potassium-titanyl-phosphate (KTP) greenlight laser
LC-MS/MS	Liquid chromatography-tandem mass spectrometry
LDH	Lactate dehydrogenase
LDL	Low-density lipoproteins
LFTs	Liver function tests
LGV	Lymphogranuloma venereum
LH	Lutenising hormone
LIFT	Ligation of intersphincteric tract
LiSWT	Low intensity shockwave therapy
LKM	Anti-liver-kidney microsomal antibody
LOH	Late-onset hypogonadism
LPS	Lipopolysaccharide
LT	Liver transplantation
LUTS	Lower urinary tract symptoms
MANW	Metabolically healthy and normal weight
MAO	Metabolically abnormal but overweight or obese
MAPK	Mitogen-activated protein kinase
MAX-PC	Memorial Anxiety Scale for Prostate Cancer
MCDI	Masculinity in Chronic Disease Inventory
MD	Mediterranean diet

mDF	Maddrey's discriminant function
MORES	Male Osteoporosis Risk Estimation Score
MELD	Model of end-stage liver disease
MetS	Metabolic syndrome
MET	Metabolic equivalent task hours
MESA	Microsurgical epididymal sperm aspiration
MHO	Metabolically healthy but overweight or obese
MicrTese	Micro testicular sperm extraction
MIS	Müllerian inhibitory substance
miRNA	Micro-RNAs
MRCP	Magnetic resonance cholangiopancreatography
MRI	Magnetic resonance imaging
MRN	Male Role Norms Inventory
MSU	Microscopy, culture and sensitivity of urine
MUSE	Medicated urethral system for erection
NIAAA	National Institute on Alcohol Abuse and Alcoholism
NIDA	National Institute on Drug Abuse
NAAT	Nucleic acid amplification test
NICE	National Institute for Health and Care Excellence
NCCN	National Comprehensive Cancer Network
NAD	Nicotinamide adenine dinucleotide
NEN	Neuroendocrine neoplasms
NICE	National Institute of Clinical Excellence
NHS	National Health Service
NO	Nitric oxide
NOA	Non-obstructive azoospermia
NOS	Nitric oxide synthase
NPV	Negative predictive value
NRT	Nicotine replacement therapy
NSAIDS	Non-steroidal anti-inflammatory drugs
NSCLC	Non-small cell lung cancer
NSGCT	Non-seminomatous germ cell tumour
NES	Nuclear signal export zone
NLS	Nuclear signal localisation
NRS	Nutritional risk screening
OAB	Overactive bladder
OS	Overall survival
PA	Para-aortic
PAE	Prostate artery embolisation
PC/PCA	Prostate cancer
PC-QOL	Prostate cancer-related quality of life scale
PCR	Polymerase chain reaction
PD	Peyronie's disease
PDQ	Peyronie's Disease Questionnaire
PDE5i	Phosphodiesterase 5 inhibitors

PHQ-9	Patient Health Questionnaire-9
PEI	Cisplatin plus etoposide plus ifosfamide
PET	Positron emission tomography
PESA	Percutaneous epididymal sperm aspiration
PFE	Pelvic floor exercise
PKC	Protein kinase C
PLAT	Platelets
$PM_{2.5}$	Particulate matter
PPI	Proton pump inhibitor
PPV	Positive predictive value
PREHAB	Prehabilitation
PREVENT	Prostate Cancer Evidence of Exercise and Nutrition Trial
PGE1	Prostaglandin E1
PGI_2	Prostacyclin
PMN	Polymorphonuclear cells
PMP	Pseudomyxoma peritonei
PTEN	Phosphatase and tensin homolog gene
PSA	Prostate-specific antigen
PSMA	Prostate-specific membrane antigen
PSCS	Post-salvage chemotherapy surgical resection
PT	Prothrombin time
PTC	Percutaneous transhepatic cholangiography
PVR	Post-void residual
PVP	Greenlight laser photovaporisation
QOL	Quality of life
ROS	Reactive oxygen species
RARHA	Reducing alcohol-related harm
RFN-NPT	Royal Free Hospital Nutritional Prioritising Tool
RNA	Retrovirus
RPLND	Retroperitoneal lymph node dissection
RR	Relative risk
RPR	Rapid plasma reagin test
RSES	Rosenberg's Self-Esteem Inventory
RT	Pelvic radiation therapy
RTC	Road traffic collision
SAAG	Serum-to-ascites albumin gradient
SBP	Spontaneous bacterial peritonitis
SCC	Squamous cell carcinoma
SCLC	Small cell lung cancer
SELECT	The Selenium and Vitamin E Cancer Prevention Trial
SERMs	Selective oestrogen receptor modulators
SF/SF12/SF36	Short Form, Short Form 12 and 36
SHBG	Sex hormone-binding globulins
SDF	Sperm DNA fragmentation
SMC	Smooth muscle cells

SSRIs	Selective serotonin reuptake inhibitor
STAI	State-Trait Anxiety Inventory
STDs	Sexually transmitted diseases
SVR	Sustained virologic response
TARN	Trauma Audit and Research Network
TACE	Transarterial chemoembolisation
TAPP	Trans-abdominal pre-peritoneal
TA	TendoAchilles
TC	Testicular cancer
TCA	Tricyclic antidepressant
T2D	Type 2 diabetes mellitus
TESA	Testicular sperm aspiration
TESE	Testicular sperm extraction
TEP	Totally extra-peritoneal
TGs	Triglycerides
THA	Total hip arthroplasty
TIPSS	Transjugular portosystemic shunt
TLR4	Toll-like receptor 4
TNF/TNF-a	Tumour necrosis factor
TNM	Tumour, node and metastasis
TPHA	T. pallidum haemagglutination assay
TPPA	T. pallidum particle agglutination test
TAMIS	Transanal minimally invasive surgery
TRAIN	Resistance training
TRAINPRO	Resistance training and protein supplementation
TRT	Testosterone replacement therapy
TRUS	Transrectal ultrasound
TT	Bound testosterone
TNT	Total neoadjuvant therapy
TURP	Transurethral resection of the prostate
VED	Vacuum erection device
VeIP	Vinblastine, ifosfamide and cisplatin
VEGF	Vascular endothelial growth factor
VEOH	Very early onset hypogonadism
VIP	Vasoactive intestinal peptide
VCA	Viral capsid antigen
WE	Wernicke's encephalopathy
WBC	White blood cells
WHO	World Health Organization
UI	Urinary incontinence
UICC	The International Union Against Cancer
UKELD	United Kingdom Model for End-Stage Liver Disease
UTI	Urinary tract infection
US	Ultrasound
UG	Usual care group

Chapter 1
The Male Reproductive System Anatomy

Georges Mjaess, Fouad Aoun, Thierry Roumeguère, and Simone Albisinni

Mastering male reproductive anatomy seems to be a necessity in order to understand normal reproduction, as well as pathology and treatment options. This chapter aims at describing the anatomy of the male reproductive system, which mainly consists of external and internal male sex organs. External sex organs are the penis, scrotum, testis and epididymis, while internal sex organs are the vas deferens, prostate gland, seminal vesicles and bulbourethral (Cowper's) glands.

1.1 The Penis

A penis measures in average 10 cm long in flaccid state and 15 cm long in erected state. The penis is constituted of three parts: (1) a fixed perineal part (root of the penis); (2) the subpubic part; and (3) a mobile part, the body of the penis, covered with skin. The erectile bodies mainly constitute the majority of the penile section: two corpora cavernosa and one corpus spongiosum, each surrounded by the tunica albuginea, and altogether surrounded by both the Buck fascia and the Colles fascia (Fig. 1.1).

The perineal part of the penis is constituted of the two corpora cavernosa whose insertion is on the anterior two-thirds of the ischio-pubic branches. The sub-pubic part of the penis is attached to the pubis by the suspensory ligament of the penis.

G. Mjaess
Faculty of Medicine, Laboratory of Biomechanics and Medical Imaging, St. Joseph University, Beirut, Lebanon

F. Aoun
Institut Jules Bordet, Brussels, Belgium

T. Roumeguère · S. Albisinni (✉)
Department of Urology, Erasme Hospital, Université Libre de Bruxelles, Bruxelles, Belgium

S. S. Goonewardene et al. (eds.), *Men's Health and Wellbeing*,
https://doi.org/10.1007/978-3-030-84752-4_1

1

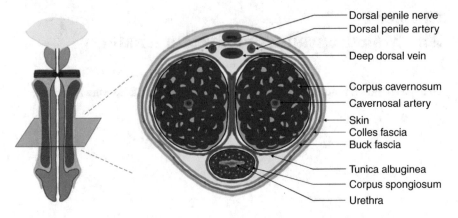

Dorsal penile nerve
Dorsal penile artery
Deep dorsal vein
Corpus cavernosum
Cavernosal artery
Skin
Colles fascia
Buck fascia
Tunica albuginea
Corpus spongiosum
Urethra

Fig. 1.1 The anatomy of the penis

The penile body is covered by a thin layer of skin coupled with a superficial leaflet (penile Dartos). The whole constitutes the penile sheath, underneath of which lies a loose connective tissue that allows for easy sliding between the sheath and the deep layers (Colles, Buck and tunica albuginea).

The corpora cavernosa are two erectile bodies consisting of vascular caverns (sinusoid spaces) lined with endothelium and surrounded by smooth muscle cells. Blood fills these spaces during erection, via the helicine arteries, aided by the relaxation of the surrounding smooth muscle cells. The two corpora cavernosa are separated by a thin permeable septum, and, joined together, form a superior gutter, in which lie the dorsal penile veins, arteries, and nerves.

The corpus spongiosum occupies the inferior gutter formed by the corpora cavernosa. It surrounds the male urethra, which ends at the top of the penile glans with the urethral meatus. The tunica albuginea which covers the corpus spongiosum is thinner than that covering the corpus cavernosum.

The penile glans is also an erectile tissue which forms the distal extremity of the penis and holds a major role in the sensitivity necessary for erection and ejaculation. It has a conical shape with a distal apex. The circumference of the base of the glans, which forms a rounded projecting border, is the corona of the glans penis or the penile crown. The terminal portion of the urethra (navicular fossa) crosses the glans and ends at its lower part through the urethral meatus. The glans is covered by the foreskin (Fig. 1.2). The frenulum is a mucous fold stretched between the lower part of the glans penis and the foreskin.

1.1.1 Vascularization

The erectile bodies and the urethra are mainly vascularized by the terminal branches of the internal pudendal artery (Fig. 1.3). The penile sheath and foreskin are vascularized by both internal and external pudendal arteries. The penile veins drain into

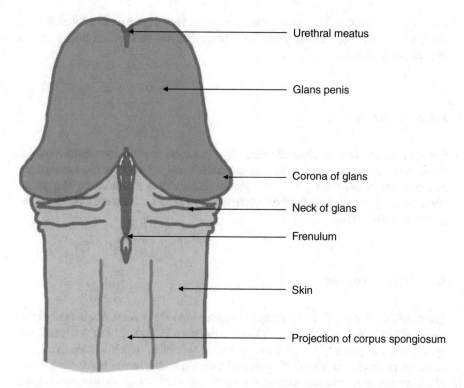

Fig. 1.2 Anatomy of the glans

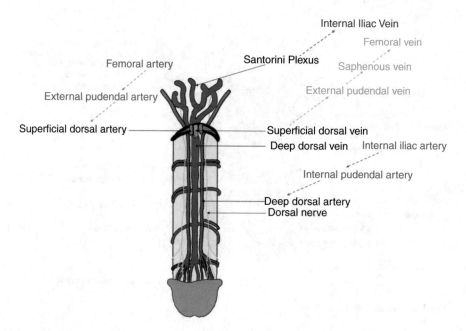

Fig. 1.3 Blood supply to the penis

the deep dorsal veins of the penis, and then in the venous plexus of Santorini or in the superficial veins, then draining into the saphenous vein which finally drains into the femoral vein.

1.1.2   Innervation

Parasympathetic and sympathetic pathways from the inferior hypogastric plexus travel in the cavernous nerves and are responsible respectively for the tumescence and detumescence of the penis. The penile sensitivity is mediated by the dorsal nerve of the penis (a branch of the pudendal nerve) and the genital branch of the genito-femoral nerve.

1.2   The Scrotum

The median raphe divides the scrotum into two separate cavities. Each cavity contains the testis, the epididymis and the vas deferens (Fig. 1.4). The testis and the epididymis are partially covered by the tunica vaginalis, which is a double-layer serosa of peritoneal origin. Tunica vaginalis covers all the lateral border and a part of the medial border of the testis and it continues cranially by the closed vestigial part of the saccus vaginalis of the peritoneum.

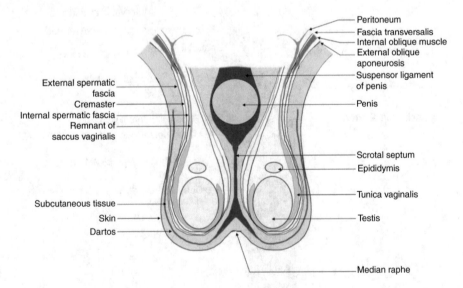

Fig. 1.4 Anatomy of the scrotum

The bursae are constituted by an evagination of the abdominal wall (all the constituent elements of this wall are thus found) from the depth to the surface:

- Internal spermatic fascia: an expansion of fascia transversalis
- Cremaster: a muscular tunica, expansion of the internal oblique and transverse muscles
- External spermatic fascia: an expansion of the external oblique muscle
- Subcutaneous tissue: an expansion of fascia superficialis containing superficial nerves and vessels
- The skin (thin and wrinkled) coupled to the dartos, which is a smooth muscle, conjunctive and elastic tissue.

1.3   Testis

The testis is a transversely flattened ovoid with a long oblique axis downwards and backwards. Its surface is smooth, and its consistency is firm and regular. It is, on average, 4 to 5 cm long, 2.5 cm thick and weighs 20 g.

It has:

- 2 edges, dorso-cranial and ventro-caudal,
- 2 poles, cranial and caudal.

It is capped, like a helmet crest, by the epididymis that extends along its dorso-cranial edge.

It is surrounded by a resistant envelope, the tunica albuginea, which sends septa inside the testicle, segmenting it into lobules ($n = 300$) that contain the seminiferous tubules, where the spermatogenesis takes place. The albuginea presents a thickening especially localized at the ventral part of the dorso-cranial edge: the mediastinum testis which contains the rete testis.

It has embryonic remains:

- Appendix testis (or sessile hydatid)
- Appendix epididymis (or pedicled hydatid)

The seminiferous tubules continue by the straight tubules, which form together the rete testis. The rete testis gives rise to the efferent ductules which join the proximal part of the epididymis (Fig. 1.5).

1.3.1   Vascularization

The testicular arteries are a pair of vessels that branch directly from the abdominal aorta. They rise anterolaterally, caudal to the renal vessels, at the level of L1-L2 vertebra (Fig. 1.6).

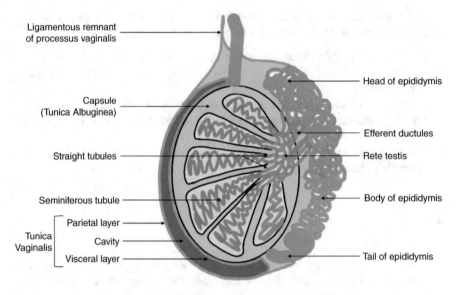

Fig. 1.5 Anatomy of the testis

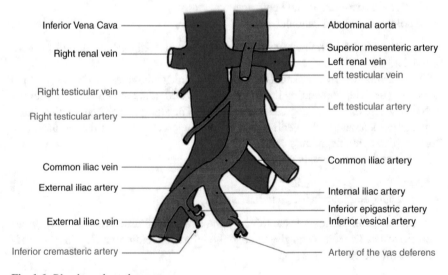

Fig. 1.6 Blood supply to the testes

The right testicular artery crosses the inferior vena cava anteriorly and trips infe-riorly and laterally, internally to the right testicular vein and the proximal ureter. On the body of the psoas muscle, it also crosses the ureter anteriorly and continues its distal course. The left testicular artery also travels internally to the testicular vein. Yet, it has a more vertical course than the right testicular artery proximally. It also crosses the left ureter anteriorly. Both testicular arteries pass lateral to the common and external iliac vessels, and as they enter the deep inguinal ring, they cross the external iliac vessels. Inside the inguinal canal, they course laterally to the vas deferens. The testicular artery in the scrotum gives rise to a branch for the epididymis before splitting into medial and lateral branches, which further divide to perforate the testis.

Vascularization of the testis is also assured by other arterial territories anastomosed with the testicular arteries: the cremasteric artery (arising from the inferior epigastric artery which in its turn arises from the external iliac artery) and the artery of the vas deferens (arising from the inferior vesical artery which in its turn arises from the anterior branch of the internal iliac artery).

As for the venous drainage, a venous network of veins called the pampiniform plexus is formed around the testis and travels cranially in order to form four branches. At the deep inguinal ring, every two branches fuse forming two testicular veins which travel alongside the testicular artery. Shortly thereafter, they fuse to form a single testicular vein. The right testicular vein drains directly into the inferior vena cava, while the left testicular vein drains perpendicularly into the left renal vein.

1.4   Epididymis

The epididymis is the place of storage and maturation of spermatozoa. It covers the posterior upper edge and the lateral facet of the test is like a helmet crest. It has an elongated shape downwards and backwards, and is formed by three segments: a head, a body and a tail. It has a 5 cm length and 1 cm width.

The head has a rounded shape and adheres to the cranial pole of the testicle. It is connected to the rete testis by the efferent ductules. The body runs along the posterior upper edge and the tail continues through the vas deferens.

1.4.1   Vascularization

The epididymis is vascularized, similarly to the testis, by the testicular, cremasteric, and vas deferens arteries.

1.5   Vas Deferens

The vas deferens is a 40 cm duct that links epididymis to ejaculatory canal, and has an external diameter of 2 mm and an interior diameter of 0.5 mm. Its origin is at the posterior superior border of the testis and constitutes a continuation of the tail of epididymis.

The vas deferens follows a sinuous pathway:

- Inside the testicular bursae, it has two portions:
  - The epididymo-testicular portion which is flexuous and medial to the epididymis at the superior border of the testis
  - The funicular portion which goes vertically into the inguinal ring
- In the inguinal canal, the vas deferens is an essential element of the spermatic cord
- At the pelvic inlet, it continues by the short iliac portion when exiting the deep inguinal ring and forms a concave curve in the fatty tissue of the Bogros space (Fig. 1.7).

- The Bogros space is the space formed by the fascia transversalis at the front, the peritoneum from top and behind, and the iliac muscle fascia at the bottom. At this level, the elements of the spermatic cord split up: the spermatic artery and the anterior spermatic plexus move away towards the posterior wall, the posterior plexus drains into the epigastric vein, and only the deferential artery accompanies the canal. The vas deferens enters in contact mainly with the epigastric artery at this level.
- Finally, in the pelvis, the vas deferens has two portions:
  - A first segment lateral to the bladder, which is located on the ceiling of the lateral part of the Retzius cavity. It goes in contact successively with the obturator pedicle, the bladder, the umbilical artery, and the ureter. This segment has a variable inclination angle depending on the bladder repletion.

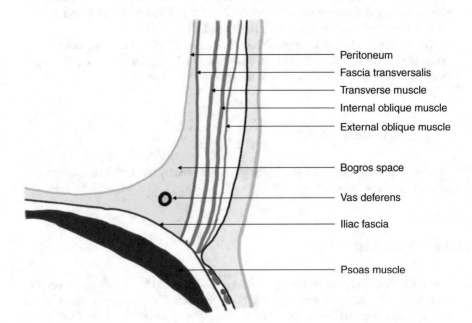

Peritoneum
Fascia transversalis
Transverse muscle
Internal oblique muscle
External oblique muscle

Bogros space

Vas deferens

Iliac fascia

Psoas muscle

Fig. 1.7 Anatomy of the Bogros Space

- A second retrovesical segment, the last portion that joins the seminal vesicle's neck, thus forming together the ejaculatory duct. Here, the vas deferens travels behind the base of the bladder, first in the genital fold formed by the peritoneum between the vesico-genital pouch and the Douglas pouch, and then in the Denonvilliers aponeurosis (Fig. 1.8).

The epididymo-testicular portion of the vas deferens is not covered by the tunica vaginalis and is mainly extra-vaginal. The tunica vaginalis only covers approximately 1 cm of the anterior facet of the initial part of the cordon.

1.6   Prostate

The prostate is located in the minor pelvis, beneath the bladder. It surrounds the prostatic urethra and is crossed by the ejaculatory ducts (Fig. 1.9). It has a chestnut shape with an upper base, a lower apex, a lateral-ventral facet and a dorsal facet. Its mean normal weight in a healthy young man is 15–20 g and it has 3 cm of length, 4 cm of width, and 2 cm of depth.

The base of the prostate is divided to an anterior slope supporting the bladder base and a posterior slope which is crossed by a median groove separating two lateral lobes (defined at the direct rectal examination). The prostatic apex and the membranous urethra are surrounded by the striated sphincter.

The prostate consists of a glandular, muscular and fibrous tissue. It is surrounded by a fibrous capsule. Its boundaries are the following:

- Anteriorly: the retropubic space (Retzius space), the venous plexus of Santorini, and the striated urethral sphincter; the prostate also is fixed anteriorly by the pubo-prostatic ligaments.

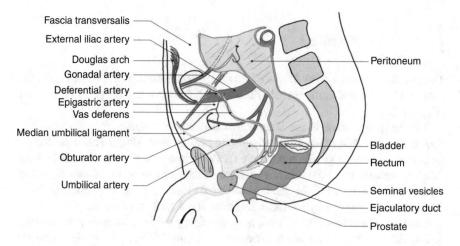

Fig. 1.8 Pelvic anatomy

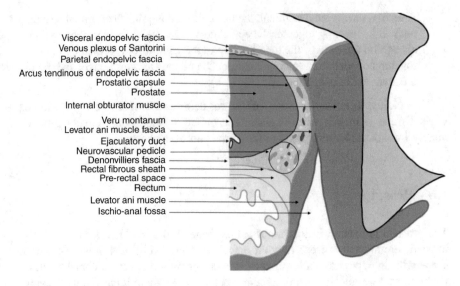

Visceral endopelvic fascia
Venous plexus of Santorini
Parietal endopelvic fascia
Arcus tendinous of endopelvic fascia
Prostatic capsule
Prostate
Internal obturator muscle
Veru montanum
Levator ani muscle fascia
Ejaculatory duct
Neurovascular pedicle
Denonvilliers fascia
Rectal fibrous sheath
Pre-rectal space
Rectum
Levator ani muscle
Ischio-anal fossa

Fig. 1.9 Prostate anatomy

- Posteriorly: the Denonvilliers's fascia and the ano-rectal angle. The posterior facet of the prostate is touchable by the direct rectal examination.
- Laterally: the endopelvic fascia and the levator ani, separating it from the obturator foramen.
- Caudally: the pelvic floor
- Cranially: the bladder.

The vasculo-nervous prostatic pedicles and the cavernous nerves pass through the postero-lateral angle of the prostate.

1.6.1 Zonal Anatomy

The prostate was divided by McNeal into five zones: the peripheral zone, the central zone, the transitional zone, anterior fibromuscular stroma, and the periurethral zone (Fig. 1.10).

The peripheral zone is the largest prostate zone accounting for approximately 70% of the prostatic gland volume in young healthy individuals. This zone constitutes the major zone in which prostatic cancer originates. Peripheral zone is separated posteriorly from the anterior rectal wall and middle rectal vessels by a thin capsule which is a part of the gland and located closest to rectum, permitting therefore to feel the prostate by the index finger on the digital rectal exam.

The central and transitional zones are collectively referred as the central prostatic gland. The transitional zone is the deepest zone, implicated in the benign prostatic

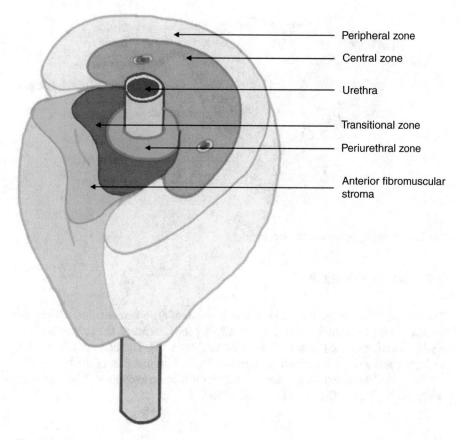

Fig. 1.10 Prostate zonal anatomy

hyperplasia, and forms 5% of the glandular prostate; whereas the central zone comprises up to 25% of the glandular tissue.

The anterior fibro-muscular stroma is a thick tissue which is contiguous with the detrusor muscle of the bladder and does not contain glandular tissue. It is thought to be with no clinical, functional or pathological importance.

Finally, periurethral zone accounts for <1% of the prostatic gland and is also implicated in the benign prostatic hyperplasia.

1.6.2   Vascularization

The prostate is vascularized by the inferior vesical artery (which gives rise to the vesiculo-deferential arteries and vesico-prostatic arteries) and the pudendal artery.

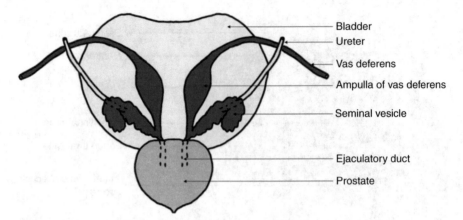

Fig. 1.11 Anatomy of the seminal vesicles

1.7 Seminal Vesicles

The seminal vesicles are a paired symmetric organ of the spermatic ducts. They are tortuous bumpy channels located in the minor pelvis, between the rectum and the bladder in front of the fascia of Denonvilliers (Fig. 1.11). The body of the seminal vesicles continues with the neck which ends in the terminal part of the bulb of the vas deferens. The length of the seminal vesicles is 6 cm in average and their greatest thickness is 15 mm. The capacity of each one is 4–6 cc.

1.8 Bulbourethral Glands

Bulbourethral glands, also called Cowper's glands, are pea-shaped glands located beneath the prostate. These glands have a length of approximately 1 cm, with ducts emptying into the urethra. They are formed by small tubules, encircled with fibrous and elastic tissue. They secrete a fluid which is added to the semen; its role consists of the lubrification and the washing out of the urethra before ejaculation.

Chapter 2
Physiology of Male Hormones

Juan Gómez Rivas, Aritz Eguibar, Jose Quesada, Mario Álvarez-Maestro, and Diego M. Carrion

2.1 Introduction to Male Sex Hormones

The testicle has two primary functions: endocrine (production of hormones) and exocrine (sperm production) 85–90% of the interior volume testicular is made up of seminiferous tubules and their germinal epithelium, place of sperm production (10–20 million gametes per day), and only 10–15% is occupied by the interstitium, where testosterone is produced (Jockenhövel and Schubert 2007).

2.2 Hypothalamic and Pituitary Hormones

The testicular function is controlled by the so-called axis hypothalamus-pituitary-testicular (Fig. 2.1). The hypothalamus gonadotropin-releasing hormone (GnRH) is secreted in the hypothalamus and stimulates hormonal production of the follicle-stimulating hormone (FSH) and luteinizing hormone (LH) by the anterior lobe of the pituitary gland (the adenohypophysis) (Hayes et al. 2001).

Numerous neurotransmitters modulate GnRH secretion and rhythm (Fig. 2.1). Alpha-adrenergic impulses stimulate GnRH secretion. Norepinephrine and prostaglandins increase hypothalamic secretion. Beta-adrenergic and dopaminergic impulses have an inhibitory action on the GnRH secretion. Endorphins, testosterone, progesterone, and prolactin, secreted in stressful situations, decrease GnRH secretion. GnRH is released by the hypothalamus in a pulsatile manner, with peaks

J. G. Rivas (✉)
Department of Urology, Clinico San Carlos University Hospital, Madrid, Spain

A. Eguibar · J. Quesada · M. Álvarez-Maestro · D. M. Carrion
Department of Urology, La Paz University Hospital, Madrid, Spain

© The Author(s), under exclusive license to Springer Nature Switzerland AG 2022
S. S. Goonewardene et al. (eds.), *Men's Health and Wellbeing*,
https://doi.org/10.1007/978-3-030-84752-4_2

every 90–120 min. This type of release is essential for the stimulatory effect of gonadotropin secretion. Continued GnRH administration would curb the pituitary discharge. The amplitude and frequency of the GnRH pulses modulate FSH and LH levels secreted by the anterior pituitary and, subsequently, the gonadal function (Hayes et al. 2001; Morales et al. 2004).

Pituitary hormones stimulate testicular functions: exocrine and endocrine. On the other hand, and due to the negative feedback process, hormones produced in the testis exert inhibitory effects on FSH secretion and LH (Table 2.1).

Pituitary LH stimulates testosterone production by Leydig cells located in the testicular interstitium by binding to specific receptors. LH release is a process discontinuous and occurs, mainly, during the night and in a pulsatile way, at intervals of about 90 min. It corresponds to the pulsatile secretion of GnRH. The available levels of this hormone will determine the amount of secretion of testosterone. But in turn, testosterone levels exert a reciprocal effect by inhibiting the LH production in the pituitary gland through two mechanisms (Table 2.2) (Vignozzi et al. 2005; Vermeulen 2003):

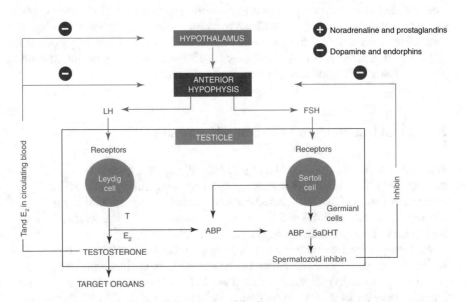

Fig. 2.1 Control of the hypothalamic-pituitary-testicular axis (*LH* luteinizing hormone, *FSH* follicle stimulant hormone, *ABD* androgen binding protein, *5a-DHT* dihydrotestosterone)

Table 2.1 Resume of the hypothalamic-pituitary-testicular axis

1. GnRH is secreted in the hypothalamus in regular amounts every 90–120 min.
2. The hypothalamus controls testicular function through GnRH by stimulating the pituitary hormones LH/FSH.
3. LH regulates and stimulates testosterone biosynthesis in Leydig cells, located in the testicular interstitium.
4. FSH stimulates spermatogenesis by acting on Sertoli cells, located in the seminiferous tubules.

Table 2.2 Hypothalamic-pituitary testicular retro-control

1. Testosterone has a weak negative feedback effect on the adenohypophysis, which results in a decrease in LH secretion. 2. On the other hand, testosterone directly inhibits GnRH secretion in the hypothalamus, causing a decrease in LH gonadotropin in the adenohypophysis, which will reduce the production of testosterone in the Leydig cells. Most of the inhibition of male hormone secretion is attributed to this feedback mechanism.	1. Testosterone exerts a depressant effect on the hypothalamic and pituitary production of gonadotropins (FSH and LH). 2. Estradiol exerts depressant effects on the hypothalamus and pituitary. 3. FSH stimulates the production of several proteins in Sertoli cells, such as inhibin, necessary for feedback control, which slows or suppresses FSH production.

A low testosterone concentration allows the hypothalamus to increase GnRH secretion, which stimulates the release of FSH and LH and thereby increases testosterone. In addition, the testicle can metabolize testosterone to estradiol through the flavouring enzymes present in the tubules and interstitium (Kaufman and Vermeulen 2005; Morley 2003). Estradiol, in physiological concentrations, also decreases the frequency and amplitude of the pulses of LH (Hayes et al. 2001).

2.3   Androgens

Male sex hormones or androgens induce the development of primary sexual characteristics in the embryo and secondary sexual characteristics in puberty. They are responsible for the general growth and protein synthesis that is reflected in the skeletal and muscular changes characteristic of the man. They are synthesized mainly in the Leydig cells of the testes, to a lesser extent in men's adrenal cortex and in women, in minute quantities, in the ovary. Like the rest of the steroid hormones, androgens are synthesized from cholesterol (Fig. 2.2).

The most important androgens are testosterone (Fig. 2.3), androstenedione, and dehydroepiandrosterone (DHEA), a precursor to the rest of the androgens and to estradiol.

The amount of testosterone is higher than the others, so it can be considered the most important testicular hormone. However, most of it is converted in the effector tissues into dihydrotestosterone (DHT), a more active hormone, as it has a greater affinity for the intracellular androgen receptor (AR) (Kelly and Jones 2013).

There is a considerable fraction of the testosterone produced that binds to albumin or sex hormone-binding globulins (SHBG) to be transported by the bloodstream, being its free concentration in serum minimal. In this way, it remains inactive until binding with its specific receptor (Heinlein and Chang 2002). The factors responsible for this transfer and how the cell causes the dissociation of the hormone-globulin complex are unknown. Still, it is believed to depend on the plasma concentration of the hormone (Heinlein and Chang 2002). Once in the cell,

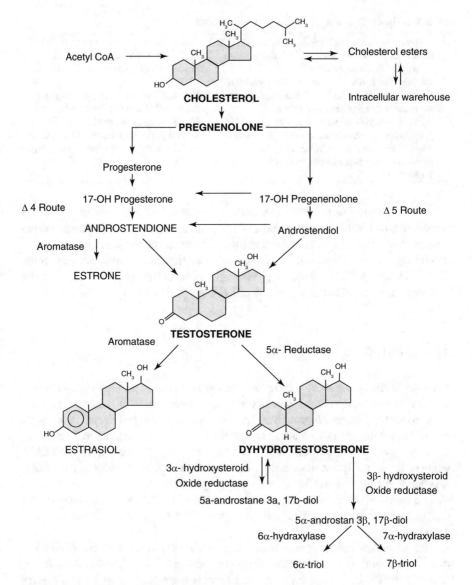

Fig. 2.2 Route of synthesis of sex hormones

the hormone binds to its receptor, located in the cytoplasm and/or in the nucleus, dimerization of the receptor is induced and its consequent activation.

AR, like those of other steroid receptors, is composed of several functional domains (Evans 1988) (Fig. 2.4):

1. The regulatory and binding domain of steroids is located in the C-terminal domain. It has several phosphorylation sites and is involved in the activation of the hormone-receptor complex.

Fig. 2.3 Structure of testosterone

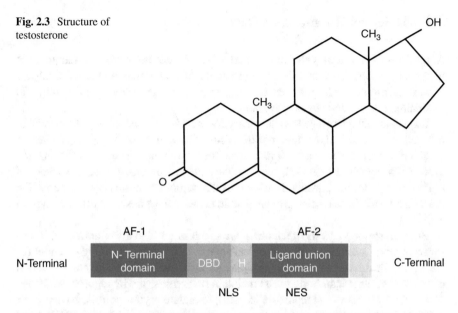

Fig. 2.4 Androgen receptors. Functional domains: A hypervariable N-terminal region, regulating transcriptional activity, a highly conserved central DNA-binding domain (DBD), a hinge region, and a long C-terminal, ligand-binding domain. Ligand-independent binding sites (AF-1, ligand-dependent (AF-2), nuclear signal localization (NLS), and nuclear signal export zone (NES) are also shown

2. The DNA-binding domain is found in the middle part and is essential for the activation of transcription. This portion is responsible for controlling which gene will be regulated by the receptor.
3. The hinge region is located between the two previous domains; it contains an important signal area for receptor movement into the nucleus after synthesis in the cytoplasm. It is a variable hydrophilic region in the different receptors.

The interaction of male sex hormones with the AR produces genomic effects, among which is the activation of the Mitogen-activated protein kinase (MAPK) and other transcription factors that induce the growth and proliferation of different cells (Geraldes et al. 2002).

There are also membrane ARs that induce non-genomic effects as they are not blocked by inhibitors of gene transcription (Gerhard and Ganz 1995; Farhat et al. 1996). Among other effects, there is an increase in the concentration of intracellular Ca^{+2} due to an increase in the formation of Inositol trisphosphate (IP3) (Estrada et al. 2000) or the phosphorylation of the Ras/Raf/extracellular signal-regulated kinase (ERK) 1/2 (Estrada et al. 2003). Through this non-genomic action, testosterone is capable of exerting a regulatory effect on vascular tone as it is capable of regulating the intracellular Ca^{+2} concentration. Likewise, testosterone can regulate the vasodilator effect of neurotransmitters, such as nitric oxide (NO) and Calcitonin gene-related peptide (CGRP) (Perusquía 2003; Isidoro et al. 2018).

2.4   Male Sex Hormones, Effects

At the sexual level, it plays a fundamental role in the development and maintenance of sexual characteristics and the male sex glands' development and functioning. As a "sex" hormone, androgens act on the central nervous system, stimulating and maintaining desire and sexual motivation.

It appears that testosterone is necessary for the normal functioning of the mechanism of ejaculation and maintenance of spontaneous erections. Its positive influence on erectile response is also known. Testosterone stimulates the activity the enzyme nitric oxide synthetase (NOS), which contributes to maintain adequate levels of nitric oxide (NO) in the smooth muscle of the corpora cavernosa of the penis. On the other hand, it has been proven that it favours the phosphodiesterase type 5 activity.

But testosterone and its metabolites are much more than a sex hormone; it performs numerous important physiological actions in the body, resulting essential for the overall health of men. Androgens play an important role in activating function cognitive; increase lean body mass; maintain bone mass (the hypogonadism is one of the main causes of osteoporosis in men); stimulate erythropoiesis; have a clear effect on lipids: improves the concentration of high-density lipoprotein (HDL) and decreases the concentration from low-density lipids (LDL); promotes cardiovascular health; even current evidence refers to an increase in life expectancy (Table 2.3).

2.5   Male Sex Hormones and Genitalia

The sex chromosomes determine if the primitive gonad is due to differentiate towards teste or ovary. Until the seventh week of gestation, the primitive gonad is common to both sexes. Subsequently, anatomical and physiological differentiation occurs that will determine the phenotype of female or male.

Testicular secretions determine the masculine character of the genitalia, both external and internal. Without this type of secretion, genetically the sex would be female, there would be no phenotypic differentiation towards male. The control of the formation of the male phenotype is due to the action of several hormones (Melmed and Jameson 2018):

1. Anti-Müllerian hormone, secreted by fetal testes, inhibits the development of Müllerian ducts (which would lead to the development of female internal genitalia).
2. Testosterone converts Wolf's ducts into the epididymis, the vessels deferens, and in the seminal vesicles.
3. DHT is later synthesized from fetal testosterone, induces the formation of the male urethra and prostate, and the fusion of the midline and elongation of the male external genitalia.

Table 2.3 Resume of the male sex hormone effects in different organs and systems

Target tissue	Active steroid	Effect
Wolff's duct	Stimulates growth and differentiation	Testosterone
External genitalia	Masculinization and growth	DHT
Urogenital sinus	Masculinization and growth	DHT
Bones	Closure of the epiphyses, anabolic effect	Estradiol/testosterone
Larynx	Growth and lengthening of the vocal cords	Testosterone/DHT
Skin	Stimulates fat production Stimulates hair growth Body and face Decreases hair growth (androgenic alopecia)	DTH
Kidneys	Stimulates the production of Erythropoietin	Testosterone/DHT
Liver	Induce enzymes, influence protein synthesis	Testosterone or DHT
Lipid metabolism	↑ HDL-cholesterol, ↓ LDL-cholesterol	Testosterone or DHT
Bone marrow	Stimulates erythropoiesis	Testosterone/DHT
Musculature	Anabolic effect	Testosterone
Testicle	Stimulates and maintains Spermatogenesis	DHT/estradiol
Prostate	Stimulates its growth and function	DHT/estradiol
Breast	Growth inhibition	Testosterone/DHT
Pituitary	Negative retro-control of the gonadotropin secretion	Testosterone/DHT
Hypothalamus	Negative retro-control of the GnRH secretion	DHT
Brain	Psychotropic effects, including on Libido	Testosterone/DHT/ Estradiol

Androgen deficiency during sexual differentiation, which occurs between weeks 9 and 14, gives rise to phenotypically intersex states, with the absence of masculinization and more or less ambiguous external genitalia. The deficits in later stages can condition abnormal development of the penis (micropenis) and abnormal testicular positioning. It is unknown how testosterone secretion is controlled in the embryo, although it seems that it can be regulated by LH and also receives influences from the placental choriogonadotropin hormone.

2.6 Testosterone at Puberty

In the prepubertal age, the levels of gonadotropins and sex hormones are low. At 6 or 7 years, the adrenarche begins, partly responsible for the growth of prepubertal and early armpit and pubic hair, but male sexual characteristics are not fully developed until puberty. Some of the organs involved in the appearance of puberty are:

1. The hypothalamic-pituitary axis.
2. The testes.

Table 2.4 Changes in puberty due to the effect of androgens

1. Thickening and pigmentation of penis and scrotum.
2. Growth of testes, penis, and scrotum.
3. Muscular development, especially in the pectoral and shoulders region.
4. Aggravation of the voice as a consequence of the elongation of the vocal cords.
5. Facilitation of the bone maturation and progressive closure of the growth cartilage.
6. Haematocrit Increase.
7. Decrease of HDL cholesterol.
8. Appearance and darkening of the axillary and pubic hair that already began in the adrenarche.
9. Psychological changes including libido increase and sexual function.
10. Maturation of the Leydig cells and beginning of the spermatogenesis.

3. The adrenal glands.
4. Other not well-known factors.

Anatomical and functional changes at puberty are predominantly a consequence of the effect of testicular androgens. These changes are summarized in Table 2.4. Complete maturation is achieved between 16 and 18 years, in 90% of cases, although the hair changes can continue until the second decade of life. If testosterone deficiency appears after birth but before puberty, virilization is poor or does not occur, this results in eunuchoid phenotypes and also a partial or total deficit of reproductive capacity (Melmed and Jameson 2018).

2.7 Testosterone in Adulthood

The effects of testosterone on adult men can be of three types:

1. Permanent and irreversible, which do not return even if there is a posterior androgenic deficiency (e.g., the severity of the voice).
2. Reversible, directly dependent on the continued secretion of androgens (e.g., influence on the production of erythropoietin, the maintenance of haemoglobin, as well as sexual function and libido).
3. Mixed (e.g., influence on spermatogenesis, consistency, and testicular size).

Over time, the sustained deficit of androgens gives rise to the manifestations known clinically as androgen deficiency syndrome in the adult male, whose predominant manifestations can be seen in Table 2.5.

2.8 Male Sex Hormones and the Cardiovascular System

Several population studies have shown a higher frequency of cardiovascular mortality in patients with low plasma testosterone levels. Studies reveal that patients with advanced cardiovascular disease have decreased plasma testosterone levels (Araujo

Table 2.5 Androgen deficiency syndrome in the adult male

1. Loss of libido
2. Erectile dysfunction.
3. Weakening of facial, axillary, and pubic hair.
4. Gynecomastia.
5. Hot flushes.
6. Decreased bone density, osteoporosis.
7. Decrease in muscle mass, body fat increase.
8. Normochromic normocytic anaemia due to erythropoietin deficiency

et al. 2005). In addition, cardiovascular diseases such as hypertension or atherosclerosis are more common in men during testosterone replacement therapy (Pongkan et al. 2016). Likewise, critical cardiovascular changes have been observed in men with prostate cancer who underwent androgen deprivation (Chou et al. 2015). All these data indicate that male sex hormones, specifically testosterone, has a fundamental role in cardiovascular health.

The basic organization of the blood vessel wall consists of three concentric layers:

1. The innermost layer, the tunica intima, is made up of one layer of endothelial cells (EC).
2. The middle layer, the tunica media, is made up of smooth muscle cells (SMC).
3. The adventitial layer, composed of collagen fibres, fibroblasts, and some sparse elastic and muscular fibres.

The tunica intima and the middle layer are separated by the internal elastic lamina, while the tunica media and the adventitial layer are separated by the external elastic lamina (Bohr et al. 1980).

The vasodilation effect of testosterone on coronary and peripheral circulation has been evaluated in animals and human trials and conclusion was that testosterone enhances the action of NO in the endothelium, inducing vasodilation (Vela Navarrete et al. 2009). The first experimental evidence was published by Yue et al. the study demonstrated the vasodilator effect of testosterone in the rabbit aorta and coronary arteries (Yue et al. 1995). Subsequently, numerous studies showed that testosterone acts as a direct vasodilator in human arteries through a non-genomic effect, independent of the ligand-dependent function of intracellular receptors and of protein transcription and synthesis (Rowell et al. 2009).

The biological actions of androgens are mainly mediated through binding to the AR, the classical genomic pathway. Androgen binding to the AR results in an AR conformational change that promotes the dissociation of chaperone proteins and facilitates receptor dimerization, nuclear transportation, phosphorylation, and deoxyribonucleic acid (DNA) binding. Upon the recruitment of co-regulators and general transcription factors, the transcription of target-genes is either induced or inhibited, leading ultimately to changes in androgen-target gene expression and cellular or biological structures and functions. This process usually takes hours before, resulting in biological changes in target cells (Cai et al. 2016).

The non-genomic effects of androgens are characterized by a rapid and reversible vasodilator effect on the vascular endothelium. The mechanism starts in the cell membrane when androgens binding to the AR on T lymphocytes, macrophages, or osteoblasts (Walker 2003), interacting between *G-protein coupled receptors* and *ion channels* (Pi et al. 2010). Most of the studies highlight the non-genomic vasodilator effect produced by testosterone; however, more recent studies have been reported increased vasorelaxation of testosterone metabolites, specifically 5β-DHT, in experimental animal models. Thus 5β-DHT is a potent androgen with a strong affinity for the intracellular AR, whereas its 5-isomer (5-DHT), which does not bind to the AR, is totally devoid of androgenic properties but is highly efficacious in producing vasorelaxation (Perusquía and Stallone 2010).

Other studies on in vitro experimental studies have found that DHEA rapidly increased the expression of endothelial NOS (eNOS) and the activity of ERK 1/2 via the nongenomic pathway, which may increase NO secretion, leading to an increased flow-mediated dilation in vivo (Williams et al. 2004).

In addition to the direct vasodilator effect of androgens on the vascular endothelium, synthesis of different factors such as prostanoids, reactive oxygen species (ROS), or protein kinase C (PKC) have been implicated as important mechanisms that contribute to the regulation of vascular tone. In this way, the loss of gonadal function increases the production of prostanoids and decreases the bioavailability of NO, which can increase both systemic vascular resistance and vascular tone (Montaño et al. 2008).

Hypertension is a significant public health problem, being a direct cause of different cardiovascular diseases such as cerebrovascular disease, coronary heart disease, kidney disease, and peripheral vascular disease. Regarding the aetiology, genetic factors, age and sex, obesity, insulin resistance, alcohol, and high salt intake have been directly related to hypertension. Regarding the pathogenesis, arterial hypertension is characterized by alterations in vascular function and structure, including endothelial dysfunction, increased vasoconstrictor responses, and increased wall-vascular lumen ratio (Briones et al. 2003).

The vascular endothelium is considered an endocrine organ involved in vasoactive, metabolic, and immune processes. It is composed of endothelial cells which responses specifically to the physical and chemical conditions of the environment. The main *endothelium-derived vasoactive substances* are described in Table 2.6.

Androgens have been shown to regulate cell proliferation and function via either a classical genomic pathway or nongenomic pathway in endothelial cells from a variety of origins (Estrada et al. 2000). Androgen diffuses into the cell, directly

Table 2.6 Endothelial-derived vasoactive factors

Vasodilator factors	Vasoconstrictor factors
Nitric Oxide (NO)	Endothelin (ET)
Endothelium-derived hyperpolarizing factor (EDHF)	Prostanoids (PGH_2, TXA_2, O_2)
Prostacyclin (PGI_2)	Angiotensin (AII)
Acetylcholine	
Bradykinin	

activating a cascade of signalling creatine kinase and MAPK. The AR ligand may upregulate Vascular Endothelial Growth Factor (VEGF) and cyclins through the genomic pathway. Androgen may also induce eNOS synthesis. These mechanisms cause cell proliferation. The effects of androgens on endothelial cells proliferation and function may be a significant factor mediating the beneficial actions of androgens in the cardiovascular system in males and may explain findings that low levels of circulating androgens are associated with increased cardiovascular morbidity and mortality in males (Perusquía 2003).

2.9   Male Sed Hormones and muscular/Bone Metabolism

In the past decades, the importance of female and male sex hormones for skeletal muscle and bone health has become recognized. Although there are multiple sex hormones, those that have been studied the most are estrogen and testosterone. Both estrogen and testosterone are present in men and women, and both hormones exert direct and indirect effects on skeletal muscle and bone. Ageing results in a highly significant loss of testosterone in women and men. Men start losing testosterone continuously throughout life, at the beginning of their third decade. Indeed, many men are hypogonadal by the eighth decade, with free testosterone levels below 320 pg/dl, the accepted minimum (Elmlinger et al. 2003). Women become postmenopausal typically by the sixth decade, thus spending approximately one-third of their lifetime in an estrogen-deficient state (Davison et al. 2005). In young men and women, there are several conditions that cause sex hormone levels to drop to nearly undetectable levels, such as trauma, spinal injury, brain injury, and bed rest. There is emerging evidence that a sedentary lifestyle and associated obesity are associated with low sex hormone levels in men. The long-term consequences of low hormone levels at a young age have yet to be clearly defined. Due to sex hormones are markedly reduced with age and life expectancy is increased, there has been recent interest in restoring hormone levels to "normal" levels in ageing men and women. As expected, bringing testosterone levels above 320 pg/dl in hypogonadal older adults has an anabolic effect on skeletal muscle. Significant gains in muscle mass and strength have been realized; however, testosterone hormone replacement in older men is not without penalty. Likewise, providing estrogen to older women has an anabolic effect on bone and possibly muscle, but there may be negative consequences of giving estrogen to women in their 60–70 s decade.

2.10   Testosterone Effects on Skeletal Muscle in Men

The profound anabolic effect of testosterone on muscle becomes evident at puberty when male gain ≈ 35% more muscle mass than females. Testosterone stimulates myoblasts and increases the number of satellite cells, which promotes protein synthesis. Once men are in their second decade, testosterone levels begin to decline, and this decrease is continuous throughout their lifetime. If serum levels of testosterone fall below 320 ng/dl, men are considered hypogonadal, a common state after 70 years. While testosterone values decline with natural ageing, many factors diminish testosterone levels at all ages, including obesity, inactivity, trauma, diet, disease, and drugs.

Skeletal muscle has many AR, and when receptors are stimulated, muscle protein synthesis occurs. AR is also responsive to Insulin-like Grow Factor-1 (IGF-1) and growth hormone, providing additional stimulation to increase muscle size. To illustrate the potent effect of testosterone, Bhasin S, et al., nearly obliterated endogenous testosterone production in normal young men with Gonadotropin-releasing hormone (GnRH); men became hypogonadal (testosterone values of 31 ng/dl) with GnRH and remained hypogonadal for 10 weeks. Before and after GnRH administration, the quantities of lean body mass and fat mass were measured. Muscle mass significantly decreased by ≈1 kg, and fat mass increased proportionately such that body weight was not different at the end of the study. This study illustrates the role of testosterone in the maintenance of standard body composition in men. In another group of men, GnRH was used to block endogenous testosterone production for 8 weeks. Various doses of testosterone were given back for the 8 weeks that GnRH was being given (Bhasin et al. 2001). Muscle mass and strength decreased in men on the low doses of testosterone but increased once testosterone levels reached a minimum of 320 pg/dl. Although none of the events were severe, the young men in this study experienced 55 adverse events, primarily prostate-specific antigen (PSA) above 4 µg/ml, hematocrit>54%, and leg oedema.

What happens when testosterone is given to older men who are already hypogonadal? Ferrando et al., addressed this issue and administered 100 mg of testosterone to six healthy men (average age: 67 year) who were hypogonadal (defined in this study as ≤480 ng/dl) to bring testosterone levels to within normal. Following 4 weeks of testosterone administration, knee extension, and flexion strength was significantly increased, and the fractional synthetic rate of quadriceps muscle protein synthesis was significantly elevated (Ferrando et al. 2002). Bhasin et al. administered graded doses of testosterone for 20 weeks to over 60 years old men (60–75 years) who were made hypogonadal following GnRH administration (Bhasin et al. 2005). The primary outcome measures were fat-free mass and maximum leg press strength. Muscle mass and strength increased in a dose-dependent manner ($r = 0.77$), the higher the dose of testosterone, the greater the increase in muscle size and strength. Decreases in fat mass also occurred and were inversely correlated with testosterone dose. The highest dose of testosterone increased muscle strength by nearly 50%, which has clear functional implications for the older man at risk for loss of independence. Unfortunately, there were 147 adverse events in this study, 12 of which were severe. Serious adverse events included haematocrit >54%, leg oedema with shortness of breath, urinary retention, prostate cancer, and haematuria with elevated PSA. Additional side effects of testosterone administration included a dramatic drop in HDL-cholesterol, which may have long-term cardiovascular consequences, and a general overall increase in PSA values (Bhasin et al. 2005).

In a recent evaluation on the safety of testosterone, Bhasin et al., concluded that "an AR modulator with anabolic properties that are free of dose-limiting adverse effects of testosterone" is needed. More recently, Sullivan et al., conducted a study in which strength training and testosterone were administered separately or together in frail old men. Testosterone was given weekly to 71 men for 12 weeks. Men trained at either 20% of one repetition maximum (1-RM) or at 80% of 1-RM. Those that performed the high-intensity strength training had a significant increase in strength of ≈25%. Men who trained at 80% of 1-RM and also received testosterone injections did not show an increase in strength over and above the strength increase shown by training alone (Sullivan et al. 2005).

2.11 Testosterone Effects on Bone in Men

A cross-sectional investigation of a large number of older men examined the relationship of declines in serum testosterone and estrogen on bone mass, fat-free mass, and muscle strength. Losses in bone mass were related to the age-related fall in endogenous testosterone and the decline in endogenous estrogen. As expected, low testosterone was correlated with poor muscle strength. Although testosterone was strongly related to muscle strength and bone, estrogen was also strongly associated with bone mineral density. Moreover, the positive relationship between testosterone and bone mineral density was independent of estrogen and bone mass, suggesting a role for both hormones and the maintenance of bone with ageing in men. The authors suggested that perhaps testosterone is aromatized to estrogen and that estrogen is responsible for the maintenance of bone mass with advancing age. It is recognized that estrogen and testosterone use different cellular pathways to inhibit osteoclastic activity and bone resorption. Perhaps hormone balance and pathway activation shift as hormone levels are altered with age (Sullivan et al. 2005).

More recently, data from the Framingham study were analyzed for 793 men who had had serum estrogen and testosterone measures taken in the early 80s. These men were followed until 1999, and the incidence of hip fracture was calculated for those with low estrogen and testosterone. The findings indicated there were no significant increased risks for hip fracture among men with low testosterone. Men with the lowest levels of estrogen had the highest incidence rates of hip fracture. In subsequent analyses, men with low estrogen and testosterone combined had the most significant risk for hip fracture (Amin et al. 2006). Bone health in men has been minimally examined and provides an ample opportunity for future inquiry (Brown 2008). In summary, falling testosterone levels with age are associated with the loss of lean muscle and bone mass. Testosterone supplementation is probably not warranted for older men due to a high incidence of detrimental effects. Inactivity is likely a major factor contributing to lower testosterone values at all ages. Exercise increases testosterone levels in young men, but it is not clear if exercise has a similar effect in older men.

Although less common than in women, osteoporosis in men still constitutes a major burden for public health. Despite the higher competing risk of mortality from other causes in older men, the remaining lifetime risk of osteoporotic fractures after 50 years may be as high as 20–25% (vs. 45–55% in women) in high-risk Caucasian populations. For hip fractures, a systematic review found that the male-female incidence ratio was also about 1:2 and was remarkably constant globally, despite greater than ten-fold variation between geographic regions. Androgens and estrogens are respectively C19 and C18 metabolites of cholesterol. The predominant gonadal androgen in men is testosterone, 95% of which is secreted by the testes. The adrenals produce the remaining 5% after conversion of the precursor DHEA. In peripheral tissues, testosterone can be converted by 5α-reductase enzymes into the more potent androgen DHT. T can also be converted into 17β-estradiol (E2) by the aromatase (CYP19A1) enzyme. The testes synthesize approximately 20% of circulating E2 in men; the remaining 80% is derived from DHEA in peripheral tissues (Kaufman and Vermeulen 2005). The circulating levels of total testosterone and E2 decrease only marginally with age in men. However, the age-related increase in SHBG is more pronounced, resulting in a more significant decrease of bioavailable or free sex steroid levels (Naessen et al. 2010). Importantly, older men generally have higher circulating E2 levels than postmenopausal women, making it plausible that E2 contributes to the conservation of the male skeleton and/or other tissues during ageing.

The effects of androgens and estrogens on male bone health can be divided into two phases: (1) peak bone mass acquisition; and (2) subsequent maintenance. The male advantage in bone strength is mainly established during the first phase by placing cortical bone further from its central axis due to greater periosteal bone formation. Young adult women achieve similar cortical thickness by limiting endosteal expansion, but this does not provide the same biomechanical advantages. Men also have greater peak trabecular bone volume due to thicker, more plate-like trabeculae. However, these gender differences are probably site-specific and require further confirmation in more prospective studies (Vanderschueren et al. 2014).

Regardless of their importance in pathophysiology, guidelines in male osteoporosis rightly point out that the role of testosterone replacement in men with late-onset hypogonadism remains controversial. Evidence supporting the clinical utility of serum sex steroid measurements for fracture prediction beyond clinical risk factors is insufficient. Further studies on sex steroid signalling in the musculoskeletal system are a high research priority because they may help in the design of additional, preferentially gender-neutral, therapeutic strategies to reach the ultimate goal of not merely preventing bone loss but also reinforcing the musculoskeletal system as a whole to prevent osteoporotic fractures in both genders (Vanderschueren et al. 2014).

With one in seven men affected over their lifetime, prostate cancer is the most prevalent solid-organ malignancy in men worldwide. Prostate cancer and treatment with androgen deprivation therapy (ADT) affects a significant number of the male population. Endocrine effects of ADT are a critical consideration in balancing the benefits and risks of treatment on long-term survival and quality of life. Muscle mass declines with ADT; however, the evidence that this correlates with a decrease in muscle strength or decreased physical performance is discordant. Cortical bone decay also occurs in association with an increase in fracture risk; hence, musculo-skeletal health optimization in men undergoing ADT is crucial. The increase in fat and loss of muscle mass are associated with a decrease in predominantly upper body strength, including maximum chest press and handgrip strength. Men undergoing ADT also report a subjective decrease in the quality of life and increased fatigue compared with controls. Objective measures of physical performance are, however, much more variable, with many studies demonstrating no significant changes in the measures of endurance, dexterity, walking speed, measures of frailty, and lower limb performance (Cheung et al. 2014).

The role of exercise and current and emerging anabolic therapies for muscle and various new strategies to prevent loss of bone mass in men undergoing ADT are discussed. Future well-designed, prospective, controlled studies are required to elucidate the effects of ADT on physical performance, which are currently lacking, and larger randomized controlled trials are required to test the efficacy of medical therapies and exercise interventions to target proven deficits and to ensure safety in men with prostate cancer (Cheung et al. 2014).

2.12   Conclusions

Understanding male hormonal physiology and the functions of testosterone, in the sexual aspect and in other systems, will allow us to understand the clinical consequences of any alteration in the hormone levels. For a correct clinical and analytical evaluation of testosterone deficiency, prior knowledge of the male hormonal pathophysiology is needed. These notions will also help to make a correct replacement treatment in patients who require it.

References

Amin S, Zhang Y, Felson DT, Sawin CT, Hannan MT, Wilson PWF, Kiel DP. Estradiol, testosterone, and the risk for hip fractures in elderly men from the framingham study. Am J Med. 2006;119:426–33.

Araujo A, Handelsman D, McKinlay J. Total testosterone as a predictor of mortality in med: results from the Massachusetts Male Aging Male Study. 2005.

Bhasin S, Woodhouse L, Casaburi R, Singh AB, Bhasin D, Berman N, Chen X, Yarasheski KE, Magliano L, Dzekov C, Dzekov J, Bross R, Phillips J, Sinha-Hikim I, Shen R, Storer TW. Testosterone dose-response relationships in healthy young men. Am J Physiol Endocrinol Metab. 2001;281:E1172–81.

Bhasin S, Woodhouse L, Casaburi R, Singh AB, Mac RP, Lee M, Yarasheski KE, Sinha-Hikim I, Dzekov C, Dzekov J, Magliano L, Storer TW. Older men are as responsive as young men to the anabolic effects of graded doses of testosterone on the skeletal muscle. J Clin Endocrinol Metab. 2005;90:678–88.

Bohr D, Somlyo A, Harvey V, Sparks J, Rhodin J. Handbook of phisiology. The cardiovacu-
lar system II: architecture of vessel wall. Bethesda, MD: American Physiological Society;
1980. p. 1–3.

Briones AM, González JM, Somoza B, Giraldo J, Daly CJ, Vila E, González MC, McGrath JC,
Arribas SM. Role of elastin in spontaneously hypertensive rat small mesenteric artery remodelling. J Physiol. 2003;552:185–95.

Brown M. Skeletal muscle and bone: effect of sex steroids and aging. Adv Physiol Educ.
2008;32:120–6.

Cai JJ, Wen J, Jiang WH, Lin J, Hong Y, Zhu YS. Androgen actions on endothelium functions and
cardiovascular diseases. J Geriatr Cardiol. 2016;13:183–96.

Cheung AS, Zajac JD, Grossmann M. Muscle and bone effects of androgen deprivation therapy:
current and emerging therapies. Endocr Relat Cancer. 2014;21:R371–94.

Chou CH, Lin CL, Lin MC, Sung FC, Kao CH. 5α-Reductase inhibitors increase acute coronary syndrome risk in patients with benign prostate hyperplasia. J Endocrinol Investig.
2015;38:799–805.

Davison SL, Bell R, Donath S, Montalto JG, Davis SR. Androgen levels in adult females: changes
with age, menopause, and oophorectomy. J Clin Endocrinol Metab. 2005;90:3847–53.

Elmlinger MW, Dengler T, Weinstock C, Kuehnel W. Endocrine alterations in the aging male. Clin
Chem Lab Med. 2003;41:934–41.

Estrada M, Liberona JL, Miranda M, Jaimovich E. Aldosterone- and testosterone-mediated intracellular calcium response in skeletal muscle cell cultures. Am J Physiol Endocrinol Metab.
2000;279:E132–9.

Estrada M, Espinosa A, Müller M, Jaimovich E. Testosterone stimulates intracellular calcium
release and mitogen-activated protein kinases via a G protein-coupled receptor in skeletal muscle cells. Endocrinology. 2003;144:3586–97.

Evans RM. The steroid and thyroid hormone receptor superfamily. Science (80-). 1988;240:889–95.

Farhat MY, Abi-Younes S, Ramwell PW. Non-genomic effects of estrogen and the vessel wall.
Biochem Pharmacol. 1996;51:571–6.

Ferrando AA, Sheffield-Moore M, Yeckel CW, Gilkison C, Jiang J, Achacosa A, Lieberman SA,
Tipton K, Wolfe RR, Urban RJ, Sheffield-Moore M, Yeckel CW. Testosterone administration to
older men improves muscle function: molecular and physiological mechanisms. 2002.

Geraldes P, Sirois MG, Bernatchez PN, Tanguay JF. Estrogen regulation of endothelial and smooth
muscle cell migration and proliferation: role of p38 and p42/44 mitogen-activated protein
kinase. Arterioscler Thromb Vasc Biol. 2002;22:1585–90.

Gerhard M, Ganz P. How do we explain the clinical benefits of estrogen? Circulation. 1995;92:5–8.
https://doi.org/10.1161/01.CIR.92.1.5.

Hayes FJ, Decruz S, Seminara SB, Boepple PA, Crowley WF. Differential regulation of gonadotropin secretion by testosterone in the human male: absence of a negative feedback effect of
testosterone on follicle-stimulating hormone secretion*. 2001.

Heinlein CA, Chang C. The roles of androgen receptors and androgen-binding proteins in nongenomic androgen actions. Mol Endocrinol. 2002;16:2181–7.

Isidoro L, Villalpando DM, Perusquía M, Ferrer M. Effects of androgens on the function of mesenteric arteries from normotensive and hypertensive rats. FASEB J. 2018;31:885.5.

Jockenhövel F, Schubert M. Chapter 1, Male hypogonadism, 2nd revise. Bremen: Uni-Med Verlag
Ag; 2007.

Kaufman JM, Vermeulen A. The decline of androgen levels in elderly men and its clinical and
therapeutic implications. Endocr Rev. 2005;26:833–76.

Kelly DM, Jones TH. Testosterone: a vascular hormone in health and disease. J Endocrinol.
2013;217:R47–71.

Melmed S, Jameson L. Capítulo 371. Hipófisis anterior: fisiología de las hormonas hipofisarias I
Harrison. Principios de Medicina Interna, 20e I AccessMedicina I McGraw-Hill Medical In:
Harrison. Principios de Medicina Interna, 20th ed. 2018.

Montaño LM, Calixto E, Figueroa A, Flores-Soto E, Carbajal V, Perusquía M. Relaxation of androgens on rat thoracic aorta: testosterone concentration dependent agonist/antagonist l-type Ca2+ channel activity, and 5β-dihydrotestosterone restricted to l-type Ca2+ channel blockade. Endocrinology. 2008;149:2517–26.

Morales A, Buvat J, Gooren LJ, Guay AT, Kaufman JM, Tan HM, Torres LO. Endocrine aspects of sexual dysfunction in men. J Sex Med. 2004;1:69–81.

Morley JE. Testosterone and behavior. Clin Geriatr Med. 2003;19:605–16.

Naessen T, Sjogren U, Bergquist J, Larsson M, Lind L, Kushnir MM. Endogenous steroids measured by high-specificity liquid chromatography-tandem mass spectrometry and prevalent cardiovascular disease in 70-year-old men and women. J Clin Endocrinol Metab. 2010;95:1889–97.

Perusquía M. Androgen-induced vasorelaxation: a potential vascular protective effect. Exp Clin Endocrinol Diabetes. 2003;111:55–9.

Perusquía M, Stallone JN. Do androgens play a beneficial role in the regulation of vascular tone? Nongenomic vascular effects of testosterone metabolites. Am J Physiol Heart Circ Physiol. 2010;298:H1301.

Pi M, Parrill AL, Quarles LD. GPRC6A mediates the non-genomic effects of steroids. J Biol Chem. 2010;285:39953–64.

Pongkan W, Chattipakorn SC, Chattipakorn N. Roles of testosterone replacement in cardiac ischemia-reperfusion injury. J Cardiovasc Pharmacol Ther. 2016;21:27–43.

Rowell KO, Hall J, Pugh PJ, Jones TH, Channer KS, Jones RD. Testosterone acts as an efficacious vasodilator in isolated human pulmonary arteries and veins: evidence for a biphasic effect at physiological and supra-physiological concentrations. J Endocrinol Investig. 2009;32:718–23.

Sullivan DH, Roberson PK, Johnson LE, Bishara O, Evans WJ, Smith ES, Price JA. Effects of muscle strength training and testosterone in frail elderly males. Med Sci Sport Exerc. 2005;37:1664–72.

Vanderschueren D, Laurent MR, Claessens F, Gielen E, Lagerquist MK, Vandenput L, Börjesson AE, Ohlsson C. Sex steroid actions in male bone. Endocr Rev. 2014;35:906–60.

Vela Navarrete R, Garcia Cardoso JV, Pardo Montero M, Jiménez Máteos-Cáceres López Farré PA. Testosterona, función endotelial, salud cardiovascular y androgenodeficiencia del varon añoso. 2009

Vermeulen A. Diagnosis of partial androgen deficiency in the aging male. Ann Endocrinol. 2003;64:109–14.

Vignozzi L, Corona G, Petrone S, Filippi S, Morelli A, Maggi M. Testosterone and sexual activity. J Endocronol Invest. 2005;28(3 Suppl):39–44. Accessed November 22, 2020

Walker WH. Nongenomic actions of androgen in sertoli cells. Curr Top Dev Biol. 2003;56:25–53.

Williams MRI, Dawood T, Ling S, Dai A, Lew R, Myles K, Funder JW, Sudhir K, Komesaroff PA. Dehydroepiandrosterone increases endothelial cell proliferation in vitro and improves endothelial function in vivo by mechanisms independent of androgen and estrogen receptors. J Clin Endocrinol Metab. 2004;89:4708–15.

Yue P, Chatterjee K, Beale C, Poole-Wilson PA, Collins P. Testosterone relaxes rabbit coronary arteries and aorta. Circulation. 1995;91:1154–60.

Chapter 3
Hypogonadism and Late Onset Hypogonadism

Paolo Verze, Raffaele Baio, Luigi Napolitano, and Vincenzo Mirone

During the process of ageing, human organism undergoes a series of morphological and functional modifications characterized by a reduced physiological efficiency and atrophy of various organs and systems. This process also involves the secretion of most hormones including testosterone. Testosterone is the main male sex hormone and the predominant circulating androgen, crucial for virilization and maintenance of the male phenotype (Mooradian et al. 1987). It modulates, either directly or via its transformation into 5α-dihydrotestosterone (a potent metabolite) or via aromatization into estradiol (E_2), important biochemical signalling pathways of human physiology. It's produced in the testicles in men, ovaries in women, and a little bit in the adrenal glands of both sexes (Hiller-Sturmhöfel and Bartke 1998). In the testicles is produced by the Leydig cells with an average secretion rate of 7 mg/day (Meikle et al. 1992). The production of testosterone increases significantly during puberty but it gradually declines as one of the processes of ageing (Feldman et al. 2002). The serum testosterone concentration gradually declines by 1.6% per year, especially after 40 years of age (Tsujimura 2013). Testosterone is an hormone with a wide range of beneficial effects on men's health and several effects on the physiology of different organs and tissues. As the most important sex male hormone, testosterone regulates: reproductive function through stimulation of Sertoli cell function, spermatogenesis, growth of primary sexual characteristics (penis, epididymis, prostate, seminal vesicle and testes) and secondary sexual characteristics (deepening of the voice, growth of facial, armpit, chest and pubic hair during

P. Verze (✉) · R. Baio
Department of Medicine, Surgery, Dentistry "Scuola medica salernitana", University of Salerno, Salerno, Italy

L. Napolitano · V. Mirone
Department of Neurosciences, Reproductive Sciences, Odontostomatology, University of Naples Federico II, Naples, Italy

puberty). During puberty an increase in testosterone levels in boys causes enlargement of the reproductive organs, increased libido, increased frequency of erection, and the growth of facial, chest, nipple and pubic hair (Birnbaum and Bertelloni 2014). During adulthood testosterone maintains male fertility, libido, mental and physical energy levels, muscle strength, general well being and prevents depression and fatigue. Testosterone is also an important anabolic steroid that promotes an increase in muscle mass, strength, density and strength of bone. It stimulates bone mineralization, muscle growth, erythropoiesis and cognitive function. Several studies suggest that testosterone deficiency contributes to obesity, insulin resistance (IR), metabolic syndrome (MetS), type 2 diabetes mellitus (T2D) (Fernández-Miró et al. 2016), cardiovascular disease (CVD) (Kloner et al. 2016), and erectile dysfunction (ED) (Elkhoury et al. 2017). The multidirectional relationship between serum testosterone and sex hormone binding globulin (SHBG) with obesity, metabolic syndrome, and type 2 diabetes is complex. Epidemiological studies support a bidirectional relationship between serum testosterone and obesity and between testosterone and the metabolic syndrome. Low serum total testosterone predicts the development of central obesity and accumulation of intra-abdominal fat (Allan and McLachlan 2010; Brand et al. 2011). Low total testosterone or SHBG levels are associated with type 2 diabetes (Corona et al. 2011). There is a significant inverse correlation between total testosterone and obesity (Kalyani et al. 2007). This includes decreased SHBG (so the proportion of bioactive free testosterone decreases), increased aromatisation of testosterone to estradiol in fat cells and cytokine-mediated inhibition of testicular steroid production (Kalyani et al. 2007). Low testosterone promotes accumulation of total and visceral fat mass with gonadotrophin inhibition (Grossmann et al. 2010). In men with visceral obesity testosterone deficiency leads to reduced lipolysis, reduced metabolic rate, visceral fat deposition and insulin resistance (Haffner 2000). MetS is characterized by central obesity, insulin resistance, dyslipidemia, and hypertension. The pathogenesis is multifactorial, but the first step is central obesity that is linked with hypertension, increased serum low-density lipoprotein (LDL), low serum high-density lipoprotein (HDL), and hyperglycemia. Central obesity is inversely related to serum testosterone level and serum total testosterone level is inversely related to insulin concentration, insulin resistance and blood pressure. Obesity is a proinflammatory state resulting in increased release and secretion of proinflammatory cytokines and adipokines, free fatty acids and estrogens from adipose tissue. These are important risk factors that may contribute to the development of metabolic syndrome, type 2 diabetes and androgen deficiency (Traish et al. 2009). Visceral fat is an active secretory tissue producing inflammatory cytokines, adipokines, biochemical modulators, and other pro-inflammatory factors: interleukin (IL)-6, IL-1b, plasminogen activator inhibitor-1, tumor necrosis factor (TNF)-a, angiotensinogen, vascular endothelial growth factor, and serum amyloid A. These factors contribute to systemic and peripheral vascular inflammation and dysfunction (Guzik et al. 2006). Free fatty acids activate nuclear factor-kB pathways resulting in increased synthesis of TNF-a that activates lipolysis, increased synthesis of IL-6 and macrophage chemoattractant protein-1. This increases recruitment of more macrophages and modulates insulin sensitivity, production of TNF-a and expression of adhesion molecules in endothelium and vascular smooth muscle cells. TNF-a also contributes to the dysregulation of insulin modulation of endothelin-1–mediated vasoconstriction and nitric oxide–mediated vasodilation, hence promoting vasoconstriction. Because of these proinflammatory agents obesity results in insulin resistance and vascular endothelial dysfunction, which are potential causal factors for increased CVD and ED. Regarding insuline resistance, it is known that low testosterone level precedes elevated fasting insulin, glucose, and haemoglobin A1c (HbA1C) values. On the cardiovascular health, low testosterone and androgen deficiency are associated with increased levels of total cholesterol, LDL, increased production of pro-inflammatory factors, increased thickness of the arterial wall, and endothelial dysfunction. In fact testosterone therapy in men improves insulin sensitivity, fasting glucose, HbA1c levels, arterial vaso-reactivity, reduces pro-inflammatory cytokines, total cholesterol, and triglyceride levels and improves endothelial function and high density lipoprotein levels. Regarding sexual function, erectile function is widely associated with the serum testosterone level. Nocturnal erection is weaker in patients with a low serum testosterone level than in men with a normal serum testosterone level. Hypoactive sexual desire and low or absent libido are often associated with ED.

3.1   Male Hypogonadism

Male hypogonadism is a disorder associated with decreased functional activity of the testes, decreased production of androgens and/or impaired sperm production (Salonia et al. 2019).

Male hypogonadism can be classified as:

Primary hypogonadism or hypergonadotropic hypogonadism: a condition given by testicles failure. It is characterized by reduced testosterone levels, increased levels of FSH and LH, impairment of spermatogenesis. Common causes of primary hypogonadism include: Klinefelter syndrome, undescended testicles, orchitis, hemochromatosis, injury to the testicles, cancer treatment.

Secondary hypogonadism or hypogonadotropic hypogonadism: It is a condition given by the alteration of the hypothalamic-pituitary axis. The testis is inadequately stimulated by gonadotropins testosterone levels are low, and FSH and LH levels are low or at the limit of normal. The main causes are hyperprolactinemia, Kallman's syndrome, pituitary, disorders, HIV/AIDS, obesity, malnutrition, systemic illness, stress, medication.

3.2 Late-Onset Hypogonadism

Late-onset Hypogonadism (LOH) is a clinical and biological syndrome associated with advancing age and characterized by typical symptoms and a deficiency in serum testosterone levels (Wang et al. 2009).

LOH is the term used to describe the decline in serum testosterone levels associated with increasing age in men above 30 years. It is a consequence of the ageing process, due to deterioration of hypothalamic-pituitary function and Leydig cell function in the testes (Nieschlag 2013). The EMAS study reported a 0.4% per annum decline in total testosterone and a 1.3% per annum decline in free testosterone (Wu et al. 2008). The incidence of biochemical hypogonadism varies from 2.1% to 12.8% (Hall et al. 2008). The incidence of low testosterone and symptoms of hypogonadism in men aged 40–79 years varies from 2% to 6% (Hall et al. 2008). There is a higher prevalence of type 2 diabetes, obesity, cardiovascular disease, osteoporosis, and anaemia in men with decreased testosterone levels (Nieschlag et al. 2006). It is characterized by diminished sexual desire, diminished erectile quality, particularly in nocturnal erections, changes in mood with concomitant decreases in intellectual activity and spatial orientation, fatigue, depression and anger, a decrease in lean body mass with associated decreases in muscle volume and strength (Zitzmann et al. 2006), a decrease in body hair and skin alterations, and decreased bone mineral density resulting in osteoporosis (Dandona and Rosenberg 2010). Several diseases and chronic co-morbidities can interfere with the HPG axis leading to the development of primary hypogonadism or, more frequently, secondary hypogonadism in adulthood. The main causes of LOH are obesity, co-morbidities and ageing. In ageing there is a deterioration of hypothalamic-pituitary (disorders of pulsed secretion of GnRH) and Leydig with reduction of the frequency and amplitude of LH pulses. SHBG levels increase with age and so the proportion of bioactive free testosterone decreases. In older men there is often an increase in aromatase activity, which metabolises testosterone to oestradiol. Several drugs can also interfere with testosterone as long-term glucocorticoid therapy methadone and tramadol that suppresses testosterone levels. Inflammatory cytokines released in states of chronic inflammation, and adipocytokines and estradiol in obesity, can suppress the hyphotalamic-pituitary-gonadal axis (HPG). In fact, the visceral fat is an active endocrine organ. Visceral fat secretes inflammatory cytokines (adipokines), pro-coagulative substances, and substances which activate the angiotensin-aldosterone system.

3.3 Diagnosis

The phenotype of the hypogonadal patient appears to be independent from the aetiology causing the problem, but it is more often affected by the age of onset of hypogonadism (during the foetal period, before puberty and after puberty) and by the extent of the deficit. When androgen deficiency develops during the foetal life (in this case we speak of "very early onset hypogonadism", VEOH), symptoms can be dramatic; in fact, if the deficit is complete, the individual will be chromosomally male but with female phenotype. Conversely, if the deficit is incomplete, we will have various defects in virilisation and ambiguous genitalia (micropenis, hypospadias, cryptorchidism) (Salonia et al. 2019; Santi and Corona 2017). Delay in puberty with an overall eunuchoidal phenotype is typical of defects manifesting in the pre- or peri-pubertal period ("early onset hypogonadism", EOH) (Salonia et al. 2019; Santi and Corona 2017). Typical anthropometric characteristics of a picture of eunuchoidism are the following:

- scant body hair;
- high-pitched voice;
- gynecomastia;
- penis of infantile size;
- underdeveloped testes (volume < 5 mL).

Those affected by this condition are sterile; in addition, there may also be behavioural disorders, which fall within what takes the name of "psychic infantilism", with impaired maturation and low desire in the sexual sphere. The patient will also have particularly long bones due to the failure to close the pituitary cartilages (with opening of the breach >5 cm compared to height). When hypogonadism occurs in adulthood ("late onset hypogonadism", LOH), symptoms can be often relatively mild, difficult to recognise and frequently confused with the ageing process (Salonia et al. 2019; Santi and Corona 2017) or with the comorbid chronic conditions. Several non-specific clinical features (such as fatigue, weakness and decreased energy, as well as sexual impairment) may be clinical manifestations. The EMAS study showed that a triad of sexual symptoms, including low libido, reduced spontaneous erections and ED, are typically associated with a decrease in testosterone serum levels. Conversely, psychological and physical symptoms were less informative (Wu et al. 2010). In severe conditions, a regression of secondary sexual characteristics can also occur. Fundamental for the correct diagnostic classification is the physical examination which must focus not only on the external genitalia but also on hair growth, the presence of gynecomastia, height and proportions of the body. We speak of eunuchoid proportions when the opening of the arms exceeds the height of the individual by at least 2 cm, since it suggests an androgen deficiency before the fusion of the epiphyses. Beard and hair growth (in the armpits, chest and pubic region) also depend on androgens; however, the changes may go unnoticed until the androgen deficiency becomes severe and prolonged. Furthermore, the ethnic strain influences the intensity of hair growth. Testicular volume is best measured with the Prader orchidometer. The length of the testicles varies between 3.5 e 5.5 cm, which corresponds to a volume of approximately 12–25 mL. Advanced age does not affect testicular volume, although it can decrease consistency. Asian men, generally, have smaller testicles than Western Europeans, regardless of body size. In view of its potential role in infertility, the presence of a varicocele must be assessed by palpation with the patient standing: it is more common on the left side. Patients with Klinefelter syndrome have a marked reduction in testicular volume (1–2 mL). In congenital hypogonadotropic hypogonadism, the testicular volume represents a good index of the degree of gonadotropic deficiency and of the possible therapeutic response. Among the first level laboratory tests, the spermiogram represents the gold standard for the evaluation of the structural and kinetic characteristics of the spermatic cells while the FSH, LH and testosterone dosage will allow to diagnose androgenic deficiency and to direct the subsequent diagnostic investigations. In the primary hypogonadism (also called hypergonadotropic), the analysis of the seminal fluid will show the presence of oligo-azoospermia. In these patients, hormonal dosage will highlight the following picture:

- Elevated gonadotropins (FSH ed. LH), due to the lack of negative feedback from the testicle on the pituitary gland;
- Drastically reduced testosterone and estrogen.

A diagnosis of primary hypogonadism requires further investigation by examining the karyotype; this test, which is expected to be XY, is also important to assess the possible presence of chromosomal alterations, although the identification of mutations is more important from a research point of view than from a diagnostic point of view. In case of normal karyotype, a testicular biopsy is indicated to differentiate genetic defects from acquired testicular changes. In patients with secondary hypogonadism (also called "hypogonadotropic"), the analysis of the seminal fluid will still show the presence of oligo-azoospermia but at the hormonal dosage there will be low values of both gonadotropins and testosterone. Furthermore, in these patients, the diagnostic procedure to confirm the presence of secondary hypogonadism must be completed by the laboratoristic study of the hypothalamus-pituitary axis (also with the determination of prolactin) and by MRI of the sella turcica, to evaluate any possible presence of tumours or congenital changes in the pituitary gland. The stimulus test with GnRH can also be performed, both to assess the extent of gonadotropic deficit and to distinguish the forms of secondary pituitary hypogonadism from hypothalamic ones. The GnRH stimulation test is performed by measuring the basal FSH and LH levels and 30 and 60 min after intravenous administration of 100 μg of GnRH. A minimum acceptable response is the doubling of LH levels and an increase of at least 50% in FSH levels. However, with the advent of extremely sensitive gonadotropin assay methods, this test is rarely used in clinical practice. A population screening for hypogonadism is not recommended. In fact, testosterone dosage should be required in subjects with signs and/or symptoms suggestive of hypogonadism. In this context, the guidelines of the Endocrine Society suggest to evaluate also subjects with a history of bone fractures for minimal trauma, in chronic therapy with corticosteroids and opiates or suffering from HIV infection, moderate-severe COPD, type 2 diabetes mellitus and severe renal failure requiring hemodialysis, conditions often associated with hypotestosteronemia. The mainstay of a LOH diagnosis includes the presence of signs and symptoms consistent with hypogonadism, coupled with biochemical evidence of low morning serum total testosterone levels on two or more occasions, measured with a reliable assay. Total

Testosterone includes the free share and the share bound to carrier proteins and can be measured by radioimmunoassay or immunometric method or by liquid chromatography and tandem mass spectrometry (LC-MS/MS). Therefore, most of the total testosterone circulates bound to SHBG or albumin; only 0.5–3% of circulating testosterone is unbound and "free". Testosterone levels show a circadian variation, which persist in ageing men (Guay et al. 2008; Travison et al. 2017). Testosterone tends to decline in the late afternoon and decreases during acute illness. Testosterone concentration in healthy young people ranges from 300 to 1000 ng/dL in most laboratories. Alterations in SHBG levels (due to ageing, obesity, diabetes mellitus, hyperthyroidism, drugs or chronic diseases or on a congenital basis) can affect total testosterone levels. Likewise, testosterone levels are potentially influenced by food intake (Gagliano-Jucá et al. 2019); hence, serum total testosterone should be measured in fasting conditions and in the morning (between 7.00 and 11.00 h). Moreover, a confirmatory measurement should always be undertaken in the case of a primary pathological value. Liquid Chromatography-Tandem Mass Spectrometry (LC-MS/MS) represents the standard and most accurate method for sex steroid evaluation; however, standardised automated platform immuno-assays for total testosterone assessment demonstrate a good correlation with LC-MS/MS (Huhtaniemi et al. 2012). Conversely, available immuno-assays are not able to provide an accurate estimation of fT; therefore, direct fT evaluation with these methods is not recommended and should be avoided (Rosner and Vesper 2010). Liquid chromatography-tandem mass spectrometry remains the standard method for fT determination. Alternatively, fT can be derived from specific mathematical calculations taking into account serum SHBG and albumin levels (Vermeulen et al. 1999). Data derived from available meta-analyses have documented that testosterone therapy is ineffective when baseline levels are above 12 nmol/L (3.5 ng/mL). Positive outcomes are documented when testosterone levels are below 12 nmol/l, being higher in symptomatic patients with more severe forms of hypogonadism (T < 8 nmol/L). Hence, 12 nmol/L should be considered as a possible cut-off to start with testosterone therapy in the presence of hypogonadal symptoms (Corona et al. 2016; Rastrelli et al. 2017). As reported above, in the presence of clinical conditions which may potentially interfere with SHBG levels, the evaluation of fT should be considered in order to better estimate androgen levels. Unfortunately, despite its potential clinical value (Boeri et al. 2017), no validated thresholds for fT are available from clinical studies and this represents an area of uncertainty; however, some data indicate that fT levels below 225 pmol/L (< 6.5 ng/dl) are associated with hypogonadal symptoms (Wu et al. 2010; Rastrelli et al. 2016; Bhasin et al. 2018; Isidori et al. 2015). The determination of LH must be performed along with prolactin (PRL) when pathological total testosterone levels are detected, in order to correctly define the underlying conditions and exclude possible organic forms. In men with reduced testosterone levels, LH levels allow to distinguish primary hypogonadism (high LH levels) from secondary hypogonadism (low or inappropriately normal levels of LH). Elevated LH levels indicate a primary defect in the testicle, while reduced or inappropriately normal levels of LH suggest an alteration at the hypothalamic-pituitary level. LH surges occur over a range of about 1 to 3 h in healthy men; therefore, as gonadotropin levels fluctuate, blood sampling should be repeated if results are unclear. FSH is less pulsatile than LH as it has a longer half-life. An isolated increase in FSH levels suggests damage to the seminiferous tubules. Inhibin B, a hormone produced by Sertoli cells which reduces FSH levels, is reduced in the presence of damage to the seminiferous tubules. The hCG stimulation test is performed by measuring testosterone levels at baseline and 24, 48, 72 and 120 h after the intramuscular administration of a single bolus of hCG with a variable dose between 1500 and 4000 IU. An alternative protocol includes 3 injections of 1500 IU of hCG in subsequent days and the measurement of testosterone levels 24 h after the last administration. An acceptable response to hCG stimulation is reaching twice the baseline testosterone concentration. In prepubertal boys, an increase in testosterone levels above 150 ng/dL indicates the presence of testicular tissue. The absence of response indicates the lack of testicular tissue or the presence of a severe deterioration in the function of Leydig cells. Measuring levels of MIS (Müllerian inhibitory substance), a peptide produced by Sertoli cells, is also used to assess the presence of testicular tissue in prepubertal boys with cryptorchidism. Due to its negative influence on libido, PRL can also be considered as first-line screening in patients with reduced sexual desire. In addition, pituitary magnetic resonance imaging (MRI) scanning, as well as other pituitary hormone evaluation, is required in the presence of specific symptoms such as visual disturbances, headache or when hyperprolactinaemia is confirmed. In addition, limited evidence suggests performing pituitary MRI also in the case of severe hypogonadism (<6 nmol/L; 1.75 ng/mL) with inadequate gonadotropin levels (Dalvi et al. 2016; Molitch 2017). Past history of surgical intervention for cryptorchidism or hypospadias must be taken into account as possible signs of congenital defects. Likewise, chronic and systemic comorbid conditions must be comprehensively investigated in every patient. Possible use of drugs potentially interfering with the HPG axis should be ruled out (such as drug-induced AR blockage, drug-induced 5-alpha-reductase activity blockage, drug-induced aromatase activity blockage). Acute illnesses are associated with the development of functional hypogonadism and the determination of serum total testosterone levels should be avoided in these conditions. Several self-reported questionnaires or structural interviews have been developed for the screening of hypogonadism. Although these case-history tools have demonstrated clinical utility in supporting the biochemical diagnosis of hypogonadism or in the assessment of testosterone therapy outcomes, their specificity remains relatively poor and they should not be used for a systematic screening of hypogonadal men (Millar et al. 2016). Since obesity is frequently associated with hypogonadism (mostly functional), the determination of body mass index (BMI) and the measurement of waist circumference are strongly recommended in all individuals. Testicular and penile size, as well as the presence of sexual secondary characteristics, can provide useful information regarding overall androgen status. In addition, upper segment/lower segment ratio (n.v. > 0.92) and arm-span to height ratio (n.v. < 1.00) can be useful to identify a eunuchoid body shape, especially in subjects with pre-pubertal hypogonadism or delayed puberty. Finally, digital rectal examination (DRE) should be performed in all subjects to exclude prostate abnormalities before testosterone therapy (any type) or to support the suspicion of hypogonadism (Rastrelli et al. 2019).

3.4 Therapy

During the neonatal period, in congenital hypogonadotropic hypogonadism, between 6 and 12 months, in the presence of cryptorchidism, orchidopexy is fundamental. In selected cases, treatment with chorionic gonadotropin (HCG) and in the case of micropenis, in order to obtain a penis enlargement, testosterone (T) replacement treatment for short periods is possible. In the case of post-pubertal hypogonadotropic hypogonadism, testosterone can be temporarily suspended, subsequently administering gonadotropins in case of desire for paternity. On the other hand, in post-pubertal hypergonadotropic hypogonadism, testosterone should be administered throughout life by monitoring the parameters of efficacy and safety (Isidori et al. 2015). In case of wish for paternity, in young adults with post-pubertal hypogonadotropic hypogonadism associated with a history of cryptorchidism and testicular volumes <4 ml, pre-treatment with extractive or recombinant FSH is recommended, at a dose of 75–150 IU every other day approximately 3 months before treatment with HCG at a dose of 1500–2000 IU 3 times a week every other day. On the other hand, in the case of hypogonadotropic hypogonadism associated with testicular volumes >4 ml, inhibin B values >60 pg/mL and no history of cryptorchidism, pretreatment with hCG for 3–6 months and subsequently treatment with FSH is recommended (Barbonetti et al. 2018).

In Klinefelter's syndrome, for the possible preservation of fertility, it is recommended at any age to search for spermatozoa by testicular sperm extraction (TESE) with possible cryopreservation (Bonomi et al. 2017). Anti-oestrogens, preventing the down-regulation of the HPG axis by oestrogens, are particularly useful in men with obesity and metabolic disorders but this is off-label treatment due to his adverse effects. In adult hypogonadism after 65 years, treatment with testosterone should not be recommended to all subjects with plasma levels of TT or better still of free T (calculated using the Vermeulen formula) lower than normal (Aversa and Morgentaler 2015). It is recommended to explain to the patient the response time to therapy (different for sexuality, body composition and improvement in quality of life, etc) which depends on the plasma levels of testosteronemia (they must remain in the lower half of the normal range, between 15 and 17 nmol/L) and which mainly depend on the formulation used. Testosterone is clinically effective only when chemically modified. There are several formulations:

- Oral products. The only preparation available for oral administration is Testosterone-undecanoate which bypasses the liver through lymphatic absorption, which depends on the lipid content of the food eaten. This results in an irregular level of Testosterone throughout the day;

- Transdermal products. These products, when used daily, provide constant testosterone levels. The patch formulation provides constant testosterone levels, mimicking the circadian rhythm; however, its use is limited by frequent allergic skin reactions. The 1 or 2% hydroalcoholic transdermal gels release Testosterone for 24 h after a single daily application. Testosterone is rapidly absorbed into the stratum corneum of the skin and forms a deposit. These gels have an excellent safety profile and are capable of normalizing Testosterone levels. Sometimes they require dose adjustments as absorption can vary from man to man;
- Injectable preparations. Testosterone injected into the muscle is absorbed directly into the blood stream. Depot hydroxyl oil formulations, on the other hand, allow a gradual release of testosterone and are divided according to their half-life. Testosterone propionate has a short half-life and requires bi-weekly administration (50 mg every 2–3 days); instead, cypionate and enanthate esters can be injected at a dosage of 25 mg every 3–4 weeks. In general, these products have an initial peak followed by a gradual decline in the following weeks. These fluctuations can cause side effects on mood. A formulation of Testosterone undecanoate, with a longer duration of action, requires an administration of 1000 mg every 12 weeks, so as to ensure a gradual and continuous release from the reservoir into the blood stream without the fluctuating side effects typical of enanthate formulations;
- Transbuccal preparations. They are still available in several countries, consisting on a sustained-release muco-adhesive buccal-testosterone-tablet requiring twice-daily application to the upper gums. The tablet does not dissolve completely in themouth and must be removed after twelve hours. This formulation has been proven to restore testosterone levels within the physiological range with minimal or transient local problems, including gum oedema, blistering and gingivitis (Rastrelli et al. 2018);
- Transnasal preparations. These preparations require administration two or three times a day using a specific metered-dose pump. The application is rapid, non-invasive and convenient.

Replacement therapy (TRT) follow-up should be performed every 3–6 months in the first year of therapy and every 12 months thereafter. The check-up includes a complete general and andrological examination with palpation of the prostate, combined with the PSA and hematocrit dosage. Metabolic tests (such as blood glucose and lipid profile measurements and liver function assessment) may be required but not mandatory. Prostate biopsy is only required in the presence of clinical suspicion (positive palpation for nodule) or for PSA levels >4 ng/mL or for PSA increases >1.4 ng/mL in the last 12 months of therapy. No studies and no meta-analyses have shown significant associations between TRT and prostate cancer incidence, PSA increases or need for prostate biopsies. Men on TRT should follow the same recommendations for prostate cancer as all other men for the same age. The vertebral-femoral examination should be performed every 1–2 years. Absolute contraindications to treatment are: breast or prostate cancer, presence of palpable lump or hardening of the prostate, PSA values >4 ng/mL, PSA values >3 ng/mL combined with high risk of prostate cancer, hematocrit >54%, severe LUTS, the presence of heart failure NYHA 3–4, the desire for short-term paternity (Colpi et al. 2018). TRT significantly improves all aspects of sexual function in men with low initial testosterone levels (<12 nmol/L) in comparative studies with placebo. The effects of TRT on ED are inversely proportional to initial testosterone levels. With more severe initial hypogonadism, the benefits of the therapy will be greater. Conversely, for baseline testosterone levels greater than 12 nmol/L, the benefits are comparable to placebo. These results are also conditioned by other factors, such as age. Hypogonadism may be the only cause of ED in the young subject who, therefore, will respond well to TRT; instead, it may only be one aspect of a multifactorial genesis of ED in the elderly, who will respond less brilliantly to TRT (Wu et al. 2008). In conclusion, a multidisciplinary approach (endocrinologist, paediatrician, internist, geriatrician, psychologist, etc.) is necessary for a correct diagnostic and therapeutic process of the male in the various ages of life (Morgentaler et al. 2016).

References

Allan CA, Mclachlan RI. Androgens and obesity. Curr Opin Endocrinol Diabetes Obes. 2010;17:224–32.

Aversa A, Morgentaler A. The practical management of testosterone deficiency in men. Nat Rev Urol. 2015;12:641–50.

Barbonetti A, Calogero AE, Balercia G, Garolla A, Krausz C, La Vignera S, Lombardo F, Jannini EA, Maggi M, Lenzi A, Foresta C, Ferlin A. The use of follicle stimulating hormone (FSH) for the treatment of the infertile man: Position statement from the italian society of andrology and sexual medicine (SIAMS). J Endocrinol Investig. 2018;41:1107–22.

Bhasin S, Brito JP, Cunningham GR, Hayes FJ, Hodis HN, Matsumoto AM, Snyder PJ, Swerdloff RS, Wu FC, Yialamas MA. Testosterone therapy in men with hypogonadism: An endocrine society clinical practice guideline. J Clin Endocrinol Metab. 2018;103:1715–44.

Birnbaum W, Bertelloni S. Sex hormone replacement in disorders of sex development. Endocr Dev. 2014;27:149–59.

Boeri L, Capogrosso P, Ventimiglia E, Cazzaniga W, Pederzoli F, Moretti D, Dehò F, Montanari E, Montorsi F, Salonia A. Does calculated free testosterone overcome total testosterone in protecting from sexual symptom impairment? Findings of a cross-sectional study. J Sex Med. 2017;14:1549–57.

Bonomi M, Rochira V, Pasquali D, Balercia G, Jannini EA, Ferlin A. Klinefelter syndrome (KS): Genetics, clinical phenotype and hypogonadism. J Endocrinol Investig. 2017;40:123–34.

Brand JS, Van Der Tweel I, Grobbee DE, Emmelot-Vonk MH, Van Der Schouw YT. Testosterone, sex hormone-binding globulin and the metabolic syndrome: A systematic review and meta-analysis of observational studies. Int J Epidemiol. 2011;40:189–207.

Colpi GM, Francavilla S, Haidl G, Link K, Behre HM, Goulis DG, Krausz C, Giwercman A. European academy of andrology guideline management of oligo-astheno-teratozoospermia. Andrology. 2018;6:513–24.

Corona G, Monami M, Rastrelli G, Aversa A, Sforza A, Lenzi A, Forti G, Mannucci E, Maggi M. Type 2 diabetes mellitus and testosterone: A meta-analysis study. Int J Androl. 2011;34:528–40.

Corona G, Giagulli VA, Maseroli E, Vignozzi L, Aversa A, Zitzmann M, Saad F, Mannucci E, Maggi M. Therapy of endocrine disease: Testosterone supplementation and body composition: Results from a meta-analysis study. Eur J Endocrinol. 2016;174:R99–116.

Dalvi M, Walker BR, Strachan MW, Zammitt NN, Gibb FW. The prevalence of structural pituitary abnormalities by MRI scanning in men presenting with isolated hypogonadotrophic hypogonadism. Clin Endocrinol. 2016;84:858–61.

Dandona P, Rosenberg MT. A practical guide to male hypogonadism in the primary care setting. Int J Clin Pract. 2010;64:682–96.

Elkhoury FF, Rambhatla A, Mills JN, Rajfer J. Cardiovascular health, erectile dysfunction, and testosterone replacement: Controversies and correlations. Urology. 2017;110:1–8.

Feldman HA, Longcope C, Derby CA, Johannes CB, Araujo AB, Coviello AD, Bremner WJ, Mckinlay JB. Age trends in the level of serum testosterone and other hormones in middle-aged men: Longitudinal results from the Massachusetts male aging study. J Clin Endocrinol Metab. 2002;87:589–98.

Fernández-Miró M, Chillarón JJ, Pedro-Botet J. Testosterone deficiency, metabolic syndrome and diabetes mellitus. Med Clin (Barc). 2016;146:69–73.

Gagliano-Jucá T, Li Z, Pencina KM, Beleva YM, Carlson OD, Egan JM, Basaria S. Oral glucose load and mixed meal feeding lowers testosterone levels in healthy eugonadal men. Endocrine. 2019;63:149–56.

Grossmann M, Gianatti EJ, Zajac JD. Testosterone and type 2 diabetes. Curr Opin Endocrinol Diabetes Obes. 2010;17:247–56.

Guay A, Miller MG, Mcwhirter CL. Does early morning versus late morning draw time influence apparent testosterone concentration in men aged > or =45 years? Data from the hypogonadism in males study. Int J Impot Res. 2008;20:162–7.

Guzik TJ, Mangalat D, Korbut R. Adipocytokines - novel link between inflammation and vascular function? J Physiol Pharmacol. 2006;57:505–28.

Haffner SM. Sex hormones, obesity, fat distribution, type 2 diabetes and insulin resistance: Epidemiological and clinical correlation. Int J Obes Relat Metab Disord. 2000;24(Suppl 2):S56–8.

Hall SA, Esche GR, Araujo AB, Travison TG, Clark RV, Williams RE, Mckinlay JB. Correlates of low testosterone and symptomatic androgen deficiency in a population-based sample. J Clin Endocrinol Metab. 2008;93:3870–7.

Hiller-Sturmhöfel S, Bartke A. The endocrine system: an overview. Alcohol Health Res World. 1998;22:153–64.

Huhtaniemi IT, Tajar A, Lee DM, O'neill TW, Finn JD, Bartfai G, Boonen S, Casanueva FF, Giwercman A, Han TS, Kula K, Labrie F, Lean ME, Pendleton N, Punab M, Silman AJ, Vanderschueren D, Forti G, Wu FC. Comparison of serum testosterone and estradiol measurements in 3174 European men using platform immunoassay and mass spectrometry; Relevance for the diagnostics in aging men. Eur J Endocrinol. 2012;166:983–91.

Isidori AM, Balercia G, Calogero AE, Corona G, Ferlin A, Francavilla S, Santi D, Maggi M. Outcomes of androgen replacement therapy in adult male hypogonadism: Recommendations from the Italian society of endocrinology. J Endocrinol Investig. 2015;38:103–12.

Kalyani RR, Gavini S, Dobs AS. Male hypogonadism in systemic disease. Endocrinol Metab Clin N Am. 2007;36:333–48.

Kloner RA, Carson C III, Dobs A, Kopecky S, Mohler ER III. Testosterone and cardiovascular disease. J Am Coll Cardiol. 2016;67:545–57.

Meikle AW, Mazer NA, Moellmer JF, Stringham JD, Tolman KG, Sanders SW, Odell WD. Enhanced transdermal delivery of testosterone across nonscrotal skin produces physiological concentrations of testosterone and its metabolites in hypogonadal men. J Clin Endocrinol Metab. 1992;74:623–8.

Millar AC, Lau ANC, Tomlinson G, Kraguljac A, Simel DL, Detsky AS, Lipscombe LL. Predicting low testosterone in aging men: a systematic review. CMAJ. 2016;188:E321–30.

Molitch ME. Diagnosis and treatment of pituitary adenomas: a review. JAMA. 2017;317:516–24.

Mooradian AD, Morley JE, Korenman SG. Biological actions of androgens. Endocr Rev. 1987;8:1–28.

Morgentaler A, Zitzmann M, Traish AM, Fox AW, Jones TH, Maggi M, Arver S, Aversa A, Chan JC, Dobs AS, Hackett GI, Hellstrom WJ, Lim P, Lunenfeld B, Mskhalaya G, Schulman CC, Torres LO. Fundamental concepts regarding testosterone deficiency and treatment: International expert consensus resolutions. Mayo Clin Proc. 2016;91:881–96.

Nieschlag S. Andrology: Male reproductive health and dysfunction. Springer Science & Business Media; 2013.

Nieschlag E, Swerdloff R, Behre HM, Gooren LJ, Kaufman JM, Legros JJ, Lunenfeld B, Morley JE, Schulman C, Wang C, Weidner W, Wu FC. Investigation, treatment, and monitoring of late-onset hypogonadism in males: Isa, Issam, and Eau recommendations. J Androl. 2006;27:135–7.

Rastrelli G, Corona G, Tarocchi M, Mannucci E, Maggi M. How to define hypogonadism? Results from a population of men consulting for sexual dysfunction. J Endocrinol Investig. 2016;39:473–84.

Rastrelli G, Corona G, Mannucci E, Maggi M. Reply to Eugenio Ventimiglia, Paolo Capogrosso, Walter Cazzaniga, Francesco Montorsi, and Andrea Salonia's letter to the editor re: Giovanni Corona, Giulia Rastrelli, Abraham Morgentaler, Alessandra Sforza, Edoardo Mannucci, Mario Maggi. Meta-analysis of results of testosterone therapy on sexual function based on international index of erectile function scores. Eur Urol. 2017;72:E162–3.

Rastrelli G, Maggi M, Corona G. Pharmacological management of late-onset hypogonadism. Expert Rev Clin Pharmacol. 2018;11:439–58.

Rastrelli G, Vignozzi L, Corona G, Maggi M. Testosterone and benign prostatic hyperplasia. Sex Med Rev. 2019;7:259–71.

Rosner W, Vesper H. Toward excellence in testosterone testing: A consensus statement. J Clin Endocrinol Metab. 2010;95:4542–8.

Salonia A, Rastrelli G, Hackett G, Seminara SB, Huhtaniemi IT, Rey RA, Hellstrom WJG, Palmert MR, Corona G, Dohle GR, Khera M, Chan YM, Maggi M. Paediatric and adult-onset male hypogonadism. Nat Rev Dis Primers. 2019;5:38.

Santi D, Corona G. Primary and secondary hypogonadism. In: Simoni M, Huhtaniemi IT, editors. *Endocrinology of the testis and male reproduction*. Cham: Springer International Publishing; 2017.

Traish AM, Feeley RJ, Guay A. Mechanisms of obesity and related pathologies: Androgen deficiency and endothelial dysfunction may be the link between obesity and erectile dysfunction. FEBS J. 2009;276:5755–67.

Travison TG, Vesper HW, Orwoll E, Wu F, Kaufman JM, Wang Y, Lapauw B, Fiers T, Matsumoto AM, Bhasin S. Harmonized reference ranges for circulating testosterone levels in men of four cohort studies in the United States and Europe. J Clin Endocrinol Metab. 2017;102:1161–73.

Tsujimura A. The relationship between testosterone deficiency and men's health. World J Mens Health. 2013;31:126–35.

Vermeulen A, Verdonck L, Kaufman JM. A critical evaluation of simple methods for the estimation of free testosterone in serum. J Clin Endocrinol Metab. 1999;84:3666–72.

Wang C, Nieschlag E, Swerdloff RS, Behre H, Hellstrom WJ, Gooren LJ, Kaufman JM, Legros JJ, Lunenfeld B, Morales A, Morley JE, Schulman C, Thompson IM, Weidner W, Wu FC. ISA, ISSAM, EAU, EAA and ASA recommendations: Investigation, treatment and monitoring of late-onset hypogonadism in males. Aging Male. 2009;12:5–12.

Wu FC, Tajar A, Pye SR, Silman AJ, Finn JD, O'neill TW, Bartfai G, Casanueva F, Forti G, Giwercman A, Huhtaniemi IT, Kula K, Punab M, Boonen S, Vanderschueren D. Hypothalamic-pituitary-testicular axis disruptions in older men are differentially linked to age and modifiable risk factors: The European male aging study. J Clin Endocrinol Metab. 2008;93:2737–45.

Wu FC, Tajar A, Beynon JM, Pye SR, Silman AJ, Finn JD, O'neill TW, Bartfai G, Casanueva FF, Forti G, Giwercman A, Han TS, Kula K, Lean ME, Pendleton N, Punab M, Boonen S, Vanderschueren D, Labrie F, Huhtaniemi IT. Identification of late-onset hypogonadism in middle-aged and elderly men. N Engl J Med. 2010;363:123–35.

Zitzmann M, Faber S, Nieschlag E. Association of specific symptoms and metabolic risks with serum testosterone in older men. J Clin Endocrinol Metab. 2006;91:4335–43.

Chapter 4
Male Factor Infertility

H. Harvey, J. Collins, and R. Jalil

4.1   Introduction

Infertility is the inability of a sexually active couple, not using contraception, to achieve a pregnancy in one year (Rowe et al. 2000). Infertility and subfertility affect an estimated 15% of couples (1 in 7), with male factor infertility solely responsible in approximately 30% of these cases (Agarwal et al. 2015). A detailed history, examination and specific investigations may help identify a treatable cause of infertility. However, 30–40% of infertile men will have an unremarkable history and examination, with normal semen analysis, and no clear cause is found (idiopathic male infertility) (Jungwirth et al. 2019).

Fertility can be impaired due to a myriad of factors including a history of infection, malignancy, endocrine disturbances, previous urological surgery, genetic abnormality or warm testicular environment (undescended testis, varicocele).

Male factor infertility can be classified in various ways. Conceptually, it can be problem with sperm production or sperm delivery. It can also be classified as a pre testicular, a testicular, or post testicular problem.

4.2   Physiology

Human spermatogenesis requires 64 days to complete, with 5–10 days of epididymal transit time (Partin et al. 2020). The hypothalamic-pituitary-gonadal axis culminates in sperm production (Fig. 4.1). Gonadotrophin releasing hormone (GnRH),

H. Harvey · J. Collins · R. Jalil (✉)
Kings College Hospital, London, UK
e-mail: hannah.harvey6@nhs.net; rozh.jalil@nhs.net

S. S. Goonewardene et al. (eds.), *Men's Health and Wellbeing*,
https://doi.org/10.1007/978-3-030-84752-4_4

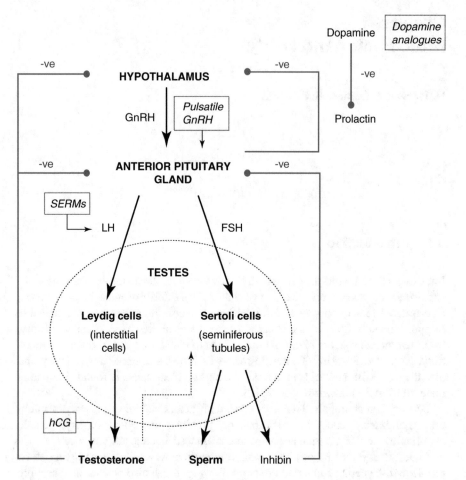

Fig. 4.1 The hypothalamic-pituitary-gonadal (HPG) axis and action of key drugs used in male infertility

released by the hypothalamus, stimulates follicle stimulating hormone (FSH) and luteinising hormone (LH) from the anterior pituitary. The cascade continues with LH acting on Leydig cells to produce testosterone, while FSH stimulates the seminiferous tubules to secrete inhibin and produce sperm. An issue at any level of this intricate pathway can lead to impaired or absent sperm production. This pathway also offers various targets for pharmacological treatment of male infertility (see Sect. 4.8).

4.3   Initial Assessment

Both partners are usually investigated in parallel to help identify any underlying and potentially treatable causes of infertility. Initial assessment requires a detailed history, which must include questions specific to male sexual development and urological history (Table 4.1).

Examination is equally important with global assessment to evaluate body mass index (BMI), muscle bulk, note of gynaecomastia, distribution of body hair and presence of secondary sexual characteristics. The penis, testes, epididymis, vas deferens and prostate should all be examined sequentially, ideally in a warm room with a chaperone. In combination these enable evaluation of sexual development, as well assessing for evidence of hypogonadism, endocrine and other medical disorders, or anatomical variations (congenital or acquired).

4.4   Semen Analysis

A semen analysis is performed on a fresh sample, delivered at body temperature, within an hour of production. For quality control purposes, the man should abstain from ejaculating for 48 h prior to the sample, but does need to have ejaculated within the preceding 7 days.

Table 4.1 Important factors in the history

Age of the patient and partner	
Reproductive history of both partners	Previous pregnancies, children, contraception use, infertility duration, previous fertility treatment
Sexual history	Previous infections, libido, erectile function, ejaculatory problems, frequency and timing of intercourse
Medical history	Diabetes mellitus, neurological conditions, respiratory illness (cystic fibrosis, primary ciliary dyskinesia), sexually transmitted infections,
Urological history	Vasectomy, trauma, testicular torsion, testicular tumour, history of undescended testes, onset of puberty, previous epididymo/orchitis/prostatitis
Surgical history	Abdominal or inguinal surgery, paediatric surgery for hypospadias or cryptorchism, orchidopexy
Social history	Alcohol intake, smoking, recreational drugs, physical activity, exogenous testosterone supplementation
Drug history	Prescribed, over the counter or recreational
Other	Cancer history, radiation or chemotherapy, family history

Table 4.2 World Health Organization normal reference values for semen analysis (Team 2010)

Parameter	Normal value
Semen volume	≥1.5 ml
pH	≥7.2
Sperm concentration	≥15 million spermatozoa/ml
Total sperm number	≥39 million spermatozoa/ejaculate
Total motility	≥40% motile
Vitality	≥58% alive
Sperm morphology (percentage of normal forms)	≥4%

An abnormal semen analysis describes an alteration in sperm concentration and/ or motility and/or morphology. The WHO normal reference values (Team 2010) are seen below (Table 4.2) and are standardised owing to the fact treatment decisions are often based on these results. The National Institute for Health and Care Excellence (NICE) guidelines recommend a repeat sample after 3 months, allowing for a repeat cycle of spermatogenesis, should any mild abnormality be detected. In cases of significant derangement, a repeat can be done in 2–4 weeks.

4.5 Definitions

There are a wide variety of terms to describe deviations from the normal semen parameters as defined by the WHO criteria.

4.5.1 Azoospermia

Azoospermia is defined as the absence of sperm in the ejaculate and is found in 10–15% of infertile males (Cocuzza et al. 2013). It can be caused by obstruction out of the testis via the vas deferens or ejaculatory ducts, or failure of spermatogenesis within the testis.

4.5.2 Oligospermia

Oligospermia describes a sperm density of less than 15 million/ml. A low sperm count can be classified as mild, moderate or severe. The underlying aetiology can be classified as pre-testicular, testicular or post testicular. Up to 25% of subfertile couples are found to have oligospermia. There is a correlation between sperm concentration and other aspects of semen quality. Both motility and morphology are usually poor with oligospermia.

4.5.3   Asthenozoospermia

Asthenozoospermia describes poor sperm motility. The sperm flagellum has a sophisticated scaffold to help drive the momentum of the sperm. A defective flagellum can therefore significantly hinder sperm motility, as seen in conditions such as primary ciliary dyskinesia. Oxidative stress is also thought to play a role, with poor resilience of the sperm membranes to reactive oxygen species (Ribas-Maynou and Yeste 2020). Cigarette smoking, infection and varicoceles have also been linked with this. Anti-sperm antibodies, formed because of a breach in the blood-testis barrier (secondary to trauma, surgery, orchitis or malignancy) are also known to impair sperm motility.

4.5.4   Teratozoospermia

Teratozoospermia describes abnormal sperm morphology and has been described as an overlooked semen parameter. Contemporary markers of sperm damage including DNA fragmentation or reactive oxygen species are linked to poorer spermatozoal morphology.

A combination of all three abnormalities is termed oligo-asthenoteratozoospermia (OAT).

4.6   Additional Investigations

Imaging of the scrotum forms part of the infertility assessment. Information will include testicular volume, the presence of any mass or varicocele, and any absence of the vas deferens. Testicular architecture can also be observed. A transrectal ultrasound (TRUS) can be used to investigate any ejaculatory duct obstruction. Vasography is an invasive test to look at the patency of the vas deferens, useful when investigating azoospermia.

Measurement of sex hormones can help identify a cause of male infertility. Follicle stimulating hormone (FSH), luteinising hormone (LH) and testosterone are key in male reproductive physiology. Testosterone and FSH act synergistically to increase the efficiency of spermatogenesis. Although endocrine causes of infertility are not common, a significant proportion of infertile men do present concomitantly with low testosterone levels. Increased prolactin levels can also be associated with sexual dysfunction and may indicate pituitary disease.

Testicular biopsy can be used diagnostically to evaluate the degree of spermatogenesis by assessing the Johnson score (a histological ten point scoring system to quantify spermatogenesis). A diagnostic biopsy is rarely done in isolation, but coupled with sperm extraction. Testicular sperm extraction (TESE) is increasingly done microscopically (microTESE) which is found to give the highest yield of sperm. This can be used for intracytoplasmic sperm injection (ICSI), which is a type of in vitro fertilisation (IVF) procedure. This is only indicated in non-obstructive cases, where it can form a cornerstone of diagnosis and treatment.

Genetic testing can be arranged to screen for a variety of conditions linked to male factor infertility, most commonly indicated in cases of severe oligospermia. This may include testing for karyotype or Y chromosome abnormalities, and other genetic conditions. The prevalence of karyotype abnormalities is inversely proportional to sperm count (2011). Chromosomal abnormalities such as Klinefelter's syndrome (47XXY) result in primary hypogonadism and variable phenotype, usually with small testicles and Leydig cell dysfunction. Y chromosome microdeletions refer to a localised region on the Y chromosome called the azoospermia factor (AZF) which is pivotal in normal spermatogenesis. These microdeletions are implicated in some cases of defective spermatogenesis, and can be further classified according to precise location. Cystic fibrosis, which can lead to congenital bilateral absence of the vas deferens (CBAVD), is screened for by testing for mutations to the cystic fibrosis transmembrane receptor (CFTR).

Sperm DNA fragmentation (SDF) testing is a newer technology which measures the quality of sperm in terms of DNA damage. Higher levels of damage are associated with poorer pregnancy rates, even in the context of in vitro fertilisation (IVF) or intracytoplasmic sperm injection (ICSI) (Kim 2018). Reactive oxygen species (ROS) have been implicated in sperm function for some time, with high levels of ROS being associated with abnormal sperm morphology. Research is ongoing and there is a lack of consensus as to who to screen in clinical practice. However, there are laboratory tests available to screen and quantify ROS (Wagner et al. 2018).

4.7 Aetiology

Table 4.3 lists some of the more common causes of male infertility. It should be noted that in patients with a testicular cause, up to 30% of cases may be idiopathic, which has ramifications for the management.

4.8 Management

The management of male infertility varies depending on the cause underlying each case, as well as the extent of impairment. Generally, management can be divided into medical therapy—typically used in pre-testicular and idiopathic cases—or assisted reproduction techniques (ART) which are more commonly used in testicular conditions and those failing to respond to medical management. In patients with a post-testicular obstructive cause, management may be either surgical or use a combination of IVF and ICSI.

Table 4.3 Classification of common causes of male infertility

Pre-testicular causes	Examples and notes
Endocrine dysfunction and hormonal abnormality	Hypogonadotropic hypogonadism/Kallmann syndrome
Systemic disease	Renal and liver failure, cystic fibrosis
Environmental factors	Hot baths
Coital issues	Erectile dysfunction, ejaculatory disorders
Drugs, alcohol, smoking or cannabis	Cannabis decreases plasma testosterone, alcohol increases the conversion of testosterone to oestradiol
Genetic abnormalities	Klinefelter's syndrome (47XXY) with an extra X chromosome is the most common genetic cause of male infertility (Hawksworth et al. 2018)
Testicular causes	**Examples and notes**
Varicoceles	Present in 15% of the adult population but in up to 50% of men presenting with fertility problems
Idiopathic	
Cryptorchism (undescended testis)	90% of patients with untreated bilateral cryptorchidism ultimately develop azoospermia (Goel et al. 2015)
Testicular injury	Trauma, post-operative, torsion
Infection	Mumps orchitis
Cancer	
Radiation or radiotherapy	
Post-testicular causes (obstruction)	**Examples and notes**
Absent vas deferens	Cystic fibrosis causing CBAVD
Obstruction at the level of epididymis, vas or ejaculatory duct	Of these the most common are epididymal obstruction or acquired obstruction of the vas deferens following vasectomy
Erectile dysfunction	
Ejaculatory disorders	Retrograde ejaculation
Infection	Prostatitis

4.8.1   Non Obstructive Azoospermia

Apart from secondary testicular failure, there is no treatment that can restore spermatogenesis in the majority of Non- obstructive azoospermia (NOA). Surgical sperm retrieval is the only treatment.

4.8.2   Hypogonadism

This describes the failure of the testis to produce a physiological level of testosterone or a normal number of spermatozoa. Primary hypogonadism is a failure of the testis and secondary hypogonadism is a failure at the level of the hypothalamus or pituitary gland (Kumar et al. 2010). Low testosterone can be linked to infertility in two ways. It has a role in decreased sperm production, and a more indirect role in reducing libido.

Hypogonadotropic hypogonadism is an uncommon, specific cause of male infertility. Gonadotrophin deficiency hinders spermatogenesis and testosterone production. It is well established that high levels of intratesticular testosterone are needed for normal spermatogenesis (Jarow and Zirkin 2005). Treatment therefore involves restoring testosterone levels, which can be done by injecting human chorionic gonadotropin (hCG) 2–3 times per week. hCG is an analogue of luteinising hormone (LH) which stimulates testicular function. The induction and maintenance of spermatogenesis is then achieved by giving follicle stimulating hormone (FSH).

Pulsatile administration of GnRH is also an effective treatment for GnRH deficiency in infertile men with hypogonadotropic hypogonadism. This is contingent on a lack of secretion from the hypothalamus but intact pituitary function such as Kallmann syndrome.

Selective oestrogen receptor modulators (SERMs), such as clomiphene citrate or tamoxifen, can be used to stimulate LH, but again rely on an intact but underactive hypothalamic-pituitary-gonadal axis. Clomiphene citrate works as an oestrogen antagonist at the level of the pituitary gland and thus stimulates the release of LH and FSH. This promotes both spermatogenesis and testosterone production. Oral clomiphene has the added benefit of not suppressing endogenous gonadotrophin secretion, which makes it superior to exogenous testosterone therapy.

Other pituitary and endocrine disorders can also cause hypogonadism. Hyperprolactinemia can be caused by pituitary tumours, medications or hypothyroidism, and leads to suppression of GnRH at the level of the pituitary. Dopamine agonist therapy (e.g. with cabergoline or bromocriptine which inhibit prolactin through negative feedback) can be used in these cases. Oestrogen excess is increasingly caused by diabetes and is known to suppress GnRH secretion, causing low FSH, LH and testosterone levels. Managing obesity is therefore paramount, as well as assessing for any other exogenous causes for high oestrogen levels.

4.8.3 Varicocele

Varicocelectomy has been the subject of much debate. Surgical varicocelectomy of clinical varicoceles has been shown to result in significant improvement of semen parameters (Baazeem et al. 2011, Esteves et al. 2016). Subclinical varicoceles are more controversial (Yamamoto et al. 1996). Treatment of varicoceles includes embolization, open, laparoscopic or microsurgical operations. A meta-analysis has shown that the microsurgery technique has a high pregnancy rate and is associated with lower recurrence rates (Cayan et al. 2009).

4.8.4   Obstructive Azoospermia

Obstructive azoospermia is commonly managed with surgical sperm retrieval. Sperm can be harvested either from the epididymis, the vas or the testis itself. Testicular sperm aspiration (TESA) is used when intratesticular obstruction is found. Microsurgical epididymal sperm aspiration (MESA) is an open surgical sperm retrieval procedure using a microscope to identify the epididymal tubules, indicated in CBAVD. Percutaneous epididymal sperm aspiration (PESA) involves aspirating fluid from the epididymis through a percutaneous approach.

4.8.5   Idiopathic Infertility

This is usually managed with assisted reproduction techniques including intra-uterine insemination (IUI) and in-vitro fertilisation (IVF). Empirical therapy has been described, although the evidence base is weak. This may consist of hormonal therapy (e.g. testosterone, anastrozole or clomiphene) or antioxidant therapy (Kumar et al. 2006, Shah and Shin 2020).

4.9   Prognosis

There are a number of prognostic factors for male infertility (Jungwirth et al. 2019), including:

- duration of infertility,
- primary versus secondary infertility (secondary infertility being when the man has successfully reproduced in the past but can no longer do so),
- results of semen analysis,
- age and fertility status of female partner.

4.10   Conclusions

Male factor infertility is a challenging topic with diverse areas for further research, as well as for the ongoing development of novel therapeutics. This is especially relevant for men with idiopathic infertility, where unlike in other idiopathic conditions, there is scant evidence for empirical therapies. Often under-recognised and continuing to bear stigma, it represents a highly important area of men's health which merits a greater discussion within the medical and general community.

References

Agarwal A, Mulgund A, Hamada A, Chyatte MR. A unique view on male infertility around the globe. Reprod Biol Endocrinol. 2015;13:37.

Baazeem A, Belzile E, Ciampi A, Dohle G, Jarvi K, Salonia A, Weidner W, Zini A. Varicocele and male factor infertility treatment: a new meta-analysis and review of the role of varicocele repair. Eur Urol. 2011;60(4):796–808.

Cayan S, Shavakhabov S, Kadioğlu A. Treatment of palpable varicocele in infertile men: a meta-analysis to define the best technique. J Androl. 2009;30(1):33–40.

Cocuzza M, Alvarenga C, Pagani R. The epidemiology and etiology of azoospermia. Clinics (Sao Paulo). 2013;68(Suppl 1):15–26.

Esteves SC, Miyaoka R, Roque M, Agarwal A. Outcome of varicocele repair in men with nonobstructive azoospermia: systematic review and meta-analysis. Asian J Androl. 2016;18(2):246–53.

Goel P, Rawat JD, Wakhlu A, Kureel SN. Undescended testicle: an update on fertility in cryptorchid men. Indian J Med Res. 2015;141(2):163–71.

Hawksworth DJ, Szafran AA, Jordan PW, Dobs AS, Herati AS. Infertility in patients with klinefelter syndrome: optimal timing for sperm and testicular tissue cryopreservation. Rev Urol. 2018;20(2):56–62.

Jarow JP, Zirkin BR. The androgen microenvironment of the human testis and hormonal control of spermatogenesis. Ann N Y Acad Sci. 2005;1061:208–20.

Jungwirth A, Diemer T, Kopa Z, Krausz C, Minhas S, Tournaye H, E. A. o. U. W. G. o. M. Infertility. EAU Guidelines on male infertility. Edn. Presented at the EAU Annual Congress Barcelona 2019. http://uroweb.org/guidelines/compilations-of-all-guidelines/, EAU Guidelines Office, Arnhem: 47. 2019

Kim GY. What should be done for men with sperm DNA fragmentation? Clin Exp Reprod Med. 2018;45(3):101–9.

Kumar R, Gautam G, Gupta NP. Drug therapy for idiopathic male infertility: rationale versus evidence. J Urol. 2006;176(4 Pt 1):1307–12.

Kumar P, Kumar N, Thakur DS, Patidar A. Male hypogonadism: symptoms and treatment. J Adv Pharm Technol Res. 2010;1(3):297–301.

Partin AW, Peters CA, Kavoussi LR, Dmockowski RR, Wein AJ. Campbell walsh wein urology. Elsevier. 2020;4096

Ribas-Maynou J, Yeste M. Oxidative stress in male infertility: causes, effects in assisted reproductive techniques, and protective support of antioxidants. Biology (Basel). 2020;9(4)

Rowe PJ, Comhaire FHH, Mahmoud TB, Ahmed MA. WHO manual for the standardized investigation and diagnosis of the infertile male. Cambridge: Cambridge University Press; 2000.

Shah T, Shin D. Empiric medical therapy for idiopathic male infertility. In: Parekattil S, Esteves S, Agarwal A, editors. Male infertility, vol. XXVII. Springer International Publishing; 2020. p. 914.

Team, W. S. a. R. H. a. R. WHO laboratory manual for the examination and processing of human semen. World Health Organization: 271; 2010.

Wagner H, Cheng JW, Ko EY. Role of reactive oxygen species in male infertility: an updated review of literature. Arab J Urol. 2018;16(1):35–43.

Yamamoto M, Hibi H, Hirata Y, Miyake K, Ishigaki T. Effect of varicocelectomy on sperm parameters and pregnancy rate in patients with subclinical varicocele: a randomized prospective controlled study. J Urol. 1996;155(5):1636–8.

Chapter 5
Erectile Dysfunction

Clare Akers, Hussain M. Alnajjar, and Asif Muneer

5.1 Introduction

Erectile dysfunction (ED) is a common male sexual disorder. Although not life threatening, it can influence psychosocial health and have a considerable impact on the quality of life of patients and their partner's (Hackett et al. 2018). ED is defined as the persistent inability to attain and/or maintain an erection sufficient to enable satisfactory sexual intercourse (NIH Consensus Conference 1993).

Normal physiological penile erection is a complex neurovascular phenomenon; it requires dilation of the penile vasculature, relaxation of corporal smooth muscle, increased intracavernosal blood flow and a normal veno-occlusive function (McMahon 2019). Any disruption to this can result in ED.

Several large epidemiological studies show a high incidence and prevalence of ED worldwide (Feldman et al. 1994; Laumann et al. 1999; Braun et al. 2000, Eardley 2013; Quilter et al. 2017). It is estimated that one third of all men will be affected by ED at some time in their life (Heidelbaugh 2010). There is a steep increase in the incidence of ED with age, usually around the 5th decade of life.

ED is subdivided into two groups:

- Primary ED: manifestation occurs within the first sexual encounter (i.e., in adolescence)
- Secondary ED: when ED occurs after a period of normal sex life.

C. Akers · H. M. Alnajjar (✉) · A. Muneer
University College London Hospital, London, UK
e-mail: Hussain.alnajjar@nhs.net; Asif.Muneer@nhs.net

© The Author(s), under exclusive license to Springer Nature Switzerland AG 2022
S. S. Goonewardene et al. (eds.), *Men's Health and Wellbeing*,
https://doi.org/10.1007/978-3-030-84752-4_5

5.2   Risk Factors

The risk factors for ED are remarkably similar to the established risk factors for cardiovascular disease (age, obesity, smoking, dyslipidaemia, diabetes mellitus, sedentary lifestyle and metabolic syndrome) (Thompson et al. 2005; Vlachopoulos et al. 2010). ED can be associated with lower urinary tract symptoms (LUTS) and benign hypertrophic hyperplasia (BPH) (Seftel et al. 2013). Rosen et al. (2004) reported that men with LUTS had an overall prevalence of ED of 49%, and 46% experienced ejaculatory disorders. Other risk factors for ED include hypogonadism, neurological disease (e.g., Parkinson's disease and multiple sclerosis), pelvic surgery (e.g., radical prostatectomy), pelvic radiotherapy, renal failure, and chronic lung disease (Hackett et al. 2018).

There are a variety of medications that can have a negative impact on erectile function (EF), refer to Table 5.1.

Various studies have demonstrated ED to be a significant cardiovascular risk factor. It can predict cardiovascular disease 3–5 years before the onset of symptoms of a cardiovascular event (Jackson et al. 2010). Therefore, men over the age of 40 with chronic ED must be considered as a potential cardiovascular risk and appropriately investigated.

5.3   Diagnosing Erectile Dysfunction

The first step in evaluating ED is always to undertake a detailed medical and sexual history from the patient and where available their partner (Salonia et al. 2020, Hackett et al. 2018). It is essential to establish a relaxed atmosphere during

Table 5.1 Common medical conditions and medications associated with erectile dysfunction

Neurogenic	Medications	Psychological
Spinal cord injury	Antidepressants	Depression
Multiple sclerosis	Excessive alcohol	Anxiety
Cerebrovascular accident (CVA)	Recreational drugs	Substance
Peripheral neuropathy	Antihypertensives	misuse
Diabetes	Antipsychotics	Bipolar
Major pelvic surgery (e.g., radical prostatectomy) or radiotherapy (pelvis or retroperitoneum)	Anticonvulsants	disorder
		Schizophrenia
		Stress
Vascular	**Hormonal**	
Diabetes	Low testosterone	
Major pelvic surgery (e.g., radical prostatectomy) or radiotherapy (pelvis or retroperitoneum)	Decreased Luteinising hormone (LH)	
Cardiovascular diseases (e.g., hypertension, coronary artery disease, peripheral vasculopathy)	High prolactin	
Obesity		
Smoking		

history-taking, which will make it easier to ask personal questions about the patient's erectile function and other aspects of their sexual history.

The history should be aimed at identifying typical risk factors which are associated with ED. It is essential to identify their past medical history such as diabetes, cardiovascular diseases, surgical procedures, and drug history e.g., antihypertensives. Enquiring about their smoking history, alcohol intake and any recreational drug misuse is essential. Medical conditions such as metabolic or endocrine disorders (e.g., diabetes and hypogonadism) should always be treated as the first step when treating ED. Modifiable risk factors, such as lifestyle factors can be addressed either prior to or during treatment (Gupta et al. 2011)

The sexual history must include information about onset (gradual or sudden) and duration of the erectile problem. Other details that should be obtained and recorded include current emotional status, current and previous sexual relationships, and any previous treatments that have been trialled. Men should be asked about duration; rigidity of morning, nocturnal and sexually stimulated erections. Difficulties with arousal, libido, ejaculation, and orgasm should also be explored (Althof et al. 2013). Where available, the sexual health status of the partner should also be obtained (Salonia et al. 2020)

The use of validated questionnaires, such as the International Index for Erectile Dysfunction (IIEF), are useful tools to assist the clinician in assessing sexual function domains (i.e., libido, orgasmic function, intercourse, erectile function, and overall satisfaction, as well as impacts of particular treatments), but they are not a substitute for a thorough medical and sexual history (Hackett et al. 2018).

5.4 Physical Examination

Every patient should have a genital examination. This may reveal unsuspected diagnoses, such as pre-malignant or malignant genital lesions, Peyronie's disease, or signs and symptoms suggestive for hypogonadism (i.e., small testes). A digital rectal exam (DRE) of the prostate is not deemed essential but should be performed in the presence of genito-urinary symptoms or secondary ejaculatory problems (Hackett et al. 2018).

Weight and waist circumference should be measured, as should blood pressure and heart rate, (if they have not already been checked in the previous three to six months) to elicit a potential cardiovascular cause (Salonia et al. 2020).

5.5 Blood Tests

The choice of investigations must be tailored to the patient's risk factors and individual circumstances. As ED can be the presenting feature of diabetes and is associated with cardiovascular disease (Vlachopoulos et al. 2010; Jackson et al. 2013), serum lipids and fasting blood glucose should be obtained in all presenting patients.

Hypogonadism is a treatable cause of ED; therefore, all men should have an early morning (between 8 and 11 am) serum testosterone measured in a fasting state. If the testosterone level is low or borderline, this should be repeated, together with prolactin and luteinising hormone (LH) (Dean et al. 2015). If these results are abnormal, discussion or referral to a specialist endocrinologist should be considered.

Serum prostate-specific antigen (PSA) should be performed if clinically indicated. Additionally, it should be measured before starting testosterone replacement, at 3–6 months post treatment, and then annually (Montorsi et al. 2006).

Although, physical examination and blood tests of most men with ED may not reveal the exact diagnosis, it does present an opportunity to detect comorbid conditions (Ghanem et al. 2013) (Fig. 5.1).

5.6 Specialist Diagnostic Tests

Most patients will not need to undergo specialist investigations as they can often be managed after taking a clear medical and sexual history. However, some reasons why patients may require specialist investigation include:

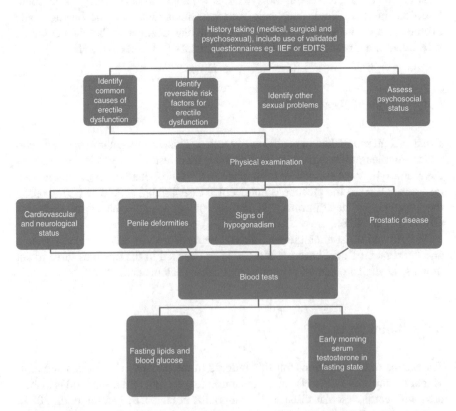

Fig. 5.1 Initial patient work up for a patient presenting with erectile dysfunction (adapted from Salonia et al. 2020)

- Young patients who have always had difficulty in obtaining and/or sustaining an erection.
- Patients who are keen to understand the aetiology of their disease.
- Those with a history of trauma.
- Where an abnormality of the penis or testes is found during examination
- Patients unresponsive to medical therapies who are considering penile prosthesis (Hackett et al. 2018)

5.7 Nocturnal Penile Tumescence Study (NPT)

Nocturnal and early morning erections are a normal physiological event in men and relate to the rapid eye movement pattern of sleep (Hackett et al. 2018). NPT is a sleep study that is used to determine nocturnal penile rigidity and tumescence. Typically, men have 3–5 erections during 8 h of sleep with an erectile event indicated by >60% rigidity, lasting for >10 min (Zou et al. 2019). Men require an overnight stay in hospital and a positive result (where strong erectile activity has been recorded), will indicate psychogenic ED (Fig. 5.2).

5.8 Intracavernous Injection Test

This test performed in outpatients involves administering an intracavernosal injection (usually prostaglandin E1) to assess penile rigidity after 10 min (Hatzichristou et al. 1999). The main use of this test is to assess penile deformities and plan surgical management. Its use as a diagnostic test for ED is limited, as a positive result will be found in patients with both normal and mild vascular disease (Hatzichristou et al. 1999).

Fig. 5.2 Rigiscan

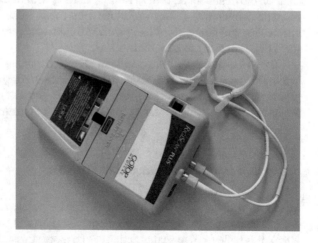

5.9 Duplex Ultrasound of the Penis

This is a radiological investigation that assesses the inflow and outflow of blood through the cavernosal arteries after an injection of a vasoactive agent (such as prostaglandin E1) (Gupta et al. 2015). Maximum systolic velocity of >25–30 cm/s and a minimum end-diastolic velocity <5 cm/s are considered normal. Indications for Doppler ultrasound include assessment of arterial integrity and veno-occlusive mechanism, and assessment for vascular anomalies, e.g. peyronie's plaque, fistula or priapism (Hackett et al. 2018)

5.10 Cavernosography

Cavernosography is rarely offered nowadays, even in specialist UK andrological centres. The sole indication being for investigating young men with primary venous aetiology (Kirby et al. 2013).

5.11 Treatment

The principal objective in the management of ED is to enable the patient or couple to enjoy a satisfactory sexual experience. This involves identifying and treating curable causes of ED, initiating lifestyle changes, risk-factor modification, and providing education and counselling to the patient and their partner (Hackett et al. 2018).

Treatment options depend on the severity of the ED, the patient and partners choice, the patient's health, and their co-morbidities. Guidelines on the management of erectile dysfunction were developed to standardise care for men with ED (Hackett et al. 2008). These guidelines illustrate a three-stage process to guide clinicians on the treatment algorithm for ED, starting with first line oral phosphodiesterase type 5 inhibitors (PDE5i), followed by second line intraurethral alprostadil/injectables, and finally third line penile prosthesis surgery for those men with end-stage ED, refractory to pharmacotherapy.

ED can be successfully treated with current treatment options, but it cannot be cured. The few exceptions to this are post-traumatic arteriogenic ED in young patients, psychogenic ED, and hormonal causes (e.g., hypogonadism) (Isidori et al. 2014; Maggi et al. 2013), which possibly could be cured with particular treatments (Fig. 5.3).

5.12 ED and Cardiovascular Disease

Most men with coronary heart disease (CHD) can safely resume sexual activity and use ED treatments following appropriate education and counselling (Jackson et al. 2013). It is believed that sexual activity is no more stressful to the heart than normal

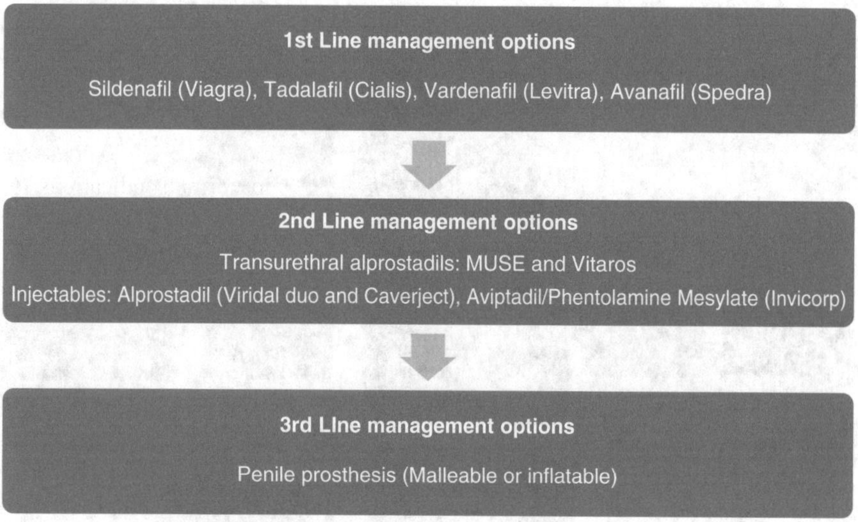

Fig. 5.3   ED treatment algorithm (adapted from Hackett et al. 2008)

daily activities, and there is no current evidence that approved ED treatments add to the overall cardiovascular risk in patients with or without diagnosed cardiovascular disease (Hackett et al. 2018).

Patients with unstable angina, recent history of stroke or myocardial infarction, unstable dysrhythmia and poorly compensated heart failure are the exception (Jackson et al. 2013). Many ED guidelines follow the Princeton III consensus guidelines for managing patients with ED and cardiovascular disease. Patients are assigned to three groups according to their risk factors: low risk, intermediate risk and high cardiac risk. For those patients at high cardiac risk, assessment and management should be supervised by a cardiologist (Nehra et al. 2012) (Fig. 5.4).

5.13   First Line Treatment

5.13.1   Oral Pharmacotherapy: PDE5 Inhibitors

PDE5i selectively inhibit PDE5 isoenzyme, increasing the amount of cyclic guanosine monophosphate (cGMP) available for smooth muscle relaxation, inducing vasodilation and increased corporal blood flow which leads to an erection.

PDE5i oral tablets are recommended as the first line treatment for men with ED as they are non-invasive and generally well tolerated. There are currently 4 PDE5 inhibitors that have been approved by the European Medicines Agency (EMA) and the U.S Food and Drug Administration (FDA): Sildenafil (Viagra), Tadalafil (Cialis), Vardenafil (Levitra) and Avanafil (Spedra). The main difference in these medications

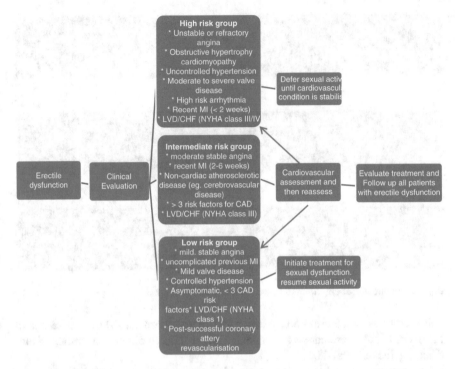

Fig. 5.4 Management algorithm and cardiovascular risk assessment for patients with ED (adapted from Princeton III Consensus conference (Nehra et al. 2012))

is that Sildenafil, Vardenafil and Avanafil are relatively short acting, with an esti-mated half-life of 4 h; whereas Tadalafil is long acting, with a half-life of 17.5 h. PDE5i are not initiators of erection and therefore require sexual stimulation to be effective (Corbin et al. 2002). Interaction of PDE5i with food, particularly fatty food, is greatest with Sildenafil (Fink et al. 2002), therefore it should be taken on an empty stomach, whereas Tadalafil is better absorbed with food. Adherence to administra-tion advice is paramount to maximise drug potential. It is recommended that patients should trial 8 doses of a PDE5i with sexual stimulation at maximum dose before being categorized as a non-responder (Hackett et al. 2018) (Tables 5.2 and 5.3).

Absolute contraindications to PDE5i are regular or intermittent use of nitrates in any form, including nicorandil, amyl nitrate "poppers", and guanyl cyclase stimula-tors, for example riociguat (Hackett et al. 2018). The use of these in combination could result in cyclic guanosine monophosphate accumulation, unpredictable falls in blood pressure and potentially devastating hypotension and death (Oliver et al. 2009). PDE5i are also contraindicated in patients who have suffered a stroke or myocardial infarction during the past 6 months, who have severe arrhythmias, unstable angina or severe heart disease (Diaconu et al. 2020). Dose adjustments may be required in those with hepatic or renal impairment or those on concomitant potent cytochrome P450 34A inhibitors such as Erythromycin, cimetidine and rito-navir (Hackett et al. 2018).

Table 5.2 Pharmaco-kinetics and safety data for different PDE5 inhibitors

Medication	Effective after	Efficacy maintained	Dose (mg)	Absorption
Sildenafil	30–60 min	Up to 12 h	25, 50 and 100 mg (starting dose 50 mg)	Take on empty stomach Not affected by alcohol
Tadalafil	30 min	>36 h	10–20 mg Daily dose 2.5–5 mg	Not affected by food or alcohol
Vardenafil	30 min	4–5 h	5, 10, 20 mg (starting dose 10 mg)	Take on empty stomach Not affected by alcohol
Avanafil	15–30 min	4–6 h	50, 100, 200 mg (starting dose 100 mg)	Not affected by food or alcohol

Table 5.3 Adverse side effects of the PDE5 inhibitors (Electronic Medicines Compendium 2017)

	Sildenafil	Tadalafil	Vardenafil	Avanafil
Very common (>1/10)	Headaches		Headache	
Common (>1/100 to >1/10)	Dizziness Visual disorders, visual colour distortion Flushing Nasal congestion Dyspepsia	Headache Flushing Nasal congestion Dyspepsia Back pain/ myalgia	Dizziness Flushing Nasal congestion Dyspepsia	Headache Flushing Nasal congestion

5.13.2 Vacuum Erection Devices (VED)

A vacuum erection device (VED) comprises of three components: a vacuum cylinder, a manual or battery-operated vacuum pump and constriction rings of various sizes.

Users begin by placing a constriction ring over the end of the vacuum cylinder. Lubricant is then applied to the penis before the clear plastic cylinder is placed over the penis and secured against the abdomen. Negative pressure is then generated using either the hand or battery-operated pump. Once, the penile tissue become engorged with blood, a constriction ring is then placed at the base of the penis which will enable the user to retain blood into the corpora.

An erection from a vacuum constriction device differs from a physiological erection because smooth muscle relaxation does not occur, as blood is simply trapped in the corpora cavernosa with the constriction ring. In some men, the size of the penis is larger than that obtained with a normal erection, and during sexual intercourse the penis may feel slightly numb, look blueish in colour or feel slightly cooler due to tissue cyanosis (Lewis and Witherington 1997)

VEDs are highly effective in inducing erections regardless of the aetiology of the ED (Levine and Dimitriou 2001; Dutta and Eid 1999), although its success depends on practice and appropriate instruction. Satisfaction rates vary considerably from 35 to 84% (Corona et al. 2017; Baltaci et al. 1995). Side effects include pain, failure to ejaculate and bruising, particularly in those on anticoagulation therapy or for those with bleeding disorders. Users should be reminded to remove the constriction ring after 30 min to avoid skin necrosis (Baltaci et al. 1995).

VED may be the treatment of choice in men who practice infrequent sexual intercourse and in those men with multiple comorbidities who require or request medication free, non-invasive management of their ED. VED are also used by men recovering from radical prostatectomy to prevent penile shortening (Hoyland et al. 2013) (Fig. 5.5).

5.14   Second Line Treatment

If first line PDE5i prove ineffective, intolerable or are contraindicated then second line treatment options should be considered.

5.14.1   Intracavernosal Injection

Intracavernous injection therapy was the first medical treatment for ED and is the most effective form of pharmacotherapy (Hackett et al. 2018). Providing the blood supply is acceptable, an excellent result can be achieved in most men as it does not require an intact nerve supply to be effective.

Intracavernosal injections are administered by the patient or their partner. However, because of the invasive nature of the treatment it is not acceptable to some

Fig. 5.5 A vacuum device with various constriction rings

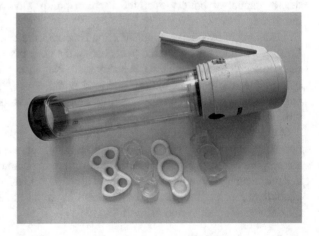

patients and their partners, and this may result in poor long-term compliance in those who do try it (Lakin et al. 1990; Porst 1996). Compliance may be a particular problem if the technique is not explained clearly and fully, thus the first injection should be administered following supervision and training in the clinical setting.

Box 5.1 Instructions on Administration of Intracavernosal Injection
- Patients may wish to bring their partners to the appointment. This should be encouraged. If a man is needle phobic or their body habitus does not allow direct visualisation of the penis, it may be appropriate to teach the partner how to inject.
- After washing hands, patients are shown how to prepare the medication. Patients may choose to video this as they are often nervous and they can be reminded that the written instructions demonstrate how to prepare the medication.
- The patient should be advised of the appropriate areas for self-injection, which is usually within the proximal third of the penis along the dorsolateral aspect of the penis.
- Visible veins are avoided to prevent bruising and patients are advised to avoid the underside (urethra) and top of the penis (to avoid nerves)
- Following administration of the medication, patients are encouraged to massage the penis whilst sitting or standing for 10 min to encourage erection.
- Erection should occur within 10–15 min and last between 30 and 60 min. This is dose dependant.
- Patients are advised to self-inject into both sides of the penis to prevent fibrosis and/or development of penile curvature.
- Intracavernosal injections should not be used more than once in 24 h and no more than 3 times in a week.

5.14.2 Alprostadil

In the UK, the first line injectable drug is alprostadil which is available as Viridal duo or Caverject. Chemically identical to prostaglandin E1 (PGE1), alprostadil targets an enzyme called adenylate cyclase, which converts adenosine triphosphate (ATP) into cyclic adenosine monophosphate (cAMP). cAMP lowers intracellular calcium, resulting in smooth muscle relaxation (Lue et al. 2004). Vasodilation occurs, which increases cavernosal arterial blood flow resulting in penile engorgement.

Alprostadil can be used in doses ranging from 2.5 to 40 mcg. The erection occurs typically 10–15 min after penile injection and frequently last 30–40 min, although the duration will be dose dependent. Adverse effects of intra-cavernous alprostadil include post injection penile pain (in up to 50% of patients) (Mulhall et al. 1999). Other complications include priapism (1%) and fibrosis (2%) (Chochina et al. 2016). Priapism is a prolonged firm erection lasting more than 4 hours without spontaneous resolution. Urgent medical attention is needed due to the potential detrimental effects on the function and viability of the corporal tissue if intervention is deferred. Systemic side effects are uncommon; the most common being mild hypotension when using higher doses (Chochina et al. 2016). Alprostadil use in ED is well tolerated and it has a good safety profile. Contraindications are few but include a history of hypersensitivity to alprostadil, a risk of priapism, severe penile deformity, or those with unstable cardiovascular disease (Hackett et al. 2018). Men with blood disorders such as multiple myeloma, leukaemia and sickle cell disease have an increased risk of priapism, therefore intracavernosal alprostadil should be used with caution or avoided (Electronic Medicines Compendium 2020b).

5.14.3 Aviptadil and Phentolamine Injection (Invicorp)

More recently Invicorp was approved and licensed in several European countries for ED. Invicorp constitutes aviptadil (Vasoactive Intestinal Peptide- VIP), which is a naturally occurring amino acid neuro-transmitter, and phentolamine, an alpha blocker. When directly injected into the corpus cavernosa, the drug relaxes the smooth muscle and erection occurs (Dinsmore and Wyllie 2008).

Invicorp is only available in one dose, 25 mcg/2 mg, therefore it is not possible to perform dose titration. Adverse events include mild or moderate facial flushing (Hackett et al. 2018), penile fibrosis and priapism.

5.14.4 Intraurethral Alprostadil

Intraurethral alprostadil is a less effective treatment than intracavernosal injection therapy. However, it is considerably less invasive, and for the needle phobic, will be the preferred method of delivery.

A formulation of alprostadil in a medicated pellet (medicated urethral system for erection, MUSE) was approved for the treatment of ED in 1997. It is available in doses of 500 mcg and 1000 mcg. Patients are told to void to moisten the urethra; the pellet is then inserted into the urethra via a small applicator. Alprostadil is delivered into the penile urethra and is absorbed through the epithelium into the venous channels of the corpus spongiosum. It reaches the vascular smooth muscle of the corpora cavernosum by retrograde flow through emissary veins (Hackett et al. 2018). The penis should be held in an upright position and massaged to encourage distribution of the drug to the penile tissues.

MUSE results in erections in approximately 30–60% of patients (Guay et al. 2000). Erection occurs within 10–15 min of administration and should last between

30 and 60 min. Side effects include penile pain (30–40%) and dizziness (2–10%). Priapism and penile fibrosis are rare (<1%). Urinary infection and urethral bleeding may occur due to poor technique.

Vitaros®, which has more recently become available, is a formulation of alprostadil intraurethral cream which is licenced at a dose of 300 mcg. A preloaded syringe is used to administer the cream into the urethra via the urethral meatus. Erection is usually achieved within 5–30 min of administration. Men are encouraged to massage the penis to initiate a response. Side effects are usually related to the local effect of topical cream, mainly genital pain.

MUSE should be avoided in men who have distal urethral stricture, and men with partners with possible or confirmed pregnancy should avoid both MUSE and Vitaros, unless with protection of a condom, due to their vasodilatory effects.

5.14.5 Psychosexual Therapy

For patients with a recognised psychological element to their ED, psychosexual counselling may be given either alone, or concurrently with another therapeutic treatment to improve patient (and partner) sexual function and satisfaction (Hackett et al. 2018)

5.15 Third Line Treatment

5.15.1 Penile Prosthesis

The surgical implantation of a penile prosthesis should be offered to all patients who have failed or declined other available treatments for erectile dysfunction (Salonia et al. 2020). They are classified as third line or 'end stage' treatment, as during placement, the corpus cavernosum smooth muscle is disrupted to make space for the cylinder, thus making the procedure irreversible (Hackett et al. 2018).

There are two types of penile prosthesis, malleable (1 piece) and inflatable penile prosthesis (2- and 3-piece devices). The 3-piece inflatable penile prosthesis may be the preferred choice for most patients due to their ability to obtain a more 'natural erection'. This device consists of three parts, 2 inflatable cylinders, a reservoir that is placed in the retzius space (extra-peritoneal) and a pump that inflates and deflates the device which is positioned within the scrotum (Wilson et al. 1999). Inflatable penile prostheses allow men the opportunity to achieve spontaneity for sexual intercourse by activating the pump 'on demand.'

Malleable penile prosthesis is the least invasive of the implant types, but they result in a constant firm penis, and men must be advised of the need to conceal the device. As malleable penile prosthesis has no working parts, they are more appropriate for someone who has limited manual dexterity, unless they have a partner who is able to manipulate the prosthesis pump.

2-piece devices have 2 cylinders and a pump in the scrotum, the reservoir is sited in the proximal part of the cylinders. Patients pump the device in order to inflate, but deflate the device by bending the implant. The 2-piece devices are usually reserved for patients who are deemed considerable risk of complications with placement of reservoir, for example those who have undergone major pelvic or abdominal surgery.

Satisfaction rates following penile prosthesis surgery are high (Levine et al. 2016; Carson et al. 2000). Pre-operative counselling at a specialist centre is essential to ensure that patients have appropriate expectations. They should be given the opportunity to see and handle the different devices. Patients should be advised that a penile prosthesis will provide girth and rigidity to the penis but no additional penile length (Hackett et al. 2018), and they do not change the patient's ability to orgasm or ejaculate. Patients should be aware of potential complications related to the surgery that include infection, erosion, and mechanical failure that may require revision surgery (Hackett et al. 2018) (Figs. 5.6, 5.7, 5.8, and 5.9).

Fig. 5.6 Malleable penile prosthesis

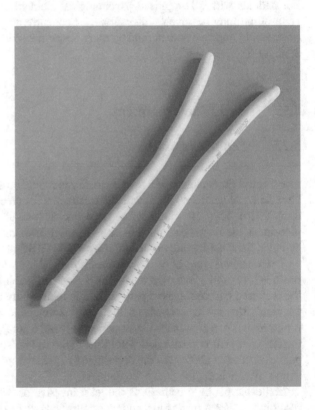

Fig. 5.7 Inflatable penile prosthesis (AMS)

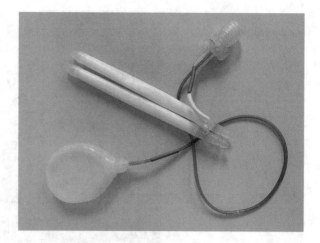

Fig. 5.8 Inflatable penile prosthesis (Coloplast titan touch Zero)

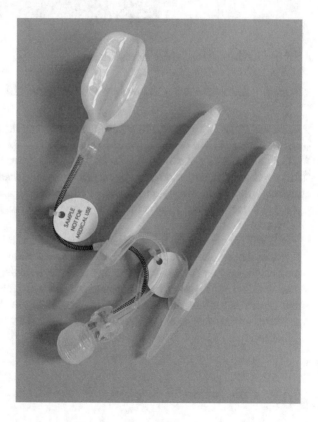

Fig. 5.9 Ambicor (two
piece penile prosthesis)

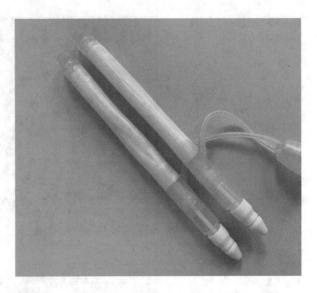

5.16   Other Available Treatment

5.16.1   Low Intensity Shockwave Therapy (LiSWT)

Since the 1980s, when it was first introduced for renal lithotripsy, shock wave therapy has been utilised over the world for different diseases, promoting regenerative effects or producing destructive effects. The shock wave carries energy, and when delivered through a medium, can be targeted non-invasively to a selected anatomic region. When low intensity shock wave treatment is applied to an organ, the shock waves interact with the tissues that have been targeted encourage angiogenesis and neovascularisation.

LiSWT for erectile dysfunction is still in its infancy. Recent meta-analyses advocate that there might be a place for LiSWT in men with mild ED wishing to avoid medical therapy and as a salvage therapy for PDE5i non-responders (Lu et al. 2017; Gao et al. 2016). However, further studies are needed to fully corroborate its potency and treatment regime, although initial reports advocate this could be a safe, non-invasive, and a well-tolerated treatment for men with mild erectile dysfunction (Hackett et al. 2018).

5.17   Conclusion

ED is one of the most common sexual disorders in men which can have a devastating psychological impact on the patient and their partner. There are several risk factors associated with ED which include diabetes, cardiovascular disease, hypertension, dyslipidaemia. ED is an early warning sign of coronary artery disease. A thorough history taking, and examination should help to identify potential causes of ED and any modifiable risk factors, including undiagnosed conditions, medications or lifestyle choices. These should be addressed either prior to commencement of treatment or in combination to improve the patient's health and erectile function.

References

Althof SE, et al. Standard operating procedures for taking a sexual history. J Sex Med. 2013;10:26.

Baltaci S, Aydos K, Kosar A, Anafarta K. Treating erectile dysfunction with a vacuum tumescence device: a retrospective analysis of acceptance and satisfaction. Br J Urol. 1995;76(6):757–60.

Braun M, Wassmer G, Klotz T, Reifenrath B, Mathers M, Engelmann U. Epidemiology of erectile dysfunction: results of the 'Cologne Male Survey'. Int J Impot Res. 2000;12(6):305–11.

Carson CC, Mulcahy JJ, Govier FE, AMS 700CX Study Group. Efficacy, safety and patient satisfaction outcomes of the AMS 700CX inflatable penile prosthesis: results of a long-term multi-center study. J Urol. 2000;164(2):376–80.

Chochina L, Naudet F, Chéhensse C, et al. Intracavernous injections in spinal cord injured men with erectile dysfunction, a systematic review and meta-analysis. Sex Med Rev. 2016;4:257–69.

Corbin JD, Francis SH, Webb DJ. Phosphodiesterase type 5 as a pharmacologic target in erectile dysfunction. Urology. 2002;60(2):4–11.

Corona G, Rastrelli G, Morgentaler A, Sforza A, Mannucci E, Maggi M. Meta-analysis of results of testosterone therapy on sexual function based on international index of erectile function scores. Eur Urol. 2017;72(6):1000–11.

Dean JD, McMahon CG, Guay AT, Morgentaler A, Althof SE, Becher EF, Bivalacqua TJ, Burnett AL, Buvat J, El Meliegy A, Hellstrom WJ. The International Society for Sexual Medicine's process of care for the assessment and management of testosterone deficiency in adult men. J Sex Med. 2015;12(8):1660–86.

Diaconu CC, Manea M, Marcu DR, Socea B, Spinu AD, Bratu OG. The erectile dysfunction as a marker of cardiovascular disease: a review. Acta Cardiologica. 2020;75(4):286–92.

Dinsmore WW, Wyllie MG. Vasoactive intestinal polypeptide/phentolamine forintracavernosal injection in erectile dysfunction. BJU Int. 2008;102(8):933–7.

Dutta TC, Eid JF. Vacuum constriction devices for erectile dysfunction: a long-term, prospective study of patients with mild, moderate, and severe dysfunction. Urology. 1999;54(5):891–3.

Eardley I. The incidence, prevalence, and natural history of erectile dysfunction. Sex Med Rev. 2013;1:3.

Electronic Medicines Compendium. Alprostadil 2020 Caverject 10 micrograms powder for solution for injection - Summary of Product Characteristics (SmPC) - (emc) (medicines.org.uk). Last updated March 2017. Accessed 6 Dec 2020.

Feldman HA, Goldstein I, Hatzichristou DG, Krane RJ, McKinlay JB. Impotence and its medical and psychosocial correlates: results of the Massachusetts Male Aging Study. J Urol. 1994;151(1):54–61.

Fink HA, Mac Donald R, Rutks IR, Nelson DB, Wilt TJ. Sildenafil for male erectile dysfunction: a systematic review and meta-analysis. Arch Int Med. 2002;162(12):1349–60.

Gao L, Qian S, Tang Z, Li J, Yuan J. A meta-analysis of extracorporeal shock wave therapy for Peyronie's disease. Int J Impot Res. 2016;28:161–6.

Ghanem HM, et al. SOP: physical examination and laboratory testing for men with erectile dysfunction. J Sex Med. 2013;10:108.

Guay AT, Perez JB, Velásquez E, Newton RA, Jacobson JP. Clinical experience with intraurethral alprostadil (MUSE®) in the treatment of men with erectile dysfunction. Eur Urol. 2000;38(6):671–6.

Gupta BP, Murad MH, Clifton MM, Prokop L, Nehra A, Kopecky SL. The effect of lifestyle modification and cardiovascular risk factor reduction on erectile dysfunction: a systematic review and meta-analysis. Arch Int Med. 2011;171(20):1797–803.

Gupta N, Herati A, Gilbert BR. Penile Doppler ultrasound predicting cardiovascular disease in men with erectile dysfunction. Curr Urol Rep. 2015;16(3):16.

Hackett G, Kell P, Ralph D, Dean J, Price D, Speakman M, Wylie K. British Society for Sexual Medicine guidelines on the management of erectile dysfunction. J Sex Med. 2008;5(8):1841–65.

Hackett G, Kirby M, Wylie K, Heald A, Ossei-Gerning N, Edwards D, Muneer A. British Society for Sexual Medicine guidelines on the management of erectile dysfunction in men—2017. J Sex Med. 2018;15(4):430–57.

Hatzichristou DG, Hatzimouratidis K, Apostolidis A, Ioannidis E, Yannakoyorgos K, Kalinderis A. Hemodynamic characterization of a functional erection. Eur Urol. 1999;36(1):60–7.

Heidelbaugh JJ. Management of erectile dysfunction. Am Fam Physician. 2010;8(3):305–12.

Hoyland K, Vasdev N, Adshead J. The use of vacuum erection devices in erectile dysfunction after radical prostatectomy. Rev Urol. 2013;15(2):67.

Isidori AM, et al. A critical analysis of the role of testosterone in erectile function: from pathophysiology to treatment-a systematic review. Eur Urol. 2014;65:99.

Jackson G, Boon N, Eardley I, Kirby M, Dean J, Hackett G, Montorsi P, Montorsi F, Vlachopoulos C, Kloner R, Sharlip I. Erectile dysfunction and coronary artery disease prediction: Evidence-based guidance and consensus. Int J Clin Pract. 2010;64(7):848–57.

Jackson G, Nehra A, Miner M, Billups KL, Burnett AL, Buvat J, Carson CC, Cunningham G, Goldstein I, Guay AT, Hackett G. The assessment of vascular risk in men with erectile dysfunction: the role of the cardiologist and general physician. Int J Clin Pract. 2013;67(11):1163–72.

Kirby M, Chapple C, Jackson G, Eardley I, Edwards D, Hackett G, Ralph D, Rees J, Speakman M, Spinks J, Wylie K. Erectile dysfunction and lower urinary tract symptoms: a consensus on the importance of co-diagnosis. Int J Clin Pract. 2013;67(7):606–18.

Lakin MM, Montague DK, Medendorp SV, Tesar L, Schover LR. Intracavernous injection therapy: analysis of results and complications. J Urol. 1990;143(6):1138–41.

Laumann EO, Paik A, Rosen RC. Sexual dysfunction in the United States: prevalence and predictors. JAMA. 1999;281(6):537–44.

Levine LA, Dimitriou RJ. Vaccum constriction and external erection devices in erectile dysfunction. Urol Clin North Am. 2001;28(2):335–42.

Levine LA, Becher E, Bella A, Brant W, Kohler T, Martinez-Salamanca JI, Trost L, Morey A. Penile prosthesis surgery: current recommendations from the International Consultation on Sexual Medicine. J Sex Med. 2016;13(4):489–518.

Lewis RW, Witherington R. External vacuum therapy for erectile dysfunction: use and results. World J Urol. 1997;15(1):78–82.

Lu Z, Lin G, Reed-Maldonado A, Wang C, Lee YC, Lue TF. Low-intensity extracorporeal shock wave treatment improves erectile function: a systematic review and meta-analysis. Eur Urol. 2017;71:223–33.

Lue TF, Giuliano F, Montorsi F, Rosen RC, Andersson KE, Althof S, Christ G, Hatzichristou D, Hirsch M, Kimoto Y, Lewis R. Summary of the recommendations on sexual dysfunctions in men. J Sex Med. 2004;1(1):6–23.

Maggi M, et al. Hormonal causes of male sexual dysfunctions and their management (hyperprolactinemia, thyroid disorders, GH disorders, and DHEA). J Sex Med. 2013;10:661.

McMahon CG. Current diagnosis and management of erectile dysfunction. Med J Aust. 2019;210(10):469–76.

Montorsi P, Ravagnani PM, Galli S, Salonia A, Briganti A, Werba JP, Montorsi F. Association between erectile dysfunction and coronary artery disease: matching the right target with the right test in the right patient. Eur Urol. 2006;50(4):721–31.

Mulhall JP, Jahoda AE, Cairney M, et al. The causes of patient dropout from penile self-injection therapy for impotence. J Urol. 1999;162:1291–4.

Nehra A, Jackson G, Miner M, Billups KL, Burnett AL, Buvat J, Carson CC, Cunningham GR, Ganz P, Goldstein I, Guay AT. The Princeton III Consensus recommendations for the management of erectile dysfunction and cardiovascular disease. In Mayo Clinic Proceedings 2012 Aug 1 (Vol. 87, No. 8, pp. 766–78). Elsevier.

NIH Consensus Conference. Impotence. NIH Consensus Development Panel on Impotence. JAMA. 1993;270(1):83–90.

Oliver JJ, Kerr DM, Webb DJ. Time-dependant interactions of the hypotensive effects of sildenafil citrate and sublingual glyceryl tinitrate. Br J Clin Pharm. 2009;67:403–12.

Porst H. The rationale for prostaglandin E1 in erectile failure: a survey of worldwide experience. The Journal of urology. 1996;155(3):802–15.

Quilter M, Hodges L, von Hurst P, Borman B, Coad J. Male sexual function in New Zealand: a population-based cross-sectional survey of the prevalence of erectile dysfunction in men aged 40–70 years. J Sex Med. 2017;14(7):928–36.

Rosen R, Altwein J, Boyle P, Kirby RS, Lukacs B, Meuleman E, O'Leary MP, Puppo P, Chris R, Giuliano F. Lower urinary tract symptoms and male sexual dysfunction: the multinational survey of the aging. male (MSAM-7). Progres en urologie: journal de l'Associationfrancaised' urologie et de la Societefrancaised'urologie. 2004;14(3):332–44.

Salonia A (Chair), Bettocchi C, Carvalho J, Corona G, Jones TH, K0061dioglu A, Martinez-Salamanca I, Minhas S (Vice-Chair), Serefoğlu EC, Verze P. Guidelines on sexual and reproductive health. Edn. presented at the EAU Annual Congress Amsterdam 2020. 978-94-92671-11-0. EAU Guidelines Office. Arnhem.

Seftel AD, De la Rosette J, Birt J, Porter V, Zarotsky V, Viktrup L. Coexisting lower urinary tract symptoms and erectile dysfunction: a systematic review of epidemiological data. Int J Clin Pract. 2013;67(1):32–45.

Thompson IM, Tangen CM, Goodman PJ, Probstfield JL, Moinpour CM, Coltman CA. Erectile dysfunction and subsequent cardiovascular disease. JAMA. 2005;294(23):2996–3002.

Vlachopoulos C, Aznaouridis K, Stefanadis C. Prediction of cardiovascular events and all-cause mortality with arterial stiffness: a systematic review and meta-analysis. J Am Coll Cardiol. 2010;55(13):1318–27.

Wilson SK, Cleves MA, Delk JR II. Comparison of mechanical reliability of original and enhanced Mentor Alpha I penile prosthesis. J Urol. 1999;162:715–8.

Zou Z, et al. The role of nocturnal penile tumescence and rigidity (NPTR) monitoring in the diagnosis of psychogenic erectile dysfunction: a review. Sex Med Rev. 2019;7:442.

Chapter 6
Benign Prostatic Hyperplasia and Lower Urinary Tract Symptoms

Luke Stroman and Ben Challacombe

Benign prostatic hyperplasia (BPH) is characterised by non-cancerous enlargement of the prostate gland, a common condition in men that increases in prevalence with age. Autopsy studies have revealed a prevalence of 50% of men aged 50–60 and 80% in men aged 80–90 (Berry et al. 1984). As the prostate enlarges it can cause obstruction to the prostatic urethra, slowing the urine flow and increasing the residual amount of urine left in the bladder after passing urine. An increase in prostatic size has been linked with worsening of symptoms, an increased risk of urinary retention and an increased need for prostate surgery (Bosch et al. 2008).

6.1 Pathophysiology

Benign prostatic hyperplasia is caused by an increased number of prostatic cells, primarily in the area of the prostate that surrounds the urethra (transition zone). Prostatic cell multiplication is increased by the effect of testosterone which either binds directly to a testosterone receptor or the enzyme 5-alpha-reductase (5AR). 5AR converts testosterone into its more potent form dihydrotestosterone which in turn binds to the testosterone receptor. Testosterone or dihydrotestosterone bound to it's receptor causes genetic and cell multiplication, increasing the size of the prostate gland.

The enlargement of the prostate gland due to BPH can obstruct the prostatic urethra and is the most common cause of bladder outlet obstruction (BOO) in men. In addition to the effect of the increased prostatic volume, the smooth muscle in the

L. Stroman · B. Challacombe (✉)
Guys and St Thomas Hospitals, London, UK
e-mail: stromanl@doctors.net.uk; benchallacombe@doctors.org.uk

© The Author(s), under exclusive license to Springer Nature
Switzerland AG 2022
S. S. Goonewardene et al. (eds.), *Men's Health and Wellbeing*,
https://doi.org/10.1007/978-3-030-84752-4_6

prostate contracting due to the effect of alpha-adrenergic muscle fibres can cause obstruction. If left untreated BOO causes thickening of the bladder wall and can cause the pressure in the bladder to increase. Back pressure from the bladder can cause ureter and renal pelvis pressures to increase subsequently, causing bilateral hydronephrosis and kidney failure due to pressure on the renal pelvis. Short term effects of BOO can include symptoms such as poor flow, hesitancy, feeling of incomplete bladder emptying and post-micturition dribbling. Long term effects can include bladder wall thickening, urinary tract infections, chronic urinary retention and renal failure.

6.2 Clinical History and Examination

Patients should be initially investigated with a full history and clinical examination. The history should focus on voiding symptoms of hesitancy, frequency, poor flow and post-micturition dribbling indicating BOO in addition to storage symptoms such as urgency to pass urine, and urge incontinence which may indicate secondary overactive bladder. An International Prostate Symptom Score (IPSS) questionnaire can be used as a validated scoring system to assess symptoms, quality of life and response to any treatments. It is important to rule out 'red flag' symptoms and signs such as visible haematuria (which could be a sign of bladder or kidney cancer or stones), bedwetting (nocturnal enuresis: a possible sign of high pressure urine retention) and lower limb neurology along with back pain (lower limb or bowel weakness could be a sign of spinal cord compression or advanced prostate cancer).

Clinical examination should be carried out, initially examining the external genitals looking for a tight foreskin or urethral meatus. A digital rectal examination (DRE) should be performed to assess prostate size and any signs of prostate cancer such as a firm, irregular or asymmetric prostate. A distended bladder should be palpated for and percussed.

6.3 Investigations

Initial investigations should include urinalysis to ensure the absence of urinary tract infection (UTI). If UTI is present this should be treated with antimicrobials and a cause investigated as both high post-void residuals and bladder stones secondary to BOO can be a cause of UTI in men. A frequency/volume bladder diary should be performed looking for volume of urine passed per void and polydipsia (increased fluid intake) which could account for polyuria (increased urine production). Renal function blood tests including serum creatinine should be performed to investigate for obstructive renal failure.

In the outpatient clinic uroflowmetry can be performed with a post-void residual (PVR) ultrasound bladder scan to assess flow and bladder emptying. Uroflowmetry should preferably be evaluated with voided volume of >150 mls. A normal urinary flow is considered to have a bell-shaped flow rate and a typical maximum flow rate (Q_{max}) of >15 mls/s, however this cannot completely rule out BOO due to physiological compensatory mechanisms. A lower threshold Q_{max} of 10 mls/s is often used to determine BOO however using this cut off has been seen to only have a 47% sensitivity and 70% specificity (Reynard et al. 1998). The reason for this is that poor flow can be due to BOO, an under-active bladder or an underfilled bladder at time of void causing reduced bladder contraction. A flat shaped uroflowmetry curve may be a sign of urethral structure and investigated with flexible cystoscopy or urethrogram. A raised PVR of >250 mls is usually an indication for treatment however one study has shown progression to hydronephrosis to be only 2% over 3 years, indicating conservative management initially to be safe but would require monitoring (Bates et al. 2003).

A prostate specific antigen (PSA) blood test to investigate for possible prostate cancer should be offered if lower urinary tract symptoms are thought to be prostatic in origin or if prostate is abnormal on DRE (NICE 2015). However, patients should be appropriately counselled about possible negative effects—PSA testing may lead to unnecessary prostate biopsy or the diagnosis of non-significant prostate cancer which may cause anxiety. Raised PSA also does not necessarily indicate prostate cancer and can be raised by either prostate cancer, large prostatic volume, inflammation, infection or trauma. A PSA of over 1.4ug/l is also a marker for BPH. Patients with a raised age-specific PSA in the absence of infection should be considered for further prostate cancer investigations in the form of prostate magnetic resonance imaging (MRI) to help delineate prostate cancer from BPH. Prostate biopsy should be considered if MRI is suspicious for prostate cancer, rapidly increasing PSA with time, raised PSA density (PSA per gram of prostate) or family history of prostate cancer (NICE 2019).

Urodynamic pressure flow studies can give a more accurate diagnosis of BOO but are invasive, using urethral and rectal pressure catheters. Urodynamics are not routinely recommended for all cases of BPH, supported by the UPSTREAM trial that concluded urodynamic assessment did not significantly affect IPSS scores after treatment or rates of BOO surgery (Lewis et al. 2020). Urodynamic studies should therefore be reserved for cases of diagnostic uncertainty, at the extremes of age, when considering revision surgery, or when there is suspected neuropathic bladder dysfunction.

Urinary tract ultrasound to image the upper urinary tract can be helpful to rule out upper tract dilatation if raised post-void residual volumes or if there are symptoms of upper tract obstruction such as loin pain. It is also important to measure prostate size and to evaluate shape in case of a protruding middle prostatic lobe. Upper tract ultrasound is not routinely indicated if renal function is normal and post-void residuals are low. In cases of normal renal function the rate of upper urinary tract dilatation has been seen to be <1% (Koch et al. 1996) (Fig. 6.1).

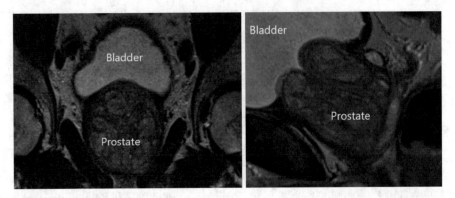

Fig. 6.1 MRI images of BPH in an obstructive 150 ml prostate. T2-weighted images showing coronal (left) and sagittal (right) views

6.4 Conservative Management

Lifestyle advice such as fluid management, reducing evening fluid intake, and reduction of bladder irritants such as caffeine or alcohol should be given to patients complaining of urgency symptoms. Bladder training has also been seen to be of benefit in patients with primarily overactive bladder symptoms. Watchful waiting without medication for BPH has been shown to be successful and a significant proportion of men with LUTS do not progress. Unmedicated men with moderate to severe LUTS secondary to BPH have been seen to a have a risk of acute urinary retention of 2% per year (McConnell et al. 1998).

6.5 Medical Management

6.5.1 Alpha-1-Adrenergic Receptor Blockers

Benign prostatic obstruction is caused in part by smooth muscle contraction through alpha-1-adrenergic receptors. Medications such as Tamsulosin, Alfuzosin and Doxazosinaim to relax prostatic smooth muscle by blocking this receptor. An improvement in symptoms of about 30% is expected with alpha blockade compared to placebo (Boyle et al. 2001). The improvement in symptoms should be immediate, although patients taking the medications for longer have been associated with a further decreased symptom score (Boyle et al. 2001). Alpha blockers can also be used to prevent urinary retention and are recommended to be given to patients with acute urine retention (without upper urinary tract dilatation) prior to catheter removal (NICE 2015).

Side effects of alpha-1-adrenergic receptor blockers include retrograde ejaculation (due to the relaxing effect on the prostate), dizziness, headache and postural hypotension—meaning it's use should be limited in patients at high risk of hypotension and falls. Alpha-blockage has been associated with intraoperative floppy iris syndrome during cataract surgery, a triad of pupil constriction, flaccid iris and iris prolapse. Patients considering cataract surgery should consider stopping alpha-blockers prior to surgery.

6.5.2 5-Alpha-Reductase Inhibitors

5-alpha-reductase inhibitors (5ARIs) prevent the conversion of testosterone to the more potent version dihydrotestosterone which happens through the enzyme 5-alpha-reductase. The action of dihydrotestosterone usually causes cell multiplication, meaning a reduction will cause less cell production and a reduction in prostatic volume, although this happens over a period of months. Examples of 5ARIs include Finasteride and Dutasteride. 5ARIs have been associated with reduction in acute urine retention when compared with placebo (McConnell et al. 1998). In addition, 5ARIs suppress vascular endothelial growth factor (VEGF) which can prevent haematuria in patients with recurrent prostatic bleeding (Foley et al. 2000). Side effects of 5ARIs include retrograde ejaculation, reduction in libido and erectile dysfunction. 5ARIs are also licenced for male pattern hair loss so can cause incidental hair regrowth.

6.5.3 Combination Medical Treatment

Alpha-1-adrenergic receptor blockers and 5ARIs can also be used together to both relax the smooth muscle and attempt to reduce prostatic volume. A randomised trial has found both an alpha-1 adrenergic receptor blocker (Doxazocin) and 5ARI (Finasteride) had an improvement in LUTS and reduction in urinary retention when compared to placebo but taking both medications in combination treatment had a significant improvement when compared to either medication separately (McConnell et al. 2003). However, there is some controversy as other trials have not shown a benefit of combination therapy against alpha blockage monotherapy (Debruyne et al. 1998). The additional side effect profile against monotherapy alone mean that most men are usually started on alpha blockage in the first instance. A combination preparation of Tamsulosin and Dutasteride (Combodart) allows men to take a single preparation.

Many men with BOO also have secondary storage urinary symptoms such as frequency and urgency which are often the most bothersome. As a result anticholinergics such as Solifenacin and Trospium can be used in conjunction with alpha blockers to relieve both sets of symptoms. A combination tablet, Vesomni, also addresses this issue.

6.6 Surgical Management

6.6.1 Transurethral Resection of the Prostate

The traditional gold standard of surgical management of BOO secondary to BPH is transurethral resection of the prostate (TURP) (Table 6.1). This procedure involves placing a cystoscope through the urethra into the bladder and resecting the prostate using diathermy, a resecting wire loop that transmits thermal energy. Resecting and removing the obstructing prostatic tissue increases urine flow through the prostatic urethra. Historically this has been done with monopolar diathermy energy that transmits the energy through the patient to a grounding pad using hypotonic

Table 6.1 New surgical techniques in BPH

Procedure	Method	Prostate size	Unique benefits and limitations
Bipolar transurethral of the prostate	Resection of prostate using thermal energy	Up to 100 ml	Widespread use and surgeon experience. Well documented results. Reduction in TUR syndrome when compared to monopolar TURP.
Greenlight laser photo vaporisation (PVP)	Vaporisation of the prostate using laser	Up to 100 ml	Similar to bipolar TURP. Some studies have suggested reduced bleeding risk compared to TURP. Safe on anticoagulants.
Holmium laser enucleation of the prostate (HOLEP)	Enucleation of the prostate using laser. Removal of the prostate using morcellator.	Any size	Can be used on large prostates >100 ml. Reduced bleeding risk. Long learning curve for the surgeon which has limited uptake. Low recurrence rates
Robotic simple prostatectomy	Robotic assisted laparoscopic excision of prostatic tissue.	Greater than 100 ml	Can be used on very large prostates. Limited to tertiary hospitals. Expensive.
Prostate Artery Embolisation	Microspheres or polyvinyl alcohol placed in prostatic arteries to occlude flow	Greater than 100 ml	Performed by interventional radiology, does not require general anaesthetic. Less improvement in Qmax than TURP but similar IPSS improvement.
Prostatic urethral lift (Urolift™)	Prostatic stapling	Up to 60 mls	Not associated with retrograde ejaculation. Can be done under local anaesthetic. Reduced improvement in Qmax and IPSS when compared with TURP.
Rezüm	Steam ablation of the prostate	*Up to 80 mls*	No evidence available that directly compares Rezüm to other BPH treatments. Little effect on ejaculatory function. May require multiple treatments. Can be done under local anaesthetic.

glycine. More recently it has been performed using bipolar diathermy and normal saline fluid that transmits energy from one side of the resecting loop to the other.

Complications of TURP include retrograde ejaculation (inability to produce ejaculate as a result of semen not being adequately ejected from the prostate) in 75% of patients, bladder neck stenosis, urethral strictures, incontinence and failure to improve urinary tract symptoms. Patients undergoing monopolar TURP are also at risk of transurethral resection (TUR) syndrome; a multifactorial syndrome characterised by dilutional hyponatraemia, hypertension secondary to absorptive fluid overload and visual disturbances from the absorption of glycine, a retinal neurotransmitter. Incidence of TUR syndrome is has been seen to be 1% in monopolar TURP but significantly less in bipolar procedures as saline is used as the irrigation fluid which is not as hypotonic as glycine (Tang et al. 2014).

TURP is typically performed on prostate glands up to 80–100 mls in volume, while alternative procedures such as holmium laser prostatectomy, open or robotic-assisted Millin's prostatectomy or prostate artery embolization may be preferred from glands >100 mls. Recent advances have seen an influx in the field of benign prostate technologies; including Urolift and Rezum, all aiming to cater for patients with various functional priorities. A table of new BPH procedures can be seen in Table 6.1.

6.6.2   Holmium Laser Enucleation of the Prostate (HOLEP)

Similar to TURP, HOLEP is performed transurethrally, but instead of resecting the prostate it uses a holmium laser to enucleate the prostate lobes in their entirety from the inside. The prostatic tissue is then pushed into the bladder where it is suctioned out using a morcellator. The prostatic tissue is not directly cut, inferring a reduced bleeding risk and allowing larger prostates can be removed using this technique. The complications of HOLEP are similar to TURP but may infer a slightly higher risk of incontinence due to complete exposure of the external sphincter to the urinary stream.

6.6.3   Greenlight Laser Photo Vaporisation (PVP)

Greenlight PVP is also performed transurethrally using a cystoscope into the bladder. The prostatic tissue is then vapourised using a laser instead of the resection of TURP. Greenlight PVP is typically performed on similar patients to TURP and has a similar side effect profile. Decision on which procedure to perform is often based on surgeon and patient preference. Outcomes in terms of IPSS, Qmax, and freedom from complications have been seen to be similar to TURP (Thomas et al. 2016).

6.6.4   Open or Robotic Simple Prostatectomy

Prior to the advent of TURP, prostatectomy for benign disease was performed using an open technique, either through a transvesical or retropubic (Millin's) approach. This is performed using a lower midline or pfannenstiel incision and adenomatous prostatic tissue enucleated. However, this is now performed rarely due to the advances in transurethral techniques such as HOLEP. More recently advances in robotic surgery have allowed for the simple prostatectomy to be performed using either retropubic or transvesical robotic approaches, although this is not offered in all centres routinely and reserved for very large prostates (>100 mls) or where HOLEP is not available.

6.6.5   Prostate Artery Embolisation

Prostate Artery Embolisation (PAE) is a non-surgical procedure carried out under local anaesthetic, sedation or general anaesthetic in the interventional radiology suite aiming to block the arteries supplying the prostate and causing shrinkage. A CT angiogram is performed prior to the procedure illustrating the prostatic vasculature. During the procedure, a percutaneous catheter is placed in the femoral artery and very small beads (microspheres) or polyvinyl alcohol placed in prostatic arteries to occlude flow. In addition to it's role in treating LUTS and retention secondary to BPH, PAE is also indicated for recurrent significant prostatic bleeding. While PAE has been associated with a reduction in flow rate and post void residuals, trials have found this reduction to be significantly less than TURP, although symptom scores have been similar (Ray et al. 2018; Insausti et al. 2020).

6.6.6   Prostatic Urethral Lift (Urolift™)

Urolift is a prostatic stapling technique that pins back the lateral lobes of the prostate as opposed to removing tissue, unlike any other technique. The unique selling point of Urolift is that it has not been associated with retrograde ejaculation, which may make it more suitable for sexually active men. However, Urolift is unlikely to work in men with very large prostates or where the is a 'ball valve' middle lobe. It may also prevent accurate assessment with prostate MRI scanning. In a randomised control trial comparing Urolift to TURP, both groups had improvements in LUTS but IPSS and Q_{max} were significantly more improved in patients that had TURP, while Urolift had better ejaculatory function preservation (Gratzke et al. 2017).

6.6.7   REZUM

Rez*ü*m uses steam injections to ablate prostate adenoma. Steam is produced from water heated outside the body using a radiofrequency generator and delivered transurethrally using a cystoscope aimed at the adenoma. Rezum is typically carried out under local anaesthetic but may require more than one treatment. It is a relatively new procedure and long-term outcomes are still undetermined. There are no randomised studies comparing Rez*ü*m to other BPH treatments.

References

Bates TS, Sugiono M, James ED. Is the conservative management of chronic retention in men ever justified? BJU International. 2003;92(6):581–3.

Berry SJ, Coffey DS, Walsh PC, Ewing LL. The development of human benign prostatic hyperplasia with age. J Urol. 1984;132:474–9.

Bosch JL, Bangma CH, Groeneveld FP, Bohnen AM. The long-term relationship between a real change in prostate volume and a significant change in lower urinary tract symptom severity in population-based men: the Krimpen study. Eur Urol. 2008;53:819–27.

Boyle P, Robertson C, Manski R, Padley RJ, Roehrborn CG. Meta-analysis of randomized trials of terazosin in the treatment of benign prostatic hyperplasia. Urology. 2001;58(5):717–22.

Debruyne FM, Jardin A, Colloi D, et al. Sustained release alfuzocin, finasteride and the combination of both in the treatment of benign prostatic hyperplasia. Eur Urol. 1998;34:169–75.

Foley SJ, Soloman LZ, Wedderburn AW, et al. A prospective study of the natural history of hematuria associated with benign prostatic hyperplasia and the effect of finasteride. J Urol. 2000;163(2):496–8.

Gratzke C, Barber N, Speakman MJ, et al. Prostatic urethral lift vs transurethral resection of the prostate: 2-year results of the BPH6 prospective, multicentre, randomized study. BJU Int. 2017;119(5):767–75.

Insausti I, de Ocáriz AS, Galbete A, et al. Randomized comparison of prostatic artery embolization versus transurethral resection of the prostate for treatment of benign prostatic hyperplasia. J Vasc Interv Radiol. 2020;31(6):882–90.

Koch WF, Ezz el Din K, de Wildt MJ, Debruyne FM, de la Rosette JJ. The outcome of renal ultrasound in the assessment of 556 consecutive patients with benign prostatic hyperplasia. J Urol. 1996;155:186–9.

Lewis AL, Young GJ, Selman LE, et al. Urodynamics tests for the diagnosis and management of bladder outlet obstruction in men: the UPSTREAM non-inferiority RCT. Health Technol Assess. 2020;24(42):1–122.

McConnell JD, Bruskewitz R, Walsh P. The effect of finasteride on the risk of acute urinary retention and the need for surgical treatment among men with benign prostatic hyperplasia. Finasteride Long-Term Efficacy and Safety Study Group. N Engl J Med. 1998;338:557–63.

McConnell JD, Roehrborn CG, Bautista OM, et al. The long-term effect of doxazosin, finasteride, and combination therapy on the clinical progression of benign prostatic hyperplasia. N Engl J Med. 2003;349:2387–98.

NICE, National Institute for Health and Care Excellence. Lower urinary tract symptoms in men: management. Clinical guideline [CG97]. https://www.nice.org.uk/guidance/cg97/chapter/1-Recommendations. [Online] June 2015.

NICE, National Institute for Health and Care Excellence. Prostate cancer: diagnosis and management. NICE guideline [NG131]. May 2019.

Ray RF, Powell J, Speakman MJ, et al. Efficacy and safety of prostate artery embolization for benign prostatic hyperplasia: an observational study and propensity-matched comparison with transurethral resection of the prostate (the UK-ROPE study). BJU Int. 2018;122(2):270–82.

Reynard JM, Yang Q, Donovan JL, et al. The ICS-'BPH' Study: uroflowmetry, lower urinary tract symptoms and bladder outlet obstruction. Br J Urol. 1998;82:619–23.

Tang Y, Li J, Pu C, Bai Y, Yuan H, Wei Q, Han P. Bipolar transurethral resection versus monopolar transurethral resection for benign prostatic hypertrophy: a systematic review and meta-analysis. J Endourol. 2014;28(9):1107–14.

Thomas JA, Tubaro A, Barber N, et al. A multicenter randomized noninferiority trial comparing GreenLight-XPS laser vaporization of the prostate and transurethral resection of the prostate for the treatment of benign prostatic obstruction: two-yr outcomes of the GOLIATH study. Eur Urol. 2016;69(1):94–102.

Chapter 7
The Impact of Urinary Incontinence on Men's Health

Julian Shah

7.1 Introduction

Urinary incontinence can be a cause of great distress for men. Men are not used to having to suffer with incontinence of urine. Male behaviour often leads to avoidance rather than to seek medical advice. Women appear to be more accustomed to incontinence and catered for whilst men are not. When did you see an advertisement for men's incontinence products?

The prevalence and aetiology of incontinence in men differs according to age and co-morbidities.

The prevalence of incontinence in men much depends upon which study is referred to. However, the condition is sufficiently common to increase the awareness of clinicians to not only how frequently male patients suffer with incontinence but also the causes and serious effects that incontinence can have on well-being. The resolution of male incontinence will have significant benefits to quality of life and even life expectancy.

7.2 Incontinence in the Elderly

The incidence of incontinence in the elderly is very common with approximately 40% of women and 10–35% of men over 80 experiencing the condition (Vaughan et al. 2018). This review reports that incontinence has effects on independence and increases the risk of depression, social isolation and falls. The prevalence in men in the USA was reported to be 24% in the community (Goode et al. 2008). In the Netherlands the prevalence of UI was 22.6% in men (Linde et al. 2017). There is

J. Shah (✉)
University College London, London, UK

© The Author(s), under exclusive license to Springer Nature
Switzerland AG 2022
S. S. Goonewardene et al. (eds.), *Men's Health and Wellbeing*,
https://doi.org/10.1007/978-3-030-84752-4_7

also an association with anxiety if patients suffer with incontinence (Bogner et al. 2002). In patients who are hospitalized the prevalence of incontinence is 22.9% in women and 16.1% in men (Junqueira and de Gouveia Santos 2017). Elderly patients who suffer with incontinence are at higher risk of loneliness but there is a strong association with other co-morbid mental health problems (Stickley et al. 2017). Strikingly urinary incontinence is also a predictor of a higher mortality (John et al. 2016).

There are a number of co-morbid conditions that are associated with incontinence in the elderly (Vaughan et al. 2018):

- Diabetes
- Obesity (Bauer et al. 2019)—higher body mass index and fat mass were associated with an increased prevalence of incontinence.
- Metabolic syndrome
- Vascular disease
- Alzheimer's, dementia, stroke, Parkinson's disease, multiple sclerosis and spinal cord injury (Grant et al. 2013).
- Heart failure—Tannenbaum and Johnell (2014) have reported that up to 50% of patients with heart failure will have lower urinary tract symptoms and incontinence. With more severe disease the incidence is three times more common. The medication used to treat heart failure can exacerbate urinary symptoms.

The increase in the incidence of obesity and diabetes in the general population has apparently increased the incidence of incontinence of urine (Markland et al. 2011). The authors showed that the incidence of incontinence had increased between the years of 2001 and 2008 because of the increased incidence of these conditions.

7.3 Incontinence in Different Age Groups

It is worthwhile considering the various ages at which males present with incontinence of urine and the causes of the incontinence.

7.4 Evolution of Continence

The natural process of continence starts in childhood and as we grow and become aware of our surroundings we are usually taught about how and when to go to the toilet and thus to avoid being incontinent. Enuresis (bedding the bed at night) can extend into adolescence and even adult life. Children usually will "grow out of it". However, it is recognized that 10% of 7 year old's wet the bed at night (von Gontard and Kuwertz-Bröking 2019) and 6% are also affected in the day. The treatment directed towards resolving incontinence should start from the age of 5 years and provides benefits to the child.

In adolescents the prevalence of nocturnal enuresis is 1–2%. The combination of nocturnal enuresis and daytime urinary incontinence is also associated with other health conditions such as constipation, obesity, chronic illness and psychological factors (von Gontard et al. 2017; Heron et al. 2017).

It is unusual for adolescents to have problems with incontinence of urine although some children will continue to suffer with an overactive bladder. This condition which is associated with detrusor overactivity (bladder instability) causes urinary incontinence. This can present with enuresis but can also present with the daytime symptoms of urgency, which may lead to urge incontinence. These symptoms are distressing for young boys and adolescents.

It is important to ensure that there is no background neurological dysfunction such as spina bifida (myelomeningocele) (Woodhouse 2005) in which incontinence is more likely to be present. There are also other conditions which are associated with urinary incontinence such as Down syndrome (Niemczyk et al. 2017), Duchenne muscular dystrophy (Morse et al. 2020), cystic fibrosis (Reichman et al. 2016) and Prader-Willi syndrome (Von Gontard et al. 2010) as examples.

By the early adult years' incontinence is usually not a common problem. Possibly post-micturition dribbling presents as incontinence. This is a condition which is common in men. Urine leaks out after voiding is completed and the penis put away. It is due to urine that pools in the bulbar urethra after voiding. It is usually a benign condition. Occasionally it can be caused by a urethral stricture. A simple free flow rate should exclude the presence of a stricture. Milking the urethra after voiding or pressure on the perineum can usually resolve this complaint.

If other forms of incontinence are present a diagnosis must be sought. Incontinence in early adult life is either due to neurological factors or the ongoing problem of bladder overactivity, which can persist throughout life. As we get into middle age incontinence can start to develop for a number of different reasons. Provided the man does not have a neurological dysfunction then the commonest cause of incontinence is bladder overactivity. With ageing this becomes more likely to be associated with bladder outflow obstruction. This can be due to a urethral stricture, bladder neck obstruction and in older men with the enlargement of the prostate. About two thirds of men with bladder outflow obstruction will develop overactive bladder contractions, which can give rise to urgency and urge incontinence.

One other cause of incontinence in men is overflow incontinence due to chronic retention of urine which is usually associated with bladder outflow obstruction or can be due to poor bladder contractility from a neurological cause (MS Spinal cord injury).

As the elderly often do suffer with multiple co-morbidities and frailty it is not surprising that incontinence can also co-exist with other conditions.

Neuropathic conditions can cause incontinence, and these can be insidious such as it may occur in multiple sclerosis or Parkinsonism or due to obvious neurological dysfunction that follows a spinal cord injury or cauda equina syndrome.

Whatever the cause the most important aspect of care is to make a primary diagnosis, which can then lead to an appropriate treatment since there is no reason why any man should be wet. Most patients can be made dry.

7.5 Symptoms of Incontinence

The principal symptoms associated with incontinence are:

1. Storage symptoms such as frequency, nocturia, urgency and urge incontinence (overactive bladder condition OAB)
2. Stress incontinence due to sphincter weakness, which can be associated with neuropathy but mostly associated with surgical intervention that follows a radical prostatectomy, HoLEP and TURP
3. Voiding dysfunction with poor flow and incontinence due to chronic retention
4. Rarely associated with fistulae such as recto-vesical or recto-prostatic fistulae or surgical intervention or radiotherapy or a combination of both.

Thus, it is important to start with a history and examination of the patient and then investigation appropriate to the nature of the condition.

After a clinic-based diagnosis (or differential diagnosis) has been made (important since this focuses the mind on the potential causes) selected investigations should be conducted.

7.6 Frequency/Volume Voided Chart

The patient should be asked to complete a frequency/volume voided chart. A record of incontinence episodes should be kept. If the patient has to wear pads he should be asked to weigh a wet pad and a dry pad. The weight change will indicate the volume of leakage. A 24-h pad collection gives the most useful information.

7.7 Urodynamic Testing

A free urine flow rate will indicate flow dysfunction. A post void residual urine will confirm retention.

The most useful investigation is a video urodynamic study in which with a combination of bladder pressure testing and X-ray screening with contrast a diagnosis can be usually confidently made.

The importance of investigation cannot be overemphasized as the only way to resolve what is clearly a cause of great distress for male is to make a diagnosis. As incontinence often leads to behavioural change and psychological distress it is important to offer and direct appropriate treatment.

7.8   Conservative Treatment of Incontinence

Naturally this will depend upon the cause.

For an elderly male with incontinence, who is frail, pads and appliances may be the most effective means of providing comfort. A Conveen sheath applied to the penis can effectively contain urinary leak provided the bladder is known to be emptying. The challenge with the Conveen is that not all elderly males have a penis suitable for the attachment of the device.

7.9   Long-Term Catheters

A long-term catheter can be offered to men who are incontinent. A urethral catheter must not be used for long-term bladder drainage. The long-term catheter due to the effects on the urethra which include inadvertent traction can cause the penis to split (iatrogenic hypospadias). Provide the patient is suitable for a visit to an operating theatre a suprapubic catheter should be placed. The catheter should be changed every 6 weeks. Daily "catheter clamping" is ideal to preserve bladder capacity. Bladder spams due to the presence of the catheter can be settled with anticholinergic medication.

If the condition has a surgical solution then this is what must be used.

7.10   Chronic Retention of Urine with Overflow

Chronic retention of urine with overflow is treated by initial catheter drainage with careful measurement of the residual urine. The cause of the failure to empty the bladder must be sought. If the cause is neuropathic then the patient would need to do intermittent self catheterisation to drain the bladder on a regular basis. High-pressure bladder storage dysfunction or bladder overactivity (confirmed by urodynamics) will need to be suppressed. The commonest cause of chronic retention of urine is bladder outflow obstruction (Abrams et al. 1978).

If the chronic retention is due to bladder outflow obstruction (usually bladder neck or prostatic outflow obstruction), then relief of the obstruction is necessary. A measurement of the volume of the residual is critical to understand whether or not the bladder will recover function. A large volume retention in the region of 2 l or more often is associated with failure to recover function in spite of surgery for obstruction and the patient may then need to do self catheterisation to empty the bladder.

Little has changed in the last 40 years with regard to the management of chronic retention. Mitchell's review in 1984 is worth reading as it provides a balanced view of the condition and its management (Mitchell 1984).

If chronic retention is due to a stricture or bladder outflow obstruction from the prostate or the bladder neck then treatment of the obstruction can help the situation dramatically returning the patient to near normal voiding. In the first instance at the time of surgery, it is worthwhile placing a SP catheter for the patient to do trials of voiding after the urethral catheter has been removed and measure the retained urine after voiding. Once voiding is restarted the residual usually falls. Once the residual falls below say 100 ml the suprapubic catheter can be used. There is no need to get "hung up" on the residual volume since much depends upon whether the residual is consistent and does not cause symptoms. A residual of 200 ml can be left if it is not causing symptoms or complications. Once the residual falls to an agreed volume the SP catheter should be removed. If the residual urine volume is consistently high then the patient should be taught to self catheterise to drain any residual urine. In some patients with large volume retention, who start voiding after surgery, intermittent catheterisation can be carried out say once a day to drain the residual as the bladder gradually recovers and when the residual falls to an accepted volume then the catheterisation can be slowly withdrawn.

7.11   Bladder Overactivity

The overall prevalence of OAB, which is common in both sexes, is 14–18% (Canadian population—Corcos et al. 2017; Wallace and Drake 2015). The condition however can be both underdiagnosed and undertreated (Burnett et al. 2020).

If the patient has symptoms of bladder overactivity (OAB) this can be based upon initial symptom history and exclusion of an inflammatory cause by collecting an MSU. The bladder must be shown to be emptying by scanning. Treatment can then start with bladder retraining and with anticholinergic medication. Many patients with this treatment will improve (Burgio et al. 2020). One medication may not suit the patient so alternatives can be tried. However, the difference between monotherapy and combination therapy is not usually large (Hsu et al. 2019). The other issue relates to compliance with treatment which is often worse than for other conditions (Dhaliwal and Wagg 2016).

Bladder retraining is much helped by enlisting the help of a continence advisor or specialist pelvic physiotherapist.

Posterior tibial nerve stimulation may benefit some patients and is a minimally invasive and safe procedure (Santos Garcia and Pereira 2018). Where facilities allow it could be a treatment worth offering to males to perhaps avoid any more invasive treatment.

If the patient fails to respond to bladder retraining and medication, then he should undergo urodynamic testing before any other treatment is used. If bladder overactivity is confirmed, then the next step is to consider Botulinum toxin injections to the bladder wall.

We are fortunate to have the availability of intravesical Botox injections, which can help patients with overactivity that does not respond to medication. The success rate of Botulinum toxin is in the region of 70% with good tolerance (Sievert et al. 2014). The patient needs to be warned about the risk of early voiding difficulty and retention after the injection. Botulinum toxin affects a change in bladder function that can lead to temporary voiding dysfunction in 10–30% of patients. Smaller doses of Botox will reduce the need to self catheterise. The patient needs to be warned preoperatively about the risk of retention and the need to self-catheterise. In some units all patients are taught self-catheterisation prior to treatment so they are familiar with it. This may be necessary for a period of time until the effect of the toxin wears off. Smaller doses of Botox produce low risk of retention but the duration of effect of Botox may be not as great with a smaller dose than with a higher dose of Botox (author's view).

7.12 Augmentation Ileo-Cystoplasty

For a patient with a high-pressure bladder overactivity whether this is due to idiopathic detrusor overactivity or a neuropathic bladder if they do not respond to Botulinum toxin or if the treatment is not as effective as was hoped, or the patient does not wish to continue with the regular treatments and needs and wishes a permanent solution then the patient can be offered an augmentation ileocystoplasty. This is an excellent procedure for resolving bladder overactivity whatever the cause and will produce satisfactory long-term benefits. Cystoplasty has to be balanced against the need to self catheterise to empty the bladder and between 50% and 100% of the patients will need to self catheterise depending upon the aetiology. Appropriate pre-operative counselling is necessary, and the patient should be taught how to self catheterise before undergoing any surgery. Long term follow up is also necessary to ensure that the patient does not develop any long-term complications. These are relatively few, except and include the risk of urinary infection and stone formation due to mucus production and infection.

7.13 Stress Incontinence

Patients who suffer with stress (sphincter weakness) urinary incontinence usually do so due to sphincter weakness associated with neurological disease due to lower motor neurone bladders with sphincter weakness incontinence or who have undergone a radical prostatectomy. Patients are also presenting after HoLEP with urinary incontinence.

Because of the increasing popularity of radical prostatectomy for cancer stress urinary incontinence due to sphincter weakness is more common.

Psychological support by nursing intervention after radical prostatectomy does improve the well being of patients (Yuan et al. 2019).

Many patients will respond well to pelvic floor exercises if instituted early, taught properly and followed-up to ensure compliance (Vrijens et al. 2015; Jalalinia et al. 2020). However there still remains some controversy as to whether PFE do have benefit at 12 months (Anderson et al. 2015). If the symptoms improve to the point at which a patient does not need to have any further management then further treatment is not necessary. Some men will manage quite happily with one pad a day as a safety pad and will not require intervention. Younger men tend to be more distressed by incontinence than older men, but not always.

If urinary incontinence persists and it is demonstrated that this due to sphincter weakness incontinence, then the patient can be offered surgical intervention. A useful review article has been published by Comiter and Dobberfuhl (2016) that discuss the pros and cons of which procedure to use.

There are three techniques currently available to treat male stress incontinence:

One is bulking agents, which do work in some patients but require particular expertise since the surgery is very much an operator dependent procedure.

The male sling was introduced to support the urethra in men post radical prostatectomy and this does work well in selected patients. Currently Meshes in the UK in men are "on hold" because of the adverse publicity that has surrounded these devices.

The "gold standard" for treatment stress urinary incontinence due to sphincter weakness is an artificial urinary sphincter (AMS800) which has been available for 40 years or more and produces an excellent outcome.

The male with post-prostatectomy incontinence will do well with an AUS800 implanted as a one-night stay procedure. The device is activated at 6 weeks. The risk of complications is around 5%. The success rate is in the region of 80%. These devices have a mean "survival" of 10 years before they need to be revised or replaced.

The artificial urinary sphincter is effective in men with incontinence due to neuropathy. If the patient with a spinal cord injury has sphincter weakness incontinence, then a bladder neck cuff is preferable to a bulbar cuff because in patients who are wheelchair bound the patient has a tendency to sit on the cuff. However, much depends upon the experience of the surgeon and the patient's morphology since if a patient is obese the operation can be difficult to perform.

Artificial urinary sphincters in patients with neuropathy do tend to well and the patient can manage satisfactorily with the device. We have to bear in mind that these devices last for about 10 years before they need revision or replacement and therefore replacing a device in a young male may have limitations in terms of the duration.

7.14 Conclusion

Incontinence in men is a distressing and seemingly common condition. There is no reason why any male should suffer with incontinence. Patients can be made continent with appropriate understanding of the condition, making an appropriate diagnosis of the cause and then applying an appropriate strategy for resolution of the problem.

References

Abrams PH, Dunn M, George N. Urodynamic findings in chronic retention of urine and their relevance to surgery. BMJ. 1978;2:1258–60.

Anderson CA, Omar MI, Campbell SE, Hunter KF, Cody JD, Glazener CMA. Conservative management for postprostatectomy urinary incontinence. Cochrane Database Syst Rev. 2015;1(1):CD001843.

Bauer SR, Grimes B, Suskind AM, Cawthon PM, Cummings S, Huang AJ. Urinary incontinence and nocturia in older men: associations with body mass, composition and strength in the health, aging and body composition study. J Urol. 2019;202(5):1015–21.

Bogner HR, Gall JJ, Swartz KL, Ford DE. Anxiety disorders and disability secondary to urinary incontinence amongst adults over the age of 50. Int J Psychiatry Med. 2002;32(2):141–54.

Burgio KL, Kraus SR, Johnson TM, Markland AD, Vaughan CP, Li P, Redden DT, Goode PS. Effectiveness of combined behavioral and drug therapy for overactive bladder symptoms in men. A randomized clinical trial. JAMA Intern Med. 2020;180(3):411.

Burnett AL, Walker DR, Feng Q, Johnston KM, Lozano-Ortega G, Nimke D, Hairston JC. Undertreatment of overactive bladder among men with lower urinary tract symptoms in the United States: A retrospective observational study. Neurourol Urodyn. 2020;39:1378–86.

Comiter CV, Dobberfuhl AD. The artificial urinary sphincter and male sling for postprostatectomy incontinence: Which patient should get which procedure? Investig Clin Urol. 2016;57:3–13.

Corcos J, MikolajPrzydacz LC, Witten J, Hickling D, Honeine C, Radomski SB, Stothers L, Wagg A. CUA guideline on adult overactive bladder. Can Urol Assoc J. 2017;11(5):E142–73.

Dhaliwal P, Wagg A. Overactive bladder: strategies to ensure treatment compliance and adherence. Clin Interv Aging. 2016;11:755–60.

Garcia MBS, Pereira JS. Electrostimulation of the posterior tibial nerve in individuals with overactive bladder: a literature review. J Phys Ther Sci. 2018;30:1333–40.

Goode PS, Burgio KL, Redden DT, Markland A, Richter HE, Sawyer P, Allman RM. Population-based study of incidence and predictors of urinary incontinence in African American and white older adults. J Urol. 2008;179(4):1449–54.

Grant RL, Drennan VM, Rait G, Petersen I, Iliffe S. First diagnosis and management of incontinence in older people with and without dementia in primary care: a cohort study using the health improvement network primary care database. PLoS Med. 2013;10(8):1–8.

Heron J, Grzeda MT, von Gontard A, Wright A, Joinson C. Trajectories of urinary incontinence in childhood and bladder and bowel symptoms in adolescence: prospective cohort study. BMJ Open. 2017;7:1–9.

Hsu FC, Weeks CE, Selph SS, Blazina I, Holmes RS, McDonagh MS. Updating the evidence on drugs to treat overactive bladder: a systematic review. Int Urogynecol J. 2019;30:1603–17.

Jalalinia SF, Raei M, VahidNaseri-Salahshour SV. The effect of pelvic floor muscle strengthening exercise on urinary incontinence and quality of life in patients after prostatectomy: a randomized clinical trial. J Caring Sci. 2020;9(1):33–8.

John G, Bardini C, Combescure C, Dällenbach P. Urinary incontinence as a predictor of death: a systematic review and meta-analysis. PLoS One. 2016;1

Junqueira JB, de Gouveia Santos VLC. Urinary incontinence in hospital patients: prevalence and associated factors. Rev Latino-Am Enfermagem. 2017;25:e2970.

Linde JM, Nijman RJM, Trzpis M, Broens PMA. Urinary incontinence in the Netherlands: prevalence and associated risk factors in adults. Neurourol Urodyn. 2017;36(6):1519–28.

Markland AD, Richter HE, Fwu C-W, Eggers P, Kusek JW. Prevalence and trends of urinary incontinence in adults in the United States, 2001 to 2008. J Urol. 2011;186(2):589–93.

Mitchell JP. Management of chronic urinary retention. Br Med J. 1984;289:515–6.

Morse CI, Higham K, Bostock EL, Jacques MF. Urinary incontinence in men with Duchenne and Becker muscular dystrophy. PLoS One. 2020;15(5):e0233527.

Niemczyk J, von Gontard A, Equit M, Medoff D, Wagner C, Curfs L. Incontinence in persons with Down Syndrome. Neurourol Urodyn. 2017;36:1550–6.

Reichman G, De Boe V, Braeckman J, Michielsen D. Incontinence in patients with cystic fibrosis. Scand J Urol. 2016;50(2):128–31.

Sievert K-D, Chapple C, Herschorn S, Joshi M, Zhou J, Nardo C, Nitti VW. Onabotulinumtoxin A 100U provides significant improvements in overactive bladder symptoms in patients with urinary incontinence regardless of the number of anticholinergic therapies used or reason for inadequate management of overactive bladder. Int J Clin Pract. 2014;68(10):1246–56.

Stickley A, Santini ZI, Koyanagi A. Urinary incontinence, mental health and loneliness among community-dwelling older adults in Ireland. BMC Urol. 2017;17:29.

Tannenbaum C, Johnell K. Managing therapeutic competition in patients with heart failure, lower urinary tract symptoms and incontinence. Drugs Aging. 2014;31:93–101.

Vaughan CP, Markland AD, Smith PP, Burgio KL, Kuchel GA, and the AGS/NIA Urinary Incontinence Conference Planning Committee and Faculty. Report and Research Agenda of the American Geriatrics Society and National Institute on aging bedside-to-bench conference on urinary incontinence in older adults: translational research agenda for a complex geriatric syndrome. Am Geriatr Soc. 2018;66(4):773–82.

von Gontard A, Kuwertz-Bröking E. The diagnosis and treatment of enuresis and functional daytime urinary incontinence. DeutschesÄrzteblatt International I DtschArztebl Int. 2019;116:279–85.

Von Gontard A, Didden R, Sinnema M, Curfs L. Urinary incontinence in persons with Prader-Willi Syndrome. BJU Int. 2010;106(11):1758–62.

von Gontard A, Cardozo L, Rantell A, Djurhuus JC. Adolescents with nocturnal enuresis and daytime urinary incontinence. How can pediatric and adult care be improved? Neurourol Urodyn. 2017;36(4):843–9.

Vrijens DMJ, Spakman JI, van Koeveringe GA, Berghmans B. Patient-reported outcome after treatment of urinary incontinence in a multidisciplinary pelvic care clinic. Int J Urol. 2015;22:1051–7.

Wallace KM, Drake MJ. Overactive bladder. F1000 Res. 2015;4:1–8.

Woodhouse CRJ. Myelomeningocele in young adults. BJUI. 2005;95:223–30.

Yuan Y, Hu Y, Cheng J-X, Ding P. Psychological nursing approach on anxiety and depression of patients with severe urinary incontinence after radical prostatectomy – a pilot study. J Int Med Res. 2019;47(11):5689–701.

Chapter 8
Metabolic Syndrome, Obesity and Cancer Risk

Giuseppe Ottone Cirulli, Alessandro Larcher, Francesco Montorsi, and Alberto Martini

8.1 General Considerations About Obesity

Obesity, defined as BMI > =30 kg/m^2, is a multifactorial disease mainly caused by the interaction of genetic and environmental factors. The increase in body fat deposits generally coincides with an increase in body weight. This results in a greater risk of comorbidities and can ultimately affect both quality of life and life expectancy (National institue of Health 2000).

Adipose tissue plays a critical role in energy homeostasis in higher organisms. It serves as the main site for energy storage in the form of triglycerides (TGs), and it also contributes to systemic glucose and lipid metabolism via its function as an endocrine organ (Fernández-Real et al. 2010).The adipocyte, a highly differentiated cell from the adipose tissue, synthesizes and releases a variety of substances that include tumour necrosis factor a (TNF-a), protein C, intercellular adhesion molecule, angiotensinogen, plasminogen activator inhibitor, adiponectin, and resistin, among others all of which fulfil different roles and are involved in the pathogenesis of chronic diseases (Guerre-Millo 2002).

The discovery of leptin, melanocortin-4, and leptin receptors and the knowledge of hormonal mediator functions that participate in keeping correct body weight have enabled further understanding of the pathogenic processes and causes of obesity (Booth et al. 2003). Among the factors that contribute to the aetiology of obesity, the genetic background inevitably overlaps with environmental factors as a necessary condition for the establishment of the disease (Aballay et al. 2013). An increase in the consumption of animal fat, added sugar, and refined foods, along with a decrease in the total intake of complex carbohydrates, fibre, fruits, and vegetables leads to increased per capita caloric availability, coupled with a social environment in which

G. O. Cirulli · A. Larcher · F. Montorsi · A. Martini (✉)
Department of Urology, San Raffaele Hospital, University Vita Salute, Milan, Italy

© The Author(s), under exclusive license to Springer Nature Switzerland AG 2022
S. S. Goonewardene et al. (eds.), *Men's Health and Wellbeing*,
https://doi.org/10.1007/978-3-030-84752-4_8

little physical activity is likely (Popkin 2001). Hence, these are not single dietary changes but interrelated multifactorial processes, often involving modifications in sociocultural and economic conditions, individual behaviour, and lifestyle (López de Blanco and Carmona 2005).

8.2   Obesity and Cancer

Many studies have explored the role of obesity and the increased risk of mortality from a number of malignancies, including cancers of the colon, pancreas, kidney, prostate, breast, and endometrium. The association between viral, bacterial, or chemical inflammation and different cancers strongly suggests that chronic inflammation is involved in the development of tumours (Lin and Karin 2007; Comba and Pasqualini 2009). Obesity is a low-inflammation state in which cytokine production from visceral adipose tissue (adipocytokines) increases, causing endothelial dysfunction and Insulin Resistance. Other authors suggest that the link between obesity and cancer is mainly due to Insulin Resistance. This link may be related to high compensatory levels of insulin. This hormone may increase cell proliferation either directly or by increasing the levels of other more potent growth factors, such as insulin-like growth factor, which is one of the growth factors responsible for the initial steps of cancer development (Aleksic et al. 2010; Walker et al. 2010). Evidence suggests, in short, that the link between obesity and cancer may involve metabolic perturbations similar to those shown to exist between Metabolic syndrome and cancer (Krukowski and West 2010).

8.3   Metabolic Syndrome: Introduction

Metabolic syndrome is an aggregation of several disorders (among which there is Obesity), which together raise the risk of an individual developing atherosclerotic cardiovascular disease, insulin resistance, diabetes mellitus, and vascular and neurological complications such as a cerebrovascular accident (Supreeya Swarup, Stat Pearls). Growing data shows the association of metabolic syndrome (MetS) or its components with cancer development and cancer-related mortality. It is also reported that MetS has a link with colorectal, breast, endometrial, pancreas, primary liver and, although controversial, prostate cancer (Uzunlulu et al. 2016).

In the Europids, MetS is diagnosed when at least three out of these five risk factors are present: central obesity (waist circumference $\geq$ 102 cm and $\geq$ 88 cm in males and females, respectively), fasting plasma glucose level $\geq$ 5.6 mmol/l (100 mg/dl) (or previous diagnosis of type 2 diabetes), triglyceride level $\geq$ 1.7 mmol/l (150 mg/ dl), HDL-cholesterol level < 1.03 mmol/l (40 mg/ dl) in men and < 1.29 mmol/l (50 mg/dl) in women (or specific therapy for these lipid abnormalities), raised blood pressure (systolic BP ≥ 130 or diastolic BP ≥ 85 mmHg) (or therapy for previously diagnosed hypertension) (Grundy et al. 2004; Bellastella et al. 2018).

Metabolic syndrome has serious implications on an individual's health and healthcare costs. It is necessary to recognize the rising prevalence of metabolic syndrome as through intervention the progression of the syndrome can be halted and potentially reversed (Burrage et al. 2018; Kim and Yi 2018; van der Pal et al. 2018).

8.4 Aetiology

The underlying aetiology of metabolic syndrome is extra weight, obesity, lack of physical activity, and genetic predisposition. The crux of the syndrome is a build-up of adipose tissue and tissue dysfunction that in turn leads to insulin resistance. Proinflammatory cytokines such as tumour necrosis factor, leptin, adiponectin, plasminogen activator inhibitor, and resistin, are released from the enlarged adipose tissue, which alters and impacts insulin handling adversely. Insulin resistance can be acquired or may be due to genetic predisposition. Impairment of the signalling pathway, insulin receptor defects, and defective insulin secretion can all contribute towards insulin resistance. Over time, the culmination of this cause development of metabolic syndrome that presents as vascular and autonomic damage (White et al. 2018).

Regarding the tumour's pathogenesis, it is suggested that in MetS insulin resistance and insulin-like growth factor 1 system play a key role. The adipokines secreted from visceral adipocytes, free fatty acids and aromatase activity contribute to this process (Catharina et al. 2018).

The distribution of body fat is also important, and it is known that upper body fat plays a strong role in developing insulin resistance. Fat accumulation can be intraperitoneal (visceral fat) or subcutaneous. Visceral fat may contribute to insulin resistance more strongly than subcutaneous fat. However, both are known to play a role in the development of metabolic syndrome. In upper body obesity, high levels of non-esterified fatty acids are released from the adipose tissue causing lipid to accumulate in other parts of the body such as liver and muscle, further perpetuating insulin resistance (Cozma et al. 2018).

8.5 Epidemiology

Epidemiological data about MetS and the risk of cancer, recently collected and analyzed by Esposito et al. (2012), demonstrated that MetS is associated with an increased risk of several common cancers in adults (colorectal, liver, pancreas, endometrial and breast postmenopausal cancers): the risk of cancer varied according to gender, geographical population and definitions of metabolic syndrome. Although the magnitude of the reported associations was small to moderate [relative risk (RR) between 1.1 and 1.6], and different between sites and across populations, these findings may have a clinical significance, given the widespread diffusion of MetS and the increased cancer mortality associated to MetS. The association of MetS and colorectal cancer was present in both sexes; liver cancer appeared to be robustly associated with MetS in men, whereas pancreas cancer was consistently associated with MetS in women (Esposito et al. 2012). The principal factors thought to be responsible for the association between MetS and cancer are central obesity and hyperglycemia.

In a population-based study, which recruited 16,667 individuals with an age higher than forty and assessed 45,828 person years in cases with MetS; it is reported that pancreas cancer risk has increased in men (standardized incidence rate 178 (114–266)) and colorectal cancer risk has increased in women (standardized incidence rate 133 (101–170)) (Russo et al. 2008).

In a meta-analysis that has evaluated 38,940 cancer cases, although difference demonstrated according to sex, population, MetS definitions used, the presence of MetS has been shown to be in association with liver (1.58, $p < 0.0001$), colorectal (1.25, $p < 0.001$) and bladder (1.10, $p = 0.013$) cancer in men and endometrial (1.61, $p = 0.001$), pancreas (1.58, $p < 0.0001$), postmenopausal breast (1.56, $p = 0.017$), rectal (1.52, $p = 0.005$) and colorectal (1.34, $p = 0.006$) cancer in women (Bellastella et al. 2018).

Cancer-related mortality is reported to be high in cases with MetS [5 del n5]. In a 14-year follow-up of 33,230 male cases (28% with MetS) aged between 20 and 88 with no known cancer at baseline, patients with MetS were associated with a 56% greater age-adjusted risk in cancer mortality. Mortality risk was 83% higher in individuals with 3 or more MetS components than those with none (Esposito et al. 2012).

8.6 Pathophysiology

Metabolic syndrome adversely influences several body systems. Insulin resistance causes microvascular damage, which predisposes a patient to endothelial dysfunction, vascular resistance, hypertension, and vessel wall inflammation. Endothelial damage can impact the homeostasis of the body causing atherosclerotic disease and development of hypertension. Furthermore, hypertension adversely affects several body functions including increased vascular resistance and stiffness causing peripheral vascular disease, structural heart disease comprising of left ventricular hypertrophy and cardiomyopathy, and leading to renal impairment.

Accumulated effects of endothelial dysfunction and hypertension due to metabolic syndrome can further result in ischemic heart disease. Endothelial dysfunction due to increased levels of plasminogen activator type 1 and adipokine levels can cause thrombogenicity of the blood and hypertension causes vascular resistance through which coronary artery disease can develop. Also, dyslipidaemia associated with metabolic syndrome can drive the atherosclerotic process leading to symptomatic ischemic heart disease (Cătoi et al. 2018; He et al. 2018).

Although every component of MetS is known to have an association with cancer development, it is still debated whether the effects of these components are additive or synergistic.

8.7   Obesity and Cancer

Being overweight and obese has been reported to be responsible for death in 14% of men and 20% of women. Epidemiological studies have demonstrated that an increase in waist circumference and/or body mass index (BMI) has an association with colon, postmenopausal breast, endometrium, oesophagus, liver, gallbladder, gastric (cardia) and kidney cancer development (Cowey and Hardy 2006). In the pathophysiology of obesity and cancer relationship, insulin resistance is found to play an essential role. Chronic elevated insulin levels related to insulin resistance increase the bioavailability of insulin-like growth factor (IGF)-1; meanwhile, an increase in endogenous oestrogen levels and bioavailability play an important role in cancer development especially in hormone-dependent cancers (Pothiwala et al. 2009).

8.8   Hyperglycaemia and Cancer

Hyperglycaemia and cancer development are liked to each other regardless of BMI (Levine et al. 1990). In a population-based study (Rapp et al. 2006) in which 140,000 adults (63,585 male, 77,228 female) were followed for an average of 8.4 years, elevation of fasting plasma glucose was found to be associated with cancer development (HR 1.20; 95% CI 1.03–1.39 in male, 1.28; 95% CI 1.08–1.53 in female), and the strongest association was observed with hepatocellular cancer in men (HR 4.58; 95% CI 1.81–11.62).

In a 10-year follow-up of 1,298,385 Korean adults (829,770 male and 468,615 female), fasting glucose levels were found to be in association with increase in pancreas, liver and renal cancer and it is reported that mortality risk was increased in individuals with plasma glucose ≥ 140 mg/dl than those with plasma glucose <90 mg/dl in all cancer types (HR 1.29; 95% CI 1.22–1.37 in males, HR 1.23; 95% CI 1.09–1.39 in females). The strongest association in both sexes were observed with pancreatic cancer (HR 1.91; 95% CI 1.52–2.41 in males, HR 2.05; 95% CI 1.43–2.93 in females) (Jee et al. 2005). It is stated that colorectal cancer risk is increased both in men and women with type 2 diabetes and this is associated with colon cell proliferation triggered by hyperinsulinemia and increase free IGF-1 levels (Yang et al. 2005; Berster and Göke 2008). It is also shown that chronic insulin treatment is associated with increased colorectal adenoma risk in type 2 diabetic patients (Chung et al. 2008).

8.9   Dyslipidaemia and Cancer

Low high density lipoprotein (HDL) cholesterol levels are reported to be correlated with increased in lung cancer incidence, and in individuals with very low HDL cholesterol ($\leq$20 mg/dl), cancer risk is 6.5 fold increased (Kucharska-Newton et al. 2008; Shor et al. 2008). High triglyceride levels have been shown to be associated with prostate cancer in men. Low-density lipoprotein cholesterol levels have been found to be associated with hematologic malignancy development (approximately 15-fold increase) (Wuermli et al. 2005).

8.10   Hypertension and Cancer

Although there is no clear data revealing a link between hypertension and cancer development, it is suggested that it is in relation with increased cancer mortality, and inhibition of apoptosis may have a role in this. In a meta-analysis that included 10 prospective studies, the mean 9–20-year follow-up of 47,119 hypertensive cases, OR for cancer mortality was found to be 1.23 (1.11–1.36) (Grossman et al. 2002).

8.11   Possible Mechanisms of MetS and Cancer Association

It is proposed that obesity, inflammation and insulin resistance are all intercon-nected and this is potentially as a result of adipose tissue hypoxemia. It is also stated that development of insulin resistance in obese individuals is associated with tumour necrosis factor alpha (TNF-α) secreted from adipose tissue impairing intracellular insulin signal cascade, elevation in free fatty acid levels, decrease in adiponectin levels and also inhibition of peroxisome proliferator-activated receptor gamma by TNF-α and interleukin (IL)-1 stimulated with nuclear factor kappa B (NF-κB) (Ye 2009).

8.12   Insulin Resistance/Hyperinsulinemia/IGF-1 System

It is accepted that hyperinsulinemia/insulin resistance is the primary mechanism responsible for many manifestations of MetS. The strongest evidence in MetS and cancer association focuses on obesity and hyperinsulinemia/insulin resistance. Insulin is a major anabolic hormone that stimulates cell proliferation and its effect on cancer cell proliferation is suggested to be with IGF-1 stimulation. Growth hor-mone is the primary stimulant for IGF-1 production in liver and insulin stimulates IGF-1 production in liver by upregulating growth hormone receptors.

Hyperinsulinemia also increases IGF-1 bioavailability by decreasing hepatic secretion of IGF-binding protein-1 and 2 (Giovannucci 2001). IGF receptor is overexpressed in breast and colon cancers and its activation activates the p21 ras/mitogen-activated protein kinase (MAPK) pathway and phosphatidylinositol-3 kinase/AKT pathway for cell proliferation (Ibrahim and Yee 2004).

Among the proliferative and anti-apoptotic characteristics of IGF, its angiogenetic effect (due to vascular endothelial growth factor (VEGF)) is also suggested to play an important role in colon, endometrium, breast and prostate cancer development (Hoeben et al. 2004). On the other hand, it is reported that hyperinsulinemia and IGF-1 inhibit sex-binding globulin synthesis in liver increasing bio-availability of sex hormones and this may have a role in hormone-dependent cancers like breast, endometrium and prostate cancer (Calle and Kaaks 2004).

8.13 Aromatase Activity

The most important oestrogen source after menopause is aromatization of androgens in adipose tissue with cytochrome p450 enzyme complex. In obese women (especially visceral obesity), an increase of oestrogen bioavailability with the effect of insulin and IGF-1 and an increase in oestrogen synthesis due to aromatization in adipose tissue are considered to be the most important factors in the development of breast and endometrium cancer in postmenopausal women (Morimoto et al. 2002).

8.14 Adipokines

Adipokines are a group of signal molecules, which play a role in appetite and energy balance, inflammation, insulin sensitivity, angiogenesis, lipid metabolism, cell proliferation and atherosclerosis. Most of these functions are postulated to be in association with MetS and cancer and to have a role in the relationship between them.

8.15 Leptin

Leptin is a hormone secreted from adipocytes, causing the inhibition of appetite with metabolic signal and increase in basal metabolism. Obese individuals have hyperleptinemia due to leptin resistance and therefore, are more sensitive to MetS. It is suggested that high leptin levels are associated with prostate, colon, breast and endometrium cancer, and the cell proliferation effect of leptin via MAPK signal may have a role in this. On the other hand, it is also suggested that leptin may stimulate angiogenesis and increase the expression of matrix metalloproteinase-2 and contribute to cancer metastasis (Brauna et al. 2011).

8.16   Adiponectin

Contrary to other hormones secreted from adipocytes, adiponectin levels are low in obese patients. It is known that adiponectin increases insulin sensitivity in muscle and liver with 5′-adenosine monophosphate-activated protein kinase (AMPK) pathway, decreases plasma free fatty acid concentration and have anti-inflammatory anti-atherosclerotic properties. It is considered that adiponectin is inversely correlated with breast, endometrium and gastric cancer risk and this is associated with its ameliorating effect on insulin resistance and its antiproliferative, antiapoptotic and anti-inflammatory effects (Kadowaki and Yamauchi 2005).

8.17   VEGF

Angiogenesis is a critical process for tumour development and metastasis. The secretion of VEGF, the most important proangiogenic factor secreted from adipocytes (only visceral adipose tissue), is stimulated by insulin, IGF-1, oestrogen, leptin, TNF-α and hypoxia (Miyazawa-Hoshimoto et al. 2003).

8.18   Proinflammatory Cytokines

After the discovery of an association between TNF-α and insulin resistance (Richardson et al. 2013), more cytokines were under focus to identify similar associations, including IL-1β, IL-6, IL-8, IL-10, macrophage inflammatory protein-1, monocyte chemoattractant protein-1. For instance, TNF-α and IL-1β activate IKKβ/NF-κB and JNK pathways in adipocytes, hepatocytes, and associated macrophages and cause insulin resistance (Shoelson et al. 2006; Richardson et al. 2013). Inflammation is related with many cancer types, including gastric, pancreas, oesophagus, liver, gall bladder and colorectal cancer. El-Omar et al. showed that carriage of multiple proinflammatory polymorphisms of IL-1B, IL-1 receptor antagonist, TNF-α, and IL-10 conferred greater risk, with ORs (and 95% CIs) of 2.8 (1.6–5.1) for one, 5.4 (2.7–10.6) for 2, and 27.3 (7.4–99.8) for 3 or 4 high-risk genotypes (El-Omar et al. 2003). Aforementioned cytokines also lead to cancer cachexia, thereby increasing cancer-related mortality. In a study by Mantovani et al. serum levels of IL-1α, IL-6, and TNFα were significantly higher in cancer patients than in healthy individuals (Mantovani et al. 2000). It is found that cytokines, reactive oxygen products and inflammatory pathways (NF-κB) cause cancer by decreasing the tumour-suppression function with increasing cell cycle and stimulation of oncogene expression (Mendonça et al. 2015).

8.19   Chronic Hyperglycemia

Chronic hyperglycemia may act as carcinogenic agent through different mechanisms (Stattin et al. 2007; Van der Heiden et al. 2009; Iwatsuki et al. 2010; Pereira et al. 2013): first, advanced glycation end-products (AGEs) may cause carcinogenic oxidative damage to DNA (Pereira et al. 2013); second, hyperglycemia promotes the epithelial mesenchymal transition phenomenon (Iwatsuki et al. 2010); third, many cancers have both an overexpression of glucose transporter (GLUT) proteins, mainly GLUT1 and GLUT3, and an enhanced activity of enzymes involved in glycolysis, including hexokinase-2 (Van der Heiden et al. 2009); finally, high glucose uptake/storage by cancer cells is associated with advanced grading, greater metastatic potential and cancer chemotherapy resistance (Stattin et al. 2007).

8.20   Damage by Endocrine Disruptors Exposure and Air Pollution

The term endocrine disruptors (EDs) refers to a series of exogenous chemical compounds spread all over the world and believed to produce deleterious effects on human health, including male and female infertility and other reproductive alterations, changes in human behaviour and neuroendocrine patterns, MetS development, and some cancers (breast, prostate and testicular cancers) (Soto and Sonnenschein 2010; Casals-Casas and Desvergne 2011). The most known and studied EDs are contained in pesticides, insecticides, plastics and food packaging, perfumes, and include phthalates, endosulfan, polycarbonate, bisphenol a, dioxins (Soto and Sonnenschein 2010; Casals-Casas and Desvergne 2011). Most EDs are small and lipophilic agents that bind nuclear receptors, leading to a subsequent alteration of genes expression (Soto and Sonnenschein 2010; Casals-Casas and Desvergne 2011). Almost all EDs show xenoestrogen or antiandrogen properties, interfering with biological effects of sex steroid hormones; in addition, other hormonal pathways, involving thyroid hormone and glucocorticoid receptors, peroxisome proliferator-activated receptors (PPARs), are well established targets of EDs (Soto and Sonnenschein 2010, Casals-Casas and Desvergne 2011).

Worldwide, exposure to small (2.5 micron in diameter) particulate matter ($PM_{2.5}$) ranks 5th in the list of the top risk factors for death, as it increases the risk for ischemic heart disease, stroke, lung cancer, lower respiratory infections and chronic obstructive pulmonary disease (Health Effects Institute 2017). The generation of a state of subclinical inflammation represents the potential mechanism linking long-term exposure to $PM_{2.5}$ to both MetS and cancer (i.e. lung cancer) (Health Effects Institute 2017).

8.21   Disorders of Circadian Clock

Recent findings support a link between circadian clocks (CCs), epigenetics and cancer (Masri et al. 2015). Among the exogenous factors that may influence CCs, daily schedules of light/darkness, sleep/awakeness and food intake can play an important entraining action (Mistlberger 2011). Circadian rhythms orchestrate biochemical and physiological processes in living organisms to respond to day/night cycle with the variations of its correlated factors, in particular with the variations of light–inhibited and darkness-stimulated melatonin secretion.

The main circadian clock gene is an endogenous, self-sustaining pacemaker, identified as a histone acetyltransferase, that operates with a periodicity of 24 h (Doi et al. 2006). Disruption in proper circadian timekeeping results in detrimental effect, which can lead to metabolic disorders, uncontrolled cell growth and cancer (Sahar and Sassone-Corsi 2012). The association between MetS and cancer is more frequent in elderly subjects, likely because with ageing, circadian desynchrony occurs at the expense of peripheral metabolic pathologies and central neurodegenerative disorders, involving the complex structures regulating circadian rhythmicity, which frequently causes sleep disturbance and reduced melatonin secretion (Liu and Chang 2017). An example of the consequences of the desynchronizing effect on circadian rhythmicity by altered schedules of light/darkness and sleep/awakeness can be observed in shift workers, which presented an elevated cancer risk, especially for breast cancer, in epidemiological studies. Through complex pathways involving both brain circuits and peripheral organs, chronic disrupted sleep promote unhealthy feeding behaviour, insulin resistance, alterations of glucose and lipids metabolism, thus causing MetS (Sahar and Sassone-Corsi 2012).

It has been speculated that the development of MetS and cancer in shift workers is greatly and negatively influenced by circadian clock disruption, also taking into account that this disruption may directly affect cell cycle, favouring an anarchic cell proliferation (Hunt and Sassone-Corsi 2007).

However, melatonin variations are not only important for shift workers, but their correlation to the time of food intake may influence the body fat and body mass index, and consequently cancer risk.

8.22   Examples of MetS and Cancer Association

8.22.1   MetS and Colorectal Cancer

Epidemiological data reveal that colorectal cancer and adenoma risk is increased in patients with MetS. In a study from Korea, it is shown that abdominal obesity causes an independent increase in precancerous lesion risk (Pothiwala et al. 2009). In another study consisting of 368,277 cases, a high BMI and excess weight have a positive relationship with colon cancer only in men and an increase in waist circumference has a positive relationship in both men and women (Pischon et al. 2006). In 'Physicians' Health Study' (22,046 male physicians), it is demonstrated that colorectal cancer risk is increased 1.4 fold in individuals with 2 or more MetS components than those who have none, and the highest risk increase was observed with obesity and diabetes (Stürmer et al. 2006).

In a study investigating the association of MetS and its components with colorectal cancer, it is found that only in men there is a direct relation with both colon and rectum cancer as shown by Pelucchi et al. (2010). Another population-based cohort study has shown that the presence of MetS increases colorectal cancer risk only in men (RR 1.78; 95% CI 1.0–3.6) (Ahmed et al. 2006). It is postulated that the increased risk of colorectal cancer in men than in women with MetS may be caused due to differences in adipose tissue distribution (Calle 2007).

8.22.2   MetS and Breast Cancer

Growing data show the association between MetS components and insulin resistance with postmenopausal breast cancer, and the presence of MetS is considered a poor prognosis indicator (Lawlor et al. 2004). Between 1983 and 2007, in 2 case-control study analyses of 3869 postmenopausal women with breast cancer and 4082 postmenopausal control cases, it is demonstrated that postmenopausal breast cancer is higher in women with MetS than those without (OR 1.75; 95% CI 1.37–2.22) and risk increases more in women older than 70 (OR 3.04; 95% CI 1.75–5.29) (Rosato et al. 2011). In relationship between breast cancer and MetS, the production of extra gonadal estrogen, increase in estrogen bioavailability due to low sex hormone binding globulin levels and mitogenic effect on both non-transformed and neoplastic breast epithelial cells triggered by insulin resistance and hyperinsulinemia are suggested to play key roles (Vona-Davis et al. 2007).

8.22.3   MetS and Pancreatic Cancer

In a cohort of 577,315 Europeans, women with MetS had an increased risk of pancreatic cancer, with a relative risk of 1.58 (95% CI 1.34–1.87) (Esposito et al. 2012). Russo et al. reported a positive association between MetS and pancreatic cancer in 43 men who were prescribed antihypertensive, lipid-lowering, and antidiabetic drugs (Russo et al. 2008). A Japanese study indicated an increased risk of pancreatic cancer (HR, 1.99; 95% CI, 1.00–3.96) among women who had ≥2 metabolic components (Inoue et al. 2009). In another study, of all metabolic components, fasting glucose and BP were significantly associated with the risk of pancreatic cancer in both men and women (Johansen et al. 2010).

Park et al. suggest that the presence of MetS and 4 or 5 metabolic components carry an approximately 40% (adjusted HR, 1.47; 95% CI, 1.19–1.81) and 60% (adjusted HR, 1.64; 95% CI, 1.06–2.51) higher risk of pancreatic cancer, respectively. In the same work, evaluating the association of each metabolic component with pancreatic cancer, no individual metabolic component alone had a statistically significant association with the risk of pancreatic cancer. This result is subtly in conflict to previous reports (Park et al. 2020).

8.22.4 MetS and Prostate Cancer

Although there are prospective studies showing the association between MetS and prostate cancer, there are also other studies reporting a decrease in prostate cancer incidence in individuals with MetS (Håheim et al. 2006). Conflicting data exists regarding the link between obesity and prostate cancer. For example, in an analysis evaluating 31 cohort and 25 case–control studies, for every 5 kg/m^2 increase in BMI, relative risk for prostate cancer was found 1.05, and a higher risk was observed in patients with progressed diseases than localized diseases (MacInnis and English 2006). On the other hand, in a population-based case–control study, it is observed that obesity (BMI $\geq$ 30 kg/m^2) is inversely related to prostate cancer (Robinson et al. 2005). Nevertheless, it has also been shown that there is a positive correlation between obesity and advanced stage or metastatic prostate cancer and that non-metastatic advanced stage prostate cancer is inversely correlated with weight loss (Rodriguez et al. 2007).

The relationship between prostate cancer and diabetes is more complex. There are reports suggesting that prostate cancer risk is moderately low in men with diabetes (Bonovas et al. 2004).

In a recent study, Liu et al. tried to investigate the pathological risk of prostate cancer (PCa) according to the obesity and metabolic status of Chinese patients undergoing radical prostatectomy (Liu et al. 2020). Among 1016 men, 551 (54.2%), 106 (10.4%), 238 (23.4%), and 121 (11.9%) were assigned to the metabolically healthy and normal weight (MHNW) group, metabolically abnormal but normal weight (MANW) group, metabolically healthy but overweight or obese (MHO) group, and metabolically abnormal but overweight or obese (MAO) group, respectively. Compared with the MHNW group, the MAO group had a significantly greater risk of a higher prostatectomy Gleason score [odds ratio (OR), 1.907; 95% confidence interval (95% CI), 1.144–3.182], pathological stage (OR, 1.606; 95% CI, 1.035–2.493), and seminal vesicle invasion (OR, 1.673; 95% CI, 1.041–2.687). In contrast, the ORs were not increased in the MHO or MANW group. They also showed that in the context of normal weight, metabolic disorders were associated with lymph node involvement. The metabolic status and body mass index were not associated with extracapsular extension or surgical margins in any of the four groups. In Liu's study, in synthesis, the MAO phenotype was associated with aggressive PCa, including a higher prostatectomy Gleason score, pathological stage, and seminal vesicle invasion and might also be associated with disease progression, showing that obesity and metabolic disorders act synergistically to increase the pathological risk of PCa.

8.22.5 MetS and Bladder Cancer

In a prospective cohort study of 580,000 people Haggstrom et al. showed that MetS was associated with a significantly increased risk of bladder cancer in men (RR 1.10, 95%CI 1.01–1.18), whereas no association was observed in women (Häggström et al. 2011). Similarly, Russo et al. in an Italian population-based study and using a "non-traditional" pharmacologically based definition of MetS (i.e. individuals simultaneously treated with hypoglycemic, antihypertensive and hypolipemics drugs) observed an increased risk of bladder cancer only in men (standardized incidence ratios – ratio between observed and expected cases [109, 95% 8 CI 82–143]) (Russo et al. 2008). As a whole, in their meta-analysis, Esposito et al. estimated that in men the presence of MetS was significantly associated with bladder cancer with a RR of 1.10 (95% CI 1.02–1.18) (Esposito et al. 2012). Little is known regarding the influence of MetS on pathological and prognostic factors of bladder cancer, and if its cumulative effect is greater or lower than its single components. A recent study investigated this correlation in 262 consecutive patients undergoing RC for muscle-invasive urothelial BCa (Cantiello et al. 2014). Cantiello et al. showed that MetS, defined according to the NCEP ATP III criteria, did not emerge as an independent predictor of the risk of both a higher pathological stage, and lymph vascular invasion and lymph node invasion; conversely, BMI, considered as a surrogate of obesity, was found to be an independent predictor of both oncological conditions. Unfortunately, no data were available on other prognostic parameters (Cantiello et al. 2014). In their study carried out with a large cohort of patients within the Me-Can project, Haggstrom et al. showed that MetS did not predict the risk of cancer-specific mortality, while they observed in male patients that an increased blood pressure was the only independent risk factor of BCa mortality (RR 1.34, 95% CI 1.06–1.69) (Häggström et al. 2011). Consequently, MetSas a whole did not seem to confer a risk of worse prognosis, but two studies are few, and further clinical studies with a large cohort of patients are required.

8.22.6 MetS and Kidney Cancer

Several studies have evaluated the association between kidney cancer risk and MetS components. Most cancers that develop in the kidneys are RCCs; thus, most of the studies are limited to RCC. From a Swedish cohort, it was reported that higher BMI and hypertension are significant risk factors of RCC for men. A meta-analysis of 22 clinical studies available in MEDLINE from 1966 to 1998 showed that increased

BMI is equally associated with increased risk of RCC in both sexes (Chow et al. 2000). The findings from studies for Caucasians, African-Americans, and Chinese showed a strong relationship between increased blood pressure and higher risk of RCC (Colt et al. 2011; Wang et al. 2012). A total of 153,852 Swedish people were analyzed in a study of the general population, and results showed that the morbidity and mortality of RCC increased in diabetic patients (Lindblad et al. 1999). Furthermore, Joh et al. reported that type 2 DM was a risk factor for RCC in women (Joh et al. 2011). However, in terms of dyslipidemia, only a few studies have been conducted and the results are inconsistent (Horiguchi et al. 2008; Ahn et al. 2009; Van Hemelrijck et al. 2012).

A recent publication by Tae Ryom Oh et al. showed how kidney cancer risk was significantly higher in patients with MetS, and there was no difference according to sex (Oh et al. 2019). The hazards ratio of kidney cancer increased with increasing number of MetS components. For patients not diagnosed with MetS but with abdominal obesity and hypertension, the likelihood of developing kidney cancer was similar to that of patients diagnosed with MetS. In the same study, interestingly, patients with improved MetS within two years had increased risk of kidney cancer compared with those without MetS. MetS, in conclusion, is an independent risk factor for kidney cancer, and the obesity and hypertension components of MetS are also powerful risk factors.

8.22.7   MetS and Thyroid Cancer

De-Tao Yin et al., in a recent meta-analysis showed that Insulin Resistance, dysglycemia, high BMI and hypertension significantly increased the thyroid cancer risk (Yin et al. 2018).

There was an increased risk for thyroid cancer for patients with insulin resistance (risk ratios [RR] = 1.59, 95% confidence interval [CI] = 1.12–2.27, $P = 0.01$), dysglycemia (RR = 1.40, 95%CI = 1.15–1.70, $P < 0.001$), high BMI (RR = 1.35, 95%CI = 1.23–1.48, $P < 0.001$) and hypertension (RR = 1.34, 95%CI = 1.22–1.47, $p < 0.001$). However, patients with dyslipidemia, both total cholesterol (RR = 1.09, 95%CI = 0.98–1.21, $P = 0.13$) and triglyceride (RR = 1.01, 95%CI = 0.91–1.12, $P = 0.82$) was not associated with thyroid cancer (Yin et al. 2018).

8.23   Prevention

8.23.1   Diet

The basic approach for cancer prevention in patients with MetS is to prevent risk factors. Lifestyle changes including weight loss and a healthy diet, especially the Mediterranean diet (MD), are known to decrease cancer risk in normal population

(Giacosa et al. 2013). Populations living in the area of Mediterranean Sea showed a decrease incidence of cancer compared with those living in regions of North Europe or US, this has been attributed to healthier dietary habits (Grosso et al. 2013). In the past decade, several reports indicated a protective role of MD towards neoplastic diseases. In particular, a meta-analysis study by Sofi and colleagues reported that MD is responsible of 6% reduction of cancer death/incidence (Sofi et al. 2010). An updated study, done on a very large cohort (335,873 subjects of the European Prospective Investigation Into Cancer and Nutrition), reported a lower cancer risk in subjects following MD (Couto et al. 2011). Nowadays, several epidemiological studies focused on the association of MD and specific type of cancers (breast and colorectal cancers). The MD contributes to the prevention of colorectal cancer through a high intake of fibre. Barera et al. have highlighted the nutraceutical effects of β-glucans, which seem to reduce low-density lipoprotein cholesterol (LDL), IL-6 and advanced glycation end-product (AGE) levels. They are also linked to colon cancer prevention (Barera et al. 2016). The MD is able to reduce gastric cancer's incidence and mortality in the South areas of selected Mediterranean country, such as France, Greece and Italy when compared with the North areas of the same countries. Higher adherence to MD lower up of 20% the incidence of all gastric cancers (Decarli et al. 1986; La Vecchia et al. 1995; Praud et al. 2014). The beneficial role of MD has also been confirmed in a large case-control study in Italy. The authors found that an enhanced adherence to MD (high consumption of fruit, vegetables and legumes) reduced risk of both oral cavity and pharynx cancer and larynx cancer (Giraldi et al. 2017). Head and neck cancers are currently the sixth cause of death in the world; however, as confirmed by Bosetti et al. a greater adherence to the MD reduces the risk of oral, laryngeal and pharyngeal cancer (by 23 and 29%, respectively) (Bosetti et al. 2003). Jacobs et al. have suggested that consumption of whole grain (bran, germ, endosperm) ≥ 4 times/week reduces the risk of cancer by 40% compared to controls, while the Continuous Update Project has determined that the intake of non-starchy vegetables and fruits lowers the risk of mouth, pharynx, larynx, oesophagus and stomach cancers (Jacobs et al. 1998; Norat et al. 2014). On the other hand, the intake of refined cereal grains (bread, pasta or rice) increase the risk of upper digestive tract, stomach, colorectal, breast and thyroid cancer (Chatenoud et al. 1999). Another pivotal food of the MD is pulses. Thanks to their content in tannins, saponins, protease inhibitor and phytic acid, legumes play an important anti-cancer role (Mudryj et al. 2014). Even the consumption of nuts was associated to the reduction of cancer's risk, in particular for those of the digestive tract. The protective role of nuts comes from their content of ellagic acid, anacardic acids, genistein, resveratrol and phytic acid (Falasca et al. 2014). The consumption of peanut products ≥2/week was associated with 58% reduction of colorectal cancer risk in Taiwan women (Yeh et al. 2006). According to IARC, ethanol was positively correlated to cancers (e.g. mouth, pharynx, larynx, oesophagus, colorectal in men, breast in pre and post-menopause). The Continuous Update Project also confirmed the carcinogen effect of alcohol in liver and colorectal cancers (WCRF 2007). However, the studies on MD show that a moderately intake of red wine in pre-menopause women reduced breast cancer risk, inhibiting the conversion of androgens to estrogens, catalyzing by aromatase (Fidanza et al. 2004; Shufelt et al. 2012). This effect is due to polyphenols content of red wine, including flavonoids (anthocyanins and flavan-3-ols) and non-flavonoids (resveratrol, cinnamates and gallic acid (Arranz et al. 2012; Di Daniele et al. 2017). In a very recent study Huang et al. showed that higher plant protein intake was associated with small reductions in risk of overall and cardiovascular disease mortality (Huang et al. 2020). Plant protein intake was inversely associated with mortality from all CVD, stroke, and "other causes combined," representing risk reductions of 5% to 12% per 1-SD intake increment and 12% to 30% per 10 g/1000 kcal intake increment in men, and 7% to 12% per 1 SD and 17% to 29% per 10 g/1000 kcal in women (all $P < .003$). By contrast, plant protein intake was not significantly associated with mortality from cancer, heart disease, respiratory disease, infections or injuries/accidents in either men or women. However, their findings provide evidence that dietary modification in choice of protein sources may influence health and longevity.

8.23.2   Vitamins and Micronutrients

The available data on vitamins and dietary supplements for chemoprevention are extensive and difficult to interpret. Multiple observational and prospective studies of the use of supplemental vitamins and minerals to prevent cancer have been disappointing (Vastag 2009). A systematic review of 38 studies found that neither vitamin C nor vitamin E supplementation was beneficial for prevention of the cancers evaluated (Coulter et al. 2006). A subsequent long-term randomized trial (mean 9.4 years treatment) in 8000 women found no evidence that supplementation with vitamin C, E, or beta-carotene (singly or in combination) decreased cancer incidence or cancer mortality (Lin et al. 2009). Additionally, two long-term observational studies, one including over 160,000 women with follow-up of approximately eight years (Neuhouser et al. 2009) and another including over 180,000 multiethnic participants with 11-year follow-up (Park et al. 2011), found no association between multivitamin use and risk of cancer.

8.23.2.1   Vitamin D

Studies of the relationship between vitamin D intake or serum levels of 25(OH) D and cancer risk have been inconsistent (Chung et al. 2011). Studies vary in regard to participants (sex, baseline serum levels), types of cancer evaluated, and dose of vitamin D. Overall, it does not appear that vitamin D supplements should be prescribed to decrease cancer risk (Freedman et al. 2010).

8.23.2.2   Calcium

Increased calcium intake has been linked to reduced risk of colorectal cancer but may be associated with an increased risk of prostate cancer. There may be a minimum level of calcium intake, around 700 mg/day, that confers protection against colorectal cancer without significantly increasing prostate cancer risk (Pietinen et al. 1999; Schuurman et al. 1999).

8.23.2.3   Folate and Other B Vitamins

Folate is present in green, leafy vegetables, fruits, cereals and grains, nuts, and meats. Folic acid, a synthetic form included in supplements, has many of the same biologic effects as folate but is more bioavailable. Folate is important in DNA synthesis, methylation, and repair, as well as in the regulation of gene expression. The role of folate or folic acid in cancer prevention is uncertain. Folate has been associated with a decreased risk for colon and other cancers, especially in individuals who consume alcohol, in observational studies. However, some randomized trials have suggested the possibility that folic acid may increase risk for cancer (Graham and Karen 2018).

8.23.2.4   Iron

Observational studies suggest that increased iron stores or dietary iron may be associated with increased risk for cancer (Knekt et al. 1994; van Asperen et al. 1995). A randomized trial conducted to evaluate the benefits of phlebotomy in patients with peripheral artery disease found a significant reduction in cancer incidence at six months (HR 0.65, 95% CI 0.43–0.97) in patients assigned to the phlebotomy group compared with controls (Zacharski et al. 2008). This finding warrants confirmation.

8.23.2.5   Vitamin E and Selenium

Men should avoid vitamin E supplementation at doses that exceed dietary intakes. Current evidence does not support a role for vitamin E supplementation in the prevention or treatment of cancers, cardiovascular disease, dementia, or infection. The relationship between prostate cancer and selenium intake and level is complex. However, there is no evidence supporting a chemopreventive effect from selenium supplementation (Klein et al. 2011). The Selenium and Vitamin E Cancer Prevention Trial (SELECT) was designed to study the role of selenium and vitamin E as agents to decrease the incidence of prostate cancer. In this trial, 35,533 men were randomly assigned to selenium (200 mcg/day), vitamin E (400 international units/day), both, or neither with appropriate placebos (Lippman et al. 2005). This trial was stopped in October 2008 after an independent data safety monitoring committee found no evidence of a decrease in the incidence of prostate cancer. Dietary supplementation with vitamin E significantly increased the incidence of prostate cancer, and there were nonsignificant increases with selenium and the combination of vitamin E plus selenium compared with placebo (HR for developing prostate cancer 1.17, 1.09, and 1.05, respectively) (Klein et al. 2011).

8.23.2.6 Others

In cases with bariatric surgery cancer-related mortality was shown to have decreased compared to all cases and matched controls (HR 0.38 and 0.40, $p < 0.001$ respectively) (Adams et al. 2007). Epidemiological studies have revealed that metformin therapy decreased cancer incidence and the risk of cancer-related mortality in diabetics when compared to those treated with sulfonylureas or other therapies. In patients using sulfonylureas or insulin, cancer-related mortality was found to be 1.3 and 1.9 fold higher respectively than metformin-using patients (Bowker et al. 2006). In a study investigating the cancer risk of glucose-lowering agents adjusted HR for cancer risk was found 1.36 (95% CI 1.19–1.54) in sulfonylurea monotherapy, 1.42 (95% CI 1.27–1.60) in insulin mono-therapy group and 0.54 (95% CI 0.43–0.66) in insulin + metformin group (Currie et al. 2009). These findings have given birth to the opinion that metformin treatment in patients with type 2 diabetes may have a cancer-preventive effect. Today, an insulin-sensitizing agent, metformin may have cancer-preventive effects independent of its hypoglycemic effect (Evans et al. 2005). It is postulated that metformin suppresses in vitro and in vivo cancer cell growth and there may be potential mechanisms like liver kinase b1/AMPK pathway activation, induction of apoptosis, inhibition of protein synthesis, decrease in insulin levels, activation of immune system and eradication of cancer stem cell (Kourelis and Siegel 2012). The benefit of metformin as an adjuvant chemotherapeutic agent and its affect as increased response to chemotherapy is still investigated. Some of the observational studies managed to prove this effect of metformin. Jiralerspong et al. showed that diabetic patients with breast cancer receiving metformin and neo-adjuvant chemo-therapy have a higher pathologic complete response rate than diabetics not receiving metformin (Jiralerspong et al. 2009). Prospective studies are warranted to clarify this effect of metformin.

As conclusion Obesity and MetS and its components are found to be associated with cancer development and mortality. The only and most efficient preventive method is still lifestyle change. Metformin and diet are promising agents for prevention and some studies show favouring effects of metformin on cancer treatment.

References

Aballay LR, Eynard AR, del Pilar Díaz M, Navarro A, Muñoz SE. Overweight and obesity: a review of their relationship to metabolic syndrome, cardiovascular disease, and cancer in South America. Nutr Rev. 2013;71(3):168–79. https://doi.org/10.1111/j.1753-4887.2012.00533.x.

Adams TD, et al. Long-term mortality after gastric bypass surgery. N Engl J Med. 2007; https://doi.org/10.1056/NEJMoa066603.

Ahmed RL, Schmitz KH, Anderson KE, Rosamond WD, Folsom AR. The metabolic syndrome and risk of incident colorectal cancer. Cancer. 2006; https://doi.org/10.1002/cncr.21950.

Ahn J, et al. Prediagnostic total and high-density lipoprotein cholesterol and risk of cancer. Cancer Epidemiol Biomark Prev. 2009; https://doi.org/10.1158/1055-9965.EPI-08-1248.

Aleksic T, et al. Type 1 insulin-like growth factor receptor translocates to the nucleus of human tumor cells. Cancer Res. 2010; https://doi.org/10.1158/0008-5472.CAN-10-0052.

Arranz S, et al. Wine, beer, alcohol and polyphenols on cardiovascular disease and cancer. Nutrients. 2012; https://doi.org/10.3390/nu4070759.

Asperen IAV, Feskens EJM, Bowles CH, Kromhout D. Body iron stores and mortality due to cancer and ischaemic heart disease: a 17-year follow-up study of elderly men and women. Int J Epidemiol. 1995; https://doi.org/10.1093/ije/24.4.665.

Barera A, et al. β-glucans: ex vivo inflammatory and oxidative stress results after pasta intake. Immun Ageing. 2016; https://doi.org/10.1186/s12979-016-0068-x.

Bellastella G, Scappaticcio L, Esposito K, Giugliano D, Maiorino MI. Metabolic syndrome and cancer: "The common soil hypothesis". Diabetes Res Clin Pract. 2018; https://doi.org/10.1016/j.diabres.2018.05.024.

Berster JM, Göke B. Type 2 diabetes mellitus as risk factor for colorectal cancer. Arch Physiol Biochem. 2008; https://doi.org/10.1080/13813450802008455.

Bonovas S, Filioussi K, Tsantes A. Diabetes mellitus and risk of prostate cancer: a meta-analysis. Diabetologia. 2004; https://doi.org/10.1007/s00125-004-1415-6.

Booth ML, et al. Change in the prevalence of overweight and obesity among young Australians, 1969-1997. Am J Clin Nutr. 2003; https://doi.org/10.1093/ajcn/77.1.29.

Bosetti C, et al. Influence of the mediterranean diet on the risk of cancers of the upper aerodigestive tract. Cancer Epidemiol Biomark Prev. 2003;

Bowker SL, Majumdar SR, Veugelers P, Johnson JA. Increased cancer-related mortality for patients with type 2 diabetes who use sulfonylureas or insulin. Diabetes Care. 2006; https://doi.org/10.2337/diacare.29.02.06.dc05-1558.

Brauna S, Bitton-Worms K, le Roith D. The link between the metabolic syndrome and cancer. Int J Biol Sci. 2011; https://doi.org/10.7150/ijbs.7.1003.

Burrage E, Marshall K, Santanam N, Chantler P. Cerebrovascular dysfunction with stress and depression. Brain Circ. 2018; https://doi.org/10.4103/bc.bc_6_18.

Calle EE. Obesity and cancer. Br Med J. 2007; https://doi.org/10.1136/bmj.39384.472072.80.

Calle EE, Kaaks R. Overweight, obesity and cancer: epidemiological evidence and proposed mechanisms. Nat Rev Cancer. 2004; https://doi.org/10.1038/nrc1408.

Cantiello F, et al. Visceral obesity predicts adverse pathological features in urothelial bladder cancer patients undergoing radical cystectomy: a retrospective cohort study. World J Urol. 2014; https://doi.org/10.1007/s00345-013-1147-7.

Casals-Casas C, Desvergne B. Endocrine disruptors: from endocrine to metabolic disruption. Annu Rev Physiol. 2011; https://doi.org/10.1146/annurev-physiol-012110-142200.

Catharina AS, et al. Metabolic syndrome-related features in controlled and resistant hypertensive subjects. Arq Bras Cardiol. 2018; https://doi.org/10.5935/abc.20180076.

Cătoi AF, et al. Metabolically healthy versus unhealthy morbidly obese: chronic inflammation, nitro-oxidative stress, and insulin resistance. Nutrients. 2018; https://doi.org/10.3390/nu10091199.

Chatenoud L, et al. Refined-cereal intake and risk of selected cancers in Italy. Am J Clin Nutr. 1999; https://doi.org/10.1093/ajcn/70.6.1107.

Chow WH, Gridley G, Fraumeni JF, Järvholm B. Obesity, hypertension, and the risk of kidney cancer in men. N Engl J Med. 2000; https://doi.org/10.1056/NEJM200011023431804.

Chung YW, et al. Insulin therapy and colorectal adenoma risk among patients with type 2 diabetes mellitus: a case-control study in Korea. Dis Colon Rectum. 2008; https://doi.org/10.1007/s10350-007-9184-1.

Chung M, Lee J, Terasawa T, Lau J, Trikalinos TA. Vitamin D with or without calcium supplementation for prevention of cancer and fractures: an updated meta-analysis for the U.S. preventive services task force. Ann Int Med. 2011; https://doi.org/10.7326/0003-4819-155-12-201112200-00005.

Colt JS, et al. Hypertension and risk of renal cell carcinoma among white and black Americans. Epidemiology. 2011; https://doi.org/10.1097/EDE.0b013e3182300720.

Comba A, Pasqualini ME. Primers on molecular pathways - lipoxygenases: their role as an oncogenic pathway in pancreatic cancer. Pancreatology. 2009;9(6):724–8. https://doi.org/10.1159/000235623.

Coulter ID, et al. Antioxidants vitamin C and vitamin E for the prevention and treatment of cancer. J Gen Intern Med. 2006; https://doi.org/10.1111/j.1525-1497.2006.00483.x.

Couto E, et al. Mediterranean dietary pattern and cancer risk in the EPIC cohort. Br J Cancer. 2011; https://doi.org/10.1038/bjc.2011.106.

Cowey S, Hardy RW. The metabolic syndrome: a high-risk state for cancer? Am J Pathol. 2006; https://doi.org/10.2353/ajpath.2006.051090.

Cozma A, et al. Determining factors of arterial stiffness in subjects with metabolic syndrome. Metab Syndr Relat Disord. 2018; https://doi.org/10.1089/met.2018.0057.

Currie CJ, Poole CD, Gale EAM. The influence of glucose-lowering therapies on cancer risk in type 2 diabetes. Diabetologia. 2009; https://doi.org/10.1007/s00125-009-1440-6.

Decarli A, la Vecchia C, Cislaghi C, Mezzanotte G, Marubini E. Descriptive epidemiology of gastric cancer in Italy. Cancer. 1986; https://doi.org/10.1002/1097-0142(19861201)58:11<2560::AID-CNCR2820581134>3.0.CO;2-U.

Di Daniele ND, et al. Impact of Mediterranean diet on metabolic syndrome, cancer and longevity. Oncotarget. 2017; https://doi.org/10.18632/oncotarget.13553.

Doi M, Hirayama J, Sassone-Corsi P. Circadian regulator clock is a histone acetyltransferase. Cell. 2006; https://doi.org/10.1016/j.cell.2006.03.033.

El-Omar EM, et al. Increased risk of noncardia gastric cancer associated with proinflammatory cytokine gene polymorphisms. Gastroenterology. 2003; https://doi.org/10.1016/S0016-5085(03)00157-4.

Esposito K, Chiodini P, Colao A, Lenzi A, Giugliano D. Metabolic syndrome and risk of cancer: a systematic review and meta-analysis. Diabetes Care. 2012; https://doi.org/10.2337/dc12-0336.

Evans JMM, Donnelly LA, Emslie-Smith AM, Alessi DR, Morris AD. Metformin and reduced risk of cancer in diabetic patients. Br Med J. 2005; https://doi.org/10.1136/bmj.38415.708634.F7.

Falasca M, Casari I, Maffucci T. Cancer chemoprevention with nuts. J Natl Cancer Inst. 2014; https://doi.org/10.1093/jnci/dju238.

Fernández-Real JM, et al. Study of caveolin-1 gene expression in whole adipose tissue and its subfractions and during differentiation of human adipocytes. Nutr Metab. 2010; https://doi.org/10.1186/1743-7075-7-20.

Fidanza F, Alberti A, Lanti M, Menotti A. Mediterranean adequacy index: correlation with 25-year mortality from coronary heart disease in the Seven Countries Study. Nutr Metab Cardiovasc Dis. 2004; https://doi.org/10.1016/S0939-4753(04)80052-8.

Freedman DM, Looker AC, Abnet CC, Linet MS, Graubard BI. Serum 25-hydroxyvitamin D and cancer mortality in the NHANES III study (1988-2006). Cancer Res. 2010; https://doi.org/10.1158/0008-5472.CAN-10-1420.

Giacosa A, et al. Cancer prevention in Europe: the mediterranean diet as a protective choice. Eur J Cancer Prev. 2013; https://doi.org/10.1097/CEJ.0b013e328354d2d7.

Giovannucci E. Insulin, insulin-like growth factors and colon cancer: a review of the evidence. J Nutr. 2001; https://doi.org/10.1093/jn/131.11.3109s.

Giraldi L, Panic N, Cadoni G, Boccia S, Leoncini E. Association between mediterranean diet and head and neck cancer: results of a large case-control study in Italy. Eur J Cancer Prev. 2017; https://doi.org/10.1097/CEJ.0000000000000277.

Graham AC, Karen ME. Accelerating the Pace of Cancer Prevention- Right Now. Cancer Prev Res. 2018;11(4):171–184. https://doi.org/10.1158/1940-6207.CAPR-17-0282.

Grossman E, Messerli FH, Boyko V, Goldbourt U. Is there an association between hypertension and cancer mortality? Am J Med. 2002; https://doi.org/10.1016/S0002-9343(02)01049-5.

Grosso G, et al. Mediterranean diet and cancer: epidemiological evidence and mechanism of selected aspects. BMC Surg. 2013; https://doi.org/10.1186/1471-2482-13-S2-S14.

Grundy SM, Brewer HB, Cleeman JI, Smith SC, Lenfant C. Definition of metabolic syndrome: report of the National Heart, Lung, and Blood Institute/American Heart Association Conference on Scientific Issues Related to Definition. Circulation. 2004; https://doi.org/10.1161/01.CIR.0000111245.75752.C6.

Guerre-Millo M. Adipose tissue hormones. J Endocrinol Investig. 2002; https://doi.org/10.1007/BF03344048.

Häggström C, et al. Metabolic syndrome and risk of bladder cancer: prospective cohort study in the metabolic syndrome and cancer project (Me-Can). Int J Cancer. 2011; https://doi.org/10.1002/ijc.25521.

Håheim LL, Wisløff TF, Holme I, Nafstad P. Metabolic syndrome predicts prostate cancer in a cohort of middle-aged Norwegian men followed for 27 years. Am J Epidemiol. 2006; https://doi.org/10.1093/aje/kwj284.

He Y, et al. Linking gut microbiota, metabolic syndrome and economic status based on a population-level analysis. Microbiome. 2018; https://doi.org/10.1186/s40168-018-0557-6.

Health Effects Institute. State of global air 2017: a special report. Boston, MA: Health Effects Institute; 2017.

Hoeben A, et al. Vascular endothelial growth factor and angiogenesis. Pharmacol Rev. 2004; https://doi.org/10.1124/pr.56.4.3.

Horiguchi A, et al. Decreased serum adiponectin levels in patients with metastatic renal cell carcinoma. Jpn J Clin Oncol. 2008; https://doi.org/10.1093/jjco/hym158.

Huang J, et al. Association between plant and animal protein intake and overall and cause-specific mortality. JAMA Intern Med. 2020; https://doi.org/10.1001/jamainternmed.2020.2790.

Hunt T, Sassone-Corsi P. Riding tandem: circadian clocks and the cell cycle. Cell. 2007; https://doi.org/10.1016/j.cell.2007.04.015.

Ibrahim YH, Yee D. Insulin-like growth factor-I and cancer risk. Growth Horm An D IGF Res. 2004; https://doi.org/10.1016/j.ghir.2004.01.005.

Inoue M, et al. Impact of metabolic factors on subsequent cancer risk: results from a large-scale population-based cohort study in Japan. Eur J Cancer Prev. 2009; https://doi.org/10.1097/CEJ.0b013e3283240460.

Iwatsuki M, et al. Epithelial-mesenchymal transition in cancer development and its clinical significance. Cancer Sci. 2010; https://doi.org/10.1111/j.1349-7006.2009.01419.x.

Jacobs DR, Marquart L, Slavin J, Kushi LH. Whole-grain intake and cancer: an expanded review and meta-analysis. Nutr Cancer. 1998; https://doi.org/10.1080/01635589809514647.

Jee SH, Ohrr H, Sull JW, Yun JE, Ji M, Samet JM. Fasting serum glucose level and cancer risk in Korean men and women. JAMA. 2005;293(2):194–202. https://doi.org/10.1001/jama.293.2.194. PMID: 15644546.

Jiralerspong S, et al. Metformin and pathologic complete responses to neoadjuvant chemotherapy in diabetic patients with breast cancer. J Clin Oncol. 2009; https://doi.org/10.1200/JCO.2009.19.6410.

Joh HK, Willett WC, Cho E. Type 2 diabetes and the risk of renal cell cancer in women. Diabetes Care. 2011; https://doi.org/10.2337/dc11-0132.

Johansen D, et al. Metabolic factors and the risk of pancreatic cancer: a prospective analysis of almost 580,000 men and women in the metabolic syndrome and cancer project. Cancer Epidemiol Biomark Prev. 2010; https://doi.org/10.1158/1055-9965.EPI-10-0234.

Kadowaki T, Yamauchi T. Adiponectin and adiponectin receptors. Endocr Rev. 2005; https://doi.org/10.1210/er.2005-0005.

Kim JY, Yi ES. Analysis of the relationship between physical activity and metabolic syndrome risk factors in adults with intellectual disabilities. J Exerc Rehabil. 2018; https://doi.org/10.12965/jer.1836302.151.

Klein EA, et al. Vitamin E and the risk of prostate cancer: the selenium and vitamin E cancer prevention trial (SELECT). JAMA. 2011; https://doi.org/10.1001/jama.2011.1437.

Knekt P, et al. Body iron stores and risk of cancer. Int J Cancer. 1994; https://doi.org/10.1002/ijc.2910560315.

Kourelis TV, Siegel RD. Metformin and cancer: New applications for an old drug. Med Oncol. 2012; https://doi.org/10.1007/s12032-011-9846-7.

Krukowski RA, West DS. Consideration of the food environment in cancer risk reduction. J Am Diet Assoc. 2010; https://doi.org/10.1016/j.jada.2010.03.026.

Kucharska-Newton AM, et al. HDL-cholesterol and the incidence of lung cancer in the Atherosclerosis Risk in Communities (ARIC) study. Lung Cancer. 2008; https://doi.org/10.1016/j.lungcan.2008.01.015.

La Vecchia C, D'Avanzo B, Negri E, Decarli A, Benichou J. Attributable risks for stomach cancer in Northern Italy. Int J Cancer. 1995; https://doi.org/10.1002/ijc.2910600603.

Lawlor DA, Smith GD, Ebrahim S. Hyperinsulinaemia and increased risk of breast cancer: findings from the British women's heart and health study. Cancer Causes Control. 2004; https://doi.org/10.1023/B:CACO.0000024225.14618.a8.

Levine W, Dyer AR, Shekelle RB, Schoenberger JA, Stamler J. Post-load plasma glucose and cancer mortality in middle-aged men and women: 12-year follow-up FINDINGS of THE chicago HEART association DETECTION project IN industry. Am J Epidemiol. 1990; https://doi.org/10.1093/oxfordjournals.aje.a115495.

Lin WW, Karin M. A cytokine-mediated link between innate immunity, inflammation, and cancer. J Clin Invest. 2007;117(5):1175–83. https://doi.org/10.1172/JCI31537.

Lin J, et al. Vitamins C and E and beta carotene supplementation and cancer risk: a randomized controlled trial. J Natl Cancer Inst. 2009; https://doi.org/10.1093/jnci/djn438.

Lindblad P, et al. The role of diabetes mellitus in the aetiology of renal cell cancer. Diabetologia. 1999; https://doi.org/10.1007/s001250051122.

Lippman SM, et al. Designing the selenium and vitamin E cancer prevention trial (SELECT). J Natl Cancer Inst. 2005; https://doi.org/10.1093/jnci/dji009.

Liu F, Chang HC. Physiological links of circadian clock and biological clock of aging. Protein Cell. 2017; https://doi.org/10.1007/s13238-016-0366-2.

Liu W, et al. Metabolically abnormal obesity increases the risk of advanced prostate cancer in Chinese patients undergoing radical prostatectomy. Cancer Manag Res. 2020; https://doi.org/10.2147/CMAR.S242193.

López de Blanco M., Carmona A. La transición alimentaria y nutricional: un reto en el siglo XXI. An. venez. nutr. 2005.

MacInnis RJ, English DR. Body size and composition and prostate cancer risk: systematic review and meta-regression analysis. Cancer Causes Control. 2006; https://doi.org/10.1007/s10552-006-0049-z.

Mantovani G, et al. Serum levels of leptin and proinflammatory cytokines in patients with advanced-stage cancer at different sites. J Mol Med. 2000; https://doi.org/10.1007/s001090000137.

Masri S, Kinouchi K, Sassone-Corsi P. Circadian clocks, epigenetics, and cancer. Curr Opin Oncol. 2015; https://doi.org/10.1097/CCO.0000000000000153.

Mendonça FM, et al. Metabolic syndrome and risk of cancer: which link? Metab Clin Exp. 2015; https://doi.org/10.1016/j.metabol.2014.10.008.

Mistlberger RE. Neurobiology of food anticipatory circadian rhythms. Physiol Behav. 2011; https://doi.org/10.1016/j.physbeh.2011.04.015.

Miyazawa-Hoshimoto S, Takahashi K, Bujo H, Hashimoto N, Saito Y. Elevated serum vascular endothelial growth factor is associated with visceral fat accumulation in human obese subjects. Diabetologia. 2003; https://doi.org/10.1007/s00125-003-1221-6.

Morimoto LM, et al. Obesity, body size, and risk of postmenopausal breast cancer: the women's health initiative (United States). Cancer Causes Control. 2002; https://doi.org/10.102 3/A:1020239211145.

Mudryj AN, Yu N, Aukema HM. Nutritional and health benefits of pulses. Appl Physiol Nutr Metab. 2014; https://doi.org/10.1139/apnm-2013-0557.

National Institute of Health. NHLBI obesity education initiative expert panel on the identification, evaluation, and treatment of overweight and obesity in adults. NHLBI Obes. Educ. Initiat. 2000.

Neuhouser ML, et al. Multivitamin use and risk of cancer and cardiovascular disease in the women's health initiative cohorts. Arch Intern Med. 2009; https://doi.org/10.1001/archinternmed.2008.540.

Norat T, Aune D, Chan D, Romaguera D. Fruits and vegetables: updating the epidemiologic evidence for the WCRF/AICR lifestyle recommendations for cancer prevention. Cancer Treat Res. 2014; https://doi.org/10.1007/978-3-642-38007-5_3.

Oh TR, et al. Metabolic syndrome resolved within two years is still a risk factor for kidney cancer. J Clin Med. 2019; https://doi.org/10.3390/jcm8091329.

Park S-Y, Murphy SP, Wilkens LR, Henderson BE, Kolonel LN. Multivitamin use and the risk of mortality and cancer incidence: the multiethnic cohort study. Am J Epidemiol. 2011; https://doi.org/10.1093/aje/kwq447.

Park SK, Oh C-M, Kim M-H, Ha E, Choi Y-S, Ryoo J-H. Metabolic syndrome, metabolic components, and their relation to the risk of pancreatic cancer. Cancer. 2020;

Pelucchi C, et al. Metabolic syndrome is associated with colorectal cancer in men. Eur J Cancer. 2010; https://doi.org/10.1016/j.ejca.2010.03.010.

Pereira CS, et al. DNA damage and cytotoxicity in adult subjects with prediabetes. Mutat Res Genet Toxicol Environ Mutagen. 2013; https://doi.org/10.1016/j.mrgentox.2013.02.002.

Pietinen P, et al. Diet and risk of colorectal cancer in a cohort of Finnish men. Cancer Causes Control. 1999; https://doi.org/10.1023/A:1008962219408.

Pischon T, et al. Body size and risk of colon and rectal cancer in the European Prospective Investigation into Cancer and Nutrition (EPIC). J Natl Cancer Inst. 2006; https://doi.org/10.1093/jnci/djj246.

Popkin BM. The nutrition transition and obesity in the developing world. J Nutr. 2001;131(3):871S–3S. https://doi.org/10.1093/jn/131.3.871s.

Pothiwala P, Jain SK, Yaturu S. Metabolic syndrome and cancer. Metab Syndr Relat Disord. 2009;

Praud D, et al. Adherence to the mediterranean diet and gastric cancer risk in Italy. Int J Cancer. 2014; https://doi.org/10.1002/ijc.28620.

Rapp K, et al. Fasting blood glucose and cancer risk in a cohort of more than 140,000 adults in Austria. Diabetologia. 2006; https://doi.org/10.1007/s00125-006-0207-6.

Richardson VR, Smith KA, Carter AM. Adipose tissue inflammation: feeding the development of type 2 diabetes mellitus. Immunobiology. 2013; https://doi.org/10.1016/j.imbio.2013.05.002.

Robinson WR, Stevens J, Gammon MD, John EM. Obesity before age 30 years and risk of advanced prostate cancer. Am J Epidemiol. 2005; https://doi.org/10.1093/aje/kwi150.

Rodriguez C, et al. Body mass index, weight change, and risk of prostate cancer in the Cancer Prevention Study II Nutrition Cohort. Cancer Epidemiol Biomark Prev. 2007; https://doi.org/10.1158/1055-9965.EPI-06-0754.

Rosato V, et al. Metabolic syndrome and the risk of breast cancer in postmenopausal women. Ann Oncol. 2011; https://doi.org/10.1093/annonc/mdr025.

Russo A, Autelitano M, Bisanti L. Metabolic syndrome and cancer risk. Eur J Cancer. 2008; https://doi.org/10.1016/j.ejca.2007.11.005.

Sahar S, Sassone-Corsi P. Regulation of metabolism: the circadian clock dictates the time. Trends Endocrinol Metab. 2012; https://doi.org/10.1016/j.tem.2011.10.005.

Schuurman AG, Van Den Brandt PA, Dorant E, Goldbohm RA. Animal products, calcium and protein and prostate cancer risk in the Netherlands Cohort Study. Br J Cancer. 1999; https://doi.org/10.1038/sj.bjc.6690472.

Shoelson SE, Lee J, Goldfine AB. Inflammation and insulin resistance. J Clin Invest. 2006; https://doi.org/10.1172/JCI29069.

Shor R, et al. Low HDL levels and the risk of death, sepsis and malignancy. Clin Res Cardiol. 2008; https://doi.org/10.1007/s00392-007-0611-z.

Shufelt C, et al. Red versus white wine as a nutritional aromatase inhibitor in premenopausal women: a pilot study. J Women's Heal. 2012; https://doi.org/10.1089/jwh.2011.3001.

Sofi F, Abbate R, Gensini GF, Casini A. Accruing evidence on benefits of adherence to the Mediterranean diet on health: an updated systematic review and meta-analysis. Am J Clin Nutr. 2010; https://doi.org/10.3945/ajcn.2010.29673.

Soto AM, Sonnenschein C. Environmental causes of cancer: endocrine disruptors as carcinogens. Nat Rev Endocrinol. 2010; https://doi.org/10.1038/nrendo.2010.87.

Stattin P, et al. Prospective study of hyperglycemia and cancer risk. Diabetes Care. 2007; https://doi.org/10.2337/dc06-0922.

Stürmer T, Buring JE, Lee IM, Gaziano JM, Glynn RJ. Metabolic abnormalities and risk for colorectal cancer in the physicians' health study. Cancer Epidemiol Biomark Prev. 2006; https://doi.org/10.1158/1055-9965.EPI-06-0391.

Supreeya Swarup, Amandeep Goyal, Yulia Grigorova, R. Z. Metabolic Syndrome. StatPearls.

Uzunlulu M, Telci Caklili O, Oguz A. Association between metabolic syndrome and cancer. Ann Nutr Metab. 2016; https://doi.org/10.1159/000443743.

van der Heiden MG, Cantley LC, Thompson CB. Understanding the warburg effect: the metabolic requirements of cell proliferation. Science. 2009; https://doi.org/10.1126/science.1160809.

van der Pal KC, et al. The association between multiple sleep-related characteristics and the metabolic syndrome in the general population: the New Hoorn study. Sleep Med. 2018; https://doi.org/10.1016/j.sleep.2018.07.022.

Van Hemelrijck M, et al. The interplay between lipid profiles, glucose, BMI and risk of kidney cancer in the Swedish AMORIS study. Int J Cancer. 2012; https://doi.org/10.1002/ijc.26212.

Vastag B. Nutrients for prevention: negative trials send researchers back to drawing board. J Natl Cancer Inst. 2009; https://doi.org/10.1093/jnci/djp073.

Vona-Davis L, Howard-Mcnatt M, Rose DP. Adiposity, type 2 diabetes and the metabolic syndrome in breast cancer. Obes Rev. 2007; https://doi.org/10.1111/j.1467-789X.2007.00396.x.

Walker MP, et al. An IGF1/insulin receptor substrate-1 pathway stimulates a mitotic kinase (cdk1) in the uterine epithelium during the proliferative response to estradiol. J Endocrinol. 2010; https://doi.org/10.1677/JOE-10-0102.

Wang G, et al. Risk factor for clear cell renal cell carcinoma in Chinese population: a case-control study. Cancer Epidemiol. 2012; https://doi.org/10.1016/j.canep.2011.09.006.

WCRF. Food, nutrition, physical activity and the prevention of cancer. The American Institute for Cancer Research (AICR); World Cancer Research Fund (WCRF UK) World Cancer Research Fund International. 2007.

White LS, Van den Bogaerde J, Kamm M. The gut microbiota: cause and cure of gut diseases. Med J Aust. 2018;209(7):312–7.

Wuermli L, et al. Hypertriglyceridemia as a possible risk factor for prostate cancer. Prostate Cancer Prostatic Dis. 2005; https://doi.org/10.1038/sj.pcan.4500834.

Yang YX, Hennessy S, Lewis JD. Type 2 diabetes mellitus and the risk of colorectal cancer. Clin Gastroenterol Hepatol. 2005; https://doi.org/10.1016/S1542-3565(05)00152-7.

Ye J. Emerging role of adipose tissue hypoxia in obesity and insulin resistance. Int J Obes. 2009; https://doi.org/10.1038/ijo.2008.229.

Yeh CC, You SL, Chen CJ, Sung FC. Peanut consumption and reduced risk of colorectal cancer in women: a prospective study in Taiwan. World J Gastroenterol. 2006; https://doi.org/10.3748/wjg.v12.i2.222.

Yin DT, et al. The association between thyroid cancer and insulin resistance, metabolic syndrome and its components: a systematic review and meta-analysis. Int J Surg. 2018; https://doi.org/10.1016/j.ijsu.2018.07.013.

Zacharski LR, et al. Decreased cancer risk after iron reduction in patients with peripheral arterial disease: results from a randomized trial. J Natl Cancer Inst. 2008; https://doi.org/10.1093/jnci/djn209.

Chapter 9
Impact of Trauma and Orthopaedics on Men's Health

M. Hefny, M. A. Weston, and K. S. Mangat

9.1 Introduction

The impact of gender on musculoskeletal health has long been established. Differences in anatomy and behaviours between the sexes both play a part in influencing a patient's susceptibility to pathology, the response to treatment and longer-term prognosis. However, gender disparities also occur at a deeper, cellular level. It is widely understood that genetics and hormone profiles can impact musculoskeletal disease but less widely acknowledged are the gender discrepancies observed in bone and soft tissue structure or the different responses to stimuli observed between male and female cells (Kinney et al. 2005). Additionally, numerous studies have demonstrated disparities in clinical outcomes and complication risks between the genders for a variety of conditions.

The differing lifestyles and functional requirements of men, particularly those involved in heavy manual work or contact sports, often alter the treatment options that the clinician may consider. Men's physical and mental health rely heavily on their level of activity. Modern men in various age groups pursue active lifestyles and engage in a variety of sports. Also, the retirement age is rising in an ageing population, with men required to be employed and active at a later stage in their lives. However, physical activity can also be a cause for musculoskeletal injuries and degenerative diseases that may hinder the ability to exercise or to maintain activity levels.

Trauma and orthopaedic surgery has evolved considerably over recent decades. The principles and methods of bone fixation, soft tissue repair/reconstruction and

M. Hefny (✉) · M. A. Weston · K. S. Mangat
Department of Trauma and Orthopaedic Surgery, South Warwickshire Orthopaedic and Research Development (SWORD), South Warwickshire NHS Foundation Trust, Warwick, UK
e-mail: Mamdouh.Hefny@swft.nhs.uk; Karanjit.Mangat@swft.nhs.uk

© The Author(s), under exclusive license to Springer Nature Switzerland AG 2022
S. S. Goonewardene et al. (eds.), *Men's Health and Wellbeing*,
https://doi.org/10.1007/978-3-030-84752-4_9

joint arthroplasty have advanced alongside improved outcomes for patients. Furthermore, increasingly accessible investigation modalities uncover more potential pathology and it becomes important to address these abnormalities, necessitating further intervention.

In this chapter, we discuss the impact of common traumatic injuries and degenerative pathologies in trauma and orthopaedic surgery on men's health.

9.2   Major Trauma

Major trauma is the main cause of disability and mortality among people under the age of 45 years. In the past, it primarily involved young adult men following road traffic collisions (RTC) and other modes of injury. However, over the past two decades, the demographics of major trauma have changed considerably in the UK. Interestingly, with increasing age men are less likely to be involved in major trauma compared to women (Fig. 9.1).

People over the age of 60 are now more commonly affected, with a prevalence of 54% (Trauma Audit Research Network (TARN) 2017). This change may be due to both the ageing population in the UK and the advancement in imaging and management protocols, improving the detection and treatment of previously undiscovered injuries (Kehoe et al. 2015).

Numerous studies have demonstrated gender differences in response to trauma and sepsis (Bone 1992). Male sex and age have been noted to be risk factors for infections and multiple organ failure after significant post-traumatic blood loss,

Fig. 9.1 Trauma Audit Research Network (TARN (2017))

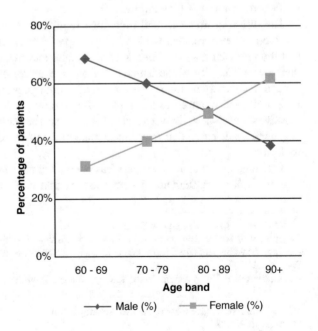

both of which carry a mortality risk in excess of 30%. The more favourable outcome in females may be explained by the protective effects of oestrogen, though the mechanism by which this occurs is not fully understood. Also, the apparent beneficial effects of oestrogen on haemorrhagic shock and sepsis have not yet been replicated consistently in clinical studies and remains an important area of research interest (Bösch et al. 2018). As such, the gender disparity itself may not be protective, rather the hormonal milieu present at the time of major trauma may influence the patient's physiological response. This may potentially prolong the patient's tolerance of permissive hypotension to ensure adequate end-organ perfusion, which could be of particular importance when transporting patients to major trauma centres.

9.3   Osteoporosis

Gender dimorphism in musculoskeletal tissue types arises during childhood development and persists into musculoskeletal degeneration in later life. Following the closure of physes (growth plates) in long bones, longitudinal growth ceases, whilst circumferential growth persists via the deposition of cortical bone. Simultaneous resorption also occurs. This constant remodelling, via a balance between resorption of old cells by osteoclasts and synthesis of new bone tissue by osteoblasts, results in the physiological synthesis of a new skeleton in each decade of life. However, the rate of bone deposition relative to resorption varies between the sexes. In men this rate is higher, resulting in more bone deposition. This causes a higher peak bone mineral density as they approach their fourth decade of life. Thereafter, bone density begins to decline, with gender discrepancies noted here too. Women demonstrate a more rapid bone resorption, which accelerates following menopause. Conversely, men experience a slower rate of resorption, which remains constant throughout life. It is for this reason that men are less susceptible to insufficiency fractures with advancing age.

Osteoporosis results from the imbalance of this physiological process with increased bone resorption and subsequent reduction in bone density. The end result is weaker bone, liable to fragility or insufficiency fractures. It is considered to be a critical health problem by the World Health Organization (WHO) (Cass et al. 2016).

Osteoporosis is a silent disease often only being revealed by a fracture. It is extremely common with a fragility fracture occurring in one in every four men over the age of 50 (Willson et al. 2015). Men are at risk of osteoporosis after the age of 65, which is over a decade later in life compared to women due to postmenopausal hormonal changes (mean age of onset is 51 years in the UK). Causes of osteoporosis in men can be primary due to genetic predisposition or secondary to medical comorbidities. For example, in prostate cancer, androgen deprivation therapy is associated with a 20% risk of fracture during the first five years of treatment. However, following fragility fractures, men are less likely to be screened for osteoporosis than women (Adler 2014). Identifying men at risk for osteoporotic fractures is essential as the risk of mortality following such injuries can be higher in men than in women (Haentjens et al. 2010).

The risk of osteoporotic fractures can be calculated using various tools. The Male Osteoporosis Risk Estimation Score (MORES) has been proposed as a gender-specific screening modality (Cass et al. 2016). However, the Fracture Risk Assessment Tool (FRAX) remains the most widely used. This provides an estimate of the ten-year probability of sustaining a major osteoporotic fracture (Compston et al. 2019). This quantifies the risk as either low, intermediate, or high. Those at low risk are recommended no prophylactic treatment, whilst those at high risk are. The intermediate group should undergo further investigation to measure bone mineral density (BMD) using dual-energy x-ray absorptiometry (DEXA) (Compston et al. 2019). Prior to starting men on treatments that may have a rapid adverse effect on their bone density, such as androgen deprivation therapy, BMD measurement using DEXA should be considered (National Institute for Health and Care Excellence (2012). Osteoporosis: assessing the risk of fragility fracture. Clinical guideline [CG146]). Recommendations for all men include healthy nutrition, regular physical activity and refraining from harmful lifestyle habits.

9.4   Disorders of the Spine

Inflammatory conditions of the spine are a clear example of why clinicians must remain mindful of gender disparities within their patients. Ankylosing Spondylitis and Diffuse Idiopathic Skeletal Hyperostosis (DISH) both occur more commonly in men. These conditions have a significant impact on the quality of life in their own right. Furthermore, they may also complicate the treatment of trauma and predispose patients to worse outcomes following injury.

Ankylosing Spondylitis is a systemic autoimmune condition that predominantly affects the spine and sacroiliac joints. Unusually for an autoimmune disorder, it largely affects men; with a male to female ratio of 4:1. It usually presents in early adulthood and its musculoskeletal manifestations include progressive spinal stiffness and kyphosis with associated sacroiliitis.

DISH is a relatively common, progressive condition, characterised by spinal stiffness, enthesopathy and chronic back pain. Its aetiology is unclear and it affects over a quarter of patients over 80 years old, although many of these remain asymptomatic. Within this age group, there is no significant gender propensity. However, male patients are more likely to experience an early onset, with a significant male gender bias (prevalence of 25% vs 15%) (Weinfeld et al. 1997) in the sixth and seventh decades of life.

Due to the loss of spine flexibility, these patients become more prone to fracture, particularly of the cervical spine. Furthermore, any cervical spine fracture is more likely to have an associated neurological injury which can prove fatal. There is an observed mortality of up to 67% in such patients sustaining a cervical spine fracture that is managed non-operatively (Whang et al. 2009). Additionally, as many major trauma patients are unable to communicate a medical history when they first present, there is a risk of iatrogenic cervical spine injury when such patients are forced into a position of neck extension by rigid spinal immobilisation. It is, therefore, vital to maintain a high suspicion of pre-existing spinal deformity and maintain immobilisation in a position that is normal for the patient.

Spinal cord injuries are life-changing events with implications on men's physical, mental and sexual health. The majority of spinal cord injuries are traumatic and occur in men under forty years of age, potentially resulting in life-changing neurological dysfunction (Aikman et al. 2018). However, non-traumatic spinal pathologies, such as Cauda Equina Syndrome, are also common. Sexual health impairment following such injuries in men can ensue, resulting in the possibility of erectile and ejaculatory dysfunction, semen abnormalities, decreased libido, anorgasmia, and dyspareunia. The extent of sexual dysfunction is dependent on the level, severity, and duration of the spinal injury. The physiology of sexual function relies on the autonomic nervous system, with the sacral nerve roots (S2–S4, parasympathetic) providing the pro-erectile vasodilating innervation to the cavernosal tissue of the penis. Sympathetic innervation from the thoracolumbar cord (T11-L2) is responsible for ejaculation, with the somatic innervation of the penis supplied by the pudendal nerve (S2–S4). (Baldo and Eardley 2006).

Most men will recover some degree of erectile function after 2 years from injury if the mechanism for erection (normal vascularity and an intact S2–4 reflex arc) is preserved. Pharmacological treatments, such as oral phosphodiesterase-5 inhibitors (sildenafil) are usually successful with up to a 70% success rate and are often the first line of therapy. Alternative options for men not responding to pharmacological treatments are specialised vacuum devices, intracavernosal injections, or surgical penile prostheses. Sex education, counselling and therapy all have important roles following spinal injury. (Hess and Hough 2012).

Ejaculation disorders, as well as semen abnormalities, are common in men, with a significant spinal injury resulting in high infertility rates. Assisted reproductive technologies, including intravaginal insemination, intrauterine insemination and in vitro fertilisation have all resulted in more favourable pregnancy rates. (Brackett 2012).

9.5 Disorders of the Lower Limb

9.5.1 Hip Fractures

Hip fractures in patients over the age of 60 years are the most common cause of trauma and orthopaedic hospital admissions. In the UK, over 75,000 hip fractures are admitted to hospitals every year. The annual expenditure is calculated to be around £2 billion. These figures are only expected to increase as the population ages and life expectancy improves. The incidence of hip fractures in women is decreasing while it is showing an uptrend in men over 84 years of age (Neuburger and Wakeman 2016). More importantly, the mortality rate in men following hip fractures is higher than in women (Guzon-Illescas et al. 2019) (see Fig. 9.2 below). The overall rate of mortality following this injury is high, 8–10% will die within the first month and around 30% within the first year. Preventative measures and treatment of risk factors in this age group are essential for primary prevention. A best practice tariff (BPT) has been applied to hospitals in the UK for safe and timely treatment of such fractures. This BPT has led to improved patient outcomes and is broadening to include femoral shaft and periprosthetic fractures of this age group of patients.

9.5.2   Lower Limb Arthritis

Knee and hip osteoarthritis are common conditions and significantly reduce the quality of life in various ways. Causes can be idiopathic, inflammatory or secondary to trauma. Pain is the chief symptom of the disease and can interfere with normal sleep patterns and day to day function. Additionally, the reduced joint range of movement can negatively impact upon activity levels.

Total Hip Arthroplasty (THA) has been described as the operation of the century (Learmonth et al. 2007). It significantly improves the quality of life with over 100,000 THAs performed annually in the UK. With more procedures being

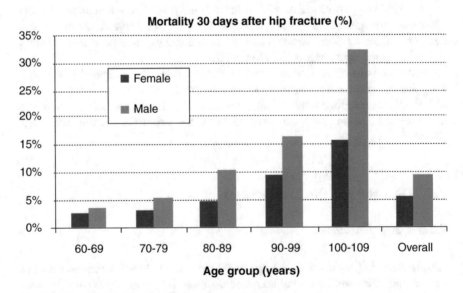

Fig. 9.2 Mortality 30 days after hip fracture by gender

undertaken every year, along with an ageing and more active population, an increasing incidence of periprosthetic fractures is being observed. With regards to THA, the relationship between gender and periprosthetic fracture risk appears inconsistent. However, the method of fixation of the femoral stem seems to show a discrepancy amongst the genders, with men having a higher risk of fracture in cemented prostheses and women a higher risk with uncemented ones (Thien et al. 2014).

In the past, patients were advised to avoid high activity levels and withdraw from sports following joint replacement surgery due to concerns about prosthesis wear and the risks of sustaining a fracture or dislocation. However, with improved implant properties and survival, as well as lack of evidence to support high failure rates in more active men, this practice has evolved (Vogel et al. 2011). An extreme recent example of this is the professional tennis player, Sir Andy Murray, who has recovered from a hip resurfacing procedure to return to top-level competition.

Osteoarthritis of the knee occurs in 1 in 20 men above the age of 45 years, compared to almost 1 in 10 women. However, this figure is rising more sharply amongst males than in females (Spitaels et al. 2020). More than 100,000 knee arthroplasty procedures are performed annually in the UK. Knee arthroplasty involves the partial or total replacement of the joint. Due to the differences in anatomy between men and women, gender-specific implants have been introduced to potentially improve outcome. These are aimed specifically towards female variation with early results not showing any significant benefit over more traditional and established prostheses (Kumar et al. 2012).

9.5.3 Achilles Tendon Rupture

The Tendo Achilles (TA) is the largest and strongest tendon in the human body. Its integral role during standing, walking, running, and jumping makes it vulnerable to repetitive strain and degenerative changes. This can eventually predispose it to failure (Del Buono et al. 2013). It is a commonly encountered debilitating injury often seen in middle-aged active men. The incidence is five times higher in men compared to women and is the most common tendon rupture in the musculoskeletal system. (Intziegianni et al. 2017).

The diagnosis of TA rupture is based on the history and clinical findings. Ultrasonography may help to confirm the diagnosis, quantify the gap between tendon ends, assess associated degenerative changes and evaluate tendon healing.

There has been a considerable debate over recent decades regarding the optimal treatment strategy, between non-operative and operative management. Recent literature has provided further insight into the benefits and risks of both options. Although several studies have demonstrated higher re-rupture rates following non-operative treatment, this appears to be a considerably lower risk than the associated complications of operative intervention, which include wound breakdown, infection, sural nerve injury, and deep venous thrombosis (DVT). Consequently, non-surgical treatment (Ochen et al. 2019) is now favoured in the majority of patients. The gold standard for non-operative treatment had previously been to immobilise in a plaster cast. However, more recently the use of functional bracing has been shown to produce similar outcomes to plaster cast treatment. This has the significant advantage of allowing safe early weight-bearing and is more cost-effective. (Costa et al. 2020).

Newly emerging biological treatments, used as adjuvants, to improve the tendon structure and function following rupture are topical and warrant further research.

9.6 Disorders of the Upper Limb, Wrist and Hand

9.6.1 Shoulder Instability

Traumatic shoulder instability is more often seen in male patients, with a relative risk of 2.55 (Brownson et al. 2015). 97% of traumatic shoulder dislocations occur anteriorly, resulting when the arm is abducted and forced into external rotation. Glenohumeral instability following dislocation is incredibly common, affecting 72–95% of patients under 20 years old (Rowe and Sakellarides 1961). Furthermore, the incidence of traumatic glenohumeral instability reduces with age; with 50% of first-time dislocations occurring between 15 and 29 years of age. A second peak in incidence occurs in the elderly but this group is disproportionately female, where diminishing muscle mass is likely to be a key factor. As such, male gender is predictive of traumatic dislocation in the young but may be considered protective against such injuries later in life. Similarly, significant gender variations are noted in recurrence rates following traumatic dislocation. Higher recurrence rates are consistently noted in young male patients between 15 years of age (86% in males versus 54% in females) and 35 years of age (29% in males versus 13% in females) (Robinson et al. 2006). It is likely that the different activities and functional requirements of men are the cause of this higher recurrence risk. Firstly, shoulder dislocations in men generally involve a higher energy trauma, either through heavy manual work, high-impact sporting activities or high-risk behaviours. As such, they may sustain greater structural damage to the joint initially. Secondly, male patients tend to return to these same activities, thereby putting their already unstable shoulder at further risk.

The final gender disparity noted in the treatment of shoulder instability is with treatment options. For the reasons described above, relating to recurrence risk and lifestyle choices, younger male patients will more frequently be offered operative management to stabilise the joint. As such, patient gender in shoulder instability can be seen to impact on the risk of sustaining this injury, the treatment offered and on the clinical outcome.

9.6.2   Distal Biceps Tendon Rupture

Rupture of the distal biceps tendon occurs relatively infrequently and is usually seen in young adults. It is rarely seen in women, with 93% of presentations in male patients. The mechanism of injury is an excessive eccentric contraction of the biceps muscle, which often results from weight-lifting or from attempting to catch a falling object, which explains why this injury usually affects the dominant arm. The use of anabolic steroid may further weaken the tendon insertion. In older, low demand patients, this injury may be managed non-operatively. However, this will inevitably result in a persistent functional deficit, with the patient losing approximately one-third of their elbow flexion strength and up to half of their supination strength. Therefore, in younger, active patients, surgical repair or reconstruction are the mainstays of treatment. Recovery after such surgery may take several months or more, which is a considerable time period for a patient to abstain from manual work or sporting activities. For this reason, patient compliance with post-operative pre-cautions can be poor and re-ruptures are occasionally seen (1.5%). More recent rehabilitation protocols now encourage little, if any, immobilisation to encourage patients to remain as compliant as possible.

9.6.3   Osteoarthritis of the Elbow

This condition is more common amongst middle-aged men, with males being affected four times more than their female counterparts (Stanley 1994). It is com-monly observed in manual workers and has a predilection for the dominant side. As a condition characterised by pain and mechanical symptoms (stiffness and locking of the joint), it can have a profound impact on a patient's functional status and abil-ity to work. Surgical management for such patients aims to promptly return them to function and relies heavily on joint preserving open or arthroscopic techniques. Total elbow arthroplasty (TEA) is reserved only for lower demand patients and is avoided in more active patients. Only 617 TEAs were performed in the UK in 2017, significantly fewer than other joint replacements.

9.6.4   Scaphoid Fractures and Wrist Instability

The scaphoid is the most commonly fractured carpal bone, with such injuries hav-ing a significant gender bias (83% occur in male patients (Dias et al. 2020)). Delayed presentation is common, with many patients believing that they have sustained a simple wrist sprain only. If left untreated, many scaphoid fractures will fail to unite, due to the tenuous blood supply to the scaphoid. Indeed, avascular necrosis of the scaphoid is a common sequela of these injuries. Furthermore, scaphoid non-union interferes with the biomechanics of the carpus, resulting in abnormal point loading of the articular surfaces and secondary degeneration may ensue, known as Scaphoid Non-union Advanced Collapse (SNAC wrist). This complication is characterised by a specific pattern of wrist joint degeneration, although its clinical manifestations are less predictable. However, at 5 years post-injury, up to 97% of patients with scaphoid fracture non-union will have radiographic signs of joint degeneration (Mack et al. 1984). Any resultant loss of wrist function represents a significant burden to the individual and also the society, with these injuries predominantly occurring in men of working age.

9.6.5 Dupuytren's Disease

Dupuytren's disease is a benign, myofibroblastic, proliferative disorder predominantly affecting the palmar fascia of the hand. It results in the formation of cords and nodules within the palm and digits, which may lead to flexion contractures of the digits which can significantly affect function as well as cosmesis. Dupuytren's disease is incurable but surgical intervention may alter progression. Dupuytren's disease most commonly involves the ulnar aspect of the hand but radial-sided disease and genital or foot involvement may be seen in aggressive disease types, known as Dupuytren's Diathesis. Dupuytren's disease has a genetic aetiology, its inheritance being autosomal dominant with variable penetrance. It affects men twice as commonly as women. Additionally, male patients tend to have a more aggressive disease with earlier onset and more rapid progression, meaning that cases presenting for surgical intervention are overwhelmingly male. Conversely, higher recurrence rates after surgery are observed in female patients. Surgery may take the form of percutaneous fasciotomy, limited palmar fasciectomy or dermo-fasciectomy with skin grafting. Digital amputation is sometimes required for more severe or recurrent deformities.

References

Adler RA. Osteoporosis in men: a review. Bone Res. 2014;2(1):14001. https://doi.org/10.1038/boneres.2014.1.

Aikman K, Oliffe JL, Kelly MT, McCuaig F. Sexual health in men with traumatic spinal cord injuries: a review and recommendations for primary health-care providers. Am J Mens Health. 2018;12(6):2044–54. https://doi.org/10.1177/1557988318790883.

Baldo O, Eardley I. Physiology of sexual function. In: Schill W-B, Comhaire F, Hargreave TB, editors. Andrology for the clinician. Berlin: Springer; 2006. p. 281–8.

Bone RC. Toward an epidemiology and natural history of SIRS (systemic inflammatory response syndrome). JAMA. 1992;268(24):3452–5. https://doi.org/10.1001/jama.1992.03490240060037.

Bösch F, Angele MK, Chaudry IH. Gender differences in trauma, shock and sepsis. Mil Med Res. 2018;5(1):35. https://doi.org/10.1186/s40779-018-0182-5.

Brackett NL. Infertility in men with spinal cord injury: research and treatment. Scientifica. 2012;2012:578257. https://doi.org/10.6064/2012/578257.

Brownson P, Donaldson O, Fox M, Rees JL, Rangan A, Jaggi A, et al. BESS/BOA patient care pathways: traumatic anterior shoulder instability. Shoulder Elbow. 2015;7(3):214–26. https://doi.org/10.1177/1758573215585656.

Cass AR, Shepherd AJ, Asirot R, Mahajan M, Nizami M. Comparison of the male osteoporosis risk estimation score (MORES) with FRAX in identifying men at risk for osteoporosis. Ann Fam Med. 2016;14(4):365. https://doi.org/10.1370/afm.1945.

Compston JE, McClung MR, Leslie WD. Osteoporosis. Lancet. 2019;393(10169):364–76. https://doi.org/10.1016/s0140-6736(18)32112-3.

Costa ML, Achten J, Wagland S, Marian IR, Maredza M, Schlüssel MM, et al. Plaster cast versus functional bracing for Achilles tendon rupture: the UKSTAR RCT. Health Technol Assess. 2020;24(8):1–86. https://doi.org/10.3310/hta24080.

Del Buono A, Chan O, Maffulli N. Achilles tendon: functional anatomy and novel emerging models of imaging classification. Int Orthop. 2013;37(4):715–21. https://doi.org/10.1007/s00264-012-1743-y.

Dias JJ, Brealey SD, Fairhurst C, Amirfeyz R, Bhowal B, Blewitt N, et al. Surgery versus cast immobilisation for adults with a bicortical fracture of the scaphoid waist (SWIFFT): a pragmatic, multicentre, open-label, randomised superiority trial. Lancet. 2020;396(10248):390–401. https://doi.org/10.1016/S0140-6736(20)30931-4.

Guzon-Illescas O, Perez Fernandez E, Crespí Villarias N, Quirós Donate FJ, Peña M, Alonso-Blas C, et al. Mortality after osteoporotic hip fracture: incidence, trends, and associated factors. J Orthop Surg Res. 2019;14(1):203. https://doi.org/10.1186/s13018-019-1226-6.

Haentjens P, Magaziner J, Colón-Emeric CS, Vanderschueren D, Milisen K, Velkeniers B, et al. Meta-analysis: excess mortality after hip fracture among older women and men. Ann Intern Med. 2010;152(6):380–90. https://doi.org/10.7326/0003-4819-152-6-201003160-00008.

Hess MJ, Hough S. Impact of spinal cord injury on sexuality: broad-based clinical practice intervention and practical application. J Spinal Cord Med. 2012;35(4):211–8. https://doi.org/10.1179/2045772312Y.0000000025.

Intziegianni K, Cassel M, Hain G, Mayer F. Gender differences of Achilles tendon cross-sectional area during loading. Sports Med Int Open. 2017;1(4):E135–E40. https://doi.org/10.1055/s-0043-113814.

Kehoe A, Smith JE, Edwards A, Yates D, Lecky F. The changing face of major trauma in the UK. Emerg Med J. 2015;32(12):911. https://doi.org/10.1136/emermed-2015-205265.

Kinney RC, Schwartz Z, Week K, Lotz MK, Boyan BD. Human articular chondrocytes exhibit sexual dimorphism in their responses to 17beta-estradiol. Osteoarthr Cartil. 2005;13(4):330–7. https://doi.org/10.1016/j.joca.2004.12.003.

Kumar V, Garg B, Malhotra R. A randomized trial comparing gender specific with gender non specific knee arthroplasty implants. Orthop Proc. 2012;94-B(SUPP_XXXVII):525. https://doi.org/10.1302/1358-992X.94BSUPP_XXXVII.EFORT2011-525.

Learmonth ID, Young C, Rorabeck C. The operation of the century: total hip replacement. Lancet. 2007;370(9597):1508–19. https://doi.org/10.1016/s0140-6736(07)60457-7.

Mack GR, Bosse MJ, Gelberman RH, Yu E. The natural history of scaphoid non-union. J Bone Joint Surg Am. 1984;66(4):504–9.

National Institute for Health and Care Excellence. Osteoporosis: assessing the risk of fragility fracture. Clinical guideline [CG146] Published date: 08 August 2012. Last updated: 07 Feb 2017. https://www.nice.org.uk/guidance/cg146.

Neuburger J, Wakeman R. Is the incidence of hip fracture increasing among older men in England? J Epidemiol Community Health. 2016;70:1049–50.

Ochen Y, Beks RB, van Heijl M, Hietbrink F, Leenen LPH, van der Velde D, et al. Operative treatment versus nonoperative treatment of Achilles tendon ruptures: systematic review and meta-analysis. BMJ. 2019;364:k5120. https://doi.org/10.1136/bmj.k5120.

Robinson CM, Howes J, Murdoch H, Will E, Graham C. Functional outcome and risk of recurrent instability after primary traumatic anterior shoulder dislocation in young patients. J Bone Joint Surg Am. 2006;88(11):2326–36. https://doi.org/10.2106/JBJS.E.01327.

Rowe CR, Sakellarides HT. Factors related to recurrences of anterior dislocations of the shoulder. Clin Orthop. 1961;20:40–8.

Spitaels D, Mamouris P, Vaes B, Smeets M, Luyten F, Hermens R, et al. Epidemiology of knee osteoarthritis in general practice: a registry-based study. BMJ Open. 2020;10(1):e031734. https://doi.org/10.1136/bmjopen-2019-031734.

Stanley D. Prevalence and etiology of symptomatic elbow osteoarthritis. J Shoulder Elb Surg. 1994;3(6):386–9. https://doi.org/10.1016/s1058-2746(09)80024-4.

Thien TM, Chatziagorou G, Garellick G, Furnes O, Havelin LI, Mäkelä K, et al. Periprosthetic femoral fracture within two years after total hip replacement: analysis of 437,629 operations in the nordic arthroplasty register association database. J Bone Joint Surg Am. 2014;96(19):e167. https://doi.org/10.2106/jbjs.m.00643.

Trauma Audit Research Network (TARN). https://www.tarn.ac.uk/. TARN Research 2017.

Vogel LA, Carotenuto G, Basti JJ, Levine WN. Physical activity after total joint arthroplasty. Sports Health. 2011;3(5):441–50. https://doi.org/10.1177/1941738111415826.

Weinfeld RM, Olson PN, Maki DD, Griffiths HJ. The prevalence of diffuse idiopathic skeletal hyperostosis (DISH) in two large American Midwest metropolitan hospital populations. Skelet Radiol. 1997;26(4):222–5. https://doi.org/10.1007/s002560050225.

Whang PG, Goldberg G, Lawrence JP, Hong J, Harrop JS, Anderson DG, et al. The management of spinal injuries in patients with ankylosing spondylitis or diffuse idiopathic skeletal hyperostosis: a comparison of treatment methods and clinical outcomes. J Spinal Disord Tech. 2009;22(2):77–85. https://doi.org/10.1097/BSD.0b013e3181679bcb.

Willson T, Nelson SD, Newbold J, Nelson RE, LaFleur J. The clinical epidemiology of male osteoporosis: a review of the recent literature. Clin Epidemiol. 2015;7:65–76. https://doi.org/10.2147/CLEP.S40966.

Chapter 10
Sexually Transmitted Diseases

Xuan Rui Sean Ong, Dominic Bagguley, Nathan Lawrentschuk, and Douglas Johnson

10.1 Introduction

Sexually transmitted diseases represent a group of syndromes caused by different pathogens that are transmitted through sexual intercourse. Despite major efforts to prevent its transmission, sexually transmitted diseases remain one of the most common global acute conditions. (Vos et al. 2016) The World Health Organisation estimated that in 2012, there were 357.4 million new cases of the four most common curable sexually transmitted diseases. (Rowley et al. 2019) They were most prevalent in adolescents and young adults and more common in women than men. However, men who have sex with men are more susceptible to some sexually transmitted diseases due to the rectal mucosa being particularly susceptible to transmission. In this particular group, rates of syphilis, gonorrhoea, chlamydia and human immunodeficiency virus continue to rise despite increases in safe sex campaigns and practices (Workowski and Bolan 2015).

Fortunately, most sexually transmitted diseases are rarely fatal. However, if left untreated, they can progress to long term complications related to fertility, organ

X. R. S. Ong · D. Bagguley
EJ Whitten Foundation Prostate Cancer Research Centre, Epworth HealthCare, Melbourne, Australia

N. Lawrentschuk
EJ Whitten Foundation Prostate Cancer Research Centre, Epworth HealthCare, Melbourne, Australia

Department of Urology, Royal Melbourne Hospital, Melbourne, Australia

D. Johnson (✉)
Department of Medicine, Royal Melbourne Hospital Melbourne, University of Melbourne, Melbourne, Australia

Internal Medicine Clinical Institute, Epworth HealthCare, Melbourne, Australia failure and psychological and social issues. It is therefore important that clinicians continue to advocate for safe sex and keep up to date with guidelines for treatment and prevention of these diseases.

This chapter aims to summarise current knowledge and guidelines pertaining to sexually transmitted diseases commonly encountered in clinical practice. It begins with outlining prevention practices pertaining to men and then will summarise the background, diagnosis and management of a select group of sexually transmitted diseases.

10.2   Prevention

Prevention of sexually transmitted diseases is multifaceted and includes: assessment of sexual health and risk behaviours, counselling and education about safe sex practices, pre exposure vaccination for vaccine preventable sexually transmitted diseases, sexually transmitted disease screening, treatment and follow up of both symptomatic and asymptomatic individuals, assessment and management of sexual partners of patients with sexually transmitted diseases and discussion about personalised risk-reduction goals (Workowski and Bolan 2015).

There are several ways to decrease the likelihood of contracting sexually transmitted diseases however abstinence from sex remains the only reliable way of avoiding transmission. Below we outline some practices specifically for men that will decrease transmission rates of sexually transmitted diseases. Vaccinations are discussed later under their disease heading.

Male Condoms   Male condoms remain the most widely available and most commonly used barrier method for the prevention of sexually transmitted diseases (Warner and Stone 2007). Studies have shown that consistent and correct use of male latex condoms is extremely effective in decreasing the risk of transmission of bacterial, viral and parasitic sexually transmitted diseases (Workowski and Bolan 2015). Breakage and slippage rates of condoms are minimal (Walsh et al. 2004) with incorrect and inconsistent use being the main reasons for failure of condoms to prevent sexually transmitted diseases (Workowski and Bolan 2015).

Male Circumcision   Male circumcision has been found to decrease rates of sexually transmitted diseases. This includes decreased rates of: HIV by 53–60%, HSV2 by 28–34%, HPV by 32–35%, trichomonas vaginalis by 48% and bacterial vaginosis by 40% (Tobian et al. 2010).

Primary Care Prevention Counselling   The Centre for Disease Control recommend that all sexually active adolescents, adults who have received a diagnosis of a sexually transmitted disease or have been diagnosed in the past, or have multiple sexual partners should have a sexual history taken and have prevention counselling (Workowski and Bolan 2015).

As mentioned previously, prevention measures do decrease risk but are not absolute. Therefore, diagnosis through microbiological testing is critical for the successful management of sexually transmitted diseases. Of note, concurrent sexually transmitted diseases are possible, particularly in at risk populations with high rates of sexually transmitted diseases such as men who have sex with men. If a sexually transmitted disease is diagnosed, we recommend performing investigations for other sexually transmitted diseases as well as contact tracing. In the following section we outline the common bacterial, viral and parasitic causes of sexually transmitted diseases in the male population.

10.3   Bacteria

10.3.1   Neisseria Gonorrhoea

Pathology: Neisseria Gonorrhoea—gram negative diplococci bacteria.

Epidemiology
The World Health Organisation estimated that in 2016 the global prevalence of gonorrhoea was 0.9% for women and 0.7% for men equalling 30.6 million cases (Rowley et al. 2019).

Transmission
Gonorrhoea is almost always sexually transmitted in adults however the risk of contracting gonorrhoea from a single sexual encounter with a positive contact is in the range of 30–70% (Sherrard 2014). Mother to neonatal (vertical) transmission can also occur.

Clinical Presentation
Men present with urethral discharge or dysuria usually around 3–10 days post contraction of the disease. Left untreated, men can experience these symptoms for many months. (Sherrard and Barlow 1996). Around 70% of women will be asymptomatic however symptomatic women present with vaginal discharge, abdominal pain or pelvic pain (Barlow and Phillips 1978). Rectal and pharyngeal infection is are often asymptomatic however can present with pain, discharge or inflammation in those areas (Sherrard 2014).

Complications usually occur when the disease has been left untreated for a significant amount of time. For men complications include: periurethral abscess, paraurethral duct infection, penile oedema, epididymitis, penile lymphangitis, (Sherrard 2014), proctitis, disseminated infection, septic arthritis and conjunctivitis (Limited 2020a). For women, complications include: inflammation of the bartholin's gland, pelvic inflammatory disease, endometritis and perihepatitis. (Sherrard 2014) HIV transmission is also facilitated by this infection (Sherrard 2014).

Diagnosis
Diagnosis of Neisseria Gonorrhoea is based on culture and nucleic acid amplification tests (NAAT). Specimens for these tests can be taken as first pass urine, urethral swab, anorectal swab, endocervical swab or pharyngeal swab. These tests are both specific and sensitive. (Papp et al. 2014) Cultures are cheaper and allow for antibiotic sensitivity tests whereas NAAT a higher sensitivity and specificity. Cultures should be taken after confirmed NAAT tests to identify resistant strains as these are becoming more common (Sherrard 2014).

For sexually active men and women, the US Preventative Services Task Force recommend considering screening in asymptomatic patients with risk factors or if they are in high epidemiological areas (new sexual partners, inconsistent condom use, previous STDs, sex workers, high prevalence areas) (Papp et al. 2014).

Management
Centre for Disease Control guidelines suggest the following regimens: (Workowski and Bolan 2015)

- Ceftriaxone 250 mg intramuscular as single dose PLUS
- Azithromycin 1 g orally as a single dose

The Australian sexually transmitted diseases management guidelines recommend a higher dose of Ceftriaxone at 500 mg intramuscular PLUS 1 gram oral azithromycin as a single dose (Alliance 2019).

Regardless of symptoms, sexual partners of patients with gonorrhoea within 60 days of onset of symptoms should be referred for testing and treatment.

10.3.2   Chlamydia Trachomatis

Pathogen: Chlamydia trachomatis serovars D-K—gram negative coccoid bacteria. Serovars A-C are a major cause of blindness whereas serovars L1-3 are associated with lymphogranuloma venereum.

Epidemiology
Chlamydia is the most common sexually transmitted bacteria with the World Health Organisation estimating global number of cases to be 127 million in people aged 15–49 years in 2016 (Rowley et al. 2019). Global prevalence in men was 2.7% in 2016 compared to 3.8% in women of the same age group (Rowley et al. 2019).

Transmission
Chlamydia is a sexually transmitted disease and can occur through vaginal, anal, oral or penile contact. It can also be transmitted from mother to neonate.

Clinical Presentation
Chlamydia remains asymptomatic in around 50% of men and 70% of women (Bébéar and De Barbeyrac 2009).

Symptomatic men usually present with manifestations of urethritis (dysuria and/ or discharge) 7–21 days after contraction. Men can also have epididymitis and proctitis. Reiters syndrome (urethritis, conjunctivitis, arthritis) or reactive arthritis have also been associated with chlamydia and is more common in males. Male fertility does not seem to be affected by infection (de Barbeyrac et al. 2006).

Symptomatic women present with mucopurulent vaginal discharge from inflammation of the adnexa. This can progress to pelvic inflammatory disease which can lead to infertility and risk of ectopic pregnancy (de Barbeyrac et al. 2006).

Diagnosis
Chlamydia is diagnosed with nucleic acid amplification tests (NAAT) from first pass urine, urethral, vaginal, rectal or oropharyngeal swabs. NAAT has a high sensitivity and specificity for chlamydia. Culture, direct immunofluorescence assays and immunosorbent assays are also available however have a lower sensitivity than NAAT (Workowski and Bolan 2015).

For sexually active asymptomatic men and women, screening should be considered especially if they are in high epidemiological areas (new sexual partners, inconsistent condom use, previous STDs, sex workers, high prevalence areas).

Management
Centre for Disease Control guidelines suggest the following regimens: (Workowski and Bolan 2015)

- Azithromycin 1 g orally as a single dose OR
- Doxycycline 100 mg orally BD for 7 days OR

  Alternate regimes

- Erythromycin base 500 mg orally 4 times a day for 7 days OR
- Levofloxacin 500 mg orally daily for 7 days OR
- Ofloxacin 300 mg orally twice a day for 7 days

Regardless of symptoms, sexual partners of patients with gonorrhoea within 60 days of onset of symptoms should be referred for testing and treatment.

10.3.3 Syphilis

Pathogen: Treponema pallidum—spiral shaped bacteria.

Epidemiology
The World Health Organisation estimated 6 millions cases of syphilis globally in people aged between 15–49 years in 2012.(Rowley et al. 2019) The global prevalence was similar between males (0.48%) and females (0.5%) in that same age group. Cases have declined since the discovery of penicillins however still remain a prominent health issue, especially in specific groups such as African-Americans, men who have sex with men and people with HIV (Baughn and Musher 2005).

Transmission

Syphilis is transmitted through direct contact with someone who has active lesions. People that have sexual contact with a positive syphilis person have been demonstrated in studies to have a 16–30% chance of becoming infected (LaFond and Lukehart 2006). However, actual transmission rates in the community are probably higher (LaFond and Lukehart 2006).

Clinical Presentation

Clinical presentation of syphilis is divided into three stages.

Primary syphilis presents with a painless ulcer (chancre) after an incubation period that can be between 10–90 days post exposure. Chancres occur at the site where treponema pallidum infiltrates the mucosa through an abrasion. Common locations for lesions are penis, labia, vagina, rectum, anal cavity and oral cavity. Untreated, these lesions will heal in 4–8 weeks (LaFond and Lukehart 2006).

Secondary syphilis is a result of proliferation of the bacteria and dissemination through blood and lymphatics. Symptoms generally appear 4–10 weeks after appearance of primary syphilis lesions. Generalised scaly, macular rash located all over the body and extending to mucous membranes is the initial sign of secondary syphilis. This can be accompanied by a systemic reaction including lethargy, fever, generalised lymphadenopathy, headache, muscles aches and pruritis. Raised white or grey lesions called condyloma latum can be found in warm moist areas of the body. Hepatitis, iritis, uveitis, arthritis and glomerulonephritis can also occur (Baughn and Musher 2005). Secondary syphilis usually resolves after 3 months. Thereafter the syphilis begins a phase lacking clinical manifestations called the latent phase.

Latent syphilis is further characterised into early latent syphilis (acquired within the preceding year) or late latent syphilis. These classifications have implications for treatment duration. During the latent phase, patients are still susceptible to intermittent seeding from the bacteria and the disease is still detectable by serological testing. Sexual transmission is less likely however the disease can still be passed from mother to fetus (LaFond and Lukehart 2006).

Tertiary syphilis is now rare due to the discovery of penicillin. Symptoms appear generally 20–40 years post contraction of the disease. Destructive granulomatous lesions called gumma can appear. This most affect bone and skin however have been reported to affect the liver, heart, brain, stomach, and upper respiratory tract (LaFond and Lukehart 2006). Cardiovascular syphilis can manifest as aortitis and complications such as aortic valve regurgitation. Neurosyphilis symptoms can range from vertigo or insomnia to seizures, cerebrovascular accidents and tabes dorsalis and optic nerve damage (Figs. 10.1, 10.2, 10.3, and 10.4).

Diagnosis

Syphilis is most commonly diagnosed by serological testing or by identification of Treponema pallidum DNA on a swab from a lesion, using nucleic acid amplification testing (NAAT) (eg polymerase chain reaction [PCR]) where available (Limited 2020b).

Fig. 10.1 Primary syphilis. Painless genital primary chancre

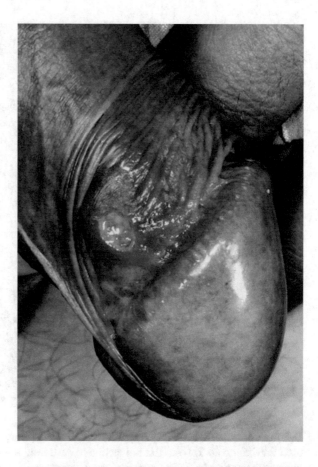

Fig. 10.2 Secondary syphilis. Macular rash (asymptomatic, non-itching)

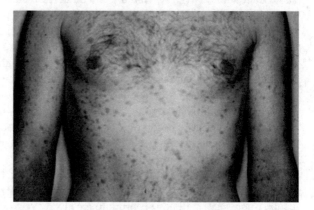

Fig. 10.3 Secondary
syphilis. Genital
condylomatalata

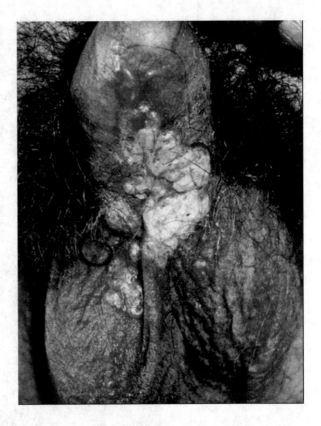

In primary and secondary syphilis, swabs or aspirates from active lesions are used for diagnosis. If NAAT is not available dark field microscopy (DFM) or direct fluorescent antibody (DFA) staining can be performed. Furthermore, serology can be used to confirm the diagnosis of active syphilis. To make a diagnosis of active syphilis using serology both: a positive nontreponemal test result (such as rapid plasma reagin [RPR] test) and a positive treponemal test result (such as treponemal enzyme immunoassay [EIA], T. pallidum particle agglutination test [TPPA] or T. pallidum haemagglutination assay [TPHA]) need to be detected.

Importantly neither test alone is sufficient for diagnosis and early in the disease serology may initially be negative. Treponemal tests may not become reactive for up to 2 weeks after chance becomes apparent and non-treponemal tests may take up to 4 weeks to become reactive. Neurosyphilis can be diagnosed from cerebrospinal fluid (Limited 2020b).

Latent syphilis can be diagnosed with serological testing. Treponemal tests (enzyme immunoassay, Treponema pallidum particle agglutination or hemagglutination, fluorescent antibody absorption) are always positive in patients with syphilis and remain positive for life (Goh 2005). These tests however can have some false positive results (Workowski and Bolan 2015). Rapid plasma reagin (RPR) tests

Fig. 10.4 Tertiary
syphilis. Gumma of
the nose

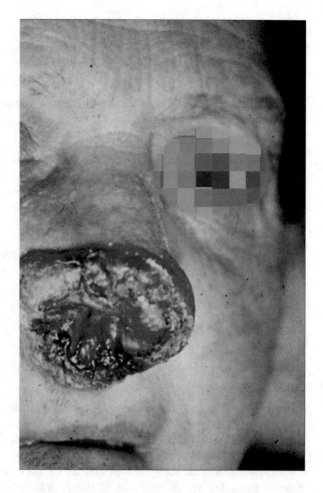

should be done to confirm syphilis after positive treponemal tests and can be used to test for reinfection or treatment monitoring.

Management
Penicillins are the mainstay of treatment for syphilis.

Centre for Disease Control guidelines suggest the following regimens: (Workowski and Bolan 2015).

For primary syphilis, secondary syphilis and early latent syphilis

- Benzathine benzylpenicillin 2.4 million international units intramuscular as a single dose for early syphilis

  For late latent syphilis

- Benzathine benzylpenicillin 2.4 million international units intramuscular weekly for 3 weeks for latent or late stage syphilis

For Neurosyphilis

- Aqueous crystalline penicillin G 18–24 million units per day, administered as 3–4 million units IV every 4 hours or continuous infusion, for 10–14 days

For penicillin allergic patients.
Penicillin is the drug of choice and careful evaluation of allergies and discussion with allergy experts is recommended.

- For penicillin allergic patients, with primary, secondary or early latent syphilis use doxycycline 100 mg orally twice a day for 14 days or tetracycline 500 mg orally four times a day for 14 days
- For penicillin allergic patients with late latent infection use doxycycline 100 mg orally twice a day for 28 days or tetracycline 500 mg orally four times a day for 28 days
- For penicillin allergic patients with neurosyphilis seek expert advice. Desensitisation after expert advice is recommended for patient with immediate penicillin hypersensitivity

10.3.4   Chancroid

Pathogen: Haemophilus ducreyi—gram negative bacterium.

Epidemiology
The global epidemiology of chancroid is poorly documented due to difficulties in diagnosis. It is not included in the World Health Organisation's estimates. In the 1990s the global prevalence of chancroid was estimated to be 7 million (Steen 2001). However, this millennia has seen a significant reduction in chancroid cases (González-Beiras et al. 2016). Europe and the US have restricted chancroid to rare sporadic cases. Transmission remains ongoing in a few countries with limited access to health services (González-Beiras et al. 2016).

Transmission
Haemophilus ducreyi is hypothesised to enter the skin and/or mucosa through abrasions present before or that occur during sexual intercourse.

Clinical Presentation
Chancroid is a form of genital ulcer disease. It usually begins as papules around 4–7 days after contraction which then evolve into pustules. These pustules then rupture finally manifesting as ulcers covered in necrotic, yellow-grey exudates. (Al-Tawfiq and Spinola 2002) Chancroid can be associated with lymphadenopathy in approximately 10–70% of cases, usually unilateral. (Morse 1989) Complications for men can include phimosis (Fig. 10.5).

Diagnosis
A clinical diagnosis of chancroid can often be mistaken with other genital ulcer disease. This is reflected in the findings of low sensitivity and specificity of physical

Fig. 10.5 Small, rounded, dirty, and nonindurated genital ulcers due to chancroid

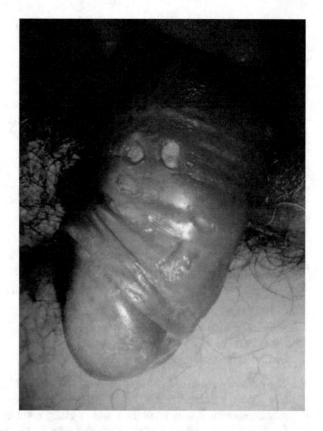

examination in identifying chancroid (Lewis 2003). A definitive diagnosis requires identification of haemophilus ducreyi on culture however since it is a fastidious organism, it is difficult to culture from genital ulcer material; often it requires a special medium that is relatively expensive not commercially available. Even when the media is used, the sensitivity is 80% (Workowski and Bolan 2015). DNA amplification using M-PCR has shown higher sensitivity and specificity for the simultaneous detection of haemophilus ducreyi, T palladium and herpes simplex virus however these are not widely available (Lewis 2000).

Management
Centre for Disease Control guidelines suggest the following regimens: (Workowski and Bolan 2015)

- Azithromycin 1 g orally as single dose OR
- Ceftriaxone 250 mg intramuscular as single dose OR
- Ciprofloxacin 500 mg orally twice a day for 3 days OR
- Erythromycin base 500 mg orally 3 times a day for 7 days

Regardless of symptoms, sexual partners of patients with chancroid within 10 days of onset of symptoms should be treated.

10.3.5   *Lymphogranuloma Venereum (LGV)*

Pathogen: Chlamydia trachomatis serovars L1, L2, L3.

Epidemiology

LGV is an endemic in tropical countries including East Africa, Southeast Asia, Latin America and the Caribbean (De Vrieze and De Vries 2014). Furthermore, the last 20 years has seen an increase in cases of LGV reported mainly in men who have sex with men usually presenting with proctitis residing in Western and Northern Europe, North America and Australia. Female cases of LGV have been reported however are rare (Peuchant et al. 2011; Verweij et al. 2012). There is no evidence of heterosexual transmission of LGV in Europe (de Vries et al. 2019).

Transmission

The transmission of LGV remains unclear (de Vries 2016). Anogenital transmission may play a part in transmission. and it has been postulated the faecal-oral route may be another mode of transmission. Asymptomatic carriers are also thought to play a role in transmission (de Vries 2016).

Clinical Presentation

The clinical presentation of LGV can be divided into genital symptoms and anorectal symptoms.

Genital symptoms can been divided into three stages:

1. The **primary stage** can begin to develop after an incubation period of 4–30 days (De Vrieze and De Vries 2014). It presents as a papule which progresses to a pustule and eventually an ulcer. Ulcers usually resolve in a short period of time (around 7 days). Mucopurulent discharge may be present if inoculation is on the cervix, urethra or rectum. As this stage is transient, it can often go untreated.
2. The **secondary stage** occurs 2–6 weeks after the primary stage (de Vries et al. 2019). Patients develop painful, usually unilateral inguinal and/or femoral lymphadenopathy which may progress to form abscesses or even rupture.
3. The **tertiary stage** occurs in patients left untreated for extended periods of time and can include chronic granulomatous inflammation, lymphoedema and elephantiasis (White 2009) (Fig. 10.6).

Anorectal symptoms are the main presentation in homosexual men with LGV. These men present with haemorrhagic proctitis, mucosal ulceration and granulomas that can manifest in pain, tenesmus, constipation and bloody discharge (De Vrieze and De Vries 2014). Inflammatory bowel disease is an important differential diagnosis. If left untreated, this can result in the development of anal strictures and the complications of it (De Vrieze and De Vries 2014) (Fig. 10.7).

Diagnosis

Where a swab or aspirate can be attained from a genital lesion, rectal lesion or lymph node specimen, LGV can be diagnosed by confirming the presence of

Fig. 10.6 LGV ulcer and
lymphadenopathy

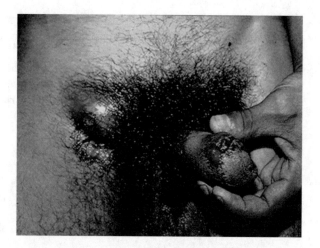

Fig. 10.7 Genital edema
following LGV

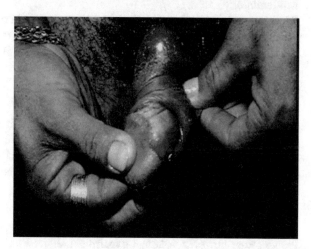

Chlamydia trachomatis on culture, direct immunofluorescence or nucleic acid
amplification test (NAAT). Subsequently, serovar specific Chlamydia trachomatis
should be identified by NAAT to complete the diagnosis (de Vries et al. 2019;
Workowski and Bolan 2015).

Management
Patients with suspected LGV should begin empirical treatment before test
results return.

Centre for Disease Control guidelines suggest the following regimens:
(Workowski and Bolan 2015)

- Doxycycline 100 mg orally twice a day for 21 days OR
- Erythromycin base 500 mg orally four times a day for 21 days

Alternate regimes for treatment of LGV exist. The Australian Guidelines recommend either doxycycline 100 mg orally twice a day for 21 days or azithromycin 1gm weekly for 3 weeks (Limited 2020c).

Sexual partners of patients with LGV within 60 days of onset of symptoms should be examined and tested for urethral, cervical, or rectal chlamydial infection. They should be empirically treated with a chlamydia regimen (azithromycin 1 g orally single dose or doxycycline 100 mg orally twice a day for 7 days) (Workowski and Bolan 2015).

10.3.6 Granuloma Inguinale (Donovanosis)

Pathogen: Klebsiella granulomatis—gram negative, rod shaped bacterium.

Epidemiology
Granuloma inguinale found rarely in developed nations however are still endemic in some tropical and developing areas including Papua New Guinea, the Caribbean, Central Australia and southern Africa. (Workowski and Bolan 2015) The disease is more prominent in lower socio-economic groups and those with poor hygiene (Velho et al. 2008).

Transmission
Granuloma inguinale is classified as a sexually transmitted disease however some reports have noted that despite being prevalent in people with high sexual activity, it relatively lower in sexual workers. Despite this, a high incidence of cervical, genital and anal lesions provide evidence in favour of the disease being sexually transmitted (Velho et al. 2008; O'Farrell 2002).

Clinical Presentation
The incubation period for granuloma inguinale has been inconsistently reported in the literature however a human experiment showed lesions presented on average around 50 days post contraction of disease (Greenblatt et al. 1939).

Lesions appear at first as a papule which then evolve into a painless, slow growing, superficial ulcer. Usually there is no regional lymphadenopathy. Ulcers have four clinical subtypes (O'Farrell 2002):

1. Ulcerogranulomatous—painless, red ulcers that bleed easily that can become extensive
2. Hypertrophic—raised ulcers with irregular edges which can be completely dry
3. Necrotic—foul smelling, tissue destroying, deep ulcers
4. Dry sclerotic lesions with fibrous or scar tissue

Extragenital spread can occur and infection can spread to the pelvis, visceral organs, bones and mouth. Secondary bacterial infections can develop and granuloma inguinale can co-exist with other sexually transmitted diseases (Workowski and Bolan 2015) (Fig. 10.8).

Fig. 10.8 Very well-
defined ulceration that
grows slowly and bleeds
readily on contact

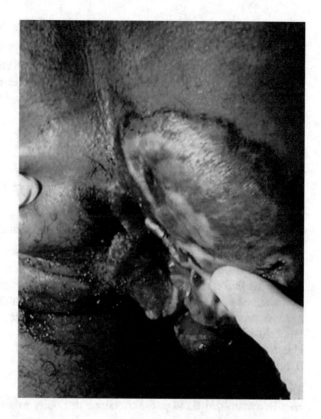

Diagnosis

The diagnosis of granuloma inguinale predominantly relies upon the visualisation of Donovan bodies on histopathological staining from biopsy or where available nucleic amplification testing. Klebsiella granulomatis is particularly hard to culture (Workowski and Bolan 2015; Velho et al. 2008; O'Farrell 2002).

Management

Patients with suspected granuloma inguinale should begin empirical treatment before test results return.

Centre for Disease Control guidelines suggest the following regimens: (Workowski and Bolan 2015)

- Azithromycin 1 g orally once per week or 500 mg orally daily for at least 3 weeks until all lesions have completely healed OR
- Doxycycline 100 mg orally twice a day for at least 3 weeks until all lesions have completely healed OR
- Ciprofloxacin 750 mg orally twice a day for at least 3 weeks until all lesions have completely healed OR
- Erythromycin base 500 mg orally four times a day or at least 3 weeks until all lesions have completely healed OR

• Trimethoprim/sulfamethoxazole 160/800 mg orally twice a day or at least 3 weeks until all lesions have completely healed

Sexual partners of patients with granuloma inguinale within 60 days of onset of symptoms should be examined and offered therapy. Empirical treatment in the absence of symptoms has not been established (Workowski and Bolan 2015).

10.4 Viruses

10.4.1 Human Immunodeficiency Virus (HIV)

Pathogen: Human immunodeficiency virus—single stranded RNA retrovirus.

Epidemiology
The World Health Organisation estimated there was 38 million people living with HIV at the end of 2019 and 690,000 deaths that year (Organisation 2020d). They have defined key groups of people at higher risk of contracting HIV as: men who have sex with men, people who inject drugs, sex workers and clients, prisoners and other closed settings, and transgender people. In regards to location, Africa have the highest prevalence of HIV and is home to two thirds of all people living with HIV (Organisation 2020d).

Transmission
HIV is transmitted through the exchange of bodily fluids including blood, breast milk, semen and vaginal secretions. The risk of transmission is increased when the infected person has a high viral load which is often in the acute stage of the infection (Maartens et al. 2014). Other sexually transmitted diseases, pregnancy and receptive anal intercourse are known to increase risk of transmission. Male circumcision reduces risk of transmission (Maartens et al. 2014). Notably, infected people cannot transmit the disease through hugging, shaking hands, food, water or even kissing. Furthermore, infected people taking antiretroviral therapy for at least 6 months with a low viral load (< 200 copies/ml) are also unlikely to pass on the disease (Organisation 2020d; Adolescents n.d.)

Clinical Presentation
The early stages of the infectious course can manifest in symptoms resembling the flu; these include fever, fatigue, myalgia and rash (Daar et al. 2008). This usually is within a month of contraction and is usually when viral load is highest. However, this syndrome is usually transient and many people with HIV do not know of their HIV status until later in the disease course.

HIV then progresses to a latent stage where it replicates at very low levels. Patients can still transmit the disease during this phase but are usually asymptomatic. At the end of this phase the viral load increases and CD4 T-cell levels decrease (Workowski and Bolan 2015).

As the CD4 T-cell count decreases to below 200cells/mm, the patient then progresses to having autoimmune immunodeficiency disease (AIDs). Patients are then more susceptible to opportunistic infections (tuberculosis, cryptococcal meningitis, severe bacterial infections) and also are at a higher risk of cancers such as lymphomas and Kaposi's sarcoma (Organisation 2020d).

Diagnosis

The diagnosis of HIV has a multi-test approach and we would recommend following the local testing algorithm at your health centre as availability of assays differs in different health settings.

Currently, where available, the preferred testing algorithm uses a fourth-generation antigen/antibody combination HIV-1/2 immunoassay (which detects both HIV-1, HIV-2 antibodies and HIV P24 antigen) plus a confirmatory HIV-1/HIV-2 antibody differentiation immunoassay. A plasma HIV RNA level can also be used as a confirmatory test and used to also track treatment response (Bernard et al. 2014; Prevention 2014, 2018).

Rapid tests are available to quickly identify if a person is HIV positive or negative. These tests require a small amount of blood (finger-prick) and results come back at least within 24 hours. They also require a confirmatory test. This test becomes reactive later in the disease course so is prone to a false negative result.

Management

There is currently no cure for HIV however with the advancements in antiretroviral therapy, people living with HIV now can control the disease and stop spread of the virus. Antiretroviral therapy generally consists of two nucleoside reverse-transcriptase inhibitors plus one of: an integrase strand transfer inhibitor (Welfare 2019), a non-nucleoside reverse transcriptase inhibitor (NNRTI), or a protease inhibitor (PI) with a pharmacokinetic (PK) enhancer (also known as a booster; the two drugs used for this purpose are cobicistat and ritonavir) (Services 2020; Adolescents n.d.).

World Health Organisation recommends first line Dolutegavir or Efavirenz plus two nucleoside reverse transcriptase inhibitors (WHO 2019). However, recommendations for HIV first line therapy change, and we recommend seeking the most up-to-date recommendations via national and international guidelines. For example, up to date guidelines can be found at the webpage Guidelines for the Use of Antiretroviral Agents in Adults and Adolescents Living with HIV available at https://clinicalinfo.hiv.gov/en/guidelines/adult-and-adolescent-arv/whats-new-guidelines.

Prevention

HIV prophylaxis is a relatively new area of research that has seen much progress the last two decades. Pre-exposure and post-exposure prophylaxis (with antiretroviral therapy) is recommended for people who have vaginal or anal sex with HIV positive people or with high risk groups. They are also recommended for people at risk of HIV infection through needle sharing.

- Pre-exposure prophylaxis can be taken daily from 2 days before until 2 days after sex (Organisation 2020d), however can be taken daily for long periods of time with no side effects. (Control 2020) The effectiveness of pre-exposure prophylaxis is reported at over 90% with very little rates of HIV resistance (Riddell et al. 2018).
- Post-exposure prophylaxis needs to be started within 72 hours of potential exposure and includes counselling, testing and a 28-day course of antiretrovirals (Organisation 2020d).

10.4.2   Human Papillomavirus (HPV)

Pathogen: Human papillomavirus—small double stranded DNA virus. There are at least 150 different types (Doorbar et al. 2012).

Epidemiology
HPV is a very common sexually transmitted disease with some publications stating that "nearly all men and women will get at least one type of HPV at some point". (Intervention 2020) The global prevalence of HPV in women is estimated to be 10.41%(Burchell et al. 2006). However, the number of cases in men is not well established in the literature. A review of 12 studies from 1991–2005 estimated the prevalence of HPV in men to be 2.3%-34.8%(Partridge and Koutsky 2006).

Transmission
HPV is transmitted through abrasions in the skin where the virus can infiltrate and track down to the basal layer. Data consistently confirms sexual intercourse to be the primary route of genital HPV infection (Burchell et al. 2006) this can include vaginal or anal intercourse. Oral-genital and digital-genital transmission have been reported however are not common (Burchell et al. 2006).

Clinical Presentation
Different strains of HPV manifest in different clinical presentations. Most HPV infections are asymptomatic. Symptomatic HPV can be divided into oncogenic and non-oncogenic strains.

Non-oncogenic strains usually present as ano-genital warts. 90% of warts are caused by HPV strain 6 or 11. Warts can also appear in the mouth, nose, conjunctiva and larynx (Workowski and Bolan 2015). Warts are usually painless but can be itchy and irritating. Non-oncogenic strains can also cause recurrent respiratory papillomatosis.

Oncogenic strains are well documented and are associated with pre-cancerous and cancerous lesions of the cervix, penis, vulva, vagina, anus and head and neck (Figs. 10.9 and 10.10).

Diagnosis
HPV can be diagnosed on clinical examination. However, if this is not possible or diagnosis is uncertain, it can be diagnosed with biopsy to exclude cancer.

Fig. 10.9 Multiple papular warts of pubic skin

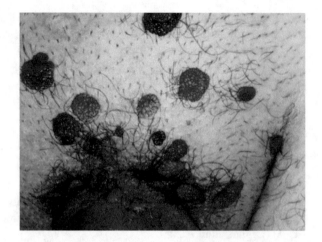

Fig. 10.10 Meatal warts

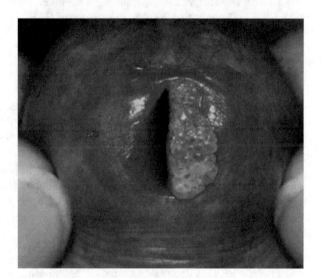

Management

Several methods can be used to eliminate external warts. Cryotherapy destroys warts through thermo-induced cell lysis. Surgical excision is an effective technique. A variety of creams, gel and ointments can be applied topically and can be effective.

Warts that are on mucosal surfaces (such as urethral meatus, anal, vaginal, cervical) can be eliminated using cryotherapy, surgical excision or topical trichloroacetic acid or bichloroacetic acid (Workowski and Bolan 2015).

No antiviral therapies are recommended (Workowski and Bolan 2015).

Prevention

The Centre for Disease Control recommends two HPV vaccinations—Gardasil 9 and Cervarix. Men are recommended to have Gardasil. Most cases of cervical cancer and most genital warts are protected against with Gardasil vaccine. Boys and girls are recommended to have two doses of the vaccine 6 months apart from aged 11–12 years (Workowski and Bolan 2015).

10.4.3   Herpes Simplex Virus (HSV)

Pathogen: Herpes simplex virus—double stranded DNA virus. This is categorised into 2 types; HSV1 and HSV2.

Epidemiology
In 2016, an estimated 3.7 billion people less than 50 years of age had HSV1 (oral or genital). Highest prevalence was in Africa (88%) and lowest was in the Americas (45%) (Organisation 2020c).

In 2016, an estimated 491 people ages between 15–49 years of age were living with HSV2 (Organisation 2020c). Prevalence of HSV2 is highest in Africa followed by the Americas. More women live with HSV2 than men (313million vs 178million).

Transmission
HSV is transmitted through epithelial mucosal cells or through interruptions in the skin then travel to nerves where they lie latent. (Kriebs 2008) Sexual transmission is most common and can occur when there is active infection or when there are no symptoms present. Interestingly, the virus can also be spread through respiratory droplets or mucocutaneous secretions if the virus is shedding (Fatahzadeh and Schwartz 2007).

The transmission from man to woman is more efficient than woman to man reflected by the fact more women have HSV than men (Organisation 2020c). Both types can affect any part of the body however HSV1 is associated with oral and genital lesions whereas HSV2 predominantly is associated with genital herpes (Kriebs 2008).

Clinical Presentation
Most people with HSV are asymptomatic (Organisation 2020c). If symptomatic, the average incubation period is 4 days. Patients then present with vesicles around the genitals, rectum or mouth. Some may a tingling sensation before the appearance of vesicles. These vesicles then rupture to leave painful ulcers. Sometimes with initial infection, systemic symptoms such as fevers, aches or lymphadenopathy. In immunocompromised patients, HSV infection can lead to more severe complications such as encephalitis or keratitis (Fig. 10.11).

Diagnosis
For symptomatic patients with lesions that can be swabbed, culture and nucleic acid amplification tests (NAAT) are available. Cultures are most accurate early on the disease process however when ulcers are crusted over, the virus can only be cultured 25% of the time (Kriebs 2008). NAAT are more sensitive (sensitivity of >95%)(Van

Fig. 10.11 Primary genital herpes, diffuse ulceration on the shaft of the penis

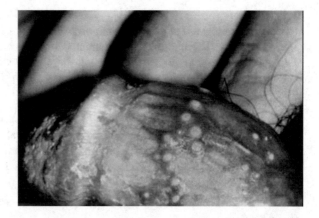

Der Pol et al. 2012). Serology based tests are available as well and have a high sensitivity and specificity (both >90%). This can be useful in patients with no symptoms or negative swab.

Management
There is no cure for herpes simplex virus however there are effective options for suppression or episodes.

For first clinical episode of genital herpes, Centre for Disease Control guidelines suggest the following regimens: (Workowski and Bolan 2015)

- Acyclovir 400 mg orally three times a day for 7–10 days OR
- Acyclovir 200 mg orally five times a day for 7–10 days OR
- Valacyclovir 1 g orally twice a day for 7–10 days OR
- Famciclovir 250 mg orally three times a day for 7–10 days

The above medications can also be considered to be used as a long term suppression agent for recurrent HSV episodes. This treatment is also likely to decrease transmission of HSV between sexual partners (Workowski and Bolan 2015).

10.4.4   Hepatitis A

Pathogen: Hepatitis A virus—unenveloped single stranded RNA virus.

Epidemiology
Approximately 1.5million cases of hepatitis A occur globally each year (Franco et al. 2012). Hepatitis A infection is one that is most common in low to middle income countries with poor sanitation and hygiene practices including many places in Africa, Asia, Central and South America. In these areas, it is estimated around 90% of children will have been infected with hepatitis A by aged 10 years (Jacobsen and Wiersma 2010).

Transmission

Hepatitis A is transmitted through fecal-oral route either by direct contact with an infected person or the ingestion of contaminated food or water. It undergoes a period of extensive shedding mainly into faeces during its 2–7 week incubation period or into the early stages of illness (Martin and Lemon 2006).

Clinical Presentation

People infected with hepatitis A commonly present with fever, nausea and vomiting, lethargy, abdominal discomfort dark urine and jaundice after a 2–7 week incubation period. Myalgia, pruritis, diarrhoea, arthralgia and a skin rash can also be present but is less common (Jeong and Lee 2010). Bloods tests will usually show raised bilirubin and liver transaminases. Patients rarely progress to fulminant hepatitis with incidence rates being reported as up to 0.5%(Jeong and Lee 2010).

Diagnosis

The diagnosis of hepatitis A is made with the presence of hepatitis A IgM antibodies in the blood. Total anti-hepatitis A antibodies can be used to identify patients with previous infection or vaccination (Workowski and Bolan 2015).

Management

Management for hepatitis A is supportive care and hospitalisation if there is clinical deterioration. The natural history of symptoms can range from days to months.

Prevention

Vaccination is the most effective way of preventing hepatitis A. They are made from formalin-inactivated hepatitis A virus. It is administered as an intramuscular injection usually 6 months apart. 94–100% of adults will have protective antibodies after 1 month post first dose and 100% of adults will have protective antibodies after the second dose. Studies have shown presence of antibodies in serum 10–20 years post vaccination with 2 doses (Workowski and Bolan 2015).

10.4.5   Hepatitis B

Pathogen: Hepatitis B virus—circular double stranded DNA virus.

Epidemiology

The World Health Organisation estimates that in 2015, there were 257 million people living with Hepatitis B and 887,000 deaths (Organisation 2020a). Prevalence is estimated to be over 8% in some parts of Africa, 2–7.9% in parts of Asia, South America and Mediterranean and less than 2% in North America, Europe and Australia (Stasi et al. 2017). The prevalence of hepatitis B has decreased since the introduction of its vaccine which is 95% effective in preventing infection.

Transmission

Hepatitis B virus is most commonly transmitted from mother to fetus or from exposure to infected blood in endemic areas (Organisation 2020a). Needle sharing transmission through tattooing, piercings, intravenous drug use, and needle-stick injuries can also occur as well as sexual transmission through bodily fluids however this has decreased now due to the vaccination. The virus has remarkable environmental resistance and can last outside of the body for 7 days during which it can still cause infection (Stasi et al. 2017).

Clinical Presentation

About 30–50% of all newly affected hepatitis B cases are symptomatic (Liang 2009) and less than 1% of infections result in fulminant hepatic failure and death (Workowski and Bolan 2015). Symptoms of acute infection include fever, lethargy, jaundice, vomiting, dark urine, arthralgia, pale stools. Symptoms usually present within 3 months of infection and can last up to 6 months (Control and Prevention 2017).

Approximately 90% of infants, 30% of children less than 5 years of age, and 2–6% of adults become chronically infected (Workowski and Bolan 2015). This happens because hepatitis B virus is not fully cleared in the acute stage. Chronic infection manifests in liver cirrhosis or hepatocellular carcinoma (HCC) in approximately 25% of those with chronic infection. Liver cirrhosis typically preceded HCC however HCC can also happen in the absence of liver cirrhosis (Tang et al. 2018). Hepatitis B accounts for at least 50% of HCC cases (Tang et al. 2018).

Diagnosis

Diagnosis of hepatitis B is based on serological markers and this can also differentiate between acute and chronic phases of the disease (Table below). Hepatitis B envelope antigen is associated with higher infectivity whereas the development of hepatitis B envelope antibodies is associated with lower infectivity.

	HBsAg	Anti-HBs	Anti-HBc
Hepatitis B acute infection	+	−	+
Resolved infection	−	+	+
Immunity from vaccination	−	+	−
Chronic infection	+	−	+

Management

There is no cure for hepatitis B and therefore treatment is supportive. The World Health Organisation recommend the use of anti-virals tenofovir and entecavir which can suppress the virus and can potentially slow the progression to liver cirrhosis and HCC (Organisation 2020a).

Prevention

The hepatitis B vaccination is a three-dose vaccination given at least 4 weeks apart as soon as possible after birth. It contains hepatitis B surface antigen and stimulates the immune system to produce surface antigen antibodies for immunity. This coverage of this vaccine has reached 85% globally in 2019 and has been effective in reducing the proportion of children less than 5 years of age chronically infected with hepatitis B from 5% to 1% (Organisation 2020a). The vaccine produces protective antibody levels in 95% of children and has been proven to last at least 20 years (Organisation 2020a).

10.4.6 Hepatitis C

Pathogen: Hepatitis C virus—enveloped RNA virus.

Epidemiology
The World Health Organisation estimates around 71 million people have hepatitis C infection globally. It is most common in eastern Mediterranean regions and in Europe where prevalence is 2.3% and 1.5% respectively (Organisation 2020b).

Transmission
Hepatitis C is a blood borne virus and is transmitted through exposure to small quantities of blood (Organisation 2020b). This most commonly happens through sharing of infected needles and blood transfusion (Shepard et al. 2005). Less commonly, transmission happens through sexual intercourse especially among men who have sex with men and those taking HIV prophylaxis (Organisation 2020b).

Clinical Presentation
About 70–80% of patients are asymptomatic during the acute phase of hepatitis C infection (Chen and Morgan 2006). When symptomatic, patient can present with lethargy, weakness, anorexia and jaundice usually 3–12 weeks after contraction of the disease (Chen and Morgan 2006). Fulminant hepatic failure is rare. During this period hepatitis C RNA is detectable in the blood.

Chronic hepatitis C infection is defined by hepatitis C RNA still being detectable 6 months after onset of acute infection (Chen and Morgan 2006) 55–85% of people will progress to this stage (Ghany et al. 2009). Increased age, men and African-Americans are more at risk of chronic infection. 10–15% of patients with chronic hepatitis C develop liver cirrhosis and this can progress to HCC (Chen and Morgan 2006).

Diagnosis
Hepatitis C is diagnosed in a two step process. Serum anti hepatitis C antibodies identify people infected with the virus previously. Then a nucleic acid test for hepatitis C RNA is performed to test for active infection (acute or chronic). People with no hepatitis C RNA but anti hepatitis C antibodies have most likely cleared the infection with a strong immune response (Organisation 2020b).

Management
Currently, the recommended treatment for hepatitis C is the use of hepatitis C virus (HCV) direct-acting antiviral (Daar, Pilcher, and Hecht) agents. The goal is to eradicate HCV RNA. A sustained virologic response (SVR) is defined as an undetectable RNA level 12 weeks following the completion of therapy. The use of DAA regimens have replaced pegylated interferon and ribavirin with sustained virological response rates of 95–99%. Recommendations for HCV first line therapy are rapidly changing, and we recommend seeking the most up to date recommendations via national and international guidelines. For example The Infectious Diseases Society of America Guidelines (available at https://www.hcvguidelines.org/) currently recommends either Glecaprevir (300 mg) /pibrentasvir (120 mg) to be taken with food for

a duration of 8 weeks or Sofosbuvir (400 mg)/velpatasvir (100 mg) for a duration of 12 weeks as first line therapy.(AASLD-IDSA 2015)

10.4.7   Mononucleosis Syndrome

Pathogen: The two most common pathogens that manifest in mononucleosis syndrome are Epstein Barr virus (EBV) and cytomegalovirus. However other lesser known pathogens (which we will not discuss here) that can mimic mononucleosis syndrome include human herpesvirus 6, HIV type 1, adenovirus, HSV type 1, streptococcus pyogenes and toxoplasma gondii (Hurt and Tammaro 2007).

Epidemiology
EBV is the most common cause of mononucleosis syndrome and it is estimated that over 90% of adults worldwide will show evidence of prior infection. (Roberts 2001) Cytomegalovirus is estimated to make up around 7% of mononucleosis syndrome cases (Hurt and Tammaro 2007). Peak age for developing the clinical disease is between 15–25 years (Roberts 2001).

Transmission
Transmission of EBV and cytomegalovirus most commonly occurs through oropharyngeal secretions and has been colloquially termed "the kissing disease". However it is also found in genital and rectal secretions and can also be transmitted through blood (Roberts 2001).

Clinical Presentation
The classic mononucleosis triad of pharyngitis, fever and generalised lymphadenopathy was first described in the 1800s and is still true today. Additionally, up to 25% of patients will have petechiae or erythema on their palate, 63% of patients can have clinically apparent splenomegaly, and isolated cases of genital ulcers (Lawee and Shafir 1983; Sisson and Glick 1998) and proctitis (Studemeister 2011) have been reported. Mononucleosis normally runs a self limiting course however it can also manifest as meningitis, encephalitis, optic neuritis, Guillian-Barre syndrome, facial nerve palsies, transverse myelitis, pancreatitis, myocarditis, hepatitis and pneumonia.

Diagnosis
As many pathogens can present as mononucleosis syndrome, a thorough history of presenting complaint, sexual history, travel history and physical examination is crucial in guiding diagnostic choices.

Definitive diagnosis of EBV is based on testing for antibodies (IgG and IgM) against viral capsid antigen (VCA), early antigens and EBV nuclear antigen proteins. PCR and EBV viral load testing can be an alternative for inconclusive antibody tests (Vouloumanou et al. 2012). Serum antibody tests and PCR are also available for detection of cytomegalovirus (Hurt and Tammaro 2007).

Management

Treatment for mononucleosis is mainly supportive (Vouloumanou et al. 2012). Paracetamol and non-steroidal inflammatory medications can be used for symptomatic control. Corticosteroids and antiviral therapy are not in routine clinical practice however may have a role in immunocompetent patients (Rafailidis et al. 2010).

10.5 Protozoa

10.5.1 Trichomoniasis

Pathogen: Trichomonas vaginalis—single cell protozoan.

Epidemiology

Trichomoniasis is estimated to be the most common non-viral sexually transmitted disease in the world (Kissinger 2015a). There is an estimated incidence of 143 million new cases of trichomoniasis globally in women and men aged between 15–49 years (Newman et al. 2015). Trichomoniasis is more common in women. In 2012, the global prevalence of trichomoniasis in men aged 15–49 years was 0.6% as compared to 5% in women (Newman et al. 2015).

Transmission

Humans are the only known host of trichomonas vaginalis. It is primarily transmitted via sexual intercourse (Kissinger 2015b).

Clinical Presentation

Around 77% of men with trichomonas vaginalis will be asymptomatic (Sena et al. 2007) Symptomatic men most commonly present with urethral discharge and irritation. Urethritis and sometime prostatitis can also ensue (Krieger et al. 1993).

For women, up to 85% are asymptomatic (Kissinger 2015b). Symptomatic women present with vaginal discharge, dysuria, vulval irritation and abdominal pain. Importantly, it has also been associated with poor birth outcomes, perinatal transmission, HIV acquisition, cervical neoplasia and herpes simplex (Kissinger 2015a).

Diagnosis

Trichomoniasis is detected using nucleic acid amplification tests of first pass urine, or meatal or high vaginal swabs. This test has a high sensitivity (95–100%) and specificity (95–100%) (Workowski and Bolan 2015). A rapid antigen detection test has also been developed. The OSOM trichomonas rapid test has results in around 10 min and a high sensitivity (82–95%) and specificity (97–100%) rate (Campbell et al. 2008).

Where these tests are not available, microscopic evaluations of wet mount preparations of genital secretions are still used. These tests are cheap and require less equipment but unfortunately have a low sensitivity (51–65%) as the accuracy of the test drops the longer the time from collection to evaluation (Workowski and Bolan 2015).

Management

Centre for Disease Control guidelines suggest the following regimens: (Workowski and Bolan 2015)

- Metronidazole 2 g orally as a single dose OR
- Tinidazole 2 g orally as a single dose OR

  Alternate regime

- Metronidazole 500 mg orally twice a day for 7 days

Retesting should be done within 2 weeks to 3 months of initial treatment to ensure resolution of disease. All sexual partners should have concurrent treatment and should abstain from sex until treatment has finished (Workowski and Bolan 2015).

References

AASLD-IDSA. Recommendations for testing, managing, and treating hepatitis C (Online). 2015. Available: https://www.hcvguidelines.org/. Accessed 28 Oct 2020.

Adolescents POAGFAA. Guidelines for the use of antiretroviral agents in adults and adolescents with HIV (Online). (n.d.) Department of Health and Human Services. Available: https://clinicalinfo.hiv.gov/sites/default/files/inline-files/AdultandAdolescentGL.pdf. Accessed 28 oct 2020.

Alliance ASH. Gonorrhoea (Online). STI management guidelines. 2019. Available: http://www.sti.guidelines.org.au/sexually-transmissible-infections/gonorrhoea. Accessed 28 Oct 2020.

Al-Tawfiq JA, Spinola SM. Haemophilus ducreyi: clinical disease and pathogenesis. Curr Opin Infect Dis. 2002;15:43–7.

Barlow D, Phillips I. Gonorrhoea in women: diagnostic, clinical, and laboratory aspects. Lancet. 1978;311:761–4.

Baughn RE, Musher DM. Secondary syphilitic lesions. Clin Microbiol Rev. 2005;18:205–16.

Bébéar C, de Barbeyrac B. Genital Chlamydia trachomatis infections. Clin Microbiol Infect. 2009;15:4–10.

Bernard M, Michele S, Laura G, Berry B, Barbara G, Kelly E, Michael A, Centers for Disease Control and Prevention. Laboratory testing for the diagnosis of HIV infection: updated recommendations. Atlanta: Centers for Disease Control and Prevention; 2014.

Burchell AN, Winer RL, de Sanjosé S, Franco EL. Epidemiology and transmission dynamics of genital HPV infection. Vaccine. 2006;24:S52–61.

Campbell L, Woods V, Lloyd T, Elsayed S, Church D. Evaluation of the OSOM Trichomonas rapid test versus wet preparation examination for detection of Trichomonas vaginalis vaginitis in specimens from women with a low prevalence of infection. J Clin Microbiol. 2008;46:3467–9.

Centers for Disease Control and Prevention. Hepatitis B: general information. South Carolina state documents depository. Atlanda: Centers for Disease Control and Prevention; 2017.

Chen SL, Morgan TR. The natural history of hepatitis C virus (HCV) infection. Int J Med Sci. 2006;3:47.

Control CFD. HIV basics (Online). Centre for Disease Control. 2020. Available: https://www.cdc.gov/hiv/basics/prep.html. Accessed 14 Sept 2020.

Daar ES, Pilcher CD, Hecht FM. Clinical presentation and diagnosis of primary HIV-1 infection. Curr Opin HIV AIDS. 2008;3:10–5.

de Barbeyrac B, Papaxanthos-Roche A, Mathieu C, Germain C, Brun JL, Gachet M, Mayer G, Bébéar C, Chene G, Hocké C. Chlamydia trachomatis in subfertile couples undergoing an in vitro fertilization program: a prospective study. Eur J Obstet Gynecol Reprod Biol. 2006;129:46–53.

de Vries HJC. The enigma of lymphogranuloma venereum spread in men who have sex with men: does ano-oral transmission plays a role? Sex Transm Dis. 2016;43(7):420–2.

De Vries H, de Barbeyrac B, de Vrieze N, Viset J, White J, Vall-Mayans M, Unemo M. 2019 European guideline on the management of lymphogranuloma venereum. J Eur Acad Dermatol Venereol. 2019;33:1821–8.

de Vrieze NHN, De Vries HJC. Lymphogranuloma venereum among men who have sex with men. An epidemiological and clinical review. Expert Rev Anti-Infect Ther. 2014;12:697–704.

Doorbar J, Quint W, Banks L, Bravo IG, Stoler M, Broker TR, Stanley MA. The biology and life-cycle of human papillomaviruses. Vaccine. 2012;30:F55–70.

Fatahzadeh M, Schwartz RA. Human herpes simplex virus infections: epidemiology, pathogenesis, symptomatology, diagnosis, and management. J Am Acad Dermatol. 2007;57:737–63.

Franco E, Meleleo C, Serino L, Sorbara D, Zaratti L. Hepatitis A: epidemiology and prevention in developing countries. World J Hepatol. 2012;4:68.

Ghany MG, Strader DB, Thomas DL, Seeff LB. Diagnosis, management, and treatment of hepatitis C: an update. Hepatology. 2009;49:1335–74.

Goh BT. Syphilis in adults. Sex Transm Infect. 2005;81:448–52.

González-Beiras C, Marks M, Chen CY, Roberts S, Mitjà O. Epidemiology of Haemophilus ducreyi infections. Emerg Infect Dis. 2016;22:1.

Greenblatt R, Dienst R, Pund E, Torpin R. Experimental and clinical granuloma inguinale. J Am Med Assoc. 1939;113:1109–16.

Hurt C, Tammaro D. Diagnostic evaluation of mononucleosis-like illnesses. Am J Med. 2007;120:911. e1–8.

Intervention CFDCA. About HPV (Online). 2020. Available: https://www.cdc.gov/hpv/parents/about-hpv.html. Accessed 4 Sept 2020.

Jacobsen KH, Wiersma ST. Hepatitis A virus seroprevalence by age and world region, 1990 and 2005. Vaccine. 2010;28:6653–7.

Jeong S-H, Lee H-S. Hepatitis A: clinical manifestations and management. Intervirology. 2010;53:15–9.

Kissinger P. Epidemiology and treatment of trichomoniasis. Curr Infect Dis Rep. 2015a;17:31.

Kissinger P. Trichomonas vaginalis: a review of epidemiologic, clinical and treatment issues. BMC Infect Dis. 2015b;15:1–8.

Kriebs JM. Understanding herpes simplex virus: transmission, diagnosis, and considerations in pregnancy management. J Midwifery Womens Health. 2008;53:202–8.

Krieger JN, Jenny C, Verdon M, Siegel N, Springwater R, Critchlow CW, Holmes KK. Clinical manifestations of trichomoniasis in men. Ann Intern Med. 1993;118:844–9.

Lafond RE, Lukehart SA. Biological basis for syphilis. Clin Microbiol Rev. 2006;19:29–49.

Lawee D, Shafir M. Solitary penile ulcer associated with infections mononucleosis. Can Med Assoc J. 1983;129:146.

Lewis DA. Diagnostic tests for chancroid. Sex Transm Infect. 2000;76:137–41.

Lewis D. Chancroid: clinical manifestations, diagnosis, and management. Sex Transm Infect. 2003;79:68–71.

Liang TJ. Hepatitis B: the virus and disease. Hepatology. 2009;49:S13–21.

Limited TG. Approach to neisseria gonorrhoea infection (Online). 2020a. Available: https://tgldcdp.tg.org.au/viewTopic?topicfile=neisseria-gonorrhoeae&guidelineName=Antibiotic&topicNavigation=navigateTopic#toc_d1e128. Accessed 28 Oct 2020.

Limited TG. Investigations for syphillis (Online). 2020b. Available: https://tgldcdp.tg.org.au/viewTopic?topicfile=syphilis&guidelineName=Antibiotic&topicNavigation=navigateTopic#toc_d1e58. Accessed 28 Oct 2020.

Limited TG. Lymphogranuloma venereum (Online). 2020c. Available: https://tgldcdp.tg.org.au/viewTopic?topicfile=infective-proctitis&guidelineName=Antibiotic#toc_d1e167. Accessed 28 Oct 2020.

Maartens G, Celum C, Lewin SR. HIV infection: epidemiology, pathogenesis, treatment, and prevention. Lancet. 2014;384:258–71.

Martin A, Lemon SM. Hepatitis A virus: from discovery to vaccines. Hepatology. 2006;43:S164–72.

Morse SA. Chancroid and Haemophilus ducreyi. Clin Microbiol Rev. 1989;2:137–57.

Newman L, Rowley J, Vander Hoorn S, Wijesooriya NS, Unemo M, Low N, Stevens G, Gottlieb S, Kiarie J, Temmerman M. Global estimates of the prevalence and incidence of four curable sexually transmitted infections in 2012 based on systematic review and global reporting. PLoS One. 2015;10:e0143304.

O'farrell N. Donovanosis. Sex Transm Infect. 2002;78:452–7.

Organisation, W. H. Hepatitis B (Online). 2020a. Available: https://www.who.int/en/news-room/fact-sheets/detail/hepatitis-b. Accessed 10 Sept 2020.

Organisation, W. H. Hepatitis C (Online). 2020b. Available: https://www.who.int/news-room/fact-sheets/detail/hepatitis-c. Accessed 10 Sept 2020.

Organisation, W. H. Herpes simplex virus (Online). World Health Organisation. 2020c. Available: https://www.who.int/news-room/fact-sheets/detail/herpes-simplex-virus.   Accessed   4 Sept 2020.

Organisation, W. H. HIV/AIDs (Online). 2020d. Available: https://www.who.int/news-room/fact-sheets/detail/hiv-aids. Accessed 10 Sept 2020.

Papp JR, Schachter J, Gaydos CA, Van der Pol B. Recommendations for the laboratory-based detection of Chlamydia trachomatis and Neisseria gonorrhoeae—2014. MMWR Recomm Rep. 2014;63:1.

Partridge JM, Koutsky LA. Genital human papillomavirus infection in men. Lancet Infect Dis. 2006;6:21–31.

Peuchant O, Baldit C, le Roy C, Trombert-Paolantoni S, Clerc M, BÉBÉAR C, de Barbeyrac B. First case of Chlamydia trachomatis L2b proctitis in a woman. Clin Microbiol Infect. 2011;17:E21–3.

Prevention CFDCA. Laboratory testing for the diagnosis of HIV infection: updated recommendations (Online). 2014. Available: https://stacks.cdc.gov/view/cdc/23447. Accessed 28 Oct 2020.

Prevention CFDCA. 2018 Quick reference guide: Recommended laboratory HIV testing algorithm for serum or plasma specimens (Online). 2018. Available: https://stacks.cdc.gov/view/cdc/50872. Accessed 28 Oct 2020.

Rafailidis PI, Mavros MN, KAPASKELIS A, FALAGAS ME. Antiviral treatment for severe EBV infections in apparently immunocompetent patients. J Clin Virol. 2010;49:151–7.

Riddell J, Amico KR, Mayer KH. HIV preexposure prophylaxis: a review. JAMA. 2018;319:1261–8.

Roberts JR. Infectious mononucleosis: epidemiology and pathophysiology. Emerg Med News. 2001;23:6–10.

Rowley J, Vander Hoorn S, Korenromp E, Low N, Unemo M, Abu-Raddad L, Chico R. Global and regional estimates of the prevalence and incidence of four curable sexually transmitted infections in 2016. WHO Bulletin. 2019, June.

Sena AC, Miller WC, Hobbs MM, Schwebke JR, Leone PA, Swygard H, Atashili J, Cohen MS. Trichomonas vaginalis infection in male sexual partners: implications for diagnosis, treatment, and prevention. Clin Infect Dis. 2007;44:13–22.

Services UDOHAH. Guidelines for the use of antiretroviral agents in adults and adolescents with HIV (Online). 2020. Available: https://aidsinfo.nih.gov/guidelines/brief-html/1/adult-and-adolescent-arv/11/what-to-start. Accessed 11 Sept 2020.

Shepard CW, Finelli L, Alter MJ. Global epidemiology of hepatitis C virus infection. Lancet Infect Dis. 2005;5:558–67.

Sherrard J. Gonorrhoea. Medicine. 2014;42:323–6.

Sherrard J, BARLOW D. Gonorrhoea in men: clinical and diagnostic aspects. Sex Transm Infect. 1996;72:422–6.

Sisson B, Glick L. Genital ulceration as a presenting manifestation of infectious mononucleosis. J Pediatr Adolesc Gynecol. 1998;11:185–7.

Stasi C, Silvestri C, Voller F. Emerging trends in epidemiology of hepatitis B virus infection. J Clin Transl Hepatol. 2017;5:272.

Steen R. Eradicating chancroid. Bull World Health Organ. 2001;79:818–26.

Studemeister A. Cytomegalovirus proctitis: a rare and disregarded sexually transmitted disease. Sex Transm Dis. 2011;38:876–8.

Tang LS, Covert E, Wilson E, Kottilil S. Chronic hepatitis B infection: a review. JAMA. 2018;319:1802–13.

Tobian AA, Gray RH, Quinn TC. Male circumcision for the prevention of acquisition and transmission of sexually transmitted infections: the case for neonatal circumcision. Arch Pediatr Adolesc Med. 2010;164:78–84.

Van der Pol B, Warren T, Taylor SN, Martens M, Jerome KR, Mena L, Lebed J, Ginde S, Fine P, Hook EW. Type-specific identification of anogenital herpes simplex virus infections by use of a commercially available nucleic acid amplification test. J Clin Microbiol. 2012;50:3466–71.

Velho PENF, Souza EMD, Belda Junior W. Donovanosis. Braz J Infect Dis. 2008;12:521–5.

Verweij SP, Ouburg S, de Vries H, Morré SA, Van Ginkel CJ, Bos H, Sebens FW. The first case record of a female patient with bubonic lymphogranuloma venereum (LGV), serovariant L2b. Sex Transm Infect. 2012;88:346–7.

Vos T, Allen C, Arora M, Barber RM, Bhutta ZA, Brown A, Carter A, Casey DC, Charlson FJ, Chen AZ. Global, regional, and national incidence, prevalence, and years lived with disability for 310 diseases and injuries, 1990–2015: a systematic analysis for the global burden of disease study 2015. Lancet. 2016;388:1545–602.

Vouloumanou EK, Rafailidis PI, Falagas ME. Current diagnosis and management of infectious mononucleosis. Curr Opin Hematol. 2012;19:14–20.

Walsh TL, Frezieres RG, Peacock K, Nelson AL, Clark VA, Bernstein L, Wraxall BG. Effectiveness of the male latex condom: combined results for three popular condom brands used as controls in randomized clinical trials. Contraception. 2004;70:407–13.

Warner L, Stone KM. Male condoms. In: Aral SO, Douglas JM, editors. Behavioral interventions for prevention and control of sexually transmitted diseases. New York: Springer; 2007.

Welfare AIOHA. Cancer in Australia 2019 (Online). Australia. 2019. Available: https://www.aihw.gov.au/reports/cancer/cancer-in-australia-2019/contents/table-of-contents. Accessed 20 June 2020.

White JA. Manifestations and management of lymphogranuloma venereum. Curr Opin Infect Dis. 2009;22:57–66.

WHO. Update of recommendations on first-and second-line antiretroviral regimens. Geneva: WHO; 2019.

Workowski KA, Bolan GA. Sexually transmitted diseases treatment guidelines, 2015. MMWR Recomm Rep. 2015;64:1.

Chapter 11
Alcohol Related Liver Disease

Sheeba Khan, Owen Cain, and Neil Rajoriya

11.1 Introduction

The last decade has seen a large increase in mortality rates associated with chronic liver disease in both men and women, a significant fraction of which is attributable to alcohol (Rehm and Shield 2019). There is now robust evidence to suggest the detrimental effects of alcohol is associated with any level of consumption (Griswold et al. 2018). Excessive alcohol consumption not only results in premature deaths in both sexes but transcends to a substantial economic and social burden (Hydes et al. 2019b). Herein, we give an overview of the health burden associated with alcohol related liver disease (ArLD). We will outline the pathophysiology of the disease, its

S. Khan (✉)
National Institute for Health Research (NIHR), Birmingham Biomedical Research Centre, University Hospitals Birmingham NHS Foundation Trust and the University of Birmingham, Birmingham, UK

Centre for Liver Research, Institute of Immunology and Immunotherapy, University of Birmingham, Birmingham, UK

The Liver Unit, Queen Elizabeth Hospital, University Hospitals Birmingham NHS Foundation Trust, Birmingham, UK
e-mail: sheebz@doctors.net.uk; Sheeba.Khan2@uhb.nhs.uk

O. Cain
Department of Histopathology, Queen Elizabeth Hospital, University Hospitals Birmingham NHS Foundation Trust, Birmingham, UK

N. Rajoriya
Centre for Liver Research, Institute of Immunology and Immunotherapy, University of Birmingham, Birmingham, UK

The Liver Unit, Queen Elizabeth Hospital, University Hospitals Birmingham NHS Foundation Trust, Birmingham, UK
e-mail: Neil.Rajoriya@uhb.nhs.uk

S. S. Goonewardene et al. (eds.), *Men's Health and Wellbeing*, https://doi.org/10.1007/978-3-030-84752-4_11

various clinical spectrums, the diagnostic tests and the treatment strategies including curative liver transplantation (LT).

11.2   The Burden of Alcoholic Related Liver Disease (ArLD)

Alcohol use is ranked as the seventh leading risk factor for both deaths and disability adjusted life years (DALYs) in 2016 (Rehm and Shield 2019). The World Health Organization (WHO) estimated three million alcohol attributable deaths globally. This reflected 5.3% of the deaths and 5.1% of DALYs worldwide in 2016 (WHO 2018a). Nearly 50% of these mortality figures were attributable to end stage liver disease (liver cirrhosis) and 10% related to liver cancer (Hepatocellular carcinoma (HCC)) (WHO 2018b). There are global and regional variations in the prevalence and mortality rates, and these are influenced by per capita alcohol consumption and socioeconomic factors such as income, education and occupational class (Liangpunsakul et al. 2016). Regions with the highest consumers of alcohol include Europe (particularly eastern Europe), Russia, Australia and Northern America (Hydes et al. 2019a). Global burden of disease study in 2017 showed the highest proportion of ArLD cirrhosis-related deaths in central Europe (44%), western Europe (41.7%) and Latin America (38.1%) (Sepanlou et al. 2020).

11.3   Genetic and Environmental Associations

Gender has a strong influence on the mortality rates. Higher mortality is seen in men and it is attributable to greater proportion involved in *high-risk* drinking behaviour (Mann et al. 2003). Women, although have a lower mortality, are more susceptible to develop ArLD at lower daily intake of alcohol and this is partially related to higher body fat and lower gastric alcohol dehydrogenase activity (ADH) (Frezza et al. 1990). The genetic susceptibility to alcohol dependence and significant alcohol related liver injury is well known (Stickel et al. 2017). Polygenic pattern of inheritance is evidenced by higher concordance in monozygotic twins compared with dizygotic twins (Hrubec and Omenn 1981). Large scale studies including genome wide association studies suggest patatin-like phospholipase domain-containing protein 3 (PNPLA3) gene to be associated with predisposition and severity of ArLD (Ali et al. 2016; Salameh et al. 2015). It represents one of the strongest single genetic modulators in ArLD.

Only 10-20% of individuals with chronic alcohol excess progress to severe forms of the disease i.e. liver cirrhosis which is partly related to the environmental and behavioural factors (Singal et al. 2018). Obesity is an important co-risk factor in heavy drinkers (Hart et al. 2010). In recent years, there has been a significant increase in the prevalence of obesity in the general population which is known to have a synergistic as well as supra-additive effect on the progression of liver disease (Boyle et al. 2018). Hence, the synergy between obesity and alcohol excess consumption is likely to accelerate not only the genesis of ArLD but also its complications. A prospective study found a combined effect between obesity and alcohol in increasing the risk for HCC (Loomba et al. 2013). Co-existence of other liver diseases such as viral hepatitis may also exacerbate the disease progression (Poynard et al. 1997).

11.4 Pathogenesis of ArLD

The pathogenesis of ArLD is multifactorial. The three key mechanisms involved, irrespective of gender are (i) ethanol induced liver injury (ii) an inflammatory immune response and (iii) disruption of intestinal barrier and dysbiosis (Dunn and Shah 2016).

11.4.1 Liver Injury

Ethanol induced liver injury is instigated through its enzymatic breakdown within hepatocytes. ADH, aldehyde hydrogenase 2 (ALDH2), catalase and cytochrome P450 2E1(CYP2E1) are key enzymes that metabolize ethanol via oxidizing pathways into various metabolites, including the key perpetrator, acetaldehyde, which is highly reactive (Zakhari and Li 2007). Covalent binding of acetaldehyde to other molecules produces hybrid compounds called "adducts" which are highly toxic and play a critical role in the development of ArLD (Tuma and Casey 2003). Adducts react with the macro-molecules such as lipids, proteins and DNA, resulting in the disruption of their structure and impairment of function and also cause DNA mutations (Setshedi et al. 2010). Reduction oxidation reactions generate Nicotinamide adenine dinucleotide (NADH), enhanced production of which causes metabolic shifts favouring formation of free fatty acids (FFA) and ultimately, contribute to development of a fatty liver—a form of chronic disease and precursor to cirrhosis (Osna et al. 2017).

Intracellular accumulation of FFA causes injury to the cell membranes and thereby instigates inflammation which eventually progresses to fibrosis and subsequent liver cirrhosis (Lieber 2004). Steps in the enzymatic breakdown of ingested ethanol are demonstrated in Fig. 11.1.

11.4.2 Immune Response

Damage associated molecular patterns (DAMPs), released from injured hepatocytes and pathogen associated molecular patterns (PAMPs), released by the bacteria and bacterial products from damaged gastrointestinal (GI) tract propagate an

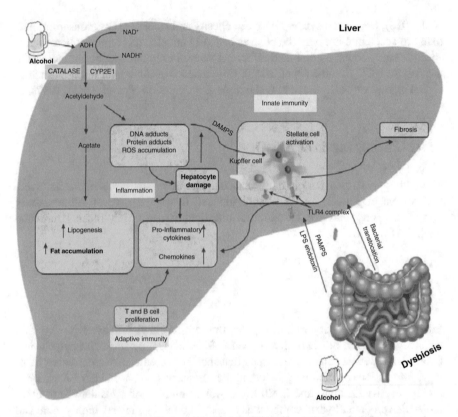

Fig. 11.1 Mechanism of ethanol induced liver injury. Ethanol ingestion results in hepatocyte damage through production of toxic metabolites including acetaldehyde, nicotinamide adenine dinucleotide and other by products. Formation of adducts results in hepatocyte damage. DAMPs released by the damaged hepatocytes and PAMPs released from the gut (through intestinal barrier disruption and dysbiosis) results in activation of innate and adaptive immunity which propagates the inflammation through release of various mediators, chemokines, activation of innate and adaptive immunity and recruitment of the inflammatory cells to the liver which eventually leads to fibrosis and cirrhosis. Alcohol Dehydrogenase (ADH); cytochrome P450 2E1 (CYP2E1); Damage associated molecular patterns (DAMPs); Deoxyribonucleic acid (DNA);Lipopolysaccharide (LPS); NAD, Nicotinamide adenine dinucleotide (NADH), Nicotinamide adenine dinucleotide-Hydr(NAD), pathogen associated molecular patterns (PAMPs); Reactive Oxidative species (ROS); Toll-like receptor 4 (TLR4)

inflammatory response which includes activation of release of pro-inflammatory cytokines and mediators, recruitment of the immune cells to the site of injury with resultant cellular dysfunction and development of ArLD (Li et al. 2019). Innate immune system comprising of Kupffer cells, hepatic stellate cells, dendritic cells, natural killer and natural killer T-lymphocytes (T-cells) are central to the initial inflammatory response to ethanol induced liver injury (Mandrekar and Ambade 2014). Activation of Kupffer cells by Lipopolysaccharide (LPS) circulating in the portal circulation via toll like receptors (TLRs) and recruitment of circulating macrophages are the key contributors to the hepatic inflammation and fibrosis (Mandrekar and Ambade 2014; Li et al. 2019). The adaptive immune system encompasses cellular and humoral subsets which are comprised of T-cells and B- lymphocytes (B-cells) respectively (Shuai et al. 2016). Alcohol excess orchestrates multiple effects on T-cells including: reduction in their number, increased T-cell activation, impaired function, altered homeostatic balance between various T-cell subsets and apoptosis of these T-cells (Pasala et al. 2015). Similar effect has been observed on B -cells including reduction in peripheral B-cell lines however, the production of immunoglobulins by the B-cells including the antibodies against liver autoantigens is enhanced (Pasala et al. 2015).

11.4.3   *Disruption of Intestinal Barrier and Dysbiosis*

Ethanol metabolites such as acetaldehyde disrupt the epithelial intestinal barrier through various mechanisms including oxidative stress, multiple signalling pathways and remodelling of cytoskeleton (Elamin et al. 2013). The disruption of the epithelial tight junctions allows translocation of luminal antigens into the portal circulation, activation of Kupffer cells within liver with subsequent activation of Tumor necrosis factor-alpha(TNF-α) and other inflammatory mediators causing hepatocellular damage (Parlesak et al. 2000; Adachi et al. 1994). Dysbiosis refers to decreased bacterial diversity and function in the gut flora (Hartmann et al. 2015). The main mechanism by which dysbiosis contributes to pathogenesis of ArLD is through bacterial translocation associated with intestinal disruption (Hartmann et al. 2015). In addition, dysbiosis also perpetuates the alteration in function of gut microbes, bile acid composition and immune dysregulation (Sarin et al. 2019).

11.5   Natural History of ArLD

ArLD represents a continuum of various clinical presentations which often overlap, these range from simple steatosis, which can progress to fibrosis and eventually develops into cirrhosis (Fig. 11.2) (Ishak et al. 1991). HCC, a complication of cirrhosis occurs in 5 to 15% cases of ArLD (Ishak et al. 1991). Binge drinking can present as a distinct presentation known as 'Alcoholic hepatitis' (AH) one of the most florid manifestations of ArLD with a high mortality (Osna et al. 2017). The deleterious effect of alcohol on the liver is threshold dependant (Kamper-Jørgensen et al. 2004) with risk for serious liver disease increasing with increasing levels of consumption beyond this threshold for an individual. Abstinence from alcohol can reverse the earlier stages of the disease and expected 5-year transplant-free survival is 60% versus 30% for those who continue to drink after developing decompensation (Parés et al. 1986).

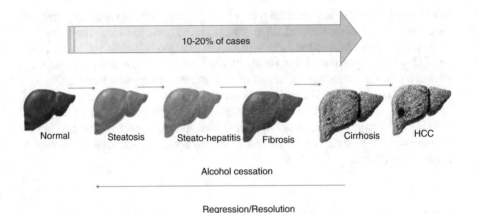

Fig. 11.2 Natural progression of ArLD. Cessation of alcohol can result in regression or resolution of liver disease depending on the stage of the disease. Genetic and environmental influences play a role in progression of liver disease. Hepatocellular carcinoma (HCC)

11.5.1   Alcoholic Steatosis

Alcoholic steatosis (fat infiltration within the liver) is the earliest and the most common presentation of ArLD (Zhou et al. 2003). It occurs in about 90% of the cases and could be seen as early as 2 weeks of heavy alcohol use (Lane and Lieber 1966). Steatosis, although reversible with abstinence, is associated with progression to cirrhosis and hence, no longer considered as a truly benign entity (Sørensen et al. 1984; Teli et al. 1995). Clinical predictors associated with increased risk of progression include female gender, continued alcohol consumption and histological features of presence of mixed droplet steatosis and megamitochondria (Fig. 11.3) (Chacko and Reinus 2016). Clinical presentation in ambulatory patients ranges from completely asymptomatic individuals with normal liver function test to incidental finding of deranged liver function test where AST and ALT do not rise above five times of upper normal limit (Sharma and Arora 2019). In hospitalised patients, hepatomegaly may be present in more than 70% of the patients (Diehl 2002).

11.5.2   Alcoholic Hepatitis

AH is a florid presentation of alcohol liver disease associated with 28-day mortality as high as 50% without treatment (Behnam Saberi et al. 2016; Drinane and Shah 2013). The typical presentation is with jaundice, abnormal liver tests and coagulopathy, leucocytosis, fever and tachycardia in individuals with a recent history of increased alcohol consumption (Frriedman 2020; Lucey et al. 2009). Features of portal hypertension may be apparent clinically such as ascites or varices due to liver

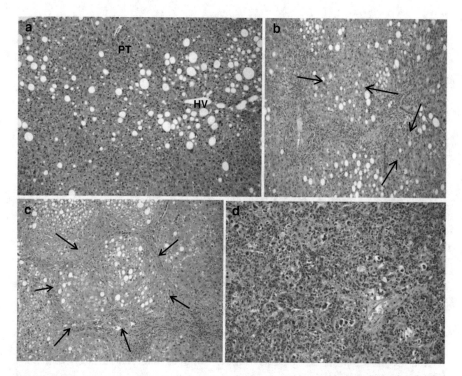

Fig. 11.3 Natural history of alcohol-related liver disease. (**a**) Early stage disease (steatosis) characterised by mild steatosis and minimal inflammation. There is no fibrosis and normal spatial relationships between portal tracts (PT) and hepatic veins (HV) are maintained. (**b**). Progressive disease (steatohepatitis) characterised by increased steatosis and inflammation, hepatocellular ballooning (difficult to appreciate at this power), and bridging fibrosis (outlined by arrows). (**c**). End stage disease (cirrhosis) characterised by bands of fibrous tissue completely encircling nodules of regenerating hepatocytes (the edge of a cirrhotic nodule is highlighted by the arrows). In this particular case there is ongoing steatohepatitis, suggestive of recent alcohol abuse. (**d**) Hepatocellular carcinoma. This tumour is arising on a background of alcohol related liver disease cirrhosis. [All four images are haematoxylin and eosin (H&E) stained sections, ×100]

inflammation leading to a rise in sinusoidal pressure/resistance (Cohen and Ahn 2009). Histology shows a constellation of lesions including steatosis, hepatocyte ballooning and an inflammatory infiltrate with polymorphonuclear neutrophils (Figs. 11.3 and 11.4) (MacSween and Burt 1986). Malnutrition and evidence of sepsis are strongly linked with increased severity and mortality (Sersté et al. 2018). Severity can be assessed by using scoring systems such as the Maddrey's discriminant function (mDF) score (Maddrey et al. 1978) and the Glasgow alcoholic hepatitis score (GAHS) score (Forrest et al. 2005) (Table 11.1). A mDF score of >32 and GAHS score > 9 indicates severe presentation associated with high mortality and warrants consideration of pharmacological intervention with steroids as per local guidelines—discussed later in chapter (Catherine Frakes Vozzo et al. 2018; Philips et al. 2019).

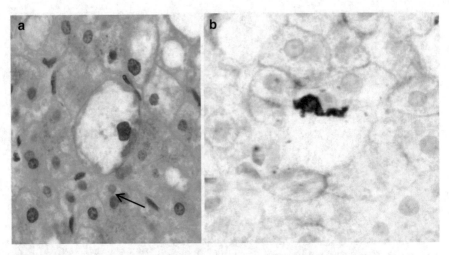

Fig. 11.4 Hepatocyte ballooning. (**a**) The histological hallmark of steatohepatitis is ballooning, characterised by cytoplasmic swelling and optical clearing of the hepatocyte cytoplasm, as seen affecting the cell in the centre of this image. There is also formation of a Mallory-Denk body (arrow), a pink 'ropy' body composed of an aggregate of intracytoplasmic filaments and proteins [H&E, ×600]. (**b**) Immunohistochemistry for keratin 8/18 can be helpful in identifying Mallory-Denk bodies, seen here as the darkly staining material in the cytoplasm of the central ballooned hepatocyte. Note the normal membranous and cytoplasmic staining of non-ballooned hepatocytes is lost, as these cytoskeletal proteins collapse down to form the Mallory-Denk body [Immunoperoxidase, ×600]

At time of writing, transplant centres in certain parts of the world consider LT an option for very carefully selected groups of patients with acute severe AH, these generally including those with severe presentation do not improve with abstinence and not showing improvement with pharmacological treatment (i.e steroid non-responders per Lille score) (Mathurin et al. 2011). These patients have a 90% mortality in 3 months if they develop additive renal dysfunction (Gianni Testino and Borro 2013).

11.5.3 Alcohol Related Cirrhosis

Cirrhosis represents the end stage on the spectrum of chronic liver disease, with often a component of reversibility with sustained abstinence of alcohol (Bonis et al. 2001). Portal hypertension represents the hallmark of cirrhosis and may be measured indirectly through hepatic venous pressure gradient (HVPG) studies across liver (Albilllos and Garcia-Tsao 2011). Portal hypertension occurs due to a combination of increased intrahepatic vascular resistance, a reduction of intrahepatic vasodilatory substances, along with increased portal vein inflow. Clinical

Table 11.1 AH severity assessment scoring systems

Score	Indication	Components	Formula			Prognostic indicators
GAHS (Forrest et al. 2005)	1. Severity staging 2. Prognostication of AH 3. Identification of patients suitable for steroid treatment	Score	1	2	3	Poor prognosis if score > 9
		Age	<50	≥50	–	
		WBC (10⁹/l)	<15	≥15	–	
		Urea(mm/l)	<5	≥5	–	
		PT ratio	<1.5	1.5-2.0	≥2	
		Bilirubin (μmol/l)	<7.3	7.3-14.6	>14.6	
mDF (Maddrey et al. 1978)	1. Severity staging of AH 2. Identification of patients suitable for steroids treatment	Bilirubin (mg/dl) Prothrombin time (sec)	4.6 × PT (PT- control PT) + total bilirubin			Score > 32 indicates poor prognosis. Patient and can be considered for steroids.
Lille score (Louvet et al. 2007)	Prognostication post treatment with steroids in AH	Bilirubin (μmol/l) Albumin (g/dl) INR	3.19-0.101 × age (yrs), +0.147 × albumin on day 0 + 0.0165 × change in bilirubin −0.206 × (renal insufficiency rated as 0 if absent and 1 if present) × 0.0065 × bilirubin day 0 – 0.0096 × INR			If >0.45 indicates lack of response to corticosteroids

GAHS Glasgow Alcoholic hepatitis score, *WBC* White blood cells, *PT* Prothrombin time, *mmol/l* millimole/litre, μmol/l micromole/litre, *mDF* Maddrey's discriminant function

manifestations of portal hypertension include the development of both ascites and varices. Cirrhosis can constitute a relatively asymptomatic phase known as *compensated cirrhosis* which can progress to a symptomatic phase of *decompensated cirrhosis* at a rate of 6 to 10% per year (Schuppan and Afdhal 2008). Common complications of decompensated cirrhosis include ascites, spontaneous bacterial peritonitis, variceal bleeding and hepatic encephalopathy (Garcia-Tsao 2016). Bacterial translocation through disrupted intestinal barrier, impaired immune response and malnutrition in cirrhotic patients predisposes patients with cirrhosis to increased bacterial infections (Riordan and Williams 2006). Various scoring systems enlisted in Table 11.2, are used in grading cirrhosis (especially pertaining to listing for LT) including the Child-Pugh (CP) score, widely used for risk stratification of cirrhosis (Infante-Rivard et al. 1987), Model of end-stage Liver disease (MELD) score and the United Kingdom model for End-stage liver disease (UKELD) score. Clinical criteria for LT assessment is dependant of local /national guidelines but should be considered if new onset decompensation of liver disease, a diagnosis of HCC, or a modified end stage liver disease (MELD) score above 21 (Crabb et al. 2020). UKELD score of 49 or above is applicable in UK.

Table 11.2 Scoring systems for risk stratification and LT referral for Liver cirrhosis

Score	Indication	Components	Formula	1	2	3	Prognostic indicators
Child Pugh (Infante-Rivard et al. 1987)	Stratification of stage of cirrhosis	Score		1	2	3	Class A: 5-6 Class B: 7-9 Class C:10-15 1 year survival rate 100%, 80% and 45% in class A, B, C cirrhosis respectively
		Bilirubin(µmol/L)		<34	34-50	>50	
		INR		<1.7	1.7-2.3	>2.3	
		Albumin(g/l)		>35	28-35	<28	
		Ascites		None	Grade 1 or 2	Grade 3	
		Encephalopathy		Absent	Mild to moderate.	Severe	
MELD/MELD-Na (Malinchoc et al. 2000; Kamath et al. 2001; Biggins et al. 2005; Kim et al. 2008)	1. Stratification of stage of cirrhosis 2. Prognostication post-TIPSS 3. Listing criteria in USA/Europe	Bilirubin (mg/dl) INR Creatinine(mg/dl) Sodium (mmol/l)	$11.2 \times in\ INR) + 9.57 \times in (creatinine) + 3.78 \times in(bilirubin) + 6.43$ $MELD\text{-}Na -[0.025 \times MELD \times (140\text{-}Na)] + 140$				Poor prognosis If >21
UKELD (Neuberger et al. 2008; Barber et al. 2011)	1. Mortality prediction in cirrhosis (UK) 2. Listing criteria	Bilirubin (µmol/l) INR Sodium (mmol/l) Creatinine (µmol/L)	$[5.395 \times in(INR) + (1.485 \times in(creatinine) + (3.13 \times in(bilirubin) - 81.565 \times in (Na, mmol/L)] + 435$				UKELD≥49 is required for listing. It indicates>9% mortality within 11 months

INR International normalised ratio, MELD Model for End stage Liver disease, Na Sodium, TIPSS Transjugular intrahepatic portosystemic shunt, UKELD United kingdom Model for End-Stage liver disease, In Natural logarithm, µmol/l micromole/litre, mmol/l millimole/litre, mg/dl milligram/decilitre

11.5.4   Hepatocellular Carcinoma (HCC)

HCC is a common solid organ neoplasm and cirrhosis is one of the major risk factors for its development and progression (Bruix et al. 2001; Herbst and Reddy 2012). It is amongst the leading causes of death in patients with cirrhosis in Europe and US (El-Serag and Mason 1999; Shaw and Shah 2011). Various predictors for HCC in cirrhosis have been proposed including high level of alpha fetoprotein (AFP), male sex, histological presence of macro regenerative nodules, dysplasia or higher proliferation index reflected in immunohistochemical studies, metabolic risk factors such as diabetes and obesity particularly increase the risk of HCC in individuals with ArLD cirrhosis (Bruix et al. 2001; Ioannou et al. 2018). A high prevalence of HCC in at risk individuals, late presentation at diagnosis and poor prognosis has led to surveillance strategies for early detection (Cabrera and Nelson 2010). Ultrasound screening for HCC in cirrhotic individuals every 6 months has been suggested with an aim to detect HCC at a stage when a potential effective treatment such as surgical resection, LT or percutaneous intervention like transarterial chemoembolization (TACE) can be offered (Bruix and Sherman 2011).

11.6   Diagnosis

Establishing an accurate history of excessive alcohol consumption is key to the diagnostic work-up. Questionnaires such as Alcohol Use Disorders Inventory Test (AUDIT) can be used to screen for hazardous drinking and alcohol dependence (Thursz et al. 2018a; b; Bush et al. 1998). In the USA, 1 alcoholic drink equivalent is described as containing 14 g (0.6floz) of pure alcohol and the National Institute on Alcohol Abuse and Alcoholism (NIAAA) defines hazardous alcohol use as more than 4 drinks on a day for men or more then 3 drinks for women (Dietary guidelines for Americans 2015-2020 2020; McCutheon 2017). Binge drinking, on the other hand, constitutes a rise in blood alcohol concentration to 0.08 mg/dl, which usually occurs after 4 drinks for women and 5 drinks for men in 2 hours (Alcoholism NIoAAa n.d.-a). Current guidance in the UK is not to consume more than 14 units a weeks on regular basis (Fig. 11.5) (UK Chief Medical Officer's Low Risk Drinking Guidelines August 2016), Among the European union member states, alcohol content in a standard unit varies from 8 to 20 g. (Commission E. n.d.)The national low risk drinking recommendations are seen in Table 11.3.

It is worth noting that alcohol consumption guidelines vary significantly across the globe (Wood et al. 2018). Analysis of the data from 83 prospective studies from 19 high income countries concluded that adoption of drinking thresholds lower than the current recommendations in most guidelines was required in order to reduce the risk of all-cause mortality (Wood et al. 2018).

Fig. 11.5 Diagrammatic representation of UK alcohol consumption guidance. 14 units of alcohol a week are equivalent to 6 glasses of wine or 6 pints of beer. ABV, Alcohol by volume; ml (millilitres)

| 6 medium glasses of wine a week. Based on 175ml 13.5%ABV | 6 pints of Beer a week. Based on 567 ml 4%ABV |

Table 11.3 National low risk alcohol consumption guidelines amongst European nations

European low-risk drinking guidance		
Country	Male	Female
Germany	Up to 24 g alcohol / per day	Up to 12 g alcohol /per day
France	No more than 10 standard drinks / week. At least 1 alcohol free day /week.	
Italy	Up to 2 drinks per day	Up to 1 drink per day
Austria	Up to 24 g of alcohol/ day	Up to 16 g of alcohol per day
Switzerland	2-3 drinks per day	1-2 drinks per day
Spain	Up to 40 g alcohol per day	Up to 20 g of alcohol per day

The good practice principles for low risk guidelines for European countries were commissioned by European Union published in 2016 by Reducing Alcohol related harm (RARHA) (Broholm et al. 2016; Commission E. n.d.)

11.6.1 Investigations

Multiple modalities of investigations are employed to aid the diagnosis of ArLD and these are based on the clinical suspicions and presentations. The aim is to (i) establish the diagnosis whilst (ii) ruling out other/ concomitant liver disease aetiologies and also (iii)establish the severity of the condition (Crabb et al. 2020). A panel of serological tests is usually combined with radiological evaluation. Additional tests including liver biopsy, portal pressure measurements and upper GI endoscopy may also be required in certain cases—the latter when portal hypertension is suspected (Heidelbaugh and Bruderly 2006; Singal et al. 2018). A general diagnostic algorithm has been outlined in Fig. 11.6.

11.6.1.1 Transient Elastography

Transient elastography, is a bedside, non-invasive technique which is performed using a Fibroscan® (Lupsor-Platon and Badea 2015). The velocity of an elastic shear wave, transmitted through an ultrasound transducer is measured and it is

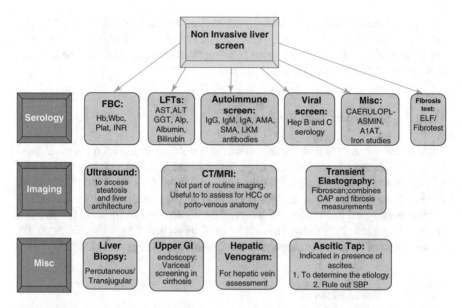

Fig. 11.6 Diagnostic Algorithm. Panel of serological tests that constitute the basic Non-invasive liver screen (Liver 2018) include: Full blood count (FBC), Haemoglobin (Hb), White blood cells (WBC), Platelets (Plat), International normalised ratio (INR). Liver function tests (LFTs): Aspartate aminotransferase (AST), Alanine aminotransferase (ALT), Gamma glutamyl transferase (GGT), Alkaline phosphatase (ALP). Immunoglobulins (Ig) G, M and A, antimitochondrial antibodies (AMA), smooth muscle antibodies(SMA), anti-liver-kidney-and microsomal antibody(LKM,) alpha 1 antitrypsin (A1AT).Imaging: Ultrasound liver is performed as part of the routine evaluation however, Computerised tomography (CT) and Magnetic resonance imaging (MRI) are reserved for specific indications. Controlled attenuation parameter (CAP) scores are used to assess fat infiltration within the liver. Liver biopsy is usually carried out when diagnosis is in doubt and performed percutaneously. Transjugular route is used when portal pressure measurements are necessary, in severe coagulopathy or if ascites present (Heidelbaugh and Bruderly 2006). Ascitic tap should be performed in all new patients with ascites, or if concerns regarding spontaneous bacterial peritonitis

directly related to the liver stiffness. This reading correlates to staging of fibrosis per Metavir score (stages F0-4) (Bedossa and Poynard 1996) and it also allows the measurement of liver steatosis through measurement of controlled attenuation parameter (CAP) score (Lupsor-Platon and Badea 2015; Foucher et al. 2006).

11.6.1.2 Imaging/Biopsy

Ultrasound is an inexpensive, widely available initial radiologic assessment of liver steatosis and mass lesions (Schwenzer et al. 2009). It also allows detection of ascites in advanced disease and splenomegaly—an indicator of portal hypertension. Doppler ultrasound liver can assess the velocity of blood flow across portal vein and aid screening of portal hypertension/ portal vein thrombosis (Di Lelio et al. 1989; Tchelepi et al. 2002). Histologic evaluation is not a standard in diagnostic process except when there is diagnostic dilemma in doubt or an accurate evaluation of staging of liver disease is required (Spycher et al. 2001). Upper GI endoscopy for upper GI variceal screening is recommended if there is a clinical suspicion of varices or in patients with fibrosis score (via transient elastography) of >20 kPa and platelets of <150,000 (de Franchis 2015).

11.6.1.3 Serological Tests for Fibrosis

Some of the widely studied serum markers of liver fibrosis in chronic liver disease include enhanced Liver Fibrosis (ELF) test, Fibrotest, Hepascore and Fibrometre (Rosenberg et al. 2004; Imbert-Bismut et al. 2001; Adams et al. 2005; Calès et al. 2018). These tests are not liver specific, the results may be affected by liver inflammation as opposed to fibrosis and hence, they fail to meet the criteria for an ideal serum biomarker. They however can be used in conjunction with Fibroscan for stratification of disease (Rockey and Bissell 2006; Papastergiou et al. 2012).

11.6.1.4 Clinical Signs

Common clinical features associated with liver cirrhosis irrespective of aetiology have been illustrated in Fig. 11.7 and may be present / absent / vary between patients depending on the clinical presentation or stage of disease (Eric Goldberg 2020).

11.7 Management of ArLD

Abstinence from alcohol is critical in the management plan hence, a multidisciplinary approach with liaison between the clinician, psychiatric and addiction support services is key in the management of ArLD (Avila et al. 2020; Stewart and Day 2003). Treatment modalities for the management of decompensated cirrhosis,including pharmacological, endoscopic, radiological or surgical are tailored per clinical presentation.(EASL Clinical Practice Guidelines for the management of patients with decompensated cirrhosis 2018) LT is the only curative treatment for patients with cirrhosis which requires careful patient selection and management (Stewart and Day 2003).

11.7.1 Ascites

Ascites is the most common complication of cirrhosis and any patient who develops ascites should have their candidacy assessed for transplantation (Runyon 2012). Diagnostic evaluation requires analysis of ascitic fluid to calculate serum-to-ascites albumin gradient (SAAG) score. It is measured by subtracting serum albumin from

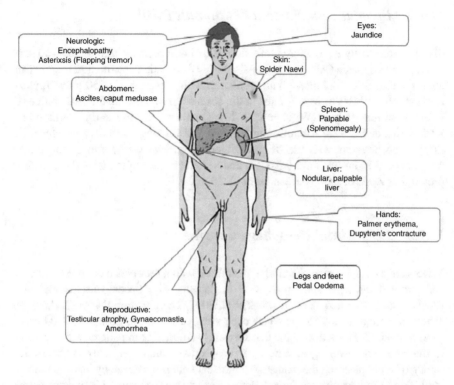

Fig. 11.7 Clinical signs of liver cirrhosis. Clinical features associated with liver cirrhosis include hepatic encephalopathy, jaundice, spider naevi, nodular liver may be palpable, splenomegaly (associated with portal hypertension), flapping tremor 'asterixsis' seen with decompensation of cirrhosis, Hypogonadism, gynecomastia, Duputren's contracture, Palmer erythema and anorexia with muscle loss (Dowd 2000)

ascitic fluid albumin and a SAAG score of >1.1 g/dl or 11/g/l has a sensitivity of 97% in indicating portal hypertensive ascites (Runyon et al. 1992; Mauer and Manzione 1988; Aithal et al. 2020). No added salt diet and titration of diuretics are the key components of medical management (Ginés et al. 1987; Santos et al. 2003). Ascites may be refractory to diuretics or patients may develop side effects of biochemical abnormalities with diuretics deemed "diuretic intolerance" (Bhogal and Sanyal 2013; Siqueira et al. 2009). There are various treatment modalities for patients who are not candidates for LT with diuretic-intolerant or refractory ascites (Aithal et al. 2020), These may include a Trans jugular portosystemic shunt (TIPSS) which aims to reduce the portal pressure via metal stent to ease decompression of the portal system, or percutaneous drainage of ascites through a large volume paracentesis (Lebrec et al. 1996; Aithal et al. 2020). Other options such as low flow continuous pump mechanisms (Alphapump ™) (Stirnimann et al. 2017) or indwelling palliative long-term ascites catheters are still being explored (Corrigan et al. 2020).

11.7.2   Spontaneous Bacterial Peritonitis (SBP)

SBP is a frequently encountered infection in cirrhotic patients with ascites and carries a poor prognosis (Alaniz and Regal 2009). Diagnostic confirmation requires an analysis of ascitic fluid through an ascitic tap with presence of >250 polymorphonuclear cells (PMN)/mm^3 in ascites in the absence of an intra-abdominal source of infection of malignancy (Wiest et al. 2012). Treatment requires early institution of intravenous antibiotics, Albumin replacement and lifelong secondary prophylaxis. Primary prophylaxis with use of an antibiotic is indicated in patients with a low protein count in ascitic fluid (<15 g/l) and secondary prophylaxis for those with a previous episode of SBP (Fernández et al. 2016).

11.7.3   Gastric and Oesophageal Varices

Varices are a complication of portal hypertension with a prevalence of 30% to 40% in compensated cirrhosis and up to 85% in patients with decompensated cirrhosis (Pagliaro and D'Amico 1994; Kovalak et al. 2007).There is high risk of mortality in setting of variceal haemorrhage which occurs at a yearly rate of 10% to 15% (Garcia-Tsao et al. 2017; Prediction of the first variceal haemorrhage in patients with cirrhosis of the liver and oesophageal varices. A prospective multicenter study 1988). In the setting of acute bleeding the mainstay of the treatment pre-endoscopy is resuscitation, intubation to protect airway, the vasoactive Terlipressin (or Octreotide in some countries) and antibiotic prophylaxis (Saner et al. 2007; Chavez-Tapia et al. 2011; Tripathi et al. 2015b). The latter 2 medications (Terlipreesin and Antibiotics) have been shown to improve survival post variceal bleeding (Carbonell et al. 2004; Zhou et al. 2018). Endoscopic modalities include injection sclerotherapy, however that has been superseded by endoscopic variceal band ligation (Laine and Cook 1995; Garcia-Pagán and Bosch 2005). In quiescent varices, non-selective beta-blockers and/or EVBL can prevent the risk of bleeding in those with medium or large varices (Garcia-Tsao et al. 2017; Tripathi et al. 2015a). Gastric variceal management differs from that of oesophageal in that injection obturation therapy is the mainstay with agents such as N Butyl-2-cyanoacrylate glue (Sarin 1997; Sarin et al. 1988; Rajoriya et al. 2011) or Thrombin(Yang et al. 2002; McAvoy et al. 2012). In the acute bleeding setting TIPSS can be considered for those patients who have ongoing bleeding despite standard treatment or pre-emptively (Neil Raoriya 2012; García-Pagán et al. 2010; Dunne et al. 2020).

11.7.4   Hepatic Encephalopathy

Hepatic Encephalopathy is a brain dysfunction caused by liver insufficiency and/or portal -systemic blood shunting; it manifests as a wide spectrum of neurological abnormalities ranging from subclinical subtle cognitive alterations to coma (Vilstrup et al. 2014). It is associated with poor quality of life for patients and associated with poor prognosis (Agrawal et al. 2015). Management involves treatment of the precipitants e.g. infection, bleeding, constipation, dehydration or medication-related, along with regular use of oral lactulose with or without Rifaximin as prophylaxis (Bass et al. 2010; Sharma et al. 2013).

11.7.5   Acute Alcoholic Hepatitis

Use of corticosteroids in selected cases of AH with DF > 32 has shown mortality benefit and can be considered (Rambaldi et al. 2008; Lucey et al. 2020). However, the recent STOPAH trial, one of the largest UK trials in hepatology showed pentoxifylline did not improve survival in patients with AH whilst prednisolone was associated with a reduction in 28-day mortality that did not reach significance and with no improvement in outcomes at 90 days or 1 year (Thursz et al. 2015). Physicians should thus follow local guidelines regarding steroids in the AH setting, as still favoured in some centres (in select candidates), especially in the absence of any other current *primetime* treatments bar good nutrition and abstinence. If steroids being used, the Lille score determines if patients should stop steroids after 1 week of initiation depending on response (Louvet et al. 2007).

11.7.6   Alcohol Withdrawal Syndrome (AWS)

AWS manifests in 24 to 72 hours of cessation of alcohol in individuals with chronic excess. The constellation of acute symptoms including anxiety, tremors, nausea, insomnia and in severe cases delirium tremens (DTs) (Leggio and Lee 2017; Physicians 2012). Severity of AWS is commonly assessed with clinical Institute withdrawal assessment for alcohol-revised (CIWA-Ar) score (Bayard et al. 2004; Sullivan et al. 1989). This is a subjective scoring system based on signs and symptoms and treatment is initiated when CIWA-Ar score is greater than 8 (Bayard et al. 2004). It is easy to use and often a good treatment algorithm for nurses to use in patients withdrawing from alcohol. Treatment protocols vary including symptom-triggered regimen, loading dose regimen and fixed dose regimen where benzodiazepines are initiated based on CIWA-Ar score, administration of loading dose every 2 hours and a fixed daily dose administered in four divided doses respectively (Kattimani and Bharadwaj 2013; WHO 2012). Landmark study by Kaim*et al* (101)) showed greater efficacy of chlordiazepoxide (a benzodiazepine) in preventing seizures and DT in AWS in comparison to other medications. Shorter acting benzodiazepines should be used in patients with established chronic liver disease (e.g cirrhosis) they are sensitive to the sedating effects which could precipitate drop in conscious level and even aspiration pneumonias (Sachdeva et al. 2015; Addolorato et al. 2016). General supportive measures include nutritional support, hydration and correction of electrolyte abnormalities especially administration of intravenous thiamine to prevent precipitation of Wernicke's encephalopathy (WE) (Bayard et al. 2004). WE is an acute neuropsychiatric condition that occurs due to depletion of thiamine (vitamin B1) in chronic alcohol users and if untreated or treated inadequately, leads to irreversible brain damage resulting in short term memory loss and ability to acquire new information (Thomson et al. 2002; Thomson and Marshall 2013).

11.7.7   Nutrition

The incidence of malnutrition in patients with cirrhosis is high and sarcopenia or loss of skeletal muscle mass is a major contributor to the adverse outcomes related to malnutrition (Dasarathy et al. 2017; Arora et al. 2012). Predisposition to malnutrition and sarcopenia is multifactorial and includes poor oral intake, anorexia, malabsorption, decreased protein synthesis and dysregulated muscle autophagy (Dasarathy et al. 2017). At present, there is no screening method available that is specific for ArLD however, there are several tools available to screen for malnutrition for patients with cirrhosis, each with their own limitations (Arora et al. 2012; Ney et al. 2020). The Royal Free Hospital Nutritional prioritisation Tool (RFN-NPT) is an easy to use, screening questionnaire that has been developed for patients with liver cirrhosis and has been found to be more sensitive than the Nutritional Risk screening (NRS) in assessing malnutrition in liver disease (Borhofen et al. 2016; Kondrup et al. 2003). Amongst micronutrients, thiamine, folate, vitamin A, vitamin C, zinc, and magnesium deficiency is seen frequently with chronic alcohol excess (125) and if not replaced adequately can lead to range of adverse clinical outcomes including: WE, anaemia, night blindness, scurvy, altered immune/ mental function and cardiovascular complications respectively. The key aspects of management include alcohol abstinence, micronutrient deficiency identification with supplementation, and general nutritional support particularly with protein supplementation and increased physical activity (Dasarathy 2016; Plauth et al. 2019).

11.7.8   Mental Health and Addiction Support

Alcohol use disorder, is central to the genesis of ArLD. It refers to an inability to stop or control alcohol use despite adverse social, occupational, or health consequences (Alcoholism NIoAAa n.d.-b). In addition, there is higher prevalence of other psychological disorders such as anxiety, depression, affective disorders, psychosis etc. in individuals with ArLD. These individuals are also, likely to develop other substance misuse and addictions (Ewusi-Mensah et al. 1983; Galbicesk 2020). Various integrated care models for patients with chronic liver disease and mental health comorbidities have been studied with observed benefits (Moriarty et al. 2007;

Ho et al. 2015; Evon et al. 2011). A UK based study demonstrated 67% reduction in hospital admissions and a reduction of emergency department attendances by 59% in complex and poorly compliant patients with ArLD, by adopting an alcohol assertive outreach team (AAOT) strategy (Hughes et al. 2013). Patient care was delivered by a multidisciplinary team comprising of medical, psychiatric, substance misuse, psychology, nursing and social work specialists. Similar multidisciplinary approach has been proposed for developing alcohol care teams (ACTs) to be rolled out in acute UK hospitals with aim to reduce mortality and improve the quality and efficiency of Alcohol care in UK through adoption of integrated pathways (Moriarty 2020).

11.7.9   Liver Transplantation

ArLD is one of the most common indications for LT in Europe and USA (Liver 2018; Fayek et al. 2016). LT for AH may contribute to this trend in the future (Im et al. 2019). Graft and patient survival after transplant for ArLD is comparable to other aetiologies of liver disease at 80 to 85%. (Liver 2018; Mathurin and Lucey 2020). Each country has stringent criteria for patient selection as well as contraindications that are applicable to patients with chronic liver disease regardless of aetiology (BASL 2012; Cholongitas and Burroughs 2012). The assessment regarding suitability and timing for transplant requires a multidisciplinary approach where, a period of abstinence balanced against risk of death associated with severity of liver disease influences the patient selection (Mathurin and Lucey 2020). The evidence for a fixed 6 month period of abstinence to predict post-transplant outcomes including compliance is contradictory (De Gottardi et al. 2007; Lucey et al. 1997) however evidence of clinical improvement associated with abstinence which may negate the need for LT is undisputed (Society 2016; Vaillant 2003). This is particularly relevant in the context of acute AH where 6 month survival rate is 30% in those who do not respond to medical treatment and could benefit from consideration for LT (Mathurin et al. 2011). A multicentre case-control study from France and Belgium showed a superior six month survival (78%) in selected acute AH patients who underwent an early liver transplant compared to the control group (24%) who received standard of care treatment (Mathurin et al. 2011). Similar results were reported by a US study with 89% survival with early liver transplant compared to 11% in matched controls (Im et al. 2016). A study from the John Hopkins Institute, USA also provided similar compelling evidence with 6-month survival in patients with acute AH reported to be 100% compared to the patients with alcohol related cirrhosis who underwent transplant after a six month period of sobriety (88%) (114).

National institute of clinical excellence (NICE) suggests a three-month period of disease evaluation prior to referral to transplant unit (Guidance 2010). A key aspect of this evaluation period is to integrate the addiction support for potential transplant candidates with an aim to minimise the risk of relapse and improve post-transplant outcomes (Matos Santana 2016). Failure to demonstrate engagement with alcohol-support or ongoing alcohol use associated with recurrent episodes of decompensation of ArLD are deemed as contraindications for LT assessment (BASL 2012). Blood testing for alcohol levels, breathalyser analysis and metabolite testing is used widely for ongoing monitoring for alcohol use as part of LT assessments (Staufer and Yegles 2016). Use of an alcohol abstinence agreement can be a useful tool to facilitate adoption of agreed code of conduct between the transplant unit and the patient. It is currently in place in all UK LT centres (Heyes et al. 2016).

11.8 Alcohol Recidivism

The relapse to hazardous levels of drinking alcohol post liver transplant is seen in about one fifth of the patients undergoing transplant for ArLD and it is associated with detrimental effects including graft rejection and graft loss (Chuncharunee et al. 2019; Dumortier et al. 2015; Rice et al. 2013). The relapse to alcohol after transplant is insidious and usually occurs between 2 to 5 years after surgery (DiMartini et al. 2010; Iruzubieta et al. 2013a). A recent meta-analysis reported any alcohol relapse and heavy alcohol relapse rates to be as high as 22% and 14% respectively (Chuncharunee et al. 2019). The effect of relapse on post-transplant outcomes is more evident in long-term with a significant reduction in 10-year survival rate in LT recipients who return to drinking compared to those who remain abstinent (Iruzubieta et al. 2013b). Poor medical compliance, increased risk of post-transplant malignancy, recurrence of cirrhosis, cardiovascular diseases and reduced long-term survival add to the poor transplant outcomes in those who relapse to hazardous alcohol use (Chuncharunee et al. 2019; Pageaux et al. 2003). Predictors for alcohol recidivism include young age, female sex, psycho-social issues, other substance misuse, short period of sobriety pre-transplant and alcohol dependence pre-transplant (Karim et al. 2010). Scoring systems proposed to predict the risk of recidivism currently lack validation and pharmacological options are limited in this area hence, psychosocial interventions are the mainstay of the pre and post-transplant management to minimise the risk of relapse (Lim and Sundaram 2018).

Acknowledgments Authors thank their colleague Professor Philip Newsome (Centre for Liver Research, Institute of Immunology and Immunotherapy, University of Birmingham) for his support.

References

Adachi Y, Bradford BU, Gao W, Bojes HK, Thurman RG. Inactivation of Kupffer cells prevents early alcohol-induced liver injury. Hepatology. 1994;20:453–60.

Adams LA, Bulsara M, Rossi E, DeBoer B, Speers D, George J, Kench J, et al. Hepascore: an accurate validated predictor of liver fibrosis in chronic Hepatitis C infection. Clin Chem. 2005;51:1867–73.

Addolorato G, Mirijello A, Barrio P, Gual A. Treatment of alcohol use disorders in patients with alcoholic liver disease. J Hepatol. 2016;65:618–30.

Agrawal S, Umapathy S, Dhiman RK. Minimal hepatic encephalopathy impairs quality of life. J Clin Exp Hepatol. 2015;5:S42–8.

Aithal GP, Palaniyappan N, China L, Härmälä S, Macken L, Ryan JM, Wilkes EA, et al. Guidelines on the management of ascites in cirrhosis. Gut. 2020;55(6):vi1–vi12. https://doi.org/10.1136/gutjnl-2020-321790.

Alaniz C, Regal RE. Spontaneous bacterial peritonitis: a review of treatment options. P & T. 2009;34:204–10.

Albilllos A, Garcia-Tsao G. Classification of cirrhosis: the clinical use of HVPG measurements. Dis Markers. 2011;31:121–8.

Alcoholism NIoAAa. Drinking levels defined. (n.d.-a), publication date: 01/01/2018.

Alcoholism NIoAAa. Alcohol use disorder. Alcohol's effect on health (n.d.-b), publication date: 01/01/2018.

Ali M, Yopp A, Gopal P, Beg MS, Zhu H, Lee W, Singal AG. A variant in PNPLA3 associated with fibrosis progression but not hepatocellular carcinoma in patients with hepatitis C virus infection. Clin Gastroenterol Hepatol. 2016;14:295–300.

Arora S, Mattina C, McAnenny C, Sullivan N, Laura M, Nina C, Gatiss G, et al. OC-075 assessment of nutritional status in patients with cirrhosis: must is not a must. Gut. 2012;61:A32.

Avila MA, Dufour J-F, Gerbes AL, Zoulim F, Bataller R, Burra P, Cortez-Pinto H, et al. Recent advances in alcohol-related liver disease (ALD): summary of a gut round table meeting. Gut. 2020;69:764.

Barber K, Madden S, Allen J, Collett D, Neuberger J, Gimson A. Elective liver transplant list mortality: development of a United Kingdom end-stage liver disease score. Transplantation. 2011;92:469–76.

Basl N. Guidelines for referral for liver transplant assessment. 2012.

Bass NM, Mullen KD, Sanyal A, Poordad F, Neff G, Leevy CB, Sigal S, et al. Rifaximin treatment in hepatic encephalopathy. N Engl J Med. 2010;362:1071–81.

Bayard M, McIntyre J, Hill KR, Woodside J Jr. Alcohol withdrawal syndrome. Am Fam Physician. 2004;69:1443–50.

Bedossa P, Poynard T. The METAVIR Cooperative Study Group. An algorithm for the grading of activity in chronic hepatitis C. Hepatology. 1996;24:289–93.

Behnam Saberi ASD, Jang Y-Y, Gurkar A, Mezey E. Current management of alcoholic hepatitis and furture therapies. J Clin Transl Hepatol. 2016;4(2):113–22.

Bhogal H, Sanyal AJ. Treatment of refractory ascites. Clin Liver Dis. 2013;2:140–2.

Biggins SW, Rodriguez HJ, Bacchetti P, Bass NM, Roberts JP, Terrault NA. Serum sodium predicts mortality in patients listed for liver transplantation. Hepatology. 2005;41:32–9.

Bonis PA, Friedman SL, Kaplan MM. Is liver fibrosis reversible? N Engl J Med. 2001;344:452–4.

Borhofen SM, Gerner C, Lehmann J, Fimmers R, Görtzen J, Hey B, Geiser F, et al. The Royal Free Hospital-nutritional prioritizing tool is an independent predictor of deterioration of liver function and survival in cirrhosis. Dig Dis Sci. 2016;61:1735–43.

Boyle M, Masson S, Anstee QM. The bidirectional impacts of alcohol consumption and the metabolic syndrome: cofactors for progressive fatty liver disease. J Hepatol. 2018;68:251–67.

Broholm K, Galluzzo L, Gandin C, Ghirini S, Ghiselli A, Jones L, Martire S, Mongan D, Montonen M, Mäkelä P, Rossi L, Sarrazin D, Scafato E, Schumacher J, Steffens R. Good practice principles for low risk drinking guidelines. Helsinki: National Institute for Health and Welfare (THL); 2016.

Bruix J, Sherman M. Management of hepatocellular carcinoma: an update. Hepatology. 2011;53:1020–2.

Bruix J, Sherman M, Llovet JM, Beaugrand M, Lencioni R, Burroughs AK, Christensen E, et al. Clinical management of hepatocellular carcinoma. Conclusions of the Barcelona-2000 EASL conference. J Hepatol. 2001;35:421–30.

Bush K, Kivlahan DR, McDonell MB, Fihn SD, Bradley KA. The AUDIT alcohol consumption questions (AUDIT-C): an effective brief screening test for problem drinking. Ambulatory care quality improvement project (ACQUIP). Alcohol use disorders identification test. Arch Intern Med. 1998;158:1789–95.

Cabrera R, Nelson DR. Review article: the management of hepatocellular carcinoma. Aliment Pharmacol Ther. 2010;31:461–76.

Calès P, Boursier J, Oberti F, Moal V, Fouchard Hubert I, Bertrais S, Hunault G, et al. A single blood test adjusted for different liver fibrosis targets improves fibrosis staging and especially cirrhosis diagnosis. Hepatol Commun. 2018;2:455–66.

Carbonell N, Pauwels A, Serfaty L, Fourdan O, Lévy VG, Poupon R. Improved survival after variceal bleeding in patients with cirrhosis over the past two decades. Hepatology. 2004;40:652–9.

Catherine Frakes Vozzo NW, Romero-Moarrero C, Fairbanks KD. Alcohol liver disease. Cleveland Clinic Center fo Continuing Education 2018.

Chacko KR, Reinus J. Spectrum of alcoholic liver disease. Clin Liver Dis. 2016;20:419–27.

Chavez-Tapia NC, Barrientos-Gutierrez T, Tellez-Avila F, Soares-Weiser K, Mendez-Sanchez N, Gluud C, Uribe M. Meta-analysis: antibiotic prophylaxis for cirrhotic patients with upper gastrointestinal bleeding – an updated Cochrane review. Aliment Pharmacol Ther. 2011;34:509–18.

Cholongitas E, Burroughs AK. The evolution in the prioritization for liver transplantation. Ann Gastroenterol. 2012;25:6–13.

Chuncharunee L, Yamashiki N, Thakkinstian A, Sobhonslidsuk A. Alcohol relapse and its predictors after liver transplantation for alcoholic liver disease: a systematic review and meta-analysis. BMC Gastroenterol. 2019;19:150.

Cohen SM, Ahn J. Review article: the diagnosis and management of alcoholic hepatitis. Aliment Pharmacol Ther. 2009;30:3–13.

Commission E. Health promotion and disease preventon knowledge gateway- alcoholic bevarages. (n.d.), publication date: 01/01/2018.

Corrigan M, Thomas R, McDonagh J, Speakman J, Abbas N, Bardell S, Thompson F, et al. Tunnelled peritoneal drainage catheter placement for the palliative management of refractory ascites in patients with liver cirrhosis. Frontline Gastroenterol. 2020;12(2):108–12.

Crabb DW, Im GY, Szabo G, Mellinger JL, Lucey MR. Diagnosis and treatment of alcohol-associated liver diseases: 2019 practice Guidance from the American Association for the Study of Liver Diseases. Hepatology. 2020;71:306–33.

Dasarathy S. Nutrition and alcoholic liver disease: effects of alcoholism on nutrition, effects of nutrition on alcoholic liver disease, and nutritional therapies for alcoholic liver disease. Clin Liver Dis. 2016;20:535–50.

Dasarathy J, McCullough AJ, Dasarathy S. Sarcopenia in alcoholic liver disease: clinical and molecular advances. Alcohol Clin Exp Res. 2017;41:1419–31.

de Franchis R. Expanding consensus in portal hypertension: report of the Baveno VI consensus workshop: stratifying risk and individualizing care for portal hypertension. J Hepatol. 2015;63:743–52.

De Gottardi A, Spahr L, Gelez P, Morard I, Mentha G, Guillaud O, Majno P, et al. A simple score for predicting alcohol relapse after liver transplantation: results from 387 patients over 15 years. Arch Intern Med. 2007;167:1183–8.

Di Lelio A, Cestari C, Lomazzi A, Beretta L. Cirrhosis: diagnosis with sonographic study of the liver surface. Radiology. 1989;172:389–92.

Diehl AM. Liver disease in alcohol abusers: clinical perspective. Alcohol. 2002;27:7–11.

Dietry Guidelines for Americans 2015–2020. 2020.

DiMartini A, Dew MA, Day N, Fitzgerald MG, Jones BL, deVera ME, Fontes P. Trajectories of alcohol consumption following liver transplantation. Am J Transplant. 2010;10:2305–12.

Dowd JG. Oxford textbook of clinical hepatology. Oxford: LWW; 2000.

Drinane MC, Shah VH. Alcoholic hepatitis: diagnosis and prognosis. Clin Liver Dis. 2013;2:80–3.

Dumortier J, Dharancy S, Cannesson A, Lassailly G, Rolland B, Pruvot FR, Boillot O, et al. Recurrent alcoholic cirrhosis in severe alcoholic relapse after liver transplantation: a frequent and serious complication. Am J Gastroenterol. 2015;110:1160–6. quiz 1167

Dunn W, Shah VH. Pathogenesis of alcoholic liver disease. Clin Liver Dis. 2016;20:445–56.

Dunne PDJ, Sinha R, Stanley AJ, Lachlan N, Ireland H, Shams A, Kasthuri R, et al. Randomised clinical trial: standard of care versus early-transjugular intrahepatic Porto-systemic shunt (TIPSS) in patients with cirrhosis and oesophageal variceal bleeding. Aliment Pharmacol Ther. 2020;52:98–106.

EASL. Clinical practice guidelines for the management of patients with decompensated cirrhosis. J Hepatol. 2018;69:406–60.

Elamin EE, Masclee AA, Dekker J, Jonkers DM. Ethanol metabolism and its effects on the intestinal epithelial barrier. Nutr Rev. 2013;71:483–99.

El-Serag HB, Mason AC. Rising incidence of hepatocellular carcinoma in the United States. N Engl J Med. 1999;340:745–50.

Eric Goldberg MC. Cirrhosis in adults: etiologies, clinical manifestations, and diagnosis. uptodate. com updated 26 Aug 2020.

Evon DM, Simpson K, Kixmiller S, Galanko J, Dougherty K, Golin C, Fried MW. A randomized controlled trial of an integrated care intervention to increase eligibility for chronic hepatitis C treatment. Am J Gastroenterol. 2011;106:1777–86.

Ewusi-Mensah I, Saunders JB, Wodak AD, Murray RM, Williams R. Psychiatric morbidity in patients with alcoholic liver disease. Br Med J (Clin Res Ed). 1983;287:1417–9.

Fayek SA, Quintini C, Chavin KD, Marsh CL. The current state of liver transplantation in the United States. Am J Transplant. 2016;16:3093–104.

Fernández J, Tandon P, Mensa J, Garcia-Tsao G. Antibiotic prophylaxis in cirrhosis: good and bad. Hepatology. 2016;63:2019–31.

Forrest EH, Evans CD, Stewart S, Phillips M, Oo YH, McAvoy NC, Fisher NC, et al. Analysis of factors predictive of mortality in alcoholic hepatitis and derivation and validation of the Glasgow alcoholic hepatitis score. Gut. 2005;54:1174–9.

Foucher J, Chanteloup E, Vergniol J, Castéra L, Le Bail B, Adhoute X, Bertet J, et al. Diagnosis of cirrhosis by transient elastography (FibroScan): a prospective study. Gut. 2006;55:403–8.

Frezza M, di Padova C, Pozzato G, Terpin M, Baraona E, Lieber CS. High blood alcohol levels in women. The role of decreased gastric alcohol dehydrogenase activity and first-pass metabolism. N Engl J Med. 1990;322:95–9.

Frriedman SL. Alcoholic hepatitis: clinical manifestations and diagnosis. uptodate.com 2020.

Galbicesk C. Drinking and drugs. Alcohol rehab guide. June ,2020.

Garcia-Pagán JC, Bosch J. Endoscopic band ligation in the treatment of portal hypertension. Nat Clin Pract Gastroenterol Hepatol. 2005;2:526–35.

García-Pagán JC, Caca K, Bureau C, Laleman W, Appenrodt B, Luca A, Abraldes JG, et al. Early use of TIPS in patients with cirrhosis and variceal bleeding. N Engl J Med. 2010;362: 2370–9.

Garcia-Tsao G. Current management of the complications of cirrhosis and portal hypertension: variceal hemorrhage, ascites, and spontaneous bacterial peritonitis. Dig Dis. 2016;34:382–6.

Garcia-Tsao G, Abraldes JG, Berzigotti A, Bosch J. Portal hypertensive bleeding in cirrhosis: risk stratification, diagnosis, and management: 2016 practice guidance by the American Association for the Study of Liver Diseases. Hepatology. 2017;65:310–35.

Gianni Testino AS, Borro P. Comment to " liver transplant for patients with alcohol liver disease": an open question. Dig Liver Dis. 2013;43(11):843–9.

Ginés P, Quintero E, Arroyo V, Terés J, Bruguera M, Rimola A, Caballería J, et al. Compensated cirrhosis: natural history and prognostic factors. Hepatology. 1987;7:122–8.

Griswold MG, Fullman N, Hawley C, Arian N, Zimsen SRM, Tymeson HD, Venkateswaran V, et al. Alcohol use and burden for 195 countries and territories, 1990-2016: a systematic analysis for the global burden of disease study 2016. Lancet. 2018;392:1015–35.

Guidance N. Alcohol- use disorders: diagnosis and management of physical complications. London: National Clinical Guidelines Centre at the Royal College of Physicians; 2010.

Hart CL, Morrison DS, Batty GD, Mitchell RJ, Davey SG. Effect of body mass index and alcohol consumption on liver disease: analysis of data from two prospective cohort studies. BMJ. 2010;340:c1240.

Hartmann P, Seebauer CT, Schnabl B. Alcoholic liver disease: the gut microbiome and liver cross talk. Alcohol Clin Exp Res. 2015;39:763–75.

Heidelbaugh JJ, Bruderly M. Cirrhosis and chronic liver failure: part I. diagnosis and evaluation. Am Fam Physician. 2006;74:756–62.

Herbst DA, Reddy KR. Risk factors for hepatocellular carcinoma. Clin Liver Dis (Hoboken). 2012;1:180–2.

Heyes CM, Schofield T, Gribble R, Day CA, Haber PS. Reluctance to accept alcohol treatment by alcoholic liver disease transplant patients: a qualitative study. Transplant Direct. 2016;2:e104.

Ho SB, Bräu N, Cheung R, Liu L, Sanchez C, Sklar M, Phelps TE, et al. Integrated care increases treatment and improves outcomes of patients with chronic hepatitis C virus infection and psychiatric illness or substance abuse. Clin Gastroenterol Hepatol. 2015;13(2005-2014):e2003.

Hrubec Z, Omenn GS. Evidence of genetic predisposition to alcoholic cirrhosis and psychosis: twin concordances for alcoholism and its biological end points by zygosity among male veterans. Alcohol Clin Exp Res. 1981;5:207–15.

Hughes NR, Houghton N, Nadeem H, Bell J, McDonald S, Glynn N, Scarfe C, et al. Salford alcohol assertive outreach team: a new model for reducing alcohol-related admissions. Frontline Gastroenterol. 2013;4:130–4.

Hydes T, Gilmore W, Sheron N, Gilmore I. Treating alcohol-related liver disease from a public health perspective. J Hepatol. 2019a;70:223–36.

Hydes T, Gilmore W, Sheron N, Gilmore I. Treating alcohol-related liver disease from a public health perspective. J Hepatol. 2019b;70:223–36.

Im G, Kim-Schluger L, Shenoy A, Schubert E, Goel A, Friedman S, Florman S, et al. Early liver transplantation for severe alcoholic hepatitis in the United States—a single-center experience. Am J Transplant. 2016;16:841–9.

Im GY, Cameron AM, Lucey MR. Liver transplantation for alcoholic hepatitis. J Hepatol. 2019;70:328–34.

Imbert-Bismut F, Ratziu V, Pieroni L, Charlotte F, Benhamou Y, Poynard T. Biochemical markers of liver fibrosis in patients with hepatitis C virus infection: a prospective study. Lancet. 2001;357:1069–75.

Infante-Rivard C, Esnaola S, Villeneuve JP. Clinical and statistical validity of conventional prognostic factors in predicting short-term survival among cirrhotics. Hepatology. 1987;7:660–4.

Ioannou GN, Green P, Lowy E, Mun EJ, Berry K. Differences in hepatocellular carcinoma risk, predictors and trends over time according to etiology of cirrhosis. PLoS One. 2018;13:e0204412.

Iruzubieta P, Crespo J, Fábrega E. Long-term survival after liver transplantation for alcoholic liver disease. World J Gastroenterol. 2013a;19:9198–208.

Iruzubieta P, Crespo J, Fábrega E. Long-term survival after liver transplantation for alcoholic liver disease. World J Gastroenterol. 2013b;19:9198–208.

Ishak KG, Zimmerman HJ, Ray MB. Alcoholic liver disease: pathologic, pathogenetic and clinical aspects. Alcohol Clin Exp Res. 1991;15:45–66.

Kamath PS, Wiesner RH, Malinchoc M, Kremers W, Therneau TM, Kosberg CL, D'Amico G, et al. A model to predict survival in patients with end-stage liver disease. Hepatology. 2001;33:464–70.

Kamper-Jørgensen M, Grønbaek M, Tolstrup J, Becker U. Alcohol and cirrhosis: dose--response or threshold effect? J Hepatol. 2004;41:25–30.

Karim Z, Intaraprasong P, Scudamore CH, Erb SR, Soos JG, Cheung E, Cooper P, et al. Predictors of relapse to significant alcohol drinking after liver transplantation. Can J Gastroenterol. 2010;24:245–50.

Kattimani S, Bharadwaj B. Clinical management of alcohol withdrawal: a systematic review. Ind Psychiatry J. 2013;22:100–8.

Kim WR, Biggins SW, Kremers WK, Wiesner RH, Kamath PS, Benson JT, Edwards E, et al. Hyponatremia and mortality among patients on the liver-transplant waiting list. N Engl J Med. 2008;359:1018–26.

Kondrup J, Rasmussen HH, Hamberg O, Stanga Z. Nutritional risk screening (NRS 2002): a new method based on an analysis of controlled clinical trials. Clin Nutr. 2003;22:321–36.

Kovalak M, Lake J, Mattek N, Eisen G, Lieberman D, Zaman A. Endoscopic screening for varices in cirrhotic patients: data from a national endoscopic database. Gastrointest Endosc. 2007;65:82–8.

Laine L, Cook D. Endoscopic ligation compared with sclerotherapy for treatment of esophageal variceal bleeding. A meta-analysis. Ann Intern Med. 1995;123:280–7.

Lane BP, Lieber CS. Ultrastructural alterations in human hepatocytes following ingestion of ethanol with adequate diets. Am J Pathol. 1966;49:593–603.

Lebrec D, Giuily N, Hadengue A, Vilgrain V, Moreau R, Poynard T, Gadano A, et al. Transjugular intrahepatic portosystemic shunts: comparison with paracentesis in patients with cirrhosis and refractory ascites: a randomized trial. French Group of Clinicians and a Group of Biologists. J Hepatol. 1996;25:135–44.

Leggio L, Lee MR. Treatment of alcohol use disorder in patients with alcohol liver disease. Am J Med. 2017;130:124–34.

Li S, Tan HY, Wang N, Feng Y, Wang X, Feng Y. Recent insights into the role of immune cells in alcoholic liver disease. Front Immunol. 2019;10:1328.

Liangpunsakul S, Haber P, McCaughan GW. Alcoholic liver disease in Asia, Europe, and North America. Gastroenterology. 2016;150:1786–97.

Lieber CS. Alcoholic fatty liver: its pathogenesis and mechanism of progression to inflammation and fibrosis. Alcohol. 2004;34:9–19.

Lim J, Sundaram V. Risk factors, scoring systems, and interventions for alcohol relapse after liver transplantation for alcoholic liver disease. Clin Liver Dis. 2018;11:105–10.

Loomba R, Yang HI, Su J, Brenner D, Barrett-Connor E, Iloeje U, Chen CJ. Synergism between obesity and alcohol in increasing the risk of hepatocellular carcinoma: a prospective cohort study. Am J Epidemiol. 2013;177:333–42.

Louvet A, Naveau S, Abdelnour M, Ramond MJ, Diaz E, Fartoux L, Dharancy S, et al. The Lille model: a new tool for therapeutic strategy in patients with severe alcoholic hepatitis treated with steroids. Hepatology. 2007;45:1348–54.

Lucey MR, Brown KA, Everson GT, Fung JJ, Gish R, Keeffe EB, Kneteman NM, et al. Minimal criteria for placement of adults on the liver transplant waiting list: a report of a national conference organized by the American Society of Transplant Physicians and the American Association for the Study of Liver Diseases. Liver Transpl Surg. 1997;3:628–37.

Lucey MR, Mathurin P, Morgan TR. Alcoholic hepatitis. N Engl J Med. 2009;360:2758–69.

Lucey MR, Im GY, Mellinger JL, Szabo G, Crabb DW. Introducing the 2019 American Association for the Study of Liver Diseases Guidance on alcohol-associated liver disease. Liver Transpl. 2020;26:14–6.

Lupsor-Platon M, Badea R. Noninvasive assessment of alcoholic liver disease using unidimensional transient elastography (Fibroscan(®)). World J Gastroenterol. 2015;21:11914–23.

MacSween RN, Burt AD. Histologic spectrum of alcoholic liver disease. Semin Liver Dis. 1986;6:221–32.

Maddrey WC, Boitnott JK, Bedine MS, Weber FL Jr, Mezey E, White RI Jr. Corticosteroid therapy of alcoholic hepatitis. Gastroenterology. 1978;75:193–9.

Malinchoc M, Kamath PS, Gordon FD, Peine CJ, Rank J, ter Borg PC. A model to predict poor survival in patients undergoing transjugular intrahepatic portosystemic shunts. Hepatology. 2000;31:864–71.

Mandrekar P, Ambade A. Immunity and inflammatory signaling in alcoholic liver disease. Hepatol Int. 2014;8(Suppl 2):439–46.

Mann RE, Smart RG, Govoni R. The epidemiology of alcoholic liver disease. Alcohol Res Health. 2003;27(3):209–19.

Mathurin P, Lucey MR. Liver transplantation in patients with alcohol-related liver disease: current status and future directions. Lancet Gastroenterol Hepatol. 2020;5:507–14.

Mathurin P, Moreno C, Samuel D, Dumortier J, Salleron J, Durand F, Castel H, et al. Early liver transplantation for severe alcoholic hepatitis. N Engl J Med. 2011;365:1790–800.

Matos Santana TE. The role of the addiction specialist in the liver transplant setting. Am J Psychiatry Resid J. 2016;11:4–5.

Mauer K, Manzione NC. Usefulness of serum-ascites albumin difference in separating transudative from exudative ascites. Another look Dig Dis Sci. 1988;33:1208–12.

McAvoy NC, Plevris JN, Hayes PC. Human thrombin for the treatment of gastric and ectopic varices. World J Gastroenterol. 2012;18:5912–7.

McCutheon V. Alcohol use as a hIGH-risk health behavior. Behavioral and Mental Health Novemeber, 2017.

Moriarty KJ. Alcohol care teams: where are we now? Frontline Gastroenterol. 2020;11:293.

Moriarty KJ, Platt H, Crompton S, Darling W, Blakemore M, Hutchinson S, Proctor D, et al. Collaborative care for alcohol-related liver disease. Clin Med. 2007;7:125.

Neil Raoriya DAG. Endoscopic Management of Oesophageal and Gastric Varices. Endoscopy of GI Tract. 2012;

Neuberger J, Gimson A, Davies M, Akyol M, O'Grady J, Burroughs A, Hudson M. Selection of patients for liver transplantation and allocation of donated livers in the UK. Gut. 2008;57:252–7.

Newsome PN, Cramb R, Davison SM, et al. Guidelines on the management of abnormal liver blood tests. Gut. 2018;67:6–19.

Ney M, Li S, Vandermeer B, Gramlich L, Ismond KP, Raman M, Tandon P. Systematic review with meta-analysis: nutritional screening and assessment tools in cirrhosis. Liver Int. 2020;40:664–73.

North Italian Endoscopic Club for the Study and Treatment of Esophageal Varices. Prediction of the first variceal hemorrhage in patients with cirrhosis of the liver and esophageal varices. A prospective multicenter study. N Engl J Med. 1988;319:983–9.

Osna NA, Donohue TM Jr, Kharbanda KK. Alcoholic liver disease: pathogenesis and current management. Alcohol Res. 2017;38:147–61.

Pageaux GP, Bismuth M, Perney P, Costes V, Jaber S, Possoz P, Fabre JM, et al. Alcohol relapse after liver transplantation for alcoholic liver disease: does it matter? J Hepatol. 2003;38:629–34.

Pagliaro L, D'Amico G. Portal hypertension in cirrhosis: natural history. In: Bosch J, Groszmann RJ, editors. Portal hypertension. Pathophysiology and treatment. Oxford: Blackwell Scientific; 1994.

Papastergiou V, Tsochatzis E, Burroughs AK. Non-invasive assessment of liver fibrosis. Ann Gastroenterol. 2012;25:218–31.

Parés A, Caballería J, Bruguera M, Torres M, Rodés J. Histological course of alcoholic hepatitis. Influence of abstinence, sex and extent of hepatic damage. J Hepatol. 1986;2:33–42.

Parlesak A, Schäfer C, Schütz T, Bode JC, Bode C. Increased intestinal permeability to macromolecules and endotoxemia in patients with chronic alcohol abuse in different stages of alcohol-induced liver disease. J Hepatol. 2000;32:742–7.

Pasala S, Barr T, Messaoudi I. Impact of alcohol abuse on the adaptive immune system. Alcohol Res. 2015;37:185–97.

Philips CAAP, Yerol PK, Rajesh S, Mahadevan P. Severe alcoholic hepatitis:current perspectives. Hepat Med. 2019;11:97–108.

Physicians R Co. Alcohol dependence and withdrawal in acute hospital. 2012.

Plauth M, Bernal W, Dasarathy S, Merli M, Plank LD, Schütz T, Bischoff SC. ESPEN guideline on clinical nutrition in liver disease. Clin Nutr. 2019;38:485–521.

Poynard T, Bedossa P, Opolon P. Natural history of liver fibrosis progression in patients with chronic hepatitis C. the OBSVIRC, METAVIR, CLINIVIR, and DOSVIRC groups. Lancet. 1997;349:825–32.

Rajoriya N, Forrest EH, Gray J, Stuart RC, Carter RC, McKay CJ, Gaya DR, et al. Long-term follow-up of endoscopic Histoacryl glue injection for the management of gastric variceal bleeding. QJM. 2011;104:41–7.

Rambaldi A, Saconato HH, Christensen E, Thorlund K, Wetterslev J, Gluud C. Systematic review: glucocorticosteroids for alcoholic hepatitis--a Cochrane Hepato-biliary group systematic review with meta-analyses and trial sequential analyses of randomized clinical trials. Aliment Pharmacol Ther. 2008;27:1167–78.

Rehm J, Shield KD. Global burden of alcohol use disorders and alcohol liver disease. Biomedicine. 2019;7:99.

Rice JP, Eickhoff J, Agni R, Ghufran A, Brahmbhatt R, Lucey MR. Abusive drinking after liver transplantation is associated with allograft loss and advanced allograft fibrosis. Liver Transpl. 2013;19:1377–86.

Riordan SM, Williams R. The intestinal flora and bacterial infection in cirrhosis. J Hepatol. 2006;45:744–57.

Rockey DC, Bissell DM. Noninvasive measures of liver fibrosis. Hepatology. 2006;43:S113–20.

Rosenberg WMC, Voelker M, Thiel R, Becka M, Burt A, Schuppan D, Hubscher S, et al. Serum markers detect the presence of liver fibrosis: a cohort study. Gastroenterology. 2004;127:1704–13.

Runyon BA. Management of Adult patients wtih ascites due to cirrhosis. Hepatology. 2012;49(6):2087–107.

Runyon BA, Montano AA, Akriviadis EA, Antillon MR, Irving MA, McHutchison JG. The serum-ascites albumin gradient is superior to the exudate-transudate concept in the differential diagnosis of ascites. Ann Intern Med. 1992;117:215–20.

Sachdeva A, Choudhary M, Chandra M. Alcohol withdrawal syndrome: benzodiazepines and beyond. J Clin Diagn Res. 2015;9:Ve01–7.

Salameh H, Raff E, Erwin A, Seth D, Nischalke HD, Falleti E, Burza MA, et al. PNPLA3 gene polymorphism is associated with predisposition to and severity of alcoholic liver disease. Am J Gastroenterol. 2015;110:846–56.

Saner FH, Canbay A, Gerken G, Broelsch CE. Pharmacology, clinical efficacy and safety of ter-lipressin in esophageal varices bleeding, septic shock and hepatorenal syndrome. Expert Rev Gastroenterol Hepatol. 2007;1:207–17.

Santos J, Planas R, Pardo A, Durández R, Cabré E, Morillas RM, Granada ML, et al. Spironolactone alone or in combination with furosemide in the treatment of moderate ascites in nonazotemic cirrhosis. A randomized comparative study of efficacy and safety. J Hepatol. 2003;39:187–92.

Sarin SK. Long-term follow-up of gastric variceal sclerotherapy: an eleven-year experience. Gastrointest Endosc. 1997;46:8–14.

Sarin SK, Sachdev G, Nanda R, Misra SP, Broor SL. Endoscopic sclerotherapy in the treatment of gastric varices. Br J Surg. 1988;75:747–50.

Sarin SK, Pande A, Schnabl B. Microbiome as a therapeutic target in alcohol-related liver disease. J Hepatol. 2019;70:260–72.

Schuppan D, Afdhal NH. Liver cirrhosis. Lancet. 2008;371:838–51.

Schwenzer NF, Springer F, Schraml C, Stefan N, Machann J, Schick F. Non-invasive assessment and quantification of liver steatosis by ultrasound, computed tomography and magnetic resonance. J Hepatol. 2009;51:433–45.

Sepanlou SG, Safiri S, Bisignano C, Ikuta KS, Merat S, Saberifiroozi M, Poustchi H, et al. The global, regional, and national burden of cirrhosis by cause in 195 countries and territories, 1990-2017: a systematic analysis for the global burden of disease study 2017. Lancet Gastroenterol Hepatol. 2020;5:245–66.

Sersté T, Cornillie A, Njimi H, Pavesi M, Arroyo V, Putignano A, Weichselbaum L, et al. The prognostic value of acute-on-chronic liver failure during the course of severe alcoholic hepatitis. J Hepatol. 2018;69:318–24.

Setshedi M, Wands JR, Monte SM. Acetaldehyde adducts in alcoholic liver disease. Oxidative Med Cell Longev. 2010;3:178–85.

Sharma P, Arora A. Clinical presentation of alcoholic liver disease and non-alcoholic fatty liver disease: spectrum and diagnosis. Transl Gastroenterol Hepatol. 2019;5

Sharma BC, Sharma P, Lunia MK, Srivastava S, Goyal R, Sarin SK. A randomized, double-blind, controlled trial comparing rifaximin plus lactulose with lactulose alone in treatment of overt hepatic encephalopathy. Am J Gastroenterol. 2013;108:1458–63.

Shaw JJ, Shah SA. Rising incidence and demographics of hepatocellular carcinoma in the USA: what does it mean? Expert Rev Gastroenterol Hepatol. 2011;5:365–70.

Shuai Z, Leung MW, He X, Zhang W, Yang G, Leung PS, Eric GM. Adaptive immunity in the liver. Cell Mol Immunol. 2016;13:354–68.

Singal AK, Bataller R, Ahn J, Kamath PS, Shah VH. ACG clinical guideline: alcoholic liver disease. Am J Gastroenterol. 2018;113:175–94.

Siqueira F, Kelly T, Saab S. Refractory ascites: pathogenesis, clinical impact, and management. Gastroenterol Hepatol. 2009;5:647–56.

Society BT UK Liver Transplant Group recommentations for alcohol-related liver disease. 2016.

Sørensen TA, Bentsen K, Eghøje K, Orholm M, Høybye G, Offersen P. Prospective evaluation of alcohol abuse and alcoholic liver injury in men as predictors of development of cirrhosis. Lancet. 1984;324:241–4.

Spycher C, Zimmermann A, Reichen J. The diagnostic value of liver biopsy. BMC Gastroenterol. 2001;1:12.

Staufer K, Yegles M. Biomarkers for detection of alcohol consumption in liver transplantation. World J Gastroenterol. 2016;22:3725–34.

Stewart SF, Day CP. The management of alcoholic liver disease. J Hepatol. 2003;38:2–13.

Stickel F, Moreno C, Hampe J, Morgan MY. The genetics of alcohol dependence and alcohol-related liver disease. J Hepatol. 2017;66:195–211.

Stirnimann G, Banz V, Storni F, De Gottardi A. Automated low-flow ascites pump for the treatment of cirrhotic patients with refractory ascites. Ther Adv Gastroenterol. 2017;10:283–92.

Sullivan JT, Sykora K, Schneiderman J, Naranjo CA, Sellers EM. Assessment of alcohol withdrawal: the revised clinical institute withdrawal assessment for alcohol scale (CIWA-Ar). Br J Addict. 1989;84:1353–7.

Tchelepi H, Ralls PW, Radin R, Grant E. Sonography of diffuse liver disease. J Ultrasound Med. 2002;21:1023–32.

Teli MR, Day CP, James OFW, Burt AD, Bennett MK. Determinants of progression to cirrhosis or fibrosis in pure alcoholic fatty liver. Lancet. 1995;346:987–90.

Thomson AD, Marshall EJ. BNF recommendations for the treatment of Wernicke's encephalopathy: lost in translation? Alcohol Alcohol. 2013;48:514–5.

Thomson AD, Cook CCH, Touquet R, Henry JA. THE Royal College of Physicians report on alcohol: guidelines for managing Wernicke's encephalopathy in THE accident and emergency department. Alcohol Alcohol. 2002;37:513–21.

Thursz MR, Richardson P, Allison M, Austin A, Bowers M, Day CP, Downs N, et al. Prednisolone or Pentoxifylline for alcoholic hepatitis. N Engl J Med. 2015;372:1619–28.

Thursz M, Gual A, Lackner C, Mathurin P, Moreno C, Spahr L, Sterneck M, et al. EASL clinical practice guidelines: management of alcohol-related liver disease. J Hepatol. 2018a;69:154–81.

Thursz M, Gual A, Lackner C, Mathurin P, Moreno C, Spahr L, Sterneck M, Cortez-Pinto H. EASL clinical practice guidelines: management of alcohol-related liver disease. J Hepatol. 2018b;69:154–81.

Tripathi D, Stanley AJ, Hayes PC, Patch D, Millson C, Mehrzad H, Austin A, et al. UK guidelines on the management of variceal haemorrhage in cirrhotic patients. Gut. 2015a;64:1680.

Tripathi D, Stanley AJ, Hayes PC, Patch D, Millson C, Mehrzad H, Austin A, et al. U.K. guidelines on the management of variceal haemorrhage in cirrhotic patients. Gut. 2015b;64:1680–704.

Tuma DJ, Casey CA. Dangerous byproducts of alcohol breakdown—focus on adducts. Alcohol Res Health. 2003;27:285.

UK Chief Medical Officer's low risk drinking guidelines. August 2016.

Vaillant GE. A 60-year follow-up of alcoholic men. Addiction. 2003;98:1043–51.

Vilstrup H, Amodio P, Bajaj J, Cordoba J, Ferenci P, Mullen KD, Weissenborn K, et al. Hepatic encephalopathy in chronic liver disease: 2014 practice guideline by the American Association for the Study of Liver Diseases and the European Association for the Study of the Liver. Hepatology. 2014;60:715–35.

WHO. Management of alcohol withdrawal. WHO mental health Gap Action Programme (mhGAP), evidence resource centre, Alcohol use disorders 2012.

WHO. Alcohol. 2018a.

WHO. Global status report on alcohol and health 2018. 2018b.

Wiest R, Krag A, Gerbes A. Spontaneous bacterial peritonitis: recent guidelines and beyond. Gut. 2012;61:297.

Wood AM, Kaptoge S, Butterworth AS, Willeit P, Warnakula S, Bolton T, Paige E, et al. Risk thresholds for alcohol consumption: combined analysis of individual-participant data for 599-912 current drinkers in 83 prospective studies. Lancet. 2018;391:1513–23.

Yang WL, Tripathi D, Therapondos G, Todd A, Hayes PC. Endoscopic use of human thrombin in bleeding gastric varices. Am J Gastroenterol. 2002;97:1381–5.

Zakhari S, Li TK. Determinants of alcohol use and abuse: impact of quantity and frequency patterns on liver disease. Hepatology. 2007;46:2032–9.

Zhou Z, Wang L, Song Z, Lambert JC, McClain CJ, Kang YJ. A critical involvement of oxidative stress in acute alcohol-induced hepatic TNF-alpha production. Am J Pathol. 2003;163:1137–46.

Zhou X, Tripathi D, Song T, Shao L, Han B, Zhu J, Han D, et al. Terlipressin for the treatment of acute variceal bleeding: a systematic review and meta-analysis of randomized controlled trials. Medicine (Baltimore). 2018;97:e13437.

Chapter 12
Recreational Drugs and Men's Health

Thilakavathi Chengodu and Nathan Lawrentschuk

12.1 Introduction

The use of recreational drugs has detrimental side effects on the lives of its users. Its users, men in this context, are often not aware and less tuned to its side effects which extend to male reproductive and sexual health.

Some of these effects include erectile dysfunction, or the inability to maintain penile erection, can be caused in the short- and long-term by nearly every drug mentioned in this chapter: including cigarettes, alcohol, cannabis, opioids, amphetamines, marijuana, cocaine, and steroids (Kulkarni et al. 2014). Other health issues include infertility, hypogonadism, Fournier's gangrene, genital mutilation syndrome, priapism, and genital infections (Haney et al. 2018).

This chapter will define what recreational drugs are and discuss its use in the context of men's health and how are health professionals able to assist in management of arising health issues for men.

T. Chengodu (✉)
EJ Whitten Foundation Prostate Cancer Research Centre, Epworth HealthCare, Richmond, VIC, Australia
e-mail: Thili.Chengodu@epworth.org.au

N. Lawrentschuk
EJ Whitten Foundation Prostate Cancer Research Centre, Epworth HealthCare, Richmond, VIC, Australia

Department of Urology, Royal Melbourne Hospital, Parkville, Australia

Department of Surgery, University of Melbourne, Parkville, Australia

© The Author(s), under exclusive license to Springer Nature Switzerland AG 2022
S. S. Goonewardene et al. (eds.), *Men's Health and Wellbeing*,
https://doi.org/10.1007/978-3-030-84752-4_12

12.2   What Are Recreational Drugs?

A formal definition of recreational drug is a *'drug (such as cocaine, marijuana, or methamphetamine) used without medical justification for its psychoactive effects often in the belief that occasional use of such a substance is not habit-forming or addictive'*(Merriam-Webster Dictionary 2021). Thus, it is a drug that is taken for pleasure and enjoyment, not for medical purposes. Recreational drugs induce an altered state of mind, by changing the perceptions, feelings, and emotions of any user. Whilst there are legal drugs for example alcohol and tobacco, most recreational drugs are illegal and used without medical supervision. There are four categories of recreational drugs at present which are analgesics, depressants, stimulants and hallucinogens, see Table 12.1.

12.3   The Impact of Drug Use on Health

The impact of drug use on an individual's health and wellbeing is extremely detrimental. It does not only affect the individual but infiltrates families, communities and becomes global (World Health Organization 2021). The far fetching consequences of recreational drugs use is now more prevalent than ever. There is a strong correlation between drug use and mental health disorders; people experiencing adverse health consequences including non-fatal overdoses, infectious diseases such as HIV and Hepatitis (World Health Organization 2021) and premature death and there is evidence of impact on social development of an individual (Schifano 2020) The 2020 National Institute on Drug Abuse (NIDA) Research Report reports that men are more likely than females to use almost all types of recreational drugs and have a higher dependence rate than women. The Australian Institute of Health and Welfare states that the consumption of alcohol and tobacco is a major cause of preventable disease and illness in Australia as is the non-medical use of pharmaceutical drugs.

Table 12.1 Categories of recreational drugs

Analgesics	Heroin; Morphine; fentanyl; Codeine
Depressants	Alcohol; barbiturates; Tranquilizers; nicotine
Stimulants	Cocaine; Methamphetamine; ecstasy
Hallucinogens	LSD (acid); Peyote (mescaline); Psilocybin (magic mushrooms), marijuana, ketamine, phencyclidine (PCP)

Adapted from (Mann 2016)

Table 12.2 Summary of drug abuse effect on health

Substance of abuse	Sexual health dysfunctions
Alcohol	Erectile dysfunction Hypogonadism Infertility Hypertension Stroke Liver damage Cancers Gout
Tobacco	Erectile dysfunction Infertility
Cocaine	Erectile dysfunction Priapism Fournier's gangrene
Marijuana	Erectile dysfunction
Heroin	Erectile dysfunction Fournier's gangrene Hypogonadism Infertility
Amphetamines	Increased susceptibility to genital herpes Penile abscesses with intra penile injection

The far-reaching consequences of illicit drug use and high consumption of alcohol and tobacco on male reproduction and sexual health is well documented and researched by health practitioners. Table 12.2 summarises some of the effects on health.

12.4 Alcohol

One of the reasons alcohol is up front is, unfortunately, the culture around alcohol use. The availability of alcohol use is enormous. And as a substance, it affects pretty much every system in the body.
–Dr Paul Grinzi, GP

Alcohol has vast consequences which is not widely known to patients but well known to health practitioners. Across most age groups around the world, it is men who have the higher rate of substance abuse within recreational drugs. The last National Health Survey conducted by Australia in 2018 still shows that more than one if 5 men still consumed more alcohol and exceeded their lifetime risk (Australian Institute of Health and Welfare 2020).

In the United Kingdom, males consumed far more alcohol on average per week compared to females in the respective age groups The Centres for Disease Control and Prevention in the United States of America state that men almost are at 2 times likely to binge drink with 22% of men reporting binge drinking and on average 5 times a month.

The harmful use of alcohol result in harm to oneself and other people, such as family members, friends, co-workers and strangers. Drinking alcohol is associated with a risk of developing health problems such as mental and behavioural disorders, including alcohol dependence, diseases such as liver cirrhosis, pancreatitis, some cancers and cardiovascular diseases and injuries resulting from violence and motor vehicle accidents (World Health Organization 2021).

In men, drinking alcohol leads to erectile dysfunction, premature ejaculation and often lower libido (Kulkarni et al. 2014; Sansone et al. 2018; Haney et al. 2018).

Consumption of alcohol excessively and over a long period affects fertility by

- lowering testosterone levels, follicle stimulating hormone, and luteinizing hormone, and raising estrogen levels, which reduce sperm production
- shrinking the testes, which can cause impotence or infertility
- changing gonadotropin release which impacts sperm production
- causing early ejaculation or decreased ejaculation
- changing the shape, size, and movement of healthy sperm

(Ajayi and Akhigbe 2020; Haney et al. 2018)

Alcohol in moderation and minimising binge drinking will be the key to better health outcomes. In Australia, the Royal Australian College of General Practitioners, (RACGP) has adopted the National Health and Medical Research (NHMRC) guidelines to reduce health risks from alcohol.

There are targeted campaigns aimed at reducing alcohol intake, for example, *'Dry July'* which is aimed at going alcohol free in the month of July to raise funds for cancer. The *'Alcohol.Think Again'* is a public education program informing Western Australian community of alcohol-related harms, ways to minimise alcohol-related risk, and raises awareness of the national low-risk drinking guidelines. This campaign focusses on the damaging effects of alcohol and promotes the NHMRC guideline on low-risk drinking levels to reduce the risk of long-term harm from alcohol.

DrinkWise (https://drinkwise.org.au/) is a non-for-profit organisation that was established by the alcohol industry of Australia in 2005. It aims to promote a safe drinking culture for all ages.

These various promotions and information sources help health practitioners to successfully navigate better health outcomes for their patients and clients affected by alcohol.

12.5   Tobacco Smoking

In Australia, men consume more tobacco products than females and is the leading cause of death and disease in Australia, killing an estimated 18,762 people annually (AIHW 2020). About 1 in 6 men still smoke and two–thirds of these smokers can be expected to die because of their tobacco use if they do not quit. The impact of smoking to the body is detrimental. It causes various health issues and leads to being at high risk for cancers. It is the same for the United Kingdom where some 78,000 people die annually according to the National Health Service (NHS) and it remains the biggest cause of death and illness. This is also the same for Asia, particularly where there is a cultural aspect attached to men smoking, making them more masculine and adult (WHO 2019). The health impacts of smoking are discussed Table 12.3.

Table 12.3 Some of the health impacts of smoking

Fertility	• The 2020 US Surgeon General's report found evidence that exposure to tobacco smoke either *in utero* or in adulthood decreased semen quality and fertility. The report found consistent evidence linking smoking to DNA damage in sperm, adversely affecting male fertility and pregnancy viability as well as anomalies in offspring.
Sexual function	• Smoking causes problems with getting or maintaining an erection due to the effects of smoking on blood flow and damage to the blood vessels of the penis
Urinary system	• Bladder cancer is inextricably linked with tobacco (around 70% cases) and with more male smokers this is significant. • Urinary symptoms of frequency, urgency and hematuria should always be taken seriously in any smoker
Muscle and bone density	• In smokers, carbon monoxide replaces some of the oxygen in the blood, making it harder for oxygen to transfer into muscle cells. Reduced oxygen in muscles causes faster tiredness. • Smoking gradually decreases your bone density in your middle and later years. This leads to low bone density increasing the risk of hip fractures.
Cardiovascular	• 30% of heart disease in those under 65 years of age is caused by smoking • Smoking temporarily raises heart rate and blood pressure, while reducing the ability of blood to carry oxygen, making the walls of the blood vessels sticky, causing a build-up of fatty deposits • Smokers have 2–3 times the risk of suffering sudden cardiac death compared with non-smokers. • Vascular diseases
Lungs	• 80% of lung cancer cases are due to smoking, making it the most common form of cancer • Cigarette smoke contains more than 7000 chemicals that interfere with the body's method of filtering air and cleaning out the lungs. More than 70 of these can cause cancer. • Cancers caused by smoking are mouth, throat, oesophagus, bladder, bowel and kidney to name a few. • COPD as well as pneumonia are other conditions leading from smoking. • Smoking also exacerbates asthma and existing lung conditions
Skin	• smokers tend to develop face wrinkles earlier than non-smokers • Smoking reduces blood flow to the skin and may damage tissues (collagen and elastin) that help keep skin looking young. • Smoking is also linked to a range of skin problems, such as psoriasis.

(Australian Institute of Health and Welfare 2020; National Institute on Drug Abuse 2020b; Royal Australian College of General Practitioners 2019; U.S Department of Health and Human Services 2020; World Health Organization 2019)

12.6   Strategies of Managing Smoking

The strategies are very simple, it is to reduce or to stop smoking entirely to cease or minimise harm from smoking (See Table 12.4).

Immediate beneficial health changes observed include drop in heart rate and blood pressure and within 12 hours, the carbon monoxide level in the blood drops a one will notice improvement in blood circulation and lung functionality increasing (AIHW 2020). Long term benefits include, reduction of risk for coronary heart disease, stroke and cancers. Of course, not forgetting that reducing or stopping smoking will reduce the chances of impotence and better-quality sperms (WHO 2019).

Health professionals from all disciplines can play an important role in supporting smoking cessation by identifying people who smoke and be able to offer them cessation referral or advice at all opportunity (National Institute on Drug Abuse 2020b). Therefore, health professionals should be proactive in discussing tobacco use in patients' or clients' life where it is relevant and when presenting with tobacco related conditions. Various interventions should be tried and be tailored to the patient's needs to ensure that patient will stay the path to smoking cessation.

Nicotine containing e-cigarettes are not first line treatment for smoking cessation as recommended by the Royal Australian College of General Practitioners (RACGP 2019). A lack of approved nicotine-containing e-cigarettes products creates ambiguity in its use as well its combined effect with alcohol and cigarettes which may lead to further adverse health effects (Jain et al. 2021). Although, it is by no means to discourage a patient if they bring this intervention up. As with all treatments, patient needs to be fully informed to the intervention through shared decision-making process in regards to the availability and efficacy of e-cigarettes.

(RACGP 2019)

It is very important and crucial that a systematic and holistic approach is taken when implementing smoking cessation intervention. This will allow the patient or client to engage, establishing a trusting relationship and stay on the path of cessation.

Table 12.4   Treatment for tobacco dependence

Nicotine replacement therapy (NRT)	Lozenges, spray, patches and gum
Medication	Varenicline / bupropion
Combination of med/NRT	Addition of either Varenicline or bupropion with NRT works well for some individuals
Cognitive Behavioural therapy	Identifying triggers Coping strategies Relaxation techniques
Hypnotherapy	Hypnosis
Motivational interview	Focused on non-confrontation Support self efficacy and optimism
Telephone support	Available 24 hours
Web services	Social media programs: Finding support groups via Facebook

12.7 Other Drugs

According to the 2019 National Drug Strategy Household Survey, 16.4% of Australians had used an illicit drug. Similar to alcohol and tobacco, there are huge implications to the health of an individual as well as the burden of disease to a community. Illicit drugs are illegal drugs that are banned from manufacture, sale or possession.

Common illicit drugs like cocaine, cannabis and ketamine will be discussed in this section. Exposure to these drugs play a vital role in male fertility. The use of recreational drugs and abuse of prescription drugs add to the complexity of care for the male with sexual and or reproductive issues. Health practitioners have to take note of any use of recreational drug use when assessing the aetiology of male fertility.

12.8 Cocaine

Cocaine stimulates the central and peripheral nervous systems thus affecting mood and behaviour. Initial use of cocaine enhances libido, however long-term affects are decreased libido and erectile as well as ejaculatory dysfunction (Blundell et al. 2018; Cosci and Chouinard 2020; National Institute on Drug Abuse 2020c). Therefore, brain function decreases leading to poor decision making and the inability to overcome the negative consequences of cocaine abuse (Table 12.5).

Reduction of blood flow in the GI tract can lead to ulcerations and chronic use of cocaine leading to appetite loss resulting in poor diet and weight loss. Due to cocaine's toxic effects on the heart, chest pains are often experienced by the users and there is increased risk of stroke (National Institute on Drug Abuse 2020c).

Another long-term effect of chronic users of cocaine is associated with the abnormalities of sperm concentration, motility and its morphology (Ajayi and Akhigbe 2020).

Table 12.5 Effects on body systems

• Reduces blood flow in the gastrointestinal tract (GI)
• Loss of appetite
• Drastic weight loss
• Malnourishment
• Increased risk of stroke
• Chest pains due to toxic effects to the heart.
• Hyperprolactinaemia and hypotestosteronaemia, which inhibit spermatogenesis

12.9   Marijuana/Cannabis

The World Health Organization (WHO) reports that cannabis is the most widely used illicit drug in the world, representing half of all drug seizures worldwide with around 2.5% of the world's population using marijuana each year (2021). It is currently the most commonly used illegal drug in the United States (NIDA 2020). The 2019 The National Survey on Drug Use and Health (NSDUH) report indicated that more than 12 million young adults (ages 18 to 25) used marijuana in the past year. AIHW reports that between 2016 and 2019, there has been an increase in the use of cannabis and it is the most widely used illicit drug in Australia.

Marijuana is often smoked in hand rolled cigarettes, in pipes and in emptied cigars and available as a brew of tea and also utilised in powder form in biscuits and cakes. Lately, vaporisers have been utilised to avoid inhaling smoke (National Institute on Drug Abuse 2020a). The effects of Marijuana can be experienced immediately after it is smoked and as it enters the bloodstream, often a pleasant euphoria, sense of relaxation, heightened sensory perception, altered sense of time and often increased appetite as well (Smith and Smith 2021).

However, there is also the effects when an individual takes too much which leads to people experiencing anxiety and fear, leading to acute psychosis (Smith and Smith 2021). In this instance the person may experience hallucinations and delusions. There is research indicating a link between the use of cannabis and mental health disorders (National Institute on Drug Abuse 2020a; Schifano 2020).

Further to the above effect is the effect of Marijuana on male fertility. Tetrahydrocannabinol (THC), the active ingredient in cannabis, reduces the serum level of LH, leading to suppression of testosterone, and spermatogenesis, resulting in oligospermia (Ajayi and Akhigbe 2020; Kulkarni et al. 2014). Marijuana can also inhibit smooth muscle relaxation of penile tissue, leading to erectile dysfunction and chronic use is reported to cause early endothelial damage, possibly predisposing men to develop vascular erectile dysfunction.

12.10   Ketamine

Ketamine is used by medical practitioners and veterinarians as an anaesthetic but this drug is now also hailed as a breakthrough for some cases of major depression due to its rapid action.

Ketamine effects are mainly around feeling happy and relaxed and having the feel of being detached from your body, often described as 'falling into a k-hole'. Due to its rapid effects, it has become a popular recreational drug. An individual may experience hallucinations and lowered sensitivity to pain to confusion, anxiety, slurred speech and sometimes vomiting (Cosci and Chouinard 2020). In large amounts, one can experience rigid muscles, convulsions and often death if medical attention is not sought immediately.

One of the consequences of using ketamine regularly and long term is developing ketamine bladder syndrome (Schifano 2020). The symptoms of this syndrome are

- Incontinence
- Urgency
- Increased bladder sensation
- Pelvic and bladder pain
- Haematuria

Apart from the bladder syndrome, there is also indication that men suffer from erectile dysfunction, a large-scale questionnaire study conducted in Taiwan reported that up to 30% of men abusing Ketamine had erectile dysfunction (Yang et al. 2018). Further to this ketamine has been reported to reduce sperm motility and its viability (Ajayi and Akhigbe 2020).

12.11 Management of Recreational Drug Use

The management of drug use requires earliest intervention possible and health professionals to be able to identify the issue at hand. The main aim of management is to ensure that the problem is diagnosed correctly and referrals made accordingly to ensure that the individual feels safe and able to trust the process. Good links with referral with the multidisciplinary team will be crucial in assessing, identification and management of any issues. Additionally any intervention need to be culturally competent to enable men of all ethnic and sexual orientation to utilise safely and without prejudice.

The models of dealing with substance abuse have evolved from strict "Rehabilitation Facilities" focused on abstinence to more nuanced and holistic approaches considering the social and contextual underlying causes of abuse. Approaches from inpatient programs over three to four weeks to transitional "dayhab" where men can enter the community with supports. These programs are tailored to men in terms of supporting them emotionally, physiologically, spiritually and socially. Important to recognise that families provide much needed support network for men experiencing problems with recreational drug use and also need caring themselves. In Australia, each state has a drug and alcohol support services tailored for men, families and young people, for example in Victoria there is Better Health Channel that is easily accessible.

Of course, government funding is lacking and financial barriers to obtaining such assistance are often cited as the main reason of poor access to such programs. In Australia, the state and territory governments fund most of the rehabilitation services, however many of these programs are run by non-government organizations and charities like Salvation Army.

12.12 Mental Health Impacts

There is a growing issue with mental health problem and recreational drug use which cannot be ignored. The prolonged use of drug affects the parts of the brain which could be the decision-making process and emotions which then later develop to mental illness. For example, an individual taking marijuana throughout his youth could increase his risk of psychosis in his adulthood. There are also trauma related events and stress that can contribute to continued drug use.

Mental illness itself may contribute to drug abuse and addiction because individuals with mental health disorders may use drugs as a form of self-medication (Substance Abuse and Mental Health Services Administration 2020) For its immediate use, there is relief from their symptoms, however, in the long run this may exacerbate their symptom. Also, the constant use of substance can lead to changes to already disrupted areas of the brain such as schizophrenia especially if there is an underlying predisposition to develop that mental illness (Schifano 2020).

It is imperative that the health system involve all levels of health care professionals who are trained in identifying the issue and able to action immediately. As well as this that there be continued government funding to ensure training is provided to build capacity work force, particularly introducing this as a training program for health professional graduates and including mental health in curriculum across health disciplines.

12.13 Conclusion

The chapter has given a brief snapshot of some of the drugs used recreationally by men. With men seeking healthcare less often than women one may ask if substance abuse is at least partially a side effect of such behaviour but also ingrained in the "risk-taking" behaviours of adolescent men who then engage in substance abuse. The lack of healthcare seeking means dealing with substance abuse is not timely.

Broader initiatives to prevent substance abuse start from empowering our children and youth in regards to the dangers and effects of any use of recreational drugs. Schools need to be empowered to develop and deliver ongoing sustainable drug education. Teachers will often be best placed to assist with delivering knowledge and guide the youth, therefore there needs to be more training and support provided to teachers in the first instance.

Research is continuously being conducted to establish definitive links to fertility although there are indications that sexual health is impacted with long term use of recreational drugs. Are we able to reverse these effects by stopping the drug use? That certainly requires more research. However, with a multitude of health professionals and more government strategies and policies in place, the general advice is to reduce the use or stop the use and continue a healthy lifestyle.

References

Ajayi AF, Akhigbe RE. The physiology of male reproduction: impact of drugs and their abuse on male fertility. Andrologia. 2020;52:e13672.

Australian Institute of Health and Welfare. Alcohol, tobacco and other drugs in Australia (Online). Australia. 2020. Available: https://www.aihw.gov.au/reports/alcohol/alcohol-tobacco-other-drugs-australia/contents/introduction. Accessed 20 Dec 2020.

Blundell MS, Dargan PI, Wood DM. The dark cloud of recreational drugs and vaping. QJM. 2018;111:145–8.

Cosci F, Chouinard G. Acute and persistent withdrawal syndromes following discontinuation of psychotropic medications. Psychother Psychosom. 2020;89:283–306.

Haney NM, Diao L, Delay K. Drugs of abuse: men's reproductive and sexual health. In: Sikka SC, Hellstorm JGW, editors. Bioenvironmental issues affecting Men's reproductive and sexual health. London: Elsevier-Academic Press; 2018.

Jain V, Rifai MA, Naderi S, Barolia R, Iqbal S, Taj M, Jia X, Merchant AT, Aronow WS, Morris PB, Virani SS. Association of Smokeless Tobacco use with the use of other illicit drugs in the United States. Am J Med. 2021;134:e15–9.

Kulkarni M, Hayden C, Kayes O. Recreational drugs and male fertility. Trends Urol Men's Health. 2014;5:5.

Mann H. What are recreational drugs? BMJ. 2016;353:i2775.

Merriam-Webster Dictionary. Recreational drug (Online). 2021. Available: https://www.merriam-webster.com/dictionary/recreational%20drug. Accessed 14 Mar 2021.

National Institute on Drug Abuse. Marijuana research report (Online). 2020a. Available: https://www.drugabuse.gov/publications/research-reports/marijuana/what-scope-marijuana-use-in-united-states. Accessed 6 Mar 2021.

National Institute on Drug Abuse. What are treatments for tobacco dependence (Online). 2020b. Available: https://www.drugabuse.gov/publications/research-reports/tobacco-nicotine-e-cigarettes/what-are-treatments-tobacco-dependence. Accessed March 9, 2021.

National Institute on Drug Abuse. What is cocaine? (Online). 2020c. Available: https://www.drugabuse.gov/publications/research-reports/cocaine/what-cocaine. Accessed 1 Mar 2021.

Royal Australian College of General Practitioners. Supporting smoking cessation: a guide for health professionals. 2nd ed. East Melbourne: Royal Australian College of General Practitioners; 2019.

Sansone A, di Dato C, de Angelis C, Menafra D, Pozza C, Pivonello R, Isidori A, Gianfrilli D. Smoke, alcohol and drug addiction and male fertility. Reprod Biol Endocrinol. 2018;16:3.

Schifano F. Coming off prescribed psychotropic medications: insights from their use as recreational drugs. Psychother Psychosom. 2020;89:274–82.

Smith GL, Smith K. Fast facts about medical cannabis and opioids: minimizing opioid use through cannabis. New York: Springer Publishing Company, LLC; 2021.

Substance Abuse and Mental Health Services Administration. 2020. Key substance use and mental health indicators in the United States: results from the 2019 national survey on drug use and health. Rockville.

U.S Department of Health and Human Services. 2020. Smoking cessation. A report of the surgeon general Atlanta, GA.

World Health Organization. WHO report on the global tobacco epidemic, 2019: offer to help quit tobacco use: executive summary (Online). 2019. Available: https://apps.who.int/iris/handle/10665/325968. Accessed 6 Feb 2021.

World Health Organization. Alcohol, drugs and addictive behaviours unit (Online). 2021. Available: https://www.who.int/teams/mental-health-and-substance-use/alcohol-drugs-and-addictive-behaviours/drugs-psychoactive/cannabis. Accessed 6 Feb 2021.

Yang SS-D, Jang M-Y, Lee K-H, Hsu W-T, Chen Y-C, Chen W-S, et al. Sexual and bladder dysfunction in male ketamine abusers: A large-scale questionnaire study. PLoS ONE 2018;13(11):e0207927. https://doi.org/10.1371/journal.pone.0207927.

Chapter 13
Benign Surgical Conditions

Vaisnavi Thirugnanasundralingam, Robert Tasevski, and Nathan Lawrentschuk

13.1 Gynaecomastia

13.1.1 Epidemiology

Gynaecomastia is the benign palpable enlargement of breast tissue in the male. True gynaecomastia results from enlargement of glandular breast tissue and is different from pseudogynaecomastia, which is the result of excess adipose accumulation in the breast region.

Asymptomatic palpable breast tissue is common in normal males, particularly in the neonate, at puberty and with increasing age. 30–40% of the general adult male population is affected by gynaecomastia (Weingertner et al. 2019). Histopathological evidence of gynaecomastia was confirmed in 40% of males in an autopsy series study, demonstrating that gynaecomastia can be a normal physiological consequence of ageing (Williams 1963).

Gynaecomastia is common in infancy, puberty, and in middle-aged to older men. In the neonate, the gynaecomastia is a result of placental transfer of the maternal oestrogens. The gynaecomastia seen during puberty occurs because the adolescent's production of oestrogen begins prior to testosterone production. In early puberty, there is

V. Thirugnanasundralingam
Department of Urology, Royal Melbourne Hospital, Melbourne, Australia

R. Tasevski (✉)
Department of General Surgery, Royal Melbourne Hospital, Melbourne, Australia

Department of Surgery, University of Melbourne, Melbourne, Australia

N. Lawrentschuk
Department of Urology, Royal Melbourne Hospital, EJ Whitten Prostate Cancer Research Centre at Epworth, Melbourne, Australia

Department of Surgery, University of Melbourne, Melbourne, Australia a period of imbalance between the amount of oestrogen compared with androgens. Finally, in the ageing male, as the production of testosterone decreases, and increasing proportions of muscle are replaced with adipose tissue, more conversion of androgens to oestrogen occurs in the peripheral adipose tissue. This, as well as the use of medications that may induce gynaecomastia, and an increase in the prevalence of co-morbid medical conditions, explains the high prevalence of this condition in older men.

In a study of 237 men with gynaecomastia, the commonest aetiologies implicated were: idiopathic (45.1%), use of anabolic steroids (13.9%), hypogonadism (11%), medication-related (7.8%), puberty (6.2%), hyperprolactinaemia (5.7%), marijuana use (3.3%), renal failure (2.1%), hyperthyroidism (2.1%) and liver impairment (1.2%) (Costanzo et al. 2018).

Gynaecomastia can also be pathologically induced by a variety of mechanisms including the use of certain medications (such as anabolic steroids, androgen deprivation therapy), organ impairment (e.g. liver impairment) and other specific pathology (such as hypogonadism) which will be explored in this section.

13.1.2 Anatomy

The male breast gland is shaped like a small 'disc' within the superficial fascia and is anchored by Cooper's ligaments to the pectoral fascia. It consists of small ducts, which coalesce into the apex of the nipple, and is surrounded by fibroadipose tissue. In adult males without gynaecomastia, the glandular tissue of the breast does not extend to beyond the areola and is not visible (Cordova and Tripoli 2020).

The nipple-areola complex is located laterally and more superior in males than in females, approximately 4–5 cm above the chest fold. This complex contains many sensory nerve endings, smooth muscle and a subareolar lymphatic plexus (Cordova and Tripoli 2020). The position of the nipple-areola complex relative to surrounding structures is of crucial importance in the aesthetics of the chest and is at risk of disruption/displacement during surgical approaches for gynaecomastia, given the glandular 'disc' of tissue is located immediately underneath this complex.

The blood supply to the breast is derived during embryological development, and hence, is similar to the segmental pattern as seen along the thoracic wall. Branches of the internal thoracic, lateral thoracic, anterior intercostal and thoraco-acromial arteries supply the breast, though there may be considerable variation between patients (Bordoni et al. 2018).

Of utmost importance in the surgical approach to gynaecomastia are the cutaneous perforators in the anterior chest wall of males. These are located in the first, second and third intercostal spaces, between the anterior axillary line and the sternum, and are branches of the internal mammary artery and the anterior intercostal arteries. These perforating cutaneous vessels provide supply to the skin overlying the anterior chest (Cordova and Tripoli 2020).

In the surgical management of gynaecomastia, important anatomy is encountered surrounding the nipple-areola complex, where incisions may be made during surgery. The perforating branches of the internal thoracic artery are the most reliable sources of blood supply to the nipple areola complex and usually pass laterally above and below this complex before anastomosing with branches of the lateral thoracic artery. Alternatively, the branches may pass obliquely around the nipple-areolar complex. Of note, a radial pattern of supply is seen in approximately 6%, where the nipple-areolar complex would be at risk of necrosis if a peri-areolar incision were to be made. (Bordoni et al. 2018).

13.1.3 Pathophysiology

Male breast tissue growth is enhanced by oestrogens and suppressed by androgens and possesses receptors for both types of hormones. Therefore, gynaecomastia may develop as a result of an absolute or relative deficiency of androgens, insufficient androgenic activity, or an increase in the level of oestrogen or oestrogenic activity (Narula and Carlson 2014).

It is crucial to understand the components of oestradiol and testosterone production, conversion and activity to comprehend the numerous ways in which an alteration in this system may lead to gynaecomastia.

- The synthesis of sex hormones is controlled by the pulsatile release of gonadotropin releasing hormone (GnRH) by the hypothalamus.
- GnRH stimulates the anterior lobe of the pituitary gland to secrete luteinizing hormone (LH) and follicle stimulating hormone (FSH) into the blood stream.
- FSH and LH stimulate Leydig cells in the testis to secrete testosterone and approximately 15% of the total oestradiol found in males.
- Both testosterone and oestradiol are carried in blood, bound to a protein called sex-hormone binding globulin (SHBG) or to albumin, and only a small percentage circulate freely in the blood. Note that testosterone binds to SHBG with greater avidity than oestradiol, this becomes important when understanding how spironolactone causes gynaecomastia.
- The "bioavailable" portion of these hormones are limited to only those bound to albumin and those circulating freely in the blood.
- The bioavailable testosterone and oestradiol enter a variety of tissues, where the aromatase enzyme complex converts some of the testosterone into oestradiol.
- In males, the aromatase enzyme complex is found in the testis, bone, skin, muscle and adipose tissues.
- The zona reticularis of the adrenal glands produce androstenedione, which is a weak androgen.
- Aromatase also converts the adrenally derived androstenedione into estrone, which can then be converted into oestradiol.
- The bioavailable testosterone and oestradiol then enter their target tissues and exert their effects.
- An enzyme called 5-alpha reductase present in skin, prostate and epididymal tissue, converts testosterone into a much more potent metabolite called dihydrotes-

tosterone. Dihydrotestosterone binds to the same receptors as testosterone, but which a much higher affinity.

In males, the effects of androgens (testosterone and dihydrotestosterone) include:

- The development of secondary sexual characteristics, such as deepening of the voice, male pattern hair growth and penile enlargement (Giagulli et al. 2011)
- Stimulation of libido (Isidori et al. 2005)
- Stimulation of bone formation and an increase in bone density (Mohamad et al. 2016)
- An increase in basal metabolic rate (Ali Abulmeaty et al. 2019)
- An increase in muscle mass (Zhou et al. 2018; Corona et al. 2016)
- Stimulation of erythropoiesis, (Bachman et al. 2014)

In males, the effects of oestrogens include:

- Glandular breast tissue development (if there is an excess of oestradiol)
- A decrease in circulating insulin levels and an increase in insulin sensitivity (Suba 2012)
- An increase in leptin sensitivity, thereby increasing the sensation of satiety (Lizcano and Guzmán 2014)
- A decrease in visceral fat deposition, switching to subcutaneous fat deposition instead (Lizcano and Guzmán 2014)
- A reduction in total cholesterol and relative low-density lipoprotein (LDL) levels (Lizcano and Guzmán 2014)
- Suppression of bone resorption and promotion of bone formation (Mohamad et al. 2016)

To summarise, there are several sources of endogenous androgen/oestradiol:

1. Testes: produces testosterone and ~ 15% of oestradiol. Much of the remainder of the body's oestradiol is converted from testosterone in peripheral tissues.
2. Adrenal Glands: Produce androstenedione; some of which is converted into oestrone by the aromatase enzyme complex, that is found in adipose tissue.
3. The aromatase enzyme complex located in testis, bone and adipose tissue converts some testosterone into oestradiol.
4. The aromatase enzyme also converts some of the oestrone derived from the adrenal glands into oestradiol.
5. Skin, prostate and epididymal tissue contain the enzyme 5-alpha reductase, which converts testosterone into the more potent dihydrotestosterone (Please see Table 13.1).

13.1.4 Histopathological Findings

The male breast is composed of simple ducts without lobule formation. If exposed to high levels of oestrogen, the male breast can develop full acinar and lobular formation which is usually only seen in female breasts.

Table 13.1 Pathophysiology of gynaecomastia

Pathophysiological category	Pathophysiological mechanism and example
Excess levels of endogenous Oestrogen Mechanisms: 1. Physiological 2. Increased GnRH secretion 3. Uninhibited increase in FSH/LH secretion 4. Uninhibited increase in testosterone and oestradiol secretion by the testes 5. Excessive production of androstenedione by adrenal glands	1. Physiological states of increased oestrogenic activity are: In the neonate, early puberty and late adulthood. 2. Hypothalamic or GnRH secreting tumour: Although exceedingly rare, a GnRH secreting tumour can cause hypogonadism in males (Ntali et al. 2014). 3. Tumour of the anterior pituitary gland: pituitary adenomas, which make up 10–15% of all intracranial masses, are a benign cause of gynaecomastia (though these patients also have other clinical manifestations). There is hyper-secretion of one or more of the pituitary hormones, which, in the case of increase FSH and LH, leads to gynaecomastia (Lake et al. 2013). Alternatively, a hyperprolactinoma may suppress GnRH secretion, producing hypogonadoropic hypogonadism and thereby causing gynaecomastia (Narula and Carlson 2014). 4. Leydig-cell or Sertoli cell testicular tumour: Germ-cell testicular tumour, ectopic hCG secreting tumour: Testicular tumours may independently secrete increased quantities of sex-hormones. The testes may be stimulated to secrete oestradiol, by means of a hCG secreting tumour. 5. Adrenocortical tumour: An adrenal tumour may produce excessive quantities of androstenedione, which is converted to oestradiol in the peripheral tissues, though this is exceedingly rare (Mihai 2015).
Excess endogenous Oestrogenic activity Mechanisms: 1. Hyperthyroidism 2. Spironolactone 3. Liver impairment	1. Hyperthyroidism: The activity of the aromatase enzyme complex may be stimulated to convert more testosterone into oestradiol. This activity is increased by hyperthyroidism, advanced age and an increase in adipose tissue (Dickson 2012). 2. Spironolactone: Increases the amount of bioavailable oestrogen. Since oestradiol binds to sex-hormone binding globulin (SHBG) with less avidity than testosterone, a competitor such as spironolactone is more likely to displace oestradiol off the SHBG and hence increase its bioavailability, causing gynaecomastia. 3. In liver impairment, hepatic degradation of oestrogens is reduced. Alcohol consumption further exacerbates gynaecomastia as ethanol can directly inhibit testosterone production (Dickson 2012).
Excess exogenous Oestrogen Mechanisms: 1. Accidental/ unintentional exposure to oestrogens 2. Exposure to oestrogen-like endocrine disrupting chemicals	1. Gynaecomastia has been found in males who have been exposed to topical vaginal oestrogen cream during coitus (Diraimondo et al. 1980). 20% of the male employees working in a factory manufacturing oral contraceptive medications developed gynaecomastia as a result of occupational exposure (Harrington et al. 1978). 2. Exogenous oestrogen-like disrupting chemicals are not oestrogens but are man-made chemicals that mimic oestrogens by binding to oestrogenic receptors. They may be found in pesticides, flame retardants and some plastic products. Exposure to these chemicals, particularly during puberty can induce a variety of disruptions to normal production of sex-hormones in both sexes (Roy et al. 2009).

(continued)

Table 13.1 (continued)

Pathophysiological category	Pathophysiological mechanism and example
Decreased endogenous androgen activity/ production Mechanisms: 1. Hypogonadism 2. Impaired androgenic receptor function/ sensitivity 3. Anabolic androgenic steroid use 4. Renal impairment	1. Primary Hypogonadism: Such as Klinefelter's Syndrome, mumps orchitis, haemochromatosis, testicular trauma can all cause gynaecomastia through testicular failure and reduced production of the sex-hormones. (Kumar et al. 2010) 2. Inherited Androgen Insensitivity: In Spinobulbar Muscular Atrophy, the abnormal presence of expanded CAG trinucleotide repeats in the gene encoding for the androgen receptor leads to androgen insensitivity. There is a relative decrease in androgenic activity and preserved oestrogenic activity and therefore, gynaecomastia can develop, (Narula and Carlson 2014). 3. Anabolic Androgenic Steroid Use: Anabolic steroids are frequently illicitly used by bodybuilders to increase muscle mass and decrease body fat. The use of anabolic androgenic steroids produces hypogonadotropic hypogonadism by means of a feedback suppression loop of the hypothalamic-pituitary-gonadal axis. There is inhibition of the pulsatile GnRH release by the hypothalamus, in response to increased levels of androgens in the blood. Upon cessation of the anabolic steroid use, this suppression may continue for some time, depending on the length of anabolic steroid use, resulting in gynaecomastia. (Rahnema et al. 2014). Previous anabolic androgenic steroid exposure was found in 20.9% of males with hypogonadism (Coward et al. 2013). 4. Renal impairment: Uraemia-associated hypogonadism found commonly in males with chronic kidney disease is multifactorial in origin but involves decreased levels of serum testosterone (due to both decreased production and increased metabolism of testosterone) and can manifest as gynaecomastia. This does not improve with dialysis. Testosterone deficiency with elevation of serum GnRH levels is present in 26–66% of males with some degree of renal impairment. (Iglesias et al. 2012)
Therapeutic androgen blockage/deprivation Mechanisms: 1. Treatment of prostate cancer 2. Treatment of benign prostatic hyperplasia/ scalp hair loss/hormone therapy in transgender women	1. Treatments for prostate cancer: Androgen deprivation therapy is often used to treat prostate cancer: Androgen receptor blockers (Bicalutamide) and GnRH analogues/agonists (Goserelin) almost always cause gynaecomastia within 6 months of commencing therapy. 2. Treatments for Benign Prostatic Hyperplasia/Scalp Hair Loss/ Transgender Women: 5-alpha reductase inhibitors such as Finasteride/Dutasteride work by inhibiting the conversion of testosterone into dihydrotestosterone which results in an overall decrease in androgenic activity. The risk of developing gynaecomastia is approximately three times higher in males exposed to 5-alpha-reductase inhibitors compared to those who had no exposure (Hagberg et al. 2017).

(continued)

Table 13.1 (continued)

Pathophysiological category	Pathophysiological mechanism and example
Medications Mechanisms: *Varies with medications*	Antibiotics: Isoniazid, ketoconazole, metronidazole, Ethionamide
	Psychiatric: Diazepam, haloperidol, phenothiazine, tricyclic antidepressants
	Hormones: Androgens, Oestrogens, anabolic steroids, growth hormone, hCG
	Cardiovascular: Verapamil, reserpine, spironolactone, Enalapril, Nifedipine, captopril, amlodipine, amiodarone, digoxin, statins, diltiazem, methyldopa
	Chemotherapeutic: Methotrexate, cyclophosphamide, alkylating agents
	Gastrointestinal: Proton pump inhibitors, ranitidine, metoclopramide, cimetidine, Domperidone
	Anti-androgens: Flutamide, finasteride, Cyproterone, Bicalutamide
	Others: Theophylline, Auranofin, Clomiphene, Penicillamine, Marijuana, Amphetamine, Heroin, Methadone, Alcohol (Fagerlund et al. 2015)

Early or 'Florid' gynaecomastia of less than 4 months is characterised by a three-layered pattern on histopathological analysis, an outer myoepithelial layer, two layers of epithelial cells and several ducts with irregular lumens, surrounded by loose connective tissue.

Fibrotic gynaecomastia is that which has been present for over 12 months and is characterised by ductal proliferation and stromal fibrosis. The fibrotic changes are permanent and require surgical excision. These changes do not regress even when the original stimulus inducing the gynaecomastia is removed. (Cordova and Tripoli 2020).

13.1.5 Clinical Presentation

There are two differing presentations of gynaecomastia. Asymptomatic and long-standing chronic gynaecomastia is usually incidentally noticed on physical examination, by which time it has been slowly growing for months or years. In contrast, painful or tender gynaecomastia generally have had a more recent onset. (Braunstein 2007).

13.1.6   History

Important information to ascertain in patients with gynaecomastia on history taking include:

- Duration of enlargement
- Any pain, discharge or other associated symptoms?
- Any other symptoms of hypogonadism?
- Constitutional symptoms such as fevers, night sweats, lethargy or unintentional weight loss
- Symptoms suggestive of thyroid-related disorders; unusual weight loss or gain, feeling hot or cold, appetite changes, palpitations
- Family history of breast cancer
- Alcohol and illicit drug use
- Anabolic steroid use, use of herbs/supplements
- Co-morbid illnesses and a detailed medication history
- Any congenital conditions, such as Klinefelter's syndrome
- Progression through childhood and puberty, and whether there were any issues during development

13.1.7   Examination

To differentiate between true gynaecomastia and pseudogynaecomastia, the examiner should bring his thumb and index finger in towards the nipple, from either side of the breast. In the case of true gynaecomastia, the examiner should feel a firm, rubbery mound of tissue surrounding the nipple-areolar complex. In pseudogynaecomastia, this rubbery 'disc', which is a manifestation of enlarged glandular tissue, does not exist, and is not felt. (Braunstein 2007).

Gynaecomastia affects both breasts in approximately 50% of patients.

Note the characteristics of the breast tissue including symmetry, size, firmness, the presence of discharge, any overlying changes to the skin or nipple. Palpate the chest wall and axilla for lymphadenopathy.

Observe whether there is any pain or tenderness associated with the enlarged breast tissue.

Examine the abdomen for masses and for features of liver impairment, such as hepatomegaly, ascites and spider naevi. A thyroid exam should be performed, looking for the presence of goitre, thyroid nodules, a tremor or hyperreflexia. Check for signs of hypercortisolaemia including the presence of central obesity, purple striae or moon face. Note the presence of secondary sexual characteristics such as facial and underarm hair.

A testicular examination should be performed, looking for testicular masses or small testicular size which may be indicative of hypogonadism.

Examination findings that raise the suspicion of cancer include the presence of a bloody discharge, skin ulceration, a firm or fixed quality to the tissue and associated lymphadenopathy. If any of these are discovered, urgent mammography, ultrasound and biopsy of the tissue should be performed. (Braunstein 2007).

13.1.8   Investigations

The most important goal in investigating gynaecomastia is to rule out malignancy. Once malignancy has been ruled out and other underlying causes have been excluded, it is safe to explore possible treatment options (Fagerlund et al. 2015).

Where there is suspicion of breast malignancy presenting as gynaecomastia, an urgent breast ultrasound, mammogram and core biopsy should be performed. Mammography has sensitivity and specificity above 90% for differentiating between benign and malignant breast conditions (Costanzo et al. 2018).

13.1.9   Blood Tests

Blood should be drawn and tested for liver function tests, urea, electrolytes and creatinine, serum luteinizing hormone, hCG, total testosterone and oestrogen levels. Some studies advocate for performing testicular tumour markers during initial investigations, which would include: serum lactate dehydrogenase (LDH), alpha-fetoprotein (AFP) and Beta-hCG, (Costanzo et al. 2018).

The finding of low serum testosterone is common and should be confirmed by testing for serum free or bioavailable testosterone. If all of the initial blood tests are within normal limits, this is considered to be idiopathic gynaecomastia, and further investigations are not required.

If any of the above are abnormal, the following panel should be tested: serum prolactin, free or bioavailable testosterone, thyroid function tests. The serum prolactin level is used to check for prolactinoma. Thyroid function tests are useful in diagnosing hyperthyroidism. Where a low free or bioavailable testosterone and an elevated LH level is found, there is likely primary testicular failure (primary hypogonadism). Where there is low free or bioavailable testosterone, but normal LH, the serum prolactin helps to differentiate between a prolactinoma (serum prolactin will be elevated) or secondary hypogonadism (serum prolactin will be normal). (Braunstein 2007).

13.1.10   Imaging

Imaging for gynaecomastia is used to look for testicular pathology, such as a testicular tumour, or to look for extra-testicular tumours such as hCG-secreting tumours or adrenocortical tumours.

Testicular ultrasound is performed where there is suspicious of testicular tumour, and are required in the following situations:

1. Elevated serum hCG in a male is concerning of a testicular germ-cell tumour, extragonadal germ-cell tumour or a hCG secreting neoplasm and requires urgent testicular ultrasound and a computed tomography scan of the chest, abdomen and pelvis to investigate.
2. Normal/decreased LH and increased Oestradiol may be a result of: a Leydig/Sertoli cell testicular tumour, Adrenocortical tumour or secondary to increased aromatase activity. In this situation, an urgent testicular ultrasound, and adrenal computed tomography scan or adrenal magnetic resonance imaging scan should be performed. (Braunstein 2007)
3. Presence of a mass on testicular examination

Breast ultrasound is useful in confirming the presence of glandular tissue in breasts, making it helpful in differentiating between pseudo-gynaecomastia and true gynaecomastia. Where there is concern for breast malignancy, however, mammography should also be performed. (Telegrafo et al. 2016).

13.1.11 Complications of Gynaecomastia

13.1.11.1 Incidental Finding of Male Breast Cancer

The role of gynaecomastia in the development of male breast cancer is uncertain. Gynaecomastia is occasionally seen alongside invasive breast cancer, however, it is much more commonly found in healthy males. (Weingertner et al. 2019) Some studies have found a correlation between gynaecomastia and male breast cancer, though this evidence is yet to become conclusive. The concern arises with newer surgical techniques such as the liposuction technique, where no pathology specimen is collected and sent for histopathological analysis, presenting a risk of having missed a case of male breast cancer.

The overall risk for malignant tumours amongst males with gynaecomastia is similar to the overall risk for the normal population. One study reported that there was no increased risk of developing breast cancer, though there was a significant increase in the risk of having testicular cancer and squamous cell skin cancer, (Olsson et al. 2002).

In a study of 452 patients who underwent surgery for gynaecomastia and histopathological analysis of the excised tissue, there were 15 incidental findings, of which 12 were ductal carcinoma in situ, 2 were atypical ductal hyperplasia and one case of infiltrating ductal carcinoma, (Senger et al. 2014). A similar study involving 5113 breasts with gynaecomastia, the overall prevalence of invasive carcinomas was 0.11% and the prevalence of in-situ carcinomas was 0.18%, (Lapid et al. 2015).

The overall lifetime risk of developing breast cancer in males is low at 0.2%. This risk increases to 3–6% in males with Klinefelter's syndrome, and in those who are related to females who developed early breast cancer. These compounding risk factors should be taken into account during initial workup of gynaecomastia. There are no specific chromosomal abnormalities associated with the development of gynaecomastia.

13.1.11.2 Breast Tenderness/Pain

In a study of 62 patients with gynaecomastia, 41.9% had associated breast pain or tenderness, 84.6% noticed nipple enlargement and 6.5% noticed nipple discharge, (Singano et al. 2017). In another study of 237 men with gynaecomastia, 51.2% had associated breast pain (Costanzo et al. 2018).

13.1.11.3 Self-Esteem/Psychological Distress

The psychological impact of gynaecomastia on patients is significant. Patients with gynaecomastia may develop self-esteem issues and as a result, may limit their social interactions, leading to worsening of psycho-emotional stress. In a study of 39 patients undergoing intervention for gynaecomastia, 94.8% reported psychological stress as a result of their gynaecomastia and one patient suffered social phobia due to their gynaecomastia (Li et al. 2012).

In a study of a cohort of 62 patients who developed gynaecomastia, 51.6% reported embarrassment, 9.7% reported associated stigma attached to the issue, and 71.0% were concerned that the gynaecomastia would have adverse effects on their health (Singano et al. 2017). In a study of 237 patients with gynaecomastia, the primary concern was related to appearance or aesthetic concerns (62.8%), followed by breast pain (Costanzo et al. 2018).

A systematic review suggested that surgical intervention is beneficial in gynaecomastia over several psychological domains including vitality, emotional discomfort, limitations due to pain and due to physical appearance (Sollie 2018).

13.1.12 Management

The mainstay of management of gynaecomastia is to treat the underlying cause. If the gynaecomastia has been present for over 12 months it is unlikely to regress even when the original stimulus is removed, and this is where surgical management is required. (Cordova and Tripoli 2020).

13.1.13   Conservative Management- Re-evaluation in 3–6 Months

For a cohort of older patients where gynaecomastia is almost certainly physiological and not bothersome to the patient, it is safe to take a watch and wait approach to investigation and treatment (Braunstein 2007). Where the gynaecomastia is bilateral, asymptomatic and incidentally discovered, after taking a history and examining the patient, it is acceptable to re-evaluate the situation in 3–6 months to see whether the changes persist.

Gynaecomastia of recent onset that is <3 cm in size has been shown to spontaneously regress in 85% of patients (Braunstein 2007). Prolonged untreated gynaecomastia becomes increasingly fibrosed, and there is increased hyalinisation of loose periductal tissue, meaning that after 12 months, it is unlikely to achieve regression of the tissue without intervention (Fagerlund et al. 2015).

Where the gynaecomastia is painful or tender, or where there has been rapid enlargement, size >4 cm and in men between the ages of 20 to 50, this conservative approach is not suitable and these patients should be thoroughly investigated (Braunstein 2007).

13.1.14   Lifestyle Modification

Obese patients should be encouraged to lose weight. Cessation or reduction of alcohol, marijuana, heroin and anabolic steroids.

13.1.15   Medication Adjustments

With the majority of medication-induced gynaecomastia, provided the gynaecomastia has been present for less than 12 months, cessation of the offending medication should result in regression of the gynaecomastia within 6 months.

13.1.16   Anti-Oestrogen Therapy

Idiopathic gynaecomastia or gynaecomastia that occurs secondary to androgen deprivation therapy can be treated with a trial of Tamoxifen 20 mg/day for up to 3 months. (Cordova and Tripoli 2020).

Aromatase inhibitors have been shown to reduce or prevent gynaecomastia by preventing the peripheral aromatisation of circulating testosterone to oestradiol and have been used with success in improving gynaecomastia and breast pain induced by androgen deprivation therapy in males with prostate cancer. In a placebo-controlled randomised trial gynaecomastia developed in 73% of patients in the placebo group, 51% of patients in the anastrazole group and only 10% of patients in the tamoxifen group. There were no significant differences in PSA response rates amongst the three groups of patients in the 6-month study duration. (Boccardo et al. 2005).

There are concerns regarding the use of anti-oestrogen therapy in patients with prostate cancer, as this could increase androgen levels by blocking the negative feedback of oestradiol on the hypothalamic-pituitary axis. More clinical trials are required to address the safety of this. (Autorino et al. 2006).

13.1.17 Prophylactic Breast Irradiation

For those patients who are being commenced on anti-androgenic medication, such as androgen receptor blockers for prostate cancer, the use of prophylactic breast irradiation is moderately effective in reducing the risk of developing gynaecomastia.

In a randomised, double-blinded, multicentre trial, 106 males with prostate cancer and no symptoms of gynaecomastia were randomised to receive either a single dose of electron beam radiotherapy or placebo "sham radiotherapy". All patients received the standard dose of Bicalutamide (an androgen receptor blocker) for 12 months. Findings demonstrated the development of gynaecomastia was 52% in the radiotherapy group and 85% in the placebo group. (Tyrrell et al. 2004).

There is also evidence of radiation treatment reducing gynaecomastia in patients who have recently developed symptoms of gynaecomastia within 6 months, though again the efficacy is moderate. Radiotherapy improved gynaecomastia in 25.9% of patients and resolved it in 7.4% of patients taking Bicalutamide for prostate cancer (Van Poppel et al. 2005). Radiotherapy-related side effects were reported by 43.9% of participants, though all of these were mild-moderate and resolved in a median duration of 5 weeks. Such adverse effects were breast/nipple erythema (36.6%), skin irritation (31.7%) and tiredness (2.4%) (Van Poppel et al. 2005).

Radiotherapy is less effective than tamoxifen in reducing/preventing gynaecomastia in patients taking androgen deprivation therapy (Perdonà et al. 2005).

13.1.18 Surgical Management

When choosing an operative approach, the surgeon should take into consideration the grade of gynaecomastia (assessed using the Simon classifications), the position of the nipple-areola complex, the amount of glandular tissue and whether there is likely to be any excess skin remaining (Özalp et al. 2018).

The grade of gynaecomastia may be assessed using the Simon classification system as follows: (Simon et al. 1973)

Grade 1: Small, visible breast enlargement—no skin redundancy
Grade 2A: Moderate breast enlargement without skin redundancy
Grade 2B: Moderate breast enlargement with skin redundancy
Grade 3: Marked breast enlargement with marked skin redundancy; i.e. pendulous female breasts

Though a number of different gynaecomastia classification systems exist, and there is no consensus as to which should be used, the Simon classification is the most commonly used.

There is no consensus as to which surgical technique is optimal for different grades of gynaecomastia, though it is generally accepted that where there is prominent skin redundancy and ptosis, optimal results are achieved through breast amputation with free nipple grafting. There is no agreement as to which technique should be utilised in moderate gynaecomastia (Grade 2A to 2B).

13.1.19   Adenectomy: The Webster Technique

The Webster technique is suitable for small or Grade 1 gynaecomastia. An incision is made along the areolar margin between the 3 o'clock and 9 o'clock positions. Glandular tissue, which is distinguished from fat due to its firmness and white colour, is removed from beneath the areola. Half a centimetre of tissue is intentionally left underneath the areolar to prevent unsightly areolar depressions. The areolar incision is closed and a drain is not usually required with this technique. (Cordova and Tripoli 2020).

13.1.20   Adenectomy: The Pull-through Technique

Suitable for Grade 1 gynaecomastia, the pull through technique involves a one centimetre incision behind the anterior axillary wall. Scissors are inseredthrough this incision to the glandular tissue, which is dissected off the fascia and 'pulled-through' and out through the incision. Again some tissue should be spared and intentionally left behind the areolar, to prevent ischaemia and areolar depression. (Cordova and Tripoli 2020).

13.1.21   Liposuction or Ultrasound Assisted Liposuction Only

The advantages of liposuction over open excisional techniques are that is does not cause nipple-areola complex necrosis or prominent scarring due to the use of a tiny incision. These techniques may be used for treatment of Grade 1 or Grade 2 gynaecomastia (Cordova and Tripoli 2020).

A study investigating liposuction for the management of gynaecomastia reported the following complications: 1% haematoma, 2% seroma, 0% nipple-areola complex necrosis, 7% persisting breast pain and 0% scarring. The recurrence rate with liposuction was found to be 0.52% amongst patients who underwent these techniques (Song et al. 2014).

13.1.22 Liposuction plus Arthroscopic Shaver Technique

The downfall of exclusively using liposuction is that glandular tissue is often left-over. Whilst open mastectomy is better at removing all glandular tissue, the resulting scar is larger than that found in liposuction techniques. The novel use of an arthroscopic shaver device, initially performed by Song et al., capitalises on the tiny incision made in liposuction techniques, by adding the use of an electrically powered endoscopic-arthroscopic cartilage shaver, thus avoiding large peri-areolar incisions and successfully removing all glandular breast tissue. (Song et al. 2014) This technique produced superior aesthetic outcomes when compared to the excision and liposuction technique, with no significant difference in the overall rate of postoperative complications in the two groups. In a study assessing 76 patients who underwent the liposuction and arthroscopic shaver technique, the rate of complications were: 2.6% seroma, 1.3% haematoma, 1.3% burn scar due to the ultrasound probe and 5.2% required re-operation due to remaining tissue. (Petty et al. 2010). This technique may be limited in cases where the breast tissue is too firm.

13.1.23 Surgical Excision (Adenectomy) with Liposuction

Where liposuction fails to remove glandular tissue, the remaining tissue can be dealt with immediately after liposuction in a combined liposuction with adenectomy/surgical excision procedure. After liposuction as described above has been completed, adenectomy may be performed by making an inferior hemispheric-areolar incision and removing any remnant glandular tissue through here.

This combination of liposuction and excision produces the most consistent results and is associated with a low rate of complications (Fagerlund et al. 2015). In an analysis of 53 patients with Grade 1 or Grade 2 gynaecomastia who underwent this technique, the rate of complications were: 0% haematomas, 0% seromas, 0% infections, 4% re-operation due to recurrence and 4% re-operation for post-operative haemorrhage (Schröder et al. 2015). The rate of nipple-areola complex necrosis in a.

13.1.24 Nipple-Sparing Mastectomy

Where there is excessive skin and a large amount of breast tissue such as that seen in Grade 3 gynaecomastia, skin-sparing techniques are not possible, as removal of the excess skin is a crucial and necessary process in achieving good clearance of the tissue and re-positioning the nipple-areolar complex to an appropriate site on the breast. In severe cases, it may not be possible to preserve the nipple, and breast amputation with free nipple transplant may be required. (Cordova and Tripoli 2020).

It is difficult to achieve very good or even good cosmesis where there is severe ptosis and excessive skin in higher Simon classifications of gynaecomastia, and the higher the grade of gynaecomastia, the higher the risk of developing operative complications, (Kasielska and Antoszewski 2013).

Overall complication rate for subcutaneous mastectomy alone was 8.5%, for subcutaneous mastectomy and skin incision it was 33.3% and for inverted-T reduction mastectomy it was 66.7%. (Kasielska and Antoszewski 2013).

13.1.25   Complications of Operative Management

13.1.25.1   Dissatisfaction

Only 62.5% of patients who underwent surgery for gynaecomastia were 'satisfied' to 'very satisfied' following the procedure (Ridha et al. 2009). Grade 3 gynaecomastia frequently requires revision surgeries to achieve satisfactory results. These revisions may include liposuction techniques alone or excisional techniques.

13.1.25.2   Sensory Changes

Altered sensation to the nipple-areola complex is a complication in 9.7% of surgeries for gynaecomastia, and occurs across a variety of liposuction and excisional techniques (Kasielska and Antoszewski 2013). Hypoaesthesia to the areola is transient in the majority of cases, but rarely may be permanent and this should be discussed as a possible risk pre-operatively.

13.1.25.3   Seromas/Wound Infection

Seroma is a complication in 3.5% of gynaecomastia surgeries. Wound infection is a complication in 0.9% of gynaecomastia surgeries (Kasielska and Antoszewski 2013). Seromas occur more frequently in surgery for higher grade gynaecomastia, and the use of a drain in these surgeries is advised (Cordova and Tripoli 2020).

13.1.25.4   Haematoma

At a rate of approximately 7.1%, post-operative haematoma is a common complication to arise from surgery for gynaecomastia, with no differences in risk observed with different surgical techniques utilised (Kasielska and Antoszewski 2013). Half of these haematomas require evacuation.

The use of a circumferential compression bandage or compression vest is recommended to avoid haematoma formation (Kasielska and Antoszewski 2013).

13.1.25.5   Nipple Necrosis

Nipple-areola complex ischemia is a serious possible complication of gynaeco-mastia surgery, particularly in the mastectomy techniques. In a study of 220 breasts that underwent nipple-sparing mastectomy, nipple ischaemia occurred in 64.1% of breasts, of which re-operation for necrosis was required in 31.3% (Ahn et al. 2018).

Factors increasing the risk of nipple-areola complex necrosis include the presence of ptosis (as seen in Grade 3 gynaecomastia) and the use of circumareolar incision. Where possible, lower peri-areolar incision is preferable over upper peri-areolar incision to preserve the vascular supply to the nipple. (Ahn et al. 2018).

13.1.25.6   Excess Skin

Secondary surgeries involving liposuction or scar revision are very common after initial surgery for severe grades of gynaecomastia, and should be discussed with patients during initial surgical planning (Cordova and Tripoli 2020).

13.1.25.7   Areolar Depression

Areolar depression is avoided by the intentional preservation of at least 0.5 cm of tissue directly behind the nipple-areolar complex (Cordova and Tripoli 2020).

13.1.25.8   Sub-Areolar Retention of Tissue

Liposuction is ineffective at removing the glandular tissue sitting just posterior to the nipple, particularly where the gynaecomastia is fibrosed, firm and established. Hence, a complication of using liposuction techniques alone is the retention of a sub-areolar nodule of tissue which may be a cause of dissatisfaction for patients.

In a study where 31 breasts underwent a liposuction-only technique, 16% had subareolar tissue retention. None of the patients who underwent ultrasound-assisted liposuction had subareolar retention of tissue, testifying to the comparative superiority of ultrasound-assisted liposuction in removing tissue. (Fruhstorfer and Malata 2003).

13.2   Inguinal Hernia

13.2.1   Pathophysiology

A hernia is a protrusion of all or part of an organ through the body wall that normally contains it. The incidence of groin hernias is common, affecting 27 out of every 100 men (Lockhart et al. 2018). The majority of these are inguinal hernias. Inguinal hernias occur 9–12 times more commonly in men than in women.

Hernias may contain omentum, mesentery or portions of small or rarely, large bowel. When the hernial contents can be returned to the body cavity by manual palpation, this is termed a 'reducible' hernia. When the hernia contents are fixed to the hernia sac, this is called an 'irreducible' or 'incarcerated' hernia. A 'strangulated' hernia is one involving vascular compromise to the hernial contents.

Although both inguinal and femoral hernias may present as a 'groin lump', inguinal hernias involve the protrusion of hernial contents into or through the inguinal canal. Femoral hernias involve the protrusion of hernial contents into the femoral canal.

In males, the inguinal canal is the route by which the testes descend from their embryological location intra-abdominally to their destination in the scrotum. The inguinal canal in males also contains the ductus deferens alongside blood vessels, nerves and lymphatic channels that supply the scrotum and testis.

There are two types of inguinal hernia, named based on the route by which they traverse the inguinal canal. Management of inguinal hernia does not differ according to whether it is of the direct or indirect type (Jenkins and O'Dwyer 2008).

A direct inguinal hernia occurs when the hernial sac protrudes through a weakness in the posterior wall of the inguinal canal (which is formed by transversalis fascia), most commonly where this wall is weakest; directly behind the superficial inguinal ring. Direct inguinal hernias more frequently occur in older males.

In contrast, an indirect inguinal hernia occurs when the hernial sac enters the inguinal canal through the deep inguinal ring to traverse the inguinal canal (alongside the ductus deferens) and exits via the superficial inguinal ring. Indirect inguinal hernias result from the embryological failure of obliteration of the pathway from the abdomen to the scrotum through which testicular descension occurs, also referred to as the 'processus vaginalis'. Hence, indirect hernias can occur in any age group, and are twice as common as direct inguinal hernias (Fitzgibbons and Forse 2015).

Indirect inguinal hernias occur lateral to the inferior epigastric vessels, whereas direct inguinal hernias occur medial to these vessels. Both types of inguinal hernia are found superior to the inguinal ligament (Fitzgibbons and Forse 2015).

13.2.2   Risk Factors for Inguinal Hernia Development

- Family history (HerniaSurge Group 2018)
- Previous contralateral hernia (HerniaSurge Group 2018)
- Male gender (Inguinal hernias occur 9–12 times more commonly in men) (Lockhart et al. 2018)

- Age; for which there are two distinct peaks; at 5 years of age for indirect inguinal hernias and between 70–80 years of age for direct inguinal hernias. (HerniaSurge Group 2018)
- Abnormal collagen metabolism (leading to a diminished collagen type 1/type 3 ratio.) (Henriksen et al. 2011)
- Previous prostatectomy, particularly open radical prostatectomy (Abe et al. 2007), (Stranne et al. 2010)
- Previous open appendicectomy (Fitzgibbons and Forse 2015)
- History of abdomino-pelvic radiation treatment (Nilsson et al. 2014)
- Low body mass index (HerniaSurge Group 2018)
- Chronic Obstructive Pulmonary Disease (Fitzgibbons and Forse 2015)
- Increased intra-abdominal pressure (Fitzgibbons and Forse 2015)
- Abdominal Aortic Aneurysm (HerniaSurge Group 2018)

Of note, occasional weight-lifting, constipation and prostatism are not noted to contribute to development of groin hernias. (HerniaSurge Group 2018).

13.2.3 Clinical Presentation

Approximately one-third of patients with inguinal hernias are asymptomatic. Most present with chronic symptoms of a groin lump associated with discomfort. Least common are acutely symptomatic patients with pain, erythema, and localised swelling.

13.2.4 Diagnosis

The majority of inguinal hernias (95%) can be diagnosed with careful history and physical examination alone. (HerniaSurge Group 2018). Inguinal hernias can be recognised as a reducible swelling superior to the inguinal ligament, (which is in contrast to femoral hernias which cause a swelling inferior to the inguinal ligament) that may be accompanied by localised pain/discomfort. In cases of uncertainty, an ultrasound can be useful in determining the presence of a hernia sac, and to define its dimensions and contents. (Light et al. 2011; Robinson et al. 2013). If groin ultrasound is negative but high suspicion remains, a dynamic CT or MRI may be of use, where the Valsalva manoeuvre is performed to better visualise what may be an occult hernia reverting to its displaced position in the hernia sac. (HerniaSurge Group 2018).

13.2.5 History

Patients may complain of a 'dragging' or heavy sensation in the groin. There may be pain or discomfort with movements that increase intra-abdominal pressure (such as sneezing, coughing or defaecating).

13.2.6 Examination

The most common physical finding is a bulge in the groin. It is more reliable to demonstrate a hernia bulge with the patient standing. A bulge can be confirmed as a hernia by placing a hand over the hernia and asking the patient to cough or perform a Valsalva manoeuvre. Smaller hernias may be detected by gently invaginating the scrotal skin with a finger through the superficial inguinal ring into the inguinal canal, attempting to palpate a localised bulge. It is important to note that reducible hernias may not be overtly visible at the time of examination, and manoeuvres such as the Valsalva manoeuvre/asking the patient to 'bear down' or cough should be performed to feel the hernial protrusion.

13.2.7 Complications of Inguinal Hernia

13.2.7.1 Pain/Discomfort

One third of patients are asymptomatic, but those who have symptoms may experience discomfort/pain on coughing, defaecating, urinating, exercise or sexual activity. There may be a burning sensation or a 'dragging' sensation. Symptoms may be relieved when lying flat or by manual reduction of the hernia. (Fitzgibbons and Forse 2015).

13.2.7.2 Incarceration

There is a 3% yearly risk of any inguinal hernia becoming incarcerated. Incarcerated hernias may remain irreducible without consequence, and do not always become strangulated (Fitzgibbons and Forse 2015).

13.2.7.3 Strangulation

When the blood supply to hernial contents is compromised, this is known as strangulation. This can lead to worsening pain, swelling, ischaemia and ultimately necrosis of the affected herniated tissues. If bowel is involved, this can lead to bowel perforation and spillage of faeculent contents into the peritoneal cavity. If there is clinical concern regarding the viability of bowel, emergency surgery is required to visualise and/or resect affected bowel and wash out the peritoneum.

13.2.7.4 BowelObstruction

All patients presenting with a bowel obstruction warrant careful groin examination to rule out a hernial cause. Bowel obstructions secondary to groin hernias rarely improve with conservative management and risk strangulation. (Fitzgibbons and Forse 2015).

13.2.8   Management of Inguinal Hernia

13.2.8.1   Conservative Management

Asymptomatic or minimally symptomatic males with inguinal hernia may be managed with 'watchful waiting', given their risk of developing a hernia-related surgical emergency is low, though it is important to note that the majority of these patients (70%) will subsequently develop pain/discomfort and eventually require surgical repair of their hernia within 5 years (Fitzgibbons and Forse 2015; HerniaSurge Group 2018).

It is important to take into account the patient's individual health status, preferences and risk factors on a case by case basis and discuss the risks and benefits of different management options, as there is no standardised 'one size fits all' approach to the management of inguinal hernias.

13.2.8.2   Surgical Management

Symptomatic inguinal hernias should be surgically managed. There is no evidence to support a 'watchful waiting' approach in these men, and further to this, there is a higher (10–20 fold increase in risk) morbidity and mortality associated with emergent repair of incarcerated or strangulated inguinal hernias compared to elective repair of uncomplicated inguinal hernia in symptomatic men (HerniaSurge Group 2018).

Acutely symptomatic inguinal hernias that warrant concern for complications such as strangulation of bowel/ bowel obstruction or peritonitis should be operated on emergently. If there is concern regarding the viability of bowel, then visualisation and/or resection of affected bowel and intra-peritoneal washout is recommended. It is also advised to avoid primary mesh repair in this cohort of patients where there is intraperitoneal contamination due to the increased risk of mesh related infection.

Principles of surgical inguinal hernia repair involve:

(a) the approximation of the defect in the abdominal wall, with the use of mesh to reinforce the defect, or, by means of
(b) non-mesh repair which is done by primary suture approximation of the abdominal wall defect.

Mesh repair is preferred in the repair of inguinal hernias as it has a lower rate of post-operative hernia recurrence. Mesh repair has an equivocal incidence of post-operative pain when compared to non-mesh repair techniques.

In cases of non-mesh repair, the Shouldice technique is the preferred method, with less risk of reoccurrence of the hernia compared to other non-mesh repair techniques. (Simons et al. 2009).

13.2.8.3 Laparo-Endoscopic Approaches

There are two well described laparo-endoscopic surgical approaches to inguinal hernia repair; Laparoscopic Trans-Abdominal Pre-Peritoneal (TAPP) and Totally Extra-Peritoneal (TEP) repair. In both techniques, mesh is placed in the pre-peritoneal plane. The differences amongst these techniques lie in the way the pre-peritoneal plane is accessed.

TAPP utilises laparoscopic ports inserted into the anterior abdominal wall to attain access to the pre-peritoneal space, through the intra-abdominal cavity. The benefit of TAPP is that it allows direct visualisation of intra-abdominal structures, so is the preferred method if there is suspicion of incarcerated bowel and in cases of recurrent hernias.

TEP utilises midline laparoscopic ports to access the pre-peritoneal space, however it does so *without* violating the abdominal cavity, and hence visualisation of abdominal structures is not possible here, making it a more technically demanding procedure. Benefits of TEP are that it can be utilised in the repair of bilateral inguinal hernias by means of the same 3–4 ports on the midline of the anterior abdominal wall. TEP is preferred when there has been previous intra-abdominal surgery complicated by adhesions, as it effectively enters the pre-peritoneal space without needing to divide complex adhesions within the peritoneal cavity.

TAPP and TEP are comparable in terms of operative times, complication risks, incidences of post-operative acute and chronic pain and hernia recurrence rates. Due to the nature of TAPP, although very rare, there is an increased incidence of vascular injury and post-operative bleeding. In contrast, in TEP, there is an increased incidence of visceral injuries. TEP, being a more technically challenging procedure than TAPP, generally takes a longer time for surgeons to become proficient in.

13.2.8.4 Open Repair

85% of inguinal hernias are repaired using an open approach (HerniaSurge Group 2018). The Lichtenstein technique which involves the onlay placement of a flat mesh in open hernia repair is the most frequently utilised technique in surgical repair of symptomatic inguinal hernias.

The Lichtenstein technique involves suturing of the mesh to the floor between transversalis fascia and the external oblique, and a slit is made in the lateral aspect of the mesh for passage of the spermatic cord. Here, the mesh is located superficial to the defect, rather than deep to the defect as seen in TAPP and TEP and hence, hernial recurrence is possible between the mesh and the defect.

13.2.9 Operative Complications

Complications after inguinal hernia repair are relatively common. The incidence is higher after emergency repairs and recurrent hernia repairs compared to elective repairs.

13.2.9.1 Short Term Complications

13.2.9.1.1 Haematoma

Average incidence of 8% reported. (Poelman et al. 2013).

13.2.9.1.2 Seroma

Average incidence of 7% reported. Does not usually require aspiration, and clinically of little significance, but can be concerning to patients who may confuse the seroma for recurrence of the hernia. (Poelman et al. 2013).

13.2.9.1.3 Wound Infections

Approximately 1% risk. Mesh infections are exceedingly rare and often do not require surgical removal of the mesh. (Poelman et al. 2013).

13.2.9.1.4 Visceral Injury

Estimated 1.8 per 1000 procedures (HerniaSurge Group 2018).

13.2.9.1.5 Vascular Injury

Estimated 0.9 per 1000 procedures (HerniaSurge Group 2018).

13.2.9.1.6 Urinary Retention

Urinary retention can occur following open or laparo-endoscopic inguinal hernia repair. The incidence is in the order of 3%, without a significant difference between approach, although it is higher when the hernia repair is performed under general anaesthesia rather than local anaesthesia (Schmedt et al. 2005).

13.2.10 Long Term Complications

13.2.10.1 Recurrence

The recurrence rate for inguinal hernias following tension free endoscopic or open repair is between 0–5%, (HerniaSurge Group 2018). Primary repair of inguinal hernia without mesh can lead to recurrence in up to 15% of patients.

13.2.10.2 Chronicpain

Due to heterogeneity in the definitions of chronic pain used by researchers, the incidence of chronic pain after repair of inguinal hernias is not well known but estimated to be as high as 25% in one randomised control trial (Langeveld et al. 2010). The chronic pain experienced following surgical repair of inguinal hernias has been attributed to damaged nerves causing neuropathic pain, formation of scar tissue or due to a reaction to prosthetic material/mesh used. In one-third of patients who experienced chronic pain post hernia repair, the pain resolved within 6 months without intervention. (Fitzgibbons and Forse 2015).

13.2.10.3 Sexual Dysfunction and Pain with Sexual Intercourse (Pain with Erection/Ejaculation)

The incidence of new onset post-operative sexual dysfunction (defined as completion of intercourse) or pain with sexual intercourse was 5.3% and 9.0% respectively, in a 2019 meta-analysis comparing minimally invasive and open repairs of inguinal hernias. The rates of pain with sexual activity was 7.4% for patients who underwent laparo-endoscopic hernia repair, compared to 12.5% of those who underwent open repair, (Ssentongo et al. 2020).

13.2.10.4 Infertility

Obstructive azoospermia (the absence of spermatozoa in the ejaculate despite normal spermatogenesis, secondary to injury to the ductus deferens) occurs in 0.03% of open bilateral inguinal hernia repairs and in 2.5% of laparo-endoscopic bilateral inguinal hernia repair surgeries involving mesh (Kordzadeh et al. 2017). There are two manners by which infertility may complicate inguinal hernia repair;

- by injury to testicular vessels causing testicular atrophy and impaired spermatogenesis (Wantz 1986)
- by injury to the ductus deferens (causing scar tissue formation or transection of the cord) resulting in obstructive azoospermia.

Both mechanisms of iatrogenic injury are thought to occur more frequently in repairs involving mesh, due to the 'foreign-body reaction' induced by the mesh, causing scar tissue formation and damage to surrounding structures (Bouchot et al. 2018). These iatrogenic injuries may also occur as a result of direct manipulation of the ductus or testicular vessels intra-operatively.

13.3   Hydrocoele

A hydrocoele is the abnormal accumulation of serous fluid typically contained within the visceral and parietal layers of the tunica vaginalis as it envelopes the testis, and/or within these layers of the spermatic cord. Hydrocoeles may, less frequently, occur in other regions within the abdomen or pelvis, and can be attributed to a diverse range of pathology (Dagur et al. 2017). Hydrocoele affects 1% of adult men, occurring most commonly in those 40 years or older (Fig. 13.1).

13.3.1   Communicating/Congenital Hydrocoele: Pathophysiology

The most common underlying pathophysiology for communicating (usually congenital) hydrocoele involves the failure of embryological closure of the processus vaginalis—an outpouching of peritoneum that develops during gestational week 12, where a physiological herniation of the parietal peritoneum occurs (processus vaginalis), allowing the testis to descend from within the abdomen, through the inguinal canal and into the scrotum.

Once testicular descent is complete, the processus vaginalis is closed off, becoming the tunica vaginalis (a bi-layered covering of the testis). Failure of closure of the processus vaginalis results in the persistence of a communicating pouch connecting the abdominal cavity to the scrotum and is associated with indirect inguinal hernia and communicating hydrocoele.

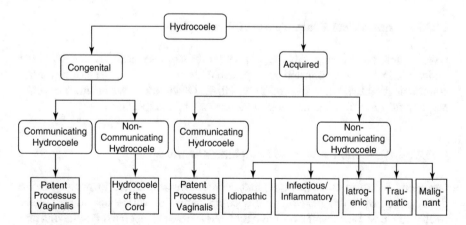

Fig. 13.1 Classification of hydrocoele

When peritoneal fluid from the intra-abdominal cavity enters the scrotum via the inguinal canal to accumulate in the space between the layers of tunica vaginalis, this is known as a communicating hydrocoele.

Congenital hydrocoeles occur in infants and are communicating hydrocoele(s) secondary to a patent processus vaginalis. These are usually asymptomatic and resolve without intervention within the first 24 months of life (Kurobe et al. 2020).

Asymptomatic communicating hydrocoeles are also found in 20% of adult men.

13.3.2   Non-communicating Hydrocoele: Pathophysiology

In contrast to communicating hydrocoele, a non-communicating hydrocoele occurs when the processus vaginalis is appropriately obliterated, and the accumulation of fluid occurs secondary to a variety of underlying pathology. These include:

13.3.2.1   Excessive Production of Serous Fluid by the Tunica Vaginalis

Serous fluid is produced by the mesothelial lining of the tunica vaginalis; excessive production of which leads to an accumulation of fluid between the two layers of tunica vaginalis. 1–2 mL of serous fluid is normally observed in the space between the two layers of tunica vaginalis. This is different to the fluid in communicating hydrocoele, which is peritoneal fluid. Excessive serous fluid production may be brought on by infection, malignancy, inflammatory conditions or testicular trauma.

13.3.2.2   Insufficient Reabsorption of Serous Fluid

Over-expression of aquaporins (AQP1) within the capillary endothelial cells of the tunica vaginalis are implicated as a possible cause of adult-onset primary non-communicating hydrocele (Hattori et al. 2014). Other causes for insufficient reabsorption of serous fluid include testicular torsion and malignancies.

13.3.2.3   Disruption to Normal Lymphatic Channels

Iatrogenic injury to testicular lymphatic channels is implicated in post-operative development of hydrocoele seen as a complication of varicocoelectomy, likely related to clamp placement over lymphatic channels intra-operatively. (Salama and Blgozah 2014), (Esposito et al. 2004).

The parietal layer of tunica vaginalis is embedded with lymphatic channels that drain the secreted serous fluid in normal conditions. However, in the endemic tropical parasitic condition Bancroftian filariasis, parasitic worms interrupt testicular lymphatic flow by residing within intrascrotal lymphatic vessels, causing accumulation of serous fluid within the layers of tunica vaginalis, (Norões and Dreyer 2010). Filariasis is the commonest cause of hydrocoele in adult men globally, though it is infrequently seen in developed countries.

13.3.2.4 OtherCauses

Less commonly, hydrocoele may occur secondary to testicular pathology, retained foreign body, trauma, infection or as a result of malignancy. Septations within the hydrocele sac seen on ultrasound, rapid evolution of the hydrocoele, inguinal or pelvic lymphadenopathy or the presence of constitutional symptoms such as unintentional weight loss and night sweats may warrant consideration of a diagnosis of malignancy.

13.3.3 Hydrocele of the Spermatic Cord

A hydrocele of the cord occurs when fluid accumulates along the spermatic cord in the inguinal canal or upper scrotum. It is caused by an abnormal closure of the processus vaginalis where the distal portion obliterates but the midportion remains patent along the cord. Typically, there are two types of cord hydroceles. Non-communicating or encysted hydroceles are more common and are characterised by fluid pooling along the length of the cord, but it does not communicate with the peritoneal cavity or tunica vaginalis. Cummunicating or funicular hydroceles communicate freely with the peritoneal cavity at the internal ring. A communicating hydrocele can enlarge with increased intra-abdominal pressure, whereas an encysted hydrocele does not.

13.3.4 Diagnosis of Hydrocoele

Hydrocoele is usually a clinical diagnosis in both adults and children. If there are other concerning symptoms or if clinical examination is undifferentiating, ultrasound may be of benefit in ruling out other groin pathology. If there is any suspicion of malignancy, or any concerning testicular irregularity seen on ultrasound, tumour markers and a CT of the chest, abdomen and pelvis should be performed. If inflammatory/infective pathology is suspected, STI screening and urine cultures should be obtained.

Most hydrocoeles in the paediatric population are diagnosed clinically, however indications for ultrasound include cases where there has been trauma to the area, suspected testicular torsion, suspicion for testicular or spermatic cord tumour or equivocal examination findings (Fourie and Banieghbal 2017).

13.3.5   History

Hydrocoeles typically present as a painless cystic scrotal mass and transilluminate. With communicating hydrocoeles, there may be a history of an increase in size during the day, and a decrease in size overnight as gravity drains the hydrocoele. Acute hydrocoeles may occur in children in conjunction with a respiratory tract infection or gastrointestinal illness, where coughing/vomiting and the associated sudden increase in intra-abdominal pressure can force peritoneal fluid through a previously undetected or small-calibre patent processus vaginalis, (Fourie and Banieghbal 2017). In this situation, hydrocoele may be painful due to sudden distention of the processus vaginalis.

13.3.6   Examination Findings

Hydrocoele presents as a tense, fluid-filled, painless and mobile mass in the upper scrotum or inguinal canal. A communicating hydrocoele may decrease in size when the patient lies flat, and may increase in size with the Valsalva manoeuvre or straining, whereas the size of a non-communicating hydrocoele is fixed.

Differentiation between hydrocoele and inguinal hernia is necessary to plan ongoing management, particularly in infants, in whom the presence of inguinal hernia may require surgical repair. Transillumination alone is not sufficient to satisfactorily differentiate between inguinal hernia and hydrocoele, as gas-filled bowel present in the scrotum can also transilluminate. If the ductus deferens is palpable above the mass, the clinician can confidently diagnose the hydrocoele and rule out inguinal hernia.

Examination should test for the presence of ipsilateral inguinal hernia, by the gentle introduction of the examiner's finger into the superficial inguinal ring. Pelvic and inguinal lymph nodes should be palpated, as enlargement may suggest underlying infectious or malignant pathology.

13.3.7   Complications of Hydrocoele

13.3.7.1   Testicular Changes and Infertility

Testicular atrophy and total arrest of spermatogenesis was observed in 8% of patients who had unilateral hydrocoele of the tunica vaginalis (Dandapat et al. 1990). In adults with large or chronic hydrocoele, the pressure of the hydrocoele exerted directly onto the testis and associated compromise of vascular supply are thought to impede spermatogenesis, resulting in infertility.

13.3.7.2   Masking/Distracting from Malignant Pathology

Benign and malignant tumours of the tunica vaginalis, testis or epididymis may present as a hydrocoele and may be overlooked.

13.3.7.3   Psychological Burden

Sexual dysfunction was found to be prevalent amongst married men with hydrocoele, presenting as a decrease in libido and difficulty sustaining erection.

13.3.7.3.1   Pain/Discomfort

Hydrocoele can rupture spontaneously or as a result of direct trauma to the tunica vaginalis, causing haematoma, oedema, inflammation and irritation of surrounding tissue. Less commonly, hydrocoele can become infected, with seeding of infection through the blood or urinary tract.

13.3.7.3.2   Development of Inguinal Hernia

In those with communicating hydrocoele, there may be development of an ipsilateral inguinal hernia, which is an absolute indication for surgical repair of the defect.

13.3.7.3.3   Calcific Changes

Calcium deposition within the testis or tunica vaginalis, as seen on ultrasound has an association with malignancy and prompts further investigation and management. However, benign formation of calculi is also observed as a result of chronic hydrocoele, and these calculi may be interpreted as calcifications on ultrasound. If suspicions remain, an inguinal approach to surgery is advised (Vahlensieck and Hesse 2010).

13.3.8   Management of Congenital/Communicating Hydrocoele

13.3.8.1   Conservative Management

The majority of asymptomatic congenital hydrocoeles resolve without intervention by 24 months of age. A resolution rate of 83.4% of hydrocoeles found in newborn boys by the age of 18 months has been reported (Osifo and Osaigbovo 2008).

A recent study found that one third of children who had hydrocoele beyond the age of 2 years experienced spontaneous resolution without intervention, and with no consequence to testicular growth. Hence, a period of watchful waiting for 6–12 months is acceptable even in children above the aye of 2, provided they are asymptomatic (Kurobe et al. 2020).

13.3.8.2 Surgical Management of Congenital/Communicating Hydrocoele

Indications for surgical repair in the paediatric population include associated ipsilateral inguinal hernia, abnormal epididymal or testicular ultrasound findings, and the presence of hydrocoele beyond 24 months of age. Unlike adults, sclerotherapy and aspiration is not a viable treatment option in the paediatric population (Fourie and Banieghbal 2017).

Communicating hydrocoeles found in adult males should be surgically corrected, due to the risk of developing an ipsilateral inguinal hernia. Communicating hydroceles are unlikely to spontaneously resolve in adults.

13.3.8.3 Open Inguinal Hernia Repair/Ligation of Patent Processus Vaginalis

Open inguinal approach is preferred in management of communicating hydrocoele to identify and ligate the patent processus vaginalis and aspirate fluid from a hydrocoele sac if required. The approach to operative management of paediatric hydrocoele is identical to operative management of paediatric inguinal hernia (Fourie and Banieghbal 2017). A transverse skin incision made in the groin, above the superficial inguinal ring. Underlying structures are gently dissected to reveal the spermatic cord, which contains the patent processus vaginalis. The spermatic cord is everted out of the incisional cavity and opened. The ductus deferens and testicular vessels are identified and preserved, and blunt dissection of the external spermatic fascia is undertaken to reveal the processus vaginalis. A small cut is made in the processus vaginalis and fluid is drained by means of digital pressure. All the cord structures are then returned to their original location.

13.3.8.4 Laparoscopic Surgery for Communicating Hydrocoele

Newer techniques utilise laparoscopic instruments to perform similar ligation of the patent processus vaginalis, with a reduction in post-operative pain and faster time to discharge. Laparascopic methods have the added benefit of allowing inspection of the contralateral inguinal region (Fourie and Banieghbal 2017).

13.3.9 Management of non-Communicating Hydrocoele

13.3.9.1 Conservative Management

Depending on clinical suspicions, urinalysis, urine culture, tumour markers and ultrasound may be indicated.

Aspiration alone is a minimally invasive method used to treat hydrocoele, but is associated with a high rate of recurrence. Aspiration of hydrocoele with administration of a sclerosing agent, such as sodium tetradecyl sulfate, is a safe, minimally invasive procedure which can be performed in the outpatient setting. Failure rate of up to 36.8%, four weeks after the first instillation of sclerosant has been reported (Musa et al. 2015). Since recurrence is common after aspiration and administration of a sclerosing agent, often more than one treatment is required. This form of management, due to its high recurrence rate, is usually reserved for those who are poor surgical candidates (Rioja et al. 2011). A study comparing techniques for managing hydrocoele found a recurrence rate of 11% amongst patients who underwent aspiration and sclerotherapy compared to a recurrence rate of 0% in those who underwent surgical management of their hydrocoele. (Roosen et al. 1991).

Asymptomatic hydrocoeles in adults do not routinely warrant treatment. The decision to manage adult hydrocoele is based on the presence of complicating factors such as discomfort, testicular atrophy and fertility issues.

13.3.9.2 Surgical Management of non-Communicating Hydrocoele

Hydrocoelectomy is gold standard in the management of adult hydrocoele and is beneficial in limiting complications such as infertility, testicular atrophy, discomfort, infection and psychological distress. It is interesting to note that 86.4% of hydrocoeles in children over the age of twelve were non-communicating hydrocele and suitable for repair via a scrotal approach (Wilson et al. 2008). Hydrocoelectomy is inappropriate if there is a suspicion of testicular malignancy and should be ruled out prior to surgery. Approaches to hydrocoelectomy involve excision of the secretory tunica vaginalis or creating a defect in the parietal layer of tunica vaginalis, exposing the serous fluid to the surrounding layers with absorptive capacity.

13.3.9.3 Open Scrotal Approaches: Jaboulay's Procedure

A midline scrotal incision is made and the contents of the scrotum are delivered out of the incision. The testis is everted through an incision made in the tunica vaginalis and much of the sac is resected, leaving only a small cuff around the border of the testis. Some surgeons leave enough sac to reflect the tunica behind the epididymis. The layers of the tunica are then sutured together for haemostasis and the scrotal contents are returned to their original position. (Rioja et al. 2011).

236

13.3.9.4 Open Scrotal Approaches: Lord's Technique

Lord's technique is an alternative approach, which differs from Jaboulay's procedure as it involves a smaller incision in the scrotal skin and hydrocoele sac, and there is minimal dissection between the layers of dartos muscle and tunica vaginalis. There is no excision of the hydrocoele sac which is plicated instead. The dissection of dartos from the tunica vaginalis performed during Jaboulay's, as well as excision of the sac, is thought to contribute to haematoma formation and is avoided in Lord's technique, (Lord 1964).

Overall recurrence rate after open hydrocoele surgery is approximately 6% (Tsai et al. 2019), with no significant difference between operative approaches. Lord's technique demonstrated a significantly lower rate of haematoma compared to Jaboulay's procedure.

13.3.9.5 Minimally Invasive Hydrocoele Surgery

Newer minimally invasive and laparoscopic techniques have been developed with success in definitively managing hydrocoele, with lower rates of scrotal hardening and smaller incisions (Saber 2015).

13.3.10 Operative Complications

13.3.10.1 Open Inguinal Approach and Ligation of Patent Processus Vaginalis (for Communicating Hydrocoele)

13.3.10.1.1 Injury to Ductus Deferens and Infertility

Injury to ductus deferens sustained during unilateral paediatric inguinal hernia repair can contribute to the development of infertility in adulthood by means of the formation of anti-sperm antibodies (Matsuda et al. 1993). The incidence of injury to ductus deferens was 0.13% in a study involving 1494 paediatric inguinal hernia repairs (Partrick et al. 1998) and 0.23% is a study of 7313 paediatric inguinal hernia repairs (Steigman et al. 1999).

13.3.10.1.2 Injury to Testicular Vessels and Infertility

Ischaemic atrophy of the testicle secondary to iatrogenic injury to the testicular vessels is a possible complication of inguinal approach surgery for hydrocoelectomy.

13.3.10.1.3   Scrotal Oedema

A common complication in all forms of hydrocoele surgery though this is usually self-resolving within months.

13.3.10.2   Open Scrotal Approach Hydrocelectomy (for Non-communicating Hydrocoele)

13.3.10.2.1   Recurrence of Hydrocoele

Recurrence rate is in the order of 9%. None of these patients require immediate re-operation, and the swelling often resolves without intervention post-operatively. (Kiddoo et al. 2004).

13.3.10.2.2   Infection

Post-operative infection has been reported in 9.3% of patients (Kiddoo et al. 2004), which is markedly higher than what is expected for a 'clean-contaminated' procedure. Antibiotic prophylaxis is recommended pre-operatively to mitigate this risk.

13.3.10.2.3   Scrotal Hardening

Scrotal hardening, thought to be due to the amount of tissue resected and handling of the hydrocoele sac is a common complication of open hydrocolectomy. It is reported in up to 24.2% of patients undergoing Jaboulay's procedure. A minimally invasive technique involving minimal handling and resection of the hydrocoele sac reported that scrotal hardening only occurred in 4.8% of patient. (Saber 2015).

13.3.10.2.4   Scrotal Oedema

Scrotal oedema is a frequently occurring post-operative complication for all forms of hydrocoele surgery, though less frequently noted in the inguinal approach, (Lashen et al. 2019) and usually self-resolving within 3 months. The incidence is about 30% (Lei et al. 2019). The oedema is thought to occur secondary to the highly reactive tissues surrounding the testis being irritated intra-operatively, and surrounding layers being irritated by leakage of hydrocoele fluid into the region.

13.3.10.2.5   Haematoma

The majority of bleeding in hydrocoele surgery is attributed to the highly vascular cut edges of the hydrocoele sac. Lower haematoma rates are reported amongst patients who undergo minimal resection of the hydrocoele sac, when compared to the extensive resection performed in Jaboulay's technique. (Saber 2015).

13.4   Varicocoele

A varicocoele is defined as the abnormal dilatation of the gonadal vein(s) (also referred to as the internal testicular veins) which forms the pampiniform plexus within the spermatic cord and scrotum. This complex network of veins is responsible for drainage of the testis.

Varicocoele affects approximately 15% of the general adult male population and is of importance in its ability to impede fertility (Clavijo et al. 2017). Varicocoele affected 35–44% of males presenting with primary infertility and 45–81% of males presenting with secondary infertility. It is the most common cause of male infertility which is surgically treatable (Kupis et al. 2015).

13.4.1   Pathophysiology

Venous drainage of the testes occurs by means of a network of veins (the pampiniform plexus), split into two bundles; the first bundle involving a meshwork of veins wrapped around the testicular artery and the second bundle consisting of veins embedded into the surrounding fat (Ergün et al. 1997). Both bundles are interconnected with small intervening veins. There are also numerous arterio-venous connections (venae comitantes) between the pampiniform plexus and testicular artery.

Whilst the pampiniform plexus functions to drain the testis, a separate cremasteric plexus drains the coverings of the testis and spermatic cord, and itself is a tributary of the inferior epigastric vein. It is important to note the anastomotic connections between the pampiniform plexus and cremasteric plexus, as these are crucial to understanding the complications implicated in varicocoelectomy.

The veins of the pampiniform plexus ultimately condense into three or four veins at the superficial inguinal ring, and further coalesce to form a single gonadal/testicular vein after exiting the deep inguinal ring. On the left, the gonadal vein usually drains into the left renal vein. On the right, the gonadal vein usually drains directly into the inferior vena cava. Though there can be variations to this, cadaveric studies demonstrate the drainage of the left gonadal vein almost always takes a longer course compared to its right-sided counterpart (Asala et al. 2001). Due to its longer course, the venous pressure is greater on the left side, predisposing formation of left-sided varicocoele rather than the much rarer isolated right-sided varicocoele (Dubin and Amelar 1977).

There are several theories as to what causes the development of varicocoele:

13.4.1.1   Increased Hydrostatic Pressure

Due to the longer course of the left gonadal vein and it's right-angled entry into the left renal vein, the venous pressures in the left gonadal vein, are higher. This is transmitted into the pampiniform plexus, where increased hydrostatic pressure causes dilatation of the veins and can contribute to formation of varicocoele.

13.4.1.2   Incompetent Valves and Variation in Drainage

Incompetent or absent valves of the internal spermatic or gonadal veins were found in one-half of men on post-mortem studies, which may contribute to pooling of venous blood in the pampiniform plexus. Further to this, the anastomotic connections that occur in some men between the veins of the pampiniform plexus and other veins that lack competent valves (such as the cremasteric plexus) may also be implicated in the development of varicocoele, (Clavijo et al. 2017). This also explains the frequent recurrence in varicocoelectomies that are performed sparing these anastomotic connections.

13.4.1.3   Extrinsic Compression

On its way to drain into the inferior vena cava, the left renal vein passes between the superior mesenteric artery and abdominal aorta and can become compressed between these structures, causing an increase in venous pressure that is then transmitted to the left pampiniform plexus, a phenomenon commonly referred to as the 'nutcracker syndrome' (Lewis et al. 2015). This can be confirmed by doppler ultrasonography if suspected.

More concerningly, external compression from a tumour, originating from the kidney/pancreas/retroperitoneum/adrenal gland can also cause varicocoele on either the left or right side. All right-sided varicocoeles require CT of the abdomen and pelvis to rule out retroperitoneal mass, however the same is also suggested for left-sided varicocole in the presence of other signs and symptoms suspicious of malignancy (Lewis et al. 2015).

13.4.2   Impairment of Fertility

The precise mechanism by which varicocoele impedes male fertility remains uncertain, though there are several factors that are thought to contribute to this complication.

13.4.2.1 Temperature

The increased temperature secondary to the reflux of 'warmer' abdominal blood causes a reduction in testosterone synthesis by Leydig cells in the testis and decreases synthetic function of the Sertoli cell, both of which are crucial in spermatogenesis.

13.4.2.2 Decreased Testicular Perfusion

Surgically induced unilateral varicocoele in monkeys demonstrated a reduction in testicular blood flow four months after the procedure. This suggests increased venous pressure even unilaterally, affects testicular blood flow bilaterally and can cause bilateral testicular atrophy, which could lead to infertility.(Harrison et al. 1983) A further study in varicocoelised rats confirmed that unilateral varicocoele causes bilateral testicular hypoxia, and induces apoptosis of germ cells (Wang et al. 2010).

13.4.2.3 Oxidative Stress

The unfavourable environment, hypoxia and possibly the backflow and accumulation of toxic metabolites from the renal/adrenal organs are thought to cause an increase in reactive oxygen species production within the testicular vessels. Reactive oxygen species can cause oxidative damage to the plasma membrane, DNA and mitochondria of spermatocytes, ultimately resulting in sperm dysfunction. This was supported by higher proportions of spermatozoa without intact DNA and with inactive mitochondria amongst men with varicocoeles than men without (Blumer et al. 2012).

13.4.3 Diagnosis of Varicocoele

13.4.3.1 History

Generally men presenting with varicocoele are asymptomatic and are often only found on routine physical examination. Some men may present with a feeling of 'fullness' or discomfort within the scrotum.

Pain secondary to varicocoeles is uncommon, but is characterised by a dull ache that worsens with activity or on standing, and improves on resting.

A thorough medical, reproductive and sexual history should be taken. Constitutional symptoms such as fevers, night sweats or weight loss may raise the possibility of a malignancy causing compression of obstruction of the left renal vein, causing the varicocoele, though this would be an infrequent cause.

13.4.3.2 Examination

Physical examination alone is usually sufficient to confidently diagnose varicocoele. The patient's scrotum should be examined and palpated in the standing position, while the Valsalva manoeuvre is performed by the patient to provoke the varicocoele. The same should be repeated when the patient is in a relaxed (seated) position.

The grading system used classifies varicocoeles according to their severity, with grade I varicocoeles only palpable with Valsalva, whereas grade III varicocoeles are large and visible whilst standing.

An isolated unilateral right-sided varicocoele is concerning, and warrants investigations looking for tumours/retro-peritoneal pathology (such as malignancy causing extrinsic compression along the venous drainage pathway).

13.4.3.3 Investigation of Varicocoele

Scrotal ultrasound is not routinely required to diagnose varicocoele in adults but is useful if examination findings were inconclusive due to variant anatomy/obesity (Lomboy and Coward 2016). Ultrasound is highly sensitive and specific in confirming the diagnosis of varicocoele, and in ruling out other inguinoscrotal pathology.

Scrotal ultrasound to estimate testicular volume is routinely performed in adolescent males with varicocoele, as a discrepancy in testicular volume of over 20% represents an indication for surgical correction (Lomboy and Coward 2016).

In men who are yet to complete a family, blood should be taken to test for androgen deficiency and other causes of infertility. Tests include total and free testosterone levels, luteinizing and follicle stimulating hormone levels, prolactin and oestrogen levels (Lomboy and Coward 2016). At least two separate semen analyses should be performed prior to consideration of surgery for varicocoele.

All men presenting with new isolated right-sided varicocoeles require CT imaging of their abdomen and pelvis looking for tumour/retroperitoneal masses that can cause varicocoele by means of extrinsic compression. CT abdomen and pelvis is also recommended for new presentations of left-sided varicocoele if there are any suspicious history or examination features, such as unintentional weight loss or haematuria (Lewis et al. 2015).

13.4.4 Management of Varicocoele

Most varicocoeles do not require intervention. Indications for surgical treatment of varicocoele include scrotal pain, testicular atrophy (particularly in children), known infertility and abnormal sperm quality. Surgical treatment of varicocoele has a role in restoring fertility in patients who have oligospermia or non-obstructive azoospermia, with demonstrated evidence of restoration of spermatogenesis. Further to this, emerging evidence suggests early repair of varicocoele can prevent future infertility and Leydig cell dysfunction. There is a duration-dependent deterioration in testicular function seen with varicocoele.

Though testicular atrophy leading to infertility is a demonstrated complication of varicocoele, there are no established criteria as to when to intervene, despite the fact that irreversible damage to testicular function can result if untreated, (Chiba et al. 2016). A discussion with the patient regarding the potential complications of varicocoele and the risks and benefits of intervention is important in guiding the decision-making process.

13.4.4.1 Percutaneous Varicocoele Embolization

This procedure can be performed in an antegrade or retrograde fashion. In the retrograde technique, an angiocatheter is used to enter the right femoral vein and threaded through to the gonadal vein, via the inferior vena cava and left renal veins. Location is confirmed with the use of venous contrast, and a sclerosing injection is injected to occlude the vein. During the antegrade approach, sclerosant is injected into an isolated vein contributing to the pampiniform plexus, after confirming anatomy using fluoroscopy (Chan 2011). Both approaches can be performed under local anaesthesia and are minimally invasive procedures. Most patients are able to return to their normal activities much sooner, as there are no muscle-splitting incisions made.

Complications of varicocoele embolization include the migration of occlusive material, flank pain, accidental arterial puncture and infection. After embolization a varicocoele recurrence rate of 12.7% has been reported. This approach is not typically complicated by hydrocoele formation (Kupis et al. 2015).

13.4.4.2 Surgical Management

13.4.4.2.1 Scrotal Approach

No longer used, the open scrotal approach involving ligation and excision of the varicosed venous plexus involves a considerable risk of damage to the testicular artery around which the pampiniform plexus is intimately wrapped. Scrotal approaches have been abandoned as the risk of compromising arterial supply to the testis is significant and can result in testicular atrophy and worsened infertility.

13.4.4.2.2 Open Non-Microsurgical Technique, 'Palomo Technique'/ Retroperitoneal Approach

The 'Palomo technique' of open non-microsurgical varicocoele repair is performed through a horizontal incision inferomedial to the anterior superior iliac spine, extending medially. Underlying fascia and muscle are split and the peritoneum retracted to expose the internal spermatic veins and artery retroperitoneally, adjacent to the ureter. At this location, the internal spermatic veins, usually two or three in number, are ligated 'high', near where they drain into the left renal vein. At this location, the testicular artery has not yet branched, and is distinctly separate from the internal spermatic veins. This technique is possible under local anaesthesia alone.

The Palomo technique is associated with a higher rate of complications and a higher recurrence rate than alternative approaches. A recurrence rate of 15% was found, and development of hydrocoele was found to be a complication in 9% of cases (Kupis et al. 2015). The high recurrence rate is likely due to the 'high ligation' of the internal spermatic veins, where there is no opportunity to ligate the periarterial plexus of fine veins (venae comitantes) along with the testicular artery that may contribute to varicocoele development in the future. The common incidence of hydrocoele as a complication of retroperitoneal varicocoelectomy is likely due to the difficulty of identifying and preserving the lymphatic structures intra-operatively.

13.4.4.2.3 Laparoscopic Technique

The advantages of the laparoscopic technique lie in easier visualisation of important structures, and hence complication rates are significantly less when this technique is utilised. In laparoscopic varicocoelectomy, the gonadal veins are ligated proximally, nearer to where they enter the left renal vein. The benefit of this is that only one or two veins require ligation (Chan 2011). Further to this, the testicular artery is easily identified and often distinct and separate from the gonadal veins, and thus is unlikely to be injured. A drawback to the laparoscopic approach is that it does not allow access for the ligation of the cremasteric veins, which can be a secondary contributor to varicocoele development.

The recurrence rate was 6.1%, and the rate of development of hydrocoele as a complication was found to be 2.84% in those patients who underwent laparoscopic varicocoele repair (Kupis et al. 2015). The comparatively lower rate of hydrocoele development is likely a result of the magnification offered by the laparoscope, allowing for identification and preservation of lymphatic channels. The recurrence of varicocoele after laparoscopic repair is thought to be due to preservation of the small arterio-venous branches that subsequently dilate and cause recurrence, or secondary to preservation of cremasteric veins that can themselves cause varicocoele.

13.4.4.2.4 Subinguinal Microsurgical Technique

With the lowest complication and recurrence rates, the microsurgical subinguinal technique for varicocoelectomy has become the standard approach in most academic institutions and can be performed under local or regional anaesthesia. The microsurgical subinguinal technique utilises 10–25x magnification of the relevant anatomy which confers unique benefits to this approach as follows:

- Access to external spermatic and gubernacular veins, which may bypass the spermatic cord and can contribute to recurrence if spared.
- Clear identification of varicocoeles and miniscule venous tributaries that have the potential to cause recurrence of varicocoele, (Chan 2011). These are likely to be missed in the absence of strong magnification.
- Visualisation of intricate anatomy, allowing identification of arteries and lymphatic vessels and hence avoiding complications related to accidental ligation of these structures, and allowing for artery-sparing varicocelectomy (Chan 2011).

The microsurgical technique boasts the lowest recurrence rate of all interventions of 1.05%, and development of hydrocoele as a complication in 0.44% (Kupis et al. 2015).

13.4.5 Complications of Varicocoelectomy Surgery

13.4.5.1 Hydrocoele

Testicular lymphatics run posteriorly to the vascular bundle, and are vulnerable to injury in varicocoelectomy surgery (Al-Said et al. 2008). Of the described surgical approaches, the lowest rate of hydrocoele was in the microsurgical subinguinal technique, due to easier intra-operative identification of lymphatic vessels.

13.4.5.2 Recurrence

Approximately 16% of men undergoing inguinal varicocoelectomy were found to have at least one external spermatic vein (Goluboff et al. 1994). External spermatic veins (usually anastomotic branches of the cremasteric vein) are generally found on the floor of the inguinal canal, travel postero-lateral to the spermatic cord and exit via the deep inguinal ring to drain into the inferior epigastric vein. Hence, suprainguinal approaches (such as that used in the Poloma technique) do not afford the opportunity to identify and ligate these tributaries, leading to recurrence of the varicocoele. Owing to the use of magnification to identify and ligate the smaller collateral veins contributing to the varicocoele, the microsurgical subinguinal varicocoelectomy had the lowest recurrence rate of all interventions.

13.4.5.3 Laparoscopic Complications

The highest rate of complications overall was attributed to the laparosopic approach, where trocar site infection and bleeding were additional complications specific only to this approach (Diegidio et al. 2011).

13.4.5.4 Ligation of the Testicular Artery

Ligation of the testicular artery, though shown to be efficacious in improving varicocele and reducing recurrence rates, may confer a risk of testicular atrophy and can impact spermatogenesis (Celigoj and Costabile 2016). Artery-sparing varicocoelectomy is preferred in patients who wish to preserve their fertility and this is best ensured using the subinguinal microsurgical approach (Celigoj and Costabile 2016). In children, however, the potential for neovascularization and compensatory hypertrophy of the remaining testicular arterial supply (the artery to ductus and the cremasteric artery) means testicular atrophy is less likely (Celigoj and Costabile 2016).

13.5 Epididymal Cyst Lesions: Spermatocoele and Epididymal Cyst

13.5.1 Anatomy and Physiology

The testis is divided by septations of its outer covering (tunica albuginea), which incompletely separates the testicular tissue into testicular lobules. Each lobule contains the tightly coiled seminiferous tubules supported by a loose connective tissue containing androgen producing Leydig cells that contribute to the surrounding testicular interstitial fluid.

The lining of the seminiferous tubules contains Sertoli cells and spermatogenic cells. Sertoli cells secrete numerous substances that nourish the developing spermatocytes. They are also responsible for the 'blood-testis barrier', the function of which is to provide a specialised micro-environment in which spermatogenesis and spermatozoal development can occur, avoiding auto-immunological sensitisation to the spermatozoa (França et al. 2016), (Cheng and Mruk 2012).

The seminiferous tubules terminate in a system of narrow straight ducts called the tubuli recti. The tubuli recti coalesce to drain into a flattened sinus called the rete testis, which is found at the thickening of the tunica albuginea, called the mediastinum testis.

Spermatozoa are produced by spermatogenic cells in the seminiferous tubules and move through these tubules, through the tubuli recti, to the rete testis and reach the efferent ducts, a series of 7–11 straight ducts that become convoluted where they fuse to form the head (caput) of the epididymis. Therefore, if the epididymis is injured distal to the caput (where the efferent ducts fuse), the result is total obstruction of the entire system on that side—an important consideration during surgeries involving this region.

The epididymis is a 6–7 metre long tightly coiled tube located on the posterolateral aspect of the testis. They have a crucial role in storing and nurturing spermatozoa until they are mature and ready for transport through the ductus deferens during ejaculation (Jones 1999). The epididymis is where the spermatozoa gain their motility function and acquire the capacity to fertilize. These changes are induced by a variety of complex maturational processes that are induced by secretions from the epididymis during their transit. Trauma, inflammation, obstruction and chemically mediated damage to the epididymis or related structures can result in impaired sperm morphology, motility and hence lead to infertility.

The epididymis also possesses a blood-epididymis barrier, formed by a pseudostratified epithelium composed of cells with multiple tight junctions delimiting the intraluminal compartment and again offering an environment where spermatozoa are sheltered from potential auto-immune sensitisation.

13.5.2   Incidence

The majority of spermatocoele/epididymal cysts are asymptomatic and often found incidentally. Scrotal ultrasound performed on 40 asymptomatic males found the incidence of spermatocoele/epididymal cyst to be 29% (Leung et al. 1984). However, between 2004 to 2015, an annual incidence of 38.5/100,000 males presented for inpatient or outpatient visits at hospitals for.

The age distribution of cystic scrotal lesions (spermatocoele/epididymal cyst or hydrocoele) has a peak mean incidence between 65 and 75 years of age. After the age of 65 years, hydrocoeles were more common than spermatocoeles/epididymal cysts (Lundström et al. 2019).

13.5.3   Pathophysiology

A spermatocoele is a cystic sac arising from the rete testis or a single efferent duct and is a result of obstruction or dilatation of the efferent ductal system. In contrast, an epididymal cyst, being a true cyst of the epididymis, is lined with epithelium and may arise from anomalous/vestigial remnants of the epididymis that do not communicate with the seminiferous tubules (Cross and Underwood 2013).

There is no way to clinically distinguish between spermatocoele and epididymal cyst, though spermatocoele fluid usually contains spermatocytes, unlike epididymal cysts. Regardless, there is no role in delineating between these two diagnoses, as management does not differ.

Spermatocoeles contain spermatozoa and interstitial fluid and as a result, usually transilluminate well. Epididymal cysts also transilluminate, but do not contain spermatozoa.(Cooper et al. 1992).

13.5.4   Pathophysiology

There is sparse conclusive evidence of the underlying risk factors that predispose to epididymal cyst and spermatocoele.

13.5.4.1   Epididymal Cyst

- Diethylstilboestrol is a synthetic oestrogen which was prescribed to pregnant females between 1938–1971 as it was thought to prevent miscarriages and premature delivery. It has been associates with an increased prevalence of reproductive tract abnormalities in females who were exposed in-utero. Males who were exposed to DES in-utero were found to have an increased prevalence of epididymal cysts, cryptorchidism and testicular inflammation or infection (Palmer et al. 2009).
- The Notch family of transmembrane receptors is involved in a signalling pathway contributing to cell fate determination and is found to be carried by reproductive cells in mice studies. Those mice who expressed the activated intracellular domain, namely, 'Notch1' were found to have hyperplasia and obstruction of the efferent ducts at the rete testis/efferent duct interface, leading to the development of epididymal cyst. The underlying process is thought to be secondary to abnormal cell proliferation of epithelial cells in the efferent ducts, with these cells causing the obstruction themselves, or by improperly fusing with the rete testis, causing dilatation.

13.5.4.2   Spermatocoele

- Agglutinated degenerated germ cells derived from the exfoliated seminiferous tubules were found in the histopathological analysis of spermatocoele in mice (Itoh et al. 1999). There was no associated inflammatory cell infiltration, granuloma or evidence of trauma found. It is possible that obstruction of an efferent duct by aggregated germ cells may initiate the formation of spermatocoele, with the release of immature germ cells into the lumen of the seminiferous tubules thought to be an effect of ageing (Itoh et al. 1999). It is unclear as to whether these findings apply to human spermatocoele development.

13.6   Clinical Presentation and Diagnosis

13.6.1   History

Spermatocoeles/epididymal cysts generally do not cause troubling symptoms and are most commonly incidentally diagnosed on physical examination or through self-examination by the patient. Rarely, when the epididymal cyst originates from an anomalous/vestigial structure, such as the appendix of the epididymis, this cyst can twist on its stalk, and cause severe acute onset testicular pain, similar to that seen in testicular torsion (Cross and Underwood 2013). Similarly, spermatocoele can also twist along its own stalk, leading to presentation as an acute scrotum (Hikosaka and Iwase 2008; Ameli et al. 2015).

Patients may complain of a palpable mass in their scrotum, or of a dull pain/ache/discomfort or 'heaviness' in their scrotum. A history of urinary tract infections, voiding difficulty, urethral discharge or fever suggests epididymitis or infectious pathology of the urinary tract and is important to consider (Coogan 2008).

Maternal exposure to DES, prior trauma/surgery to the scrotum and the presence of any familial risk factors for malignancy should be determined (Coogan 2008). Any concerning symptoms suggesting malignancy, including unintentional weight loss, unexplained fatigue or night sweats should warrant urgent urological referral and investigation for malignancy.

13.6.2   Examination

Physical examination should be performed by examining the 'normal' testis first. On physical examination, spermatocoele/epididymal cysts present as a palpable, transilluminable smooth cystic structure originating from the epididymis or on the superior aspect of the testis. Note that this contrasts with hydrocoele, in which the fluid surrounds the testis itself.

The borders of the spermatocoele/epididymal cyst should be palpably separate from the testis itself. Of the patients who have pain associated with the cyst, it is localised to the epididymis, and can be elicited on palpation of the epididymis.

13.6.3   Investigations

Ultrasound is a useful method of confirming diagnosis of spermatocoele/epididymal cyst, particularly when the cyst is large and difficult to differentiate from hydrocoele or varicocoele. Ultrasound cannot differentiate between spermatocoele and epididymal cyst.

Spermatocoele and epididymal cyst will present on ultrasound as a hypoechoic well circumscribed simple cystic structure originating from the epididymis.

Urine culture and screening for sexually transmitted disorders should be performed if epididymitis is suspected (Coogan 2008).

13.6.4 Complications of Spermatocoele/Epididymal Cyst

13.6.4.1 Infertility

There is limited conclusive evidence as to whether the presence of a spermatocoele/epididymal cyst can impact fertility, though there is a higher incidence of spermatocoele/epididymal cysts amongst males with infertility compared to fertile males (Kauffman et al. 2011).

The surgical interventions to treat these conditions pose a significant risk of iatrogenic injury to the epididymis and related structures, which in itself can cause infertility. Hence, the decision to intervene on patients with epididymal cyst/spermatocoele to treat infertility, or where the preservation of fertility is desired, is a serious undertaking. Such patients warrant referral to a urologist who is familiar with these techniques and should be made aware of the potential risk of losing fertility as a result of these interventions.

A case-control study recruited 67 males presenting with infertility, and 24 males presenting for elective vasectomy to undergo scrotal ultrasound. There was no statistically significant difference in the incidence of epididymal cysts amongst the infertile men when compared to the fertile men, which was between 66.7–73.1%. This study did not find a statistically significant risk of infertility as a result of epididymal cyst.(Weatherly et al. 2018).

13.6.4.2 Epididymitis

Epididymitis is the painful infection of the epididymis and can occur secondary to epididymal cysts. The epididymis may be overtly swollen and exquisitely tender, presenting as an acute scrotum. Chronic epididymitis lasting longer than 6 weeks may also occur. The majority of epididymal infections resolve with appropriate antibiotics, scrotal ice, scrotal elevation and anti-inflammatory medication. When the acute phase of the illness has settled, epididymectomy or surgical removal of the cyst generally has favourable outcomes (Partin et al. 2021).

13.6.4.3 Torsion

A rarer presentation of spermatocoele or epididymal cyst is acute onset unilateral testicular pain, concerning of testicular torsion. Due to their stalk like paratesticular attachment, both epididymal cyst and spermatocoele can twist on their stalk and present as an acute scrotum (Hikosaka and Iwase 2008; Ameli et al. 2015).

13.6.5 Management of Spermatocoele/Epididymal Cyst

The rate of procedural or surgical intervention, for either spermatocoele/epididymal cyst or hydrocoele is approximately 20% (Lundström et al. 2019).

The mean age of men seeking surgical management of spermatocoele is 46 years. The most common reason to seek intervention was to alleviate pain and the bothersome sensation of a mass in the scrotum. The time between diagnosis of spermatocoele and eventual surgical management was 48 months, with patients largely tolerating symptoms for some time after diagnosis had been made. The average size of spermatocoeles on excision was found to be 4.2 cm in diameter. Males with the largest spermatocoeles were on average 10 years older than those with the smallest spermatocoeles, suggesting these grow over time (Walsh et al. 2007).

Spermatocoeles/epididymal cysts, where intervention is sought in males who wish to preserve their fertility, are best approached using the microsurgical technique to cyst excision, as it has demonstrably lower rates of iatrogenic damage to the epididymis.

13.6.6 Conservative Management

13.6.6.1 No Intervention

Asymptomatic spermatocoele/epididymal cyst does not warrant any intervention.

Pain, discomfort, infertility or infection are the indications for which to consider definitive intervention. Men of reproductive age should be counselled regarding the risk of developing obstructive azoospermia as a result of intervention, due to damage to the epididymis.

13.6.6.2 Aspiration and Sclerotherapy for Spermatocoele/ Epididymal Cyst

Similar to hydrocoele management, spermatocoele/epididymal cyst can also be treated by means of aspiration and instillation of a sclerosant. Various sclerosants have been used, with varying rates of success, including phenol (48% success after one treatment), polidocanal (59% success after one treatment), doxycycline (84–85% success after one treatment) and alcohol (71% success after one treatment) (Low et al. 2020).

Aspiration and sclerosis can be performed in the outpatient setting, under local anaesthesia. A needle-catheter is punctured into the cyst after infiltration of local anaesthetic. The needle is removed, leaving the catheter tip within the cyst. Fluid is aspirated and sent for laboratory analysis if required. Following aspiration, further local anaesthetic is injected into the empty cyst cavity and the sclerosing agent is instilled. Following instillation, the area is massaged to maximise exposure of the lining of the cyst wall to the sclerosant, before removing the catheter and applying pressure to the puncture site to ensure haemostasis (Low et al. 2020).

Many patients require more than one treatment. One study found that 85% of patients with spermatocoele/epididymal cyst who underwent a maximum of two aspiration and sclerotherapy treatments had no recurrence of their symptoms on follow up at 29 months, with a post–initial treatment failure rate of 26.5%.(Low et al. 2020). Another reported a post-initial treatment failure rate of 46% using Polidocanol as the sclerosant (Jahnson et al. 2011). Another study involving 69 patients who underwent aspiration and sclerosis for either spermatocoele/epididymal cyst or hydrocoele, found a success rate in 98% of patients, with an average of 1.3 treatments. This study also found a reduction in spermatozoa concentration and motility 6 months following the treatments, though a return to baseline parameters was found at 12 months follow up. (Shan et al. 2011).

The procedure is usually well tolerated, with pain being the most commonly reported intra-procedural complication. 17.4% of patients reported intra-operative pain, of which most reported as 'mild' pain. (Shan et al. 2011). 2.6% of patients reported pain on instillation of the alcohol, described as 'stinging, burning, or being kicked in the groin' (Low et al. 2020). Despite this, 100% of surveyed patients were happy to undergo the procedure again.

Overall complication rates were 5.6% in and included superficial scrotal abscess at the puncture site requiring incision and drainage, scrotal infection requiring oral antibiotics and scrotal haematomas. A rare reported complication is focal scrotal skin necrosis, a result of leakage of the sclerosant into the scrotal wall. This requires surgical excision of the necrosed skin and re-approximation of tissue under general anaesthesia. (Low et al. 2020).

Compared to surgical management, aspiration and sclerosis has a lower rate of complications, but also a higher rate of failure. It is common to require more than one treatment. Despite this, patients reported high satisfaction rates with the procedure. Generally, aspiration and sclerosis of spermatocoele/epididymal cyst is a safe option for patients who are unable to undergo general anaesthesia, prefer outpatient management or those who do not wish to undergo the riskier surgical options.

13.6.6.3   Operative Management for Spermatocoele/Epididymal Cyst

Intervention for spermatocoele or epididymal cyst is performed if there is associated pain, discomfort, infection or infertility (Celigoj and Costabile 2016).

It is important to counsel the patient pre-operatively regarding the risk that the surgery may not bring significant relief of their symptoms, and to discuss the potential for infertility as a result of epididymal injury (Zahalsky et al. 2004).

13.6.6.4   Nonmicrosurgical Spermatocoelectomy

Spermatocoelectomy is approached with a small incision on the midline raphe of the scrotal sac. The spermatocoele/epididymal cyst is excised without opening the cyst by means of sharp dissection off the surrounding epididymis and testis. Sharp dissection is recommended to spare the delicate epididymis and testis from excessive mobilisation and force. Once haemostasis is ensured, the edges of epididymis from which the spermatocoele/epididymal cyst was excised are approximated and sutured. A portion of tunica can be used to close the defect (Rioja et al. 2011).

Dense adhesions arising from the spermatocoele may complicate the procedure, in which case partial or total epididymectomy is recommended as an alternative (Rioja et al. 2011).

13.6.6.5   Microsurgical Epididymal Cystectomy/Spermatocoelectomy

A scrotal incision is made and the parietal layer of tunica vaginalis is opened to expose the testis, epididymis and spermatocoele/epididymal cyst. An operative microscope with 10–25x magnification is utilised for visualisation of the smaller structures. The tunica vaginalis is opened with microscissors, near the border of the cyst and testis, where the visceral layer of tunica vaginalis is separated from the underlying structures due to the cyst. The aim here is to avoid puncturing the cyst itself and more importantly, to avoid damage to the epididymis. Using blunt and sharp microscissor dissection, the cyst is separated from surrounding structures, and removed intact. Bipolar electrocautery is used carefully, as required. If a spermatocoele is found, it is followed through to the single efferent duct from which is has originated. Doing so affords the surgeon visual confirmation that no other structures will be damaged during ligation of this 'stalk'. The isolated efferent duct is suture-ligated and the spermatocoele is removed intact. Prior to closure, the epididymal bed should be examined to confirm whether there has been damage to the tubules. The tunica of the epididymis is reapproximated to the tunica albuginea and the opened tunica vaginalis is closed after returning the testis and epididymis to its physiological position. The scrotum is closed in layers. The excised cyst can be sent for histopathological analysis to confirm diagnosis of epididymal cyst or spermatocoele.(Kauffman et al. 2011) (Hou et al. 2018).

Benefits of the microsurgical approach testify to the utility of intra-operative magnification to avoid damage to surrounding structures.

In patients who had infertility and abnormal sperm parameters prior to intervention, this microsurgical technique demonstrated the return of normal sperm count, motility and morphology after the microsurgical intervention, in those patients who had infertility and abnormal sperm parameters prior to intervention. There was also no evidence of post-surgical decrease in sperm count, sperm motility or derangement of sperm morphology. (Kauffman et al. 2011; Hou et al. 2018).

13.7 Operative Complications in Spermatocoele/Epididymal Cyst Surgery

13.7.1 Damage to the Epididymis, which Can Lead to Infertility

Direct injury to the delicate epididymis or surrounding structures is of particular concern in surgery for treatment of spermatocoele/epididymal cyst, owing to the proximity of these structures to the cyst that is being excised. Damage may occur during cyst dissection/excision, as a result of heat/electrocautery or, in the case of aspiration and sclerosis, as a result of leakage of sclerosant to surrounding tissues.

Histopathological evidence of damage to surrounding epididymis can be identified in 17% of analysed specimens after spermatocoelectomy. The true rate of epididymal injury is thought to be higher as a result of unidentified electrocautery induced damage. (Zahalsky et al. 2004) In contrast, there was no histopathological evidence of epididymal injury in 36 specimens who underwent the microsurgical approach as management of spermatocoele, (Kauffman et al. 2011). In the treatment of patients with spermatocoele/epididymal cyst who wish to preserve their fertility, the microsurgical approach appears to be the safest treatment option.

13.7.1.1 Scrotal Haematoma

As with any scrotal procedure/surgery, haematoma is a potential post-operative complication, though in most cases is self-limiting and does not require additional surgical management. Accumulated haematoma requires drainage only if it becomes infected or if it continues to enlarge (Celigoj and Costabile 2016).

13.7.1.2 Hydrocoele

Hydrocoele is a possible complication in any scrotal surgery where lymphatics surrounding the testis are damaged. With regards to epididymal surgery, the incidence of post-operative hydrocoele is lowest in the microsurgical approach to spermatocoele/epididymal cyst removal, as this allows visualisation and intentional sparing of the lymphatics.

13.7.1.3 Chronic Scrotal Pain

The incidence of chronic scrotal pain following vasectomy is approximately 2%, and is likely similar after spermatocoele/epididymal cyst surgery. (Partin et al. 2021) There is sparse research in this domain but should be included in pre-operative counselling for patients.

13.8 Benign Prostatic Hyperplasia

Benign Prostatic Hyperplasia (BPH) refers to the non-neoplastic enlargement of the prostate gland, that commonly occurs in males over the age of 50 years. It is one of several causes of lower urinary tract symptoms (such as urinary frequency, urinary urgency, nocturia, straining to void) in males. BPH can also cause bladder outlet obstruction (BOO), which, left untreated, can lead to renal impairment.

The risk of developing BPH rises linearly with age, with almost all men who live to the age of 90 years having evidence of microscopic BPH (Berry et al. 1984). Despite this, not all men will develop obstructive lower urinary tract symptoms. A variety of conservative, medical and surgical approaches are available to manage this extremely common condition.

13.8.1 Anatomy and Physiology

Situated at the base of the bladder and surrounding the urethra, the prostate is a male urogenital organ that produces secretions that contribute to seminal fluid. Where it surrounds the urethra, enlargement of the prostate can restrict the flow of urine, causing obstructive symptoms and compensatory hypertrophy of the detrusor muscle of the bladder.

The prostate is comprised of both fibromuscular and glandular tissue, found in different divisions of the gland. The central region, comprised of the prostatic urethra, contains muscular tissue (the internal urethral sphincter) that contracts during urination, preventing the retrograde flow of urine into the prostatic ducts. In contrast, the role of the glandular prostatic tissue is characterised by its secretory contribution to the seminal fluid, of which the prostate specific antigen (PSA) is a component.

The prostate gland is divided into lobes and is surrounded by a periprostatic capsule. The rectum being its posterior relation is useful during physical examination. The transition zone of the prostate consists of two separate glands located immediately adjacent to the internal urethral sphincter. This is where BPH changes begin to develop.

13.8.2 Pathophysiology

Whilst often referred to as 'benign prostatic hypertrophy', this is a misnomer, as the underlying pathophysiology is increased cell proliferation (hyperplasia), rather than increased cell size (hypertrophy).

Hyperplasia is an increase in cell proliferation, leading to an increased number of cells of a given type, and can be a normal physiological process to a specific stimulus. In benign prostatic hyperplasia, there is an increase in either stromal cell or epithelial cell proliferation, or, a decrease in the frequency of physiological programmed cell death (apoptosis).

A stromal cell is a connective tissue cell (in the prostate, mainly smooth muscle cells) that supports the parenchymal function of its related organ. In the adult, stromal cells possessing androgen receptors regulate the growth, death and differentiation of epithelial cells, by secretion of "andromedins". Such andromedins include fibroblast growth factor, epidermal growth factor and insulin-like growth factor. Stromal cells also express oestrogen receptors which control the secretion of stromal factors that stimulate prostate epithelial cells. (Hägglöf and Bergh 2012).

Luminal epithelial cells of the prostate express androgen receptors, which, when activated, maintain the survival of these cells. In contrast, basal epithelial prostate cells that have their androgen receptors activated suppress cell proliferation. Hence there is a crucial balance of stromal-epithelial cell interactions, of which a disruption is thought to contribute to benign prostatic hyperplasia. (Hägglöf and Bergh 2012).

13.8.2.1 Prostate Tissue Remodelling, Inflammation and Autoimmunity

BPH is caused by altered stromal cell function resulting in stromal cell and epithelial cell proliferation. The ratio of stromal to epithelial area rises from 3:1 in a normal prostate to 5:1 seen in hyperplastic prostate tissue. Proliferation of stromal cells is the predominant pathology.

This remodelling occurs in the transition zone and peri-urethral region. Remodelling changes evident in the transition zone consists of altered secretions by the luminal epithelial cells leading to calcification, clogged ducts and inflammation, and enlargement of basal epithelial cells. The induction of inflammation in the prostate flows on to activate cytokines, growth factors and radical oxygen species which have numerous pro-inflammatory and pro-hyperplastic effects (Untergasser et al. 2005).

The induction of pro-inflammatory changes is thought to initiate the phenotypical changes to stromal cells that causes their hyperplasia. Nearly all BPH specimens contain inflammatory infiltrates, without evidence of bacterial or foreign antigenic material. Auto-reactive T Lymphocytes have been found to recognise prostate secretory products such as Prostate Specific Antigen (PSA) and Kallikrein-2, suggesting an autoimmune component to BPH. Chronically activated T Lymphocytes further propagate the inflammatory process, leading to further stromal cell proliferation. Ultimately, as a result of the inflammatory changes, there is a disruption to the barrier formed by the tight junctions between prostate epithelial cells. This allows entry of autoantigens to PSA and Kallikrein-2 to be released into the stromal compartment. In normal prostate tissue, secreted PSA is stored within the ducts, resulting in a low PSA concentration detected when measured in serum. With disruption to the epithelial barrier, PSA can leak into the vascular stromal tissue to enter systemic circulation—causing elevated serum PSA concentration as seen in BPH. PSA leaked into the stromal tissue further encourages inflammatory changes as more autoimmune reactivity takes place. (Isaacs 2008).

13.8.2.2 Prostate Stem Cell Reactivation

In prepubertal boys, the size of the prostate is small, weighing around 2 g. During puberty, the prostate grows to 20 g, corresponding to the rise in serum testosterone during puberty (Vickman et al. 2020). After the ages of 18–20 years are reached, the size of the prostate remains steady, balanced by the rates of physiological cell proliferation and apoptosis, until the age of 50. From 50 years onwards, the average size of the prostate gland almost universally enlarges gradually.

Stem cells are those which are able to proliferate without external stimuli and have the capacity to differentiate into a variety of cell types. After embryogenesis, a portion of stem cells remain through to adulthood, and are able to self-renew and differentiate into a small number of cell types. During adulthood, these cells are dormant, and are contained in a specific micro-environment where their proliferation and death are tightly regulated. With regards to the prostate, there are epithelial stem cells and stromal stem cells. A sub-class of epithelial stem cells require andromedin secretion by stromal cells to proliferate, and a portion of these also require andromedin secretion to survive. (Isaacs 2008).

Reactivation of previously dormant stromal stem cells in the prostate and their subsequent secretion of andromedins activates proliferation of epithelial stem cells.

13.8.2.3 Role of Androgens

Though not thought to initiate BPH, androgens play a role in supporting the underlying hyperplasia, and this relationship is effectively targeted by use of 5-alpha reductase inhibition, which causes involution of the prostate by reducing serum and intraprostatic concentrations of dihydrotestosterone. (Goldenberg et al. 2009).

The androgen receptor is crucial for maintenance of prostate size in adult males and is primarily expressed by the luminal epithelial cells.(Vickman et al. 2020) Testosterone depletion, as a result of surgical or chemical castration has been shown to cause a decrease in the size of the prostate, and a loss of luminal epithelial cells by means of apoptosis (Vickman et al. 2020). There is also evidence demonstrating the regeneration of prostate size following androgen supplementation, further demonstrating this relationship (Isaacs 2008).

Though classically thought to exclusively affect epithelial cells, androgen deprivation therapy also decreases andromedin production by stromal cells, which is how it ultimately decreases the proliferation of luminal epithelial cells. There is also a loss of vasoactive endothelial growth factor that is usually secreted by stromal cells, which results in a global decrease in prostate vascularity (Vickman et al. 2020; Isaacs 2008).

With the recent discovery of stromal cell androgen receptor targets, future studies may identify potential therapeutic applications that may be effective in BPH.

13.8.3 Risk Factors

13.8.3.1 Genetic/Familial Component

Younger patients with the largest sized prostates at time of prostatectomy were more likely to have a first-degree relative with BPH (Sanda et al. 1994). Analysis demonstrated this was likely to be an autosomal dominant pattern of inheritance.

13.8.3.2 Diet, Physical Activity and Lifestyle Factors

Physical activity can decrease the low grade chronic inflammatory reaction that contributes to BPH and is evidenced by studies showing that a lack of physical activity is correlated with worsening and progressive lower urinary tract symptoms (Calogero et al. 2019).

Evidence that has been found linking specific lifestyle factors to the development of BPH, are not conclusive, but reflect potential for a research focus (Jacobsen 2007). Tobacco smoking has been associated with worse lower urinary tract symptoms in patients with BPH (Calogero et al. 2019). Nicotine is thought to worsen these symptoms by increasing the sympathetic nervous system output, and thereby increasing tonicity of the detrusor muscle. Men with hypertension had larger prostates and worse lower urinary tract symptoms than those without hypertension (Calogero et al. 2019). This is also likely secondary to activation of the underlying chronic inflammatory process.

Poor glycaemic control in diabetic patients has been shown to worsen BPH outcomes. Strong evidence correlates larger insulin dose requirements with larger prostate volumes, even after adjustment for other metabolic factors (Calogero et al. 2019). It is thought that insulin is an independent risk factor for BPH as it stimulates prostate growth by modulating the effects of insulin-like growth factor.

13.8.4 Clinical Presentation and Diagnosis

BPH is a histological diagnosis and can manifest in one of three ways; benign prostatic enlargement (BPE), lower urinary tract symptoms (LUTS) and/or bladder outlet obstruction (BOO).

It is important to recognise that BPH is not the exclusive cause of lower urinary tract symptoms and bladder outlet obstruction, and other urological and nonurological conditions can also contribute to these presentations.

LUTS are a frequent presentation, and is found in 51.3% of adult males (Irwin et al. 2006). The patient typically reports gradually worsening symptoms over months or years.

Less frequently, due to the slowly progressive nature of BPH, patients may become accustomed to the lower urinary tract symptoms and may not have any complaints. These patients may present with acute urinary retention, or with uraemia secondary to renal failure. Other less frequent presentations include macroscopic haematuria, urinary incontinence or recurrent urinary tract infections. (Foo 2019).

When consulting with the patient, the aim is to identify the presence and severity of lower urinary tract symptoms and bladder outlet obstruction. It is important to note that the size of the enlarged prostate does not correlate with symptom severity. A 35 cc prostate causing significant periurethral mass effect may produce worse symptoms than a 100 cc prostate with enlarged lateral lobes.

The type of intervention and the urgency at which it is required predominantly depends on the severity of symptoms, presence of renal impairment, general health of the patient and patient preferences.

13.8.4.1 History

Lower urinary tract symptoms can be separated into storage (daytime urinary frequency, nocturia, urinary urgency, incontinence), voiding (slow urine stream, needing to strain to void, spraying of urine stream, intermittent stream, terminal dribbling), and post-micturition (feeling of incomplete emptying, post-micturition dribbling) symptoms.

The International Prostate Symptom Score questionnaire (IPSS) is useful in quantifying a patient's lower urinary tract symptoms and monitoring treatment efficacy.

The occurrence of any sexual symptoms, such as difficulty in initiating or maintaining erection, ejaculatory issues along with a thorough sexual history should be discussed.

The primary focus of the history is to ascertain the likely underlying cause of the symptoms. A thorough general medical history is required, with focus on conditions that may cause LUTS. These may include diabetes, endocrinological conditions, neurologic disease and medication use. A thorough urological history is also required, screening for atypical features which may warrant a different investigative pathway. Risk factors for urological cancers should be questioned about, particularly smoking, exposure to industrial paints/dyes/rubber and petrochemical exposure.

13.8.4.2 Examination

13.8.4.2.1 Examination of the Urogenital System

The suprapubic area should be inspected for an overdistended or painful bladder. The penis should be inspected to evaluate for presence of meatal stenosis, phimosis or lesions. The testes and epididymis should be examined for palpable tenderness which could be suggestive of infection.

13.8.4.2.2 Digital Rectal Examination

The normal prostate has an anterior, posterior, medial and two lateral lobes. The medial and two lateral lobes are the most prominent, and are the lobes affected by BPH. During digital rectal examination, the index finger should be inserted into the rectum, with the fingertip directed anteriorly. The examiner should be able to feel two relatively firm lobes with a distinct sulcus between each lobe.

Digital rectal examination is helpful when the examiner is sufficiently experienced. The detection of nodular changes, induration, irregularity, loss of the sulcus or possessing a 'fixed' component to the pelvic sidewall are concerning for prostatic malignancy.

The positive predictive value for accurately detecting prostate cancer by digital rectal examination alone is approximately 9% (Bozeman et al. 2005).

The digital rectal examination can be used opportunistically to test external anal sphincter tone and function, which can aid in diagnosis of neurological causes of these symptoms.

13.8.4.3 Investigations

13.8.4.3.1 Midstream Urinalysis, Culture and Cytology

Urinalysis demonstrating presence of erythrocytes warrants further investigation, though it may be simply the result of BPH. Particularly where other risk factors such as heavy smoking or exposure to petrochemicals/dyes, erythrocytes in the urine may represent underlying urothelial malignancy.

Urinary nitrates are usually produced by bacteria, with Escherichia coli being responsible for the majority. Nitrates in the urine are abnormal and may occur as a result of chronically high post-void residual volumes, urethral stricture disease, bladder diverticulum, patent urachal cyst/remnant, renal calculi, bladder calculi and other such urological conditions. Finding nitrates in the urine should prompt further urological investigations.

Urine cytology should be tested in males presenting with LUTS or macroscopic haematuria, particularly if they have a smoking history.

13.8.4.3.2 Routine Blood Tests +/− Serum PSA

Baseline blood tests should be performed to assess the general health of the patient, and importantly to detect any renal impairment.

Testing serum PSA should be considered in males presenting with LUTS who have a life expectancy beyond ten years (Partin et al. 2021).

The serum prostate specific antigen (PSA) level is not required to make a diagnosis of BPH, and a raised PSA can be caused by prostate cancer, recent ejaculation, prostate infarction, prostatic inflammation and even bike riding! Testing PSA should be done only after discussion with the patient regarding the risks, benefits and necessity for further investigations if abnormal. The free PSA:total PSA ratio can be useful in distinguishing between BPH and prostate malignancy, but is not conclusive.

13.8.4.3.3   Uroflowmetry

This is a non-invasive test providing quantitative information of the function and patency of the bladder and the bladder outlet. The parameters measured include the 'peak urinary flow rate', voided volume, and the pattern of flow. A normal peak urinary flow rate is >10 mL/sec, and 90% of males with a peak urinary flow rate less than 10 mL/sec have some form of bladder outlet obstruction (Nitti 2005).

Note that a low peak urinary flow rate can occur as a result of either impaired detrusor contractility or bladder outlet obstruction, and basic uroflowmetry cannot distinguish between these two totally different pathologies (Chancellor et al. 1991). Further to this, a normal peak urinary flow rate does not necessarily exclude bladder outlet obstruction (Gerstenberg et al. 1982). Despite its limitations, uroflowmetry, in conjunction with other investigative methods is a minimally invasive and effective tool in quantifying the patient's symptoms and differentiating between various urological pathologies. Where more clarification is required, a formal urodynamics study, though more invasive, can provide more information.

13.8.4.3.4   Post-Void Residual Volume

The post-void residual volume measures the quantity of urine remaining in the bladder at the end of micturition, usually by a bedside ultrasound device applied to the anterior pelvic wall. There is no consensus as to what constitutes an abnormal post-void residual volume, though volumes greater than 300 mL are concerning (Asimakopoulos et al. 2016). There is conflicting research as to whether patients with higher post-void residual volumes are at risk of worse outcomes, though it is suggested that those patients who have high post-void residual volumes are closely observed if they choose to have conservative approaches to manage their BPH (Partin et al. 2021).

13.8.4.4   Additional Investigations

13.8.4.4.1   Ultrasound of the Kidneys, Ureters and Bladder

Imaging of the upper renal tracts is recommended for patients presenting with LUTs who also have renal impairment, elevated post-void residual volumes or a history significant for haematuria, recurrent UTIs, renal calculi or prior urological procedures. In this instance, an ultrasound of the kidney, ureters and bladder is recommended as the initial investigation. (Partin et al. 2021).

13.8.4.4.2 Urodynamic Studies

The invasive urodynamic test, is the gold standard for assessment of LUTS pathophysiology, and unlike basic uroflowmetry, it can differentiate between bladder outlet obstruction, detrusor overactivity, detrusor underactivity and low bladder compliance (Nitti 2005). This test involves the insertion of a specialised urinary catheter that can measure pressures and a transducer probe into the rectum.

The urodynamic study provides a wealth of information including: bladder capacity, the compliance of the bladder, the competency of the urethral sphincter, when the patient perceives the sensation of bladder filling, when the patient perceives the desire to void, the stability of the detrusor muscle and the effect of intra-abdominal pressure on voiding function (Nitti 2005).

There are two 'phases' involved in urodynamic studies, namely the filling and storage phase or 'cystometrogram', followed by the voiding phase or 'voiding pressure flow study'.

Involuntary detrusor contractions whilst the bladder is being filled suggests an overactive detrusor muscle, which can occur as a result of bladder outlet obstruction. Essentially, bladder outlet obstruction is characterised by a 'high-pressure', 'low-flow' state.

13.8.5 Complications of Benign Prostatic Hyperplasia

13.8.5.1 Bladder Decompensation

Bladder outlet obstruction (BOO), a consequence of BPH (and other urological conditions), can lead to several compensatory structural and functional changes in the detrusor muscle of the bladder. These changes occur as a response to overcome the resistance to bladder emptying, and lead to a thickened, overactive detrusor muscle. Hypertrophic changes of the detrusor muscle, including hypertrophic changes to the smooth muscle cells and increased collagenisation of the extracellular matrix have been noted. Hypertrophy of neurons of the bladder wall, and activation of previously 'silent' neuronal pathways have been implicated in the irritative urinary symptoms and detrusor overactivity seen in BOO. These changes are induced by the mechanical stress perceived by the chronically distended bladder wall in BOO, which induces the molecular signalling leading to these morphological and functional changes involved in the bladder remodelling that occurs. (Mirone et al. 2007).

Cystoscopic bladder changes that may be seen as a result of longstanding BPH include trabeculation of the bladder wall and diverticula. Biopsies of trabeculated, obstructed bladders show dense connective tissue deposition and may also show age-related fibrotic changes (Partin et al. 2021).

The concerning question today is whether there is a 'point of no return', where delayed intervention of BPH may lead to irreversible loss of bladder function.

(Presicce et al. 2018) This area is yet to be thoroughly investigated, and would be best researched by means of longitudinal, prospective randomised studies comparing outcomes for early and deferred interventions, (Presicce et al. 2018).

13.8.5.2   Acute Urinary Retention

Acute urinary retention (AUR) is a frequent acute presentation of BPH and a distressing event for the patient. AUR is defined as the acute inability to urinate, with worsening urinary urgency and pain as the bladder fills.

This is managed with urinary catheterisation to urgently decompress the bladder, and a 'trial of void' is conducted once the patient's clinical state has been sufficiently optimised. A proportion of these patients will remain in retention after the catheter is removed and will require lifelong catheterisation or another intervention for BPH. Precipitants for AUR include urinary tract infection, prostate infection, sexual activity, immobility, sympathomimetic/anticholinergic medications, antihistamines and anaesthesia—all of which can further narrow an already constricted prostatic urethra in a patient with BPH (Stimson and Fihn 1990; Roehrborn et al. 2000). AUR can also occur spontaneously in a patient with BPH, though the aetiology for this is unclear (Kaplan Steven et al. 2008).

13.8.5.3   Urinary Incontinence

Over distension of the bladder in BPH can lead to overflow incontinence. Detrusor instability as a result of BPH can result in urge incontinence.

13.8.5.4   Urinary Tract Infections

Theoretically, there ought to be an increase in the frequency of urinary tract infections, due to the poor emptying of the bladder causing retention and stasis of urine in patients with BPH. However, there is sparse conclusive research on this area to affirm this theory. (Partin et al. 2021).

13.8.5.5   Renal Impairment and Upper Tract Retention

BPH can cause bladder outlet obstruction, which can lead to pressurised retention of urine in the upper urinary tracts, leading to bilateral hydroureteronephrosis and a deterioration in renal function. An average of 13.6% of patients presenting for transurethral resection of prostate had evidence of renal failure (McConnell et al. 1994). Renal impairment secondary to BPH can remain undetected until the patient is in renal failure, as the majority of patients are asymptomatic. Renal impairment is correlated with an increased risk of surgical complication compared to those with normal renal function (Holtgrewe et al. 1989). The degree and progression of renal impairment is likely compounded by comorbid diabetes, hypertension and other factors (Hong et al. 2010).

13.8.5.6 Haematuria

Patients with BPH may develop macroscopic haematuria, with or without clots. This is likely due to irritation of a hyper-vascular and enlarged prostate. As previously mentioned, the increased activity of vascular endothelial growth factor in BPH likely contributes to its hypervascularity. 5 alpha reductase inhibitors are an effective treatment for prostatic haematuria, as it reduces vascular endothelial growth factor activity (DiPaola et al. 2001).

13.8.5.7 Bladder Calculi

The prevalence of bladder calculi is 8 times higher amongst men with a histological diagnosis of BPH (3.4%) compared to controls (0.4), although there is no observed increase in renal or ureteric calculi (Grosse 1990). The overall rate of developing bladder calculi is low. One study demonstrated that only 1 of 276 patients with BPH with moderate symptoms developed a bladder stone at 3-year follow up (Wasson et al. 1995). Screening for bladder calculi is only indicated where there are relevant symptoms, such as haematuria (Partin et al. 2021).

13.8.6 Management

Acute urinary retention accounts for up to 30% of the indications for surgical management of BPH (Partin et al. 2021). The majority of patients seek intervention to improve their quality of life and for symptom reduction.

13.8.6.1 Conservative Management- Watchful Waiting

Suitable patients with minimal symptoms may prefer a watchful waiting approach. There is a cohort of patients with BPH, for whom symptoms remain relatively stable throughout their lifetime, without need for surgery, and without development of acute urinary retention. In a 15-year longitudinal study, this cohort was 50% of included patients, who showed stable symptoms scores throughout the study period (Fukuta et al. 2012).

A trial compared surgical management of BPH (transurethral resection of prostate) with a watchful waiting approach. (Wasson et al. 1995) Patients in the watchful waiting approach received lifestyle advice and were monitored for the development of acute urinary retention, elevated post-void residual volume, development of bladder calculi, renal failure and raised symptom scores. Though the patients who received surgical management demonstrated improvement in symptomology, only 17% of patients in the watchful waiting cohort experienced treatment failure, which in the majority of cases, was secondary to elevated post-void residual volumes or a raised symptom score. This was over an average follow-up period of 2.8 years.

Lifestyle advice and reassurance should be provided, and symptoms should be monitored yearly, to offer timely intervention as required. Lifestyle advice includes avoidance of coffee, alcohol and medications that may worsen their lower urinary tract symptoms. A daily fluid intake between 1500–2000 mL is appropriate. It can be helpful to restrict fluid intake when most inconvenient (such as during long journeys), and cessation of fluid intake within two hours of sleeping, to minimise nocturia. Double-voiding should be encouraged, and constipation should be avoided.

Watchful waiting is a safe alternative for patients who are less bothered by their symptoms.

13.8.6.2 Medical Therapy

Medical therapy is an effective treatment modality for patients bothered by LUTS, who do not require imminent surgery. This cohort of patients includes those with all of: mild-moderate IPSS scores, no evidence of renal impairment, reasonable post-void residual volumes (<300 mL), a normal urinary peak flow rate and no imminent indications for surgery (Goonewardene et al. 2018).

Surgical treatment is indicated by the development of acute urinary retention, recurrent urinary tract infections, renal impairment, development of bladder calculi and recurrent haematuria. In the absence of such complications, medical management may be suitable, and should be tailored to the patient's symptoms.

Medical therapy should take into account the patient's symptoms and their severity, prostate volume, post-void residual volume and patient preferences. Alpha-adrenergic receptor blockers are usually first line for most patients with LUTS secondary to BPH (Please see Table 13.2).

For patients whose main symptoms are related to voiding, such as urinary hesitance, slow/weak stream, terminal dribbling, intermittency, overflow incontinence or chronic urinary retention, suitable medications include alpha1-blockers (a1B) and 5-alpha reductase inhibitors (5ARI).

For patients whose main symptoms are those related to storage, such as urinary frequency, urgency, nocturia and urge incontinence, optimal medications include muscarinic receptor antagonists or Beta-3 receptor agonists.

Medical therapy should aim to provide symptom relief, reduce the degree of bladder outlet obstruction, reduce urinary retention, reverse detrusor overactivity, prevent renal impairment and prevent progression of the disease. Periodically assessing the patient's symptoms by means of the IPSS questionnaire is a useful way to monitor treatment efficacy.

Table 13.2 Summary of drugs used to treat benign prostatic hyperplasia

Drug Class	Mechanism of action	Side effects	Uses and limitations
Alpha- blockers Drug name: Tamsulosin, Alfuzosin, Prazocin, silodosin, Phenoxybenzamine (each has a different selectivity for the different alpha adrenergic receptors, and a different half-life)	In the context of BPH, the LUTS are caused by narrowing of the urethral lumen by the enlarged prostate and also by contraction of the bladder neck and prostatic smooth muscle cells which further exacerbates the obstructive symptoms (Russo and Morgia 2018). The prostates of patients with BPH have been found to have a greater proportion of smooth muscle cells compared to those without BPH, on histopathological studies (Shapiro et al. 1992). Alpha 1 and alpha 2 receptors are expressed in the lower urinary tract, but the main receptor involved in prostate and bladder neck smooth muscle contraction is the alpha 1 receptor, and this is the target of alpha-blocker therapy (Walden et al. 1999). Treatment with alpha-blockers is effective in preventing symptom deterioration, and in improving IPSS scores, quality of life and peak urinary flow rate (Roehrborn 2006).	Side effects depend on the uroselectivity of the medication utilised. Where a highly uroselective alternative is chosen, fewer patients experienced dizziness, hypotension and dry mouth (Dong et al. 2009). Side effects include: • Postural hypotension • Syncope • Dizziness • Vertigo • Somnolence • Aesthenia • Anejaculation Patients who are already taking antihypertensive medications should have these reviewed and adjusted prior to commencing an alpha-blocker.	Uses: • Bothersome LUTS • Predominantly voiding dysfunction Limitations: • Do not reduce the risk of developing acute urinary retention but do reduce the risk of surgery. • Cardiovascular and neurological side effects • Anejaculation as a side effect

(continued)

Table 13.2 (continued)

Drug Class	Mechanism of action	Side effects	Uses and limitations
5-alpha-reductase inhibitors Drug name: Dutasteride, finasteride	The presence of androgens such as testosterone and dihydrotestosterone (DHT) is crucial for maintenance of prostate size. DHT is a metabolite of testosterone, and has a tenfold greater potency for androgen receptor activation compared with testosterone (Russo and Morgia 2018). DHT is synthesised by the enzymes type 1 and type 2 5-alpha-reductase. Type 2 5-alpha-reductase is found in tissues of the prostate itself, whereas type 1 5-alpha reductase is present in organs other than the prostate, including in the skin. 5-alpha-reductase inhibitors block the conversion of testosterone to DHT and limit the proliferation of prostate stromal cells. (Russo and Morgia 2018) Treatment with 5-alpha-reductase inhibitors decrease prostate volume, increased peak urinary flow rate and improved LUTS, though it typically takes 6–12 months of treatment to achieve maximal effect (Russo and Morgia 2018). Finasteride has also been shown to suppress intraprostatic vascular endothelial growth factor, and other growth-promoting agents, hence reducing the vascularity of the prostate. Finasteride proves to be an effective option in reducing the recurrence of haematuria secondary to hyper-vascular prostate enlargement seen in some patients with BPH (Foley et al. 2000).	Sexual side effects such as decreased libido, erectile dysfunction, ejaculatory dysfunction and gynaecomastia may occur as a result of therapy, but are reversible for most patients. 15% of patients taking finasteride and 7% of patients taking placebo reported new sexual side effects within 12 months of commencing therapy, however there was no difference amongst these groups in subsequent years of therapy (Wessells et al. 2003). Patients with a baseline, pre-treatment low serum testosterone level were found to be more likely to suffer from sexual side effects related to 5-alpha-reductase inhibition (Traish et al. 2014). Depression and worsening symptoms of anxiety have been implicated in 5-alpha-reductase inhibition (Traish et al. 2014). There is a risk of failing to detect, and even of encouraging the proliferation of certain subtypes of higher Gleason grade prostate cancers with the use of 5-alpha-reductase inhibitors. This is thought to be due to the unpredictable and variable responses that different subtypes of prostate cancer cells have shown towards androgens (Traish et al. 2014).	Uses: • LUTS secondary to BPH • Macroscopic haematuria secondary to a hypervascular prostate • "Pre-surgical prostate shrinkage" Limitations: • Sexual side effects • Risk of undetected prostate cancer; risk of progression of undetected prostate cancer • Depression and anxiety • 6–12 months of therapy is required before maximal effect is seen

Drug Class	Mechanism of action	Side effects	Uses and limitations
Antimuscarinic agents Drug name: Solifenacin, Oxybutinin, Fesoterodine	Muscarinic receptors are involved in detrusor muscle contraction during voiding, and antimuscarinic agents reduce the contractility of the detrusor muscle. Antimuscarinic agents are effective in reducing urinary urgency, urinary frequency and IPSS scores (Russo and Morgia 2018).	Side effects of antimuscarinic agents are related to unwanted blockade of muscarinic receptors outside the urinary tract. Side effects and their incidences over 12 months of treatment with solifenacin were: dry mouth (12.4%), constipation (5.2%), dyspepsia (2.7%), hypertension (2.4%), urinary tract infection (2.3%) and nasopharyngitis (1%) (Drake et al. 2015).	Uses: • Predominantly storage type lower urinary tract symptoms Limitations: • Anticholinergic side effects • Risk of urinary retention
Beta-3 agonists Drug name: Mirabegron	Beta-3 receptors are expressed by the detrusor muscle and urothelium, and sparsely expressed elsewhere in the body, making these an ideal therapeutic target for urological conditions. Stimulation of Beta-3 receptors induces relaxation of the detrusor muscle. By binding to Beta-3 receptors in the bladder, Mirabegron induces relaxation of the detrusor muscle.	Mirabegron produces anticholinergic side effects similar to the antimuscarinic agents. The most common side effect over a 12 month period of treatment with Mirabegron was dry mouth (2.8%). Though there is a theoretical risk of precipitating acute urinary retention amongst males with BPH, this has not been substantiated (Nitti et al. 2013).	Uses: • Predominantly storage type lower urinary tract symptoms • Overactive bladder Limitations: • Risk of urinary retention • Anticholinergic side effects

13.8.6.3 Surgical Management

Surgical management provides a faster and more significant improvement in lower urinary tract symptoms compared to medical therapy. It is interesting to note that with the advent of medical therapy to treat BPH, the age at which patients ultimately seek surgical intervention for BPH is increasing. Hence, patients seeking surgical treatment are older and have more co-morbidities compared to the same cohort of patients prior to the widespread use of medical BPH therapy.

13.8.6.4 Transurethral Resection of the Prostate

Transurethral resection of the prostate (TURP) is the gold standard for surgery to manage BPH.

Indications for TURP include:

- Acute urinary retention; often the sign of an 'end-stage bladder'.
- Recurrent macroscopic haematuria
- Bladder calculi/bladder diverticulae
- Recurrent urinary tract infections
- The development of bilateral hydroureteronephrosis and renal impairment

Generally, patients should be trialled on medical management prior to surgical intervention. Ultimately the decision to proceed to surgery is one that should be made after considering the patient's situation and preferences.

TURP can be utilised in the management of prostates of up to 100–150 cc in size.

A resectoscope inserted through the urethra is used to shave those portions of obstructive prostatic tissue encircling the urethra. Current runs through an electrified loop of wire on the end of the resectoscope, through the tissue and patient, and returns via the electrode in the grounding pad. This is known as a 'monopolar TURP'. Due to the unidirectional nature of the current used, a non-ionic irrigant (usually glycine) is continually flushed through the resectoscope and into the lower urinary tract to prevent dispersion of the electric current. Due to their hypo-osmolarity, when these irrigants are systemically absorbed, can cause electrolyte disturbances, called 'TUR syndrome'.

Great care is taken when resecting tissue, to avoid damaging the external urethral sphincter, by limiting resection to tissue that is proximal to the easily identifiable landmark; the verumontanum. The internal urethral sphincter is usually resected, rendering it incompetent. The depth of resection is guided by the visualisation of the fibrous capsule of the prostate or bladder neck. Areas of bleeding can be coagulated with carefully placed swipes of the wire. Often, continual resection of the bleeding area until the prostatic capsule is reached is required to adequately control bleeding.

Resected bladder chips are collected and sent for histopathological analysis. The bladder is emptied before confirming haemostasis, to ensure that hydrostatic pressure is not unknowingly tamponading any venous bleeding. An indwelling catheter is placed into the bladder and may be placed on traction to control bleeding at the bladder neck.

Bipolar TURP is a newer modality, where the inflow and outflow of current is through the resection wire alone. This allows for the use of iso-osmolar saline in place of the non-ionic irrigants, required for monopolar TURP. As a result, TUR syndrome is a much less frequent complication where bipolar TURP has been adapted.

13.8.6.5 Holmium Laser Enucleation of the Prostate (HoLEP)

HoLEP is a newer advancement in the management of BPH and utilises a laser fiber engaged with a holmium-yttrium-aluminum-garnet (holmium: YAG) laser, which acts through a photothermal mechanism. This laser is capable of vaporising soft tissue, cutting through prostatic tissue and producing effective haemostasis at the same time. Unlike for TURP, HoLEP does not require pre-operative cessation of anticoagulants or even anti-platelet medication, making it accessible to older and co-morbid patients in whom TURP is contra-indicated. (Kuo et al. 2003).

HoLEP is associated with negligible risks of haematuria, transfusion and infection. The majority of patients are discharged within 24–48 h and almost always go home without an indwelling catheter in place. Emerging research suggests HoLEP is a promising alternative to TURP, particularly for elderly and co-morbid patients. (Kuo et al. 2003).

The holmium:YAG laser fiber is delivered via a resectoscope to the prostate. Incisions are made down to the prostatic capsule, at the 'five o'clock and seven o'clock' positions, with the laser fibre. The energy of the laser fibre is increased, and the median lobe of the prostate is liberated. The entire prostate adenoma is dissected from the prostatic capsule, similar to stripping the peel off an orange. Ultimately, the median and lateral lobes of the prostate are freed from the capsule and pushed into the bladder. The liberated prostate tissue that was pushed into the bladder is morcellated and collected and sent for histopathological analysis. (Kuo et al. 2003).

HoLEP is not limited to prostates of a certain size, and can be used as an alternative to open prostatectomy in very large prostates, with no increase in the risk of bleeding or transfusion (Ahyai et al. 2010).

13.8.6.6 Potassium-Titanyl-Phosphate (KTP) Greenlight Laser Vaporization of the Prostate

The KTP laser is derived from the 1064 nm wavelength Ng:YAG. The laser beam is passed through a KTP crystal, that halves the wavelength whilst doubling the frequency, called 'Greenlight'. At these specifications, the energy is preferentially absorbed by haemoglobin and not by water, resulting in simultaneously highly effective coagulation and tissue vaporisation. The depth of penetration by this laser is short at 0.8 mm. (Welliver et al. 2017).

The laser fibre is introduced through a specialised scope where the 'aiming beam' can be intentionally focussed on the target areas of the prostate. The energy vaporises the prostate from 'inside to out'. The surgeon terminates firing of the laser when fibres of the prostatic capsule are seen. (Welliver et al. 2017) Occasionally, bleeding of the mucosa may obstruct visualisation, and can lead to perforation of the prostatic capsule. If bleeding is not able to be controlled, the procedure is converted to TURP. Overall however, the risk of significant bleeding in KTP is exceedingly low (Bachmann et al. 2012; Ahyai et al. 2010).

Similar toHoLEP, patients can remain on anticoagulant and antiplatelet medication and have KTP. The risk of transfusion and significant bleeding is exceedingly rare in laser procedures of the prostate. (Bachmann et al. 2012) Unlike HoLEP, tissue samples are not collected and sent for histopathological analysis, therefore there is an extremely small cohort of patients in whom a diagnosis of prostate cancer may be missed.

The most commonly occurring intraoperative complications of KTP are: bleeding (3–13%), capsule perforation (3.1–5%) and conversion to TURP (1.8–8%) (Bachmann et al. 2012).

The most common long-term complications of KTP are: retrograde ejaculation (14.3%), recurrence requiring re-operation (2–11%), and urethral stricture (3.5–6%) (Bachmann et al. 2012).

KTP is limited to prostates less than 100 cc in size, and has a high re-intervention rate, with 18% of patients requiring repeat surgery for management of BPH within 6 months of the original KTP procedure (Ahyai et al. 2010).

13.8.6.7   Prostatectomy: Open, Laparoscopic or Robotic

Prostatectomy is generally reserved for those patients with prostate volumes greater than 150 cc, or those who also have large bladder calculi.

Transvesical Open Prostatectomy is performed with the patient in supine positioning, with legs apart. A short midline incision is made between the umbilicus and pubis and muscles are separated to enter the cavity. The dome of the bladder is freed of peritoneum and brought to the surface of the incision. A 4 cm longituninal incision is made into the bladder wall and suction is used to empty the bladder of urine. Sutures are used to suspend the bladder wall on either edge of the incision. Electrocautery is used to incise the mucosa of the bladder neck around the urethral meatus. A dissecting plane is created between the adenomatous prostatic tissue and the surrounding prostatic capsule with scissors. Once dissection is completed, the attachment at the distal urethra is dissected off the prostatic capsule, and the adenoma is freed. Absorbable sutures are used to re-approximate the bladder neck to the peripheral prostate, though a large fossa remains where the adenoma was removed. The bladder is closed ensuring a 'watertight seal'. An indwelling catheter remains in the bladder until the bladder neck anastomosis has healed and become 'watertight'. This approach is particularly useful where there are large bladder calculi or diverticulae, which can be managed simultaneously. (Tubaro and de Nunzio 2006).

Prostatectomy may also be performed retropubically or extraperitoneally, with use of laparoscopic instruments or a robotic console.

The incidence of intraoperative and perioperative transfusion requiring haemorrhage is 27% in open prostatectomy, which is significantly higher than other management options. Hospital stay ranges from 6–10 days and the rate of urinary tract infection is between 6–8%. (Tubaro and de Nunzio 2006).

Late complications of prostatectomy include bladder neck stenosis (8%) and urethral structures. Erectile dysfunction occurs in 3–5% of men after prostatectomy. Retrograde ejaculation is very common, occurring in 80–90% of patients after prostatectomy. Overall recurrence of symptoms or failure is the lowest out of all interventions, at 0%. (Tubaro and de Nunzio 2006).

With the advent of safer management options for BPH, prostatectomy is reserved for those patients with massive prostates >150 cc or those with concomitant bladder pathology that can be simultaneously managed (bladder calculi/bladder diverticulae) (Tubaro and de Nunzio 2006).

13.8.7 Complications of Operative Management of BPH

13.8.7.1 Short-Term Complications of Operative Management of BPH

13.8.7.1.1 Infection/Urosepsis

13.8.7.1.1.1 Monopolar/Bipolar TURP

A multicentre study of 857 patients who underwent TURP revealed a post-TURP infection incidence of 21.6%, of which 19.3% were urinary tract infections and 2.3% were bacteraemia/septic shock (Vivien et al. 1998).

Risk factors for septic shock after TURP include preoperative bacteriuria and a duration of surgery beyond 70 minutes (Vivien et al. 1998).

Intra-operative antibiotic coverage for TURP should include at least a fluoroquinolone or trimethoprim/sulfamethoxazole. All patients should be screened for bacteriuria and treated with sensitive antibiotics to ensure sterile urine by the day of surgery. Those patients with an indwelling catheter/foreign material in the urinary tract may require a prolonged course of antibiotics to achieve sterile urine. (Partin et al. 2021).

13.8.7.1.1.2 Laser Procedures

The incidence of urinary tract infection in patients who underwent HoLEP is 7.8%. (Shigemura et al. 2013). The incidence of urinary tract infection in patients who underwent KTP is 4.6%. (Pfitzenmaier et al. 2008) The incidence of urosepsis in HoLEP and KTP is negligible. (Ahyai et al. 2010).

13.8.7.1.1.3 Prostatectomy

There were no statistically significant differences in the number of post-operative infectious complications following robotic prostatectomy or laparoscopic prostatectomy. The incidence of post-operative infections after robotic or laparoscopic prostatectomy is approximately 1.1%, (Marra et al. 2019). The incidence of urinary tract infection in open prostatectomy is 12.5% (Rao et al. 2013).

13.8.7.2 Incidental Histological Finding of Prostate Cancer

Shavings of prostatic tissue are sent for histopathologic analysis and for a subgroup of patients, this may identify previously undetected and undiagnosed prostate cancer.

The risk of incidentally finding Gleason 7 (or higher) grade prostate cancer as a result of TURP is 0.026% (1 in 382 TURPs), provided the patient's PSA was less than 4 ng/mL.(Meeks Joshua et al. 2013).

A study conducted over 13 years, that overlapped the introduction of the serum PSA prostate cancer screening method demonstrated a reduction in the amount of prostate cancer that was incidentally diagnosed at TURP as prostate cancer screening become more frequently utilised. 23% of BPH specimens were incidentally found to have T1 prostate cancers prior to the PSA screening era and this reduced to 7% of specimens found to have T1 prostate cancers after PSA screening was established (Tombal et al. 1999).

It may be concerning that prostate cancer diagnoses are being missed where prostate tissue is not sent for analysis following treatments such as KTP. It is estimated that in a pool of 60,000 laser vaporizations, only 163 clinically significant cancers would be missed, provided these patients have undergone adequate PSA screening pre-operatively (Meeks Joshua et al. 2013).

Despite the decreasing frequency of incidentally found prostate cancer as PSA screening is utilised, there is a need to counsel patients pre-operatively of this risk when histopathological samples are collected and sent during surgical procedures for BPH. In KTP and other treatments where tissue is not sent for analysis, PSA screening is suggested, if the patient is agreeable to this.

13.8.7.3 TUR Syndrome

Mild to moderate degrees of TUR syndrome occur in 1–8% of patients undergoing TURP (Hahn 2006). Severe TUR syndrome has a mortality of 25% but is rare. It manifests predominantly with neurological symptoms (92%), cardiovascular symptoms (54%) and visual disturbances (42%) (Hahn 2006).

The incidence of TUR syndrome is negligible in bipolar TURP (0.0%) compared to monopolar TURP (0.8%) (Ahyai et al. 2010).

The systemic absorption of non-ionic irrigation solution used in monopolar TURP can cause fluid overload and various fluid and electrolyte shifts—commonly hypo-osmolar or iso-osmolar hyponatraemia.

Glycine 1.5% in water is an example of one such fluid and is, in itself, hypotonic. During resection, an average of 1500 mL of irrigant fluid is absorbed through the exposed vascular bed of the prostate. Absorption of the irrigation fluid induces osmotic diuresis by the kidneys, resulting in an absolute loss of sodium from the body. If a large amount of fluid is absorbed, the renal tubular cells swell rendering the kidneys anuric. Glycine and sorbitol solutions in particular cause an abnormal intracellular uptake of water, which further compounds the fluid shifts and electrolyte disturbances (Hahn 2006).

Manifestations of TUR syndrome include a 'prickling or burning sensation of the face and neck', hypotension, bradycardia, poor urine output, blurred vision, nausea, vomiting, confusion, headache and depressed consciousness (Hahn 2006).

Most concerningly, the rapid development of hyponatraemia can precipitate seizures. Cellular oedema may result from glycine or sorbitol solution, which, in the brain, can result in cerebral herniation, cardiorespiratory collapse and death. (Hahn 2006).

Factors increasing the risk of TUR syndrome are: (O'Donnell and Foo 2009)

- Resection time > 1 hour
- Large blood loss
- Low venous pressure (in a hypotensive or hypovolaemic patient)
- Perforation of the prostatic capsule (allows rapid absorption of irrigation via the extracapsular prostatic venous plexus)
- Bladder perforation (allows rapid absorption of a large volume of irrigant directly into the peritoneal cavity)

13.8.7.4 Resection of the Ureteric Orifices

Injury to the ureteric orifice(s) is an infrequent complication of TURP but may occur where a large median lobe obstructs adequate visualisation of the ureteric orifices. In this situation, the median lobe should be resected in the midline (which is clear of the ureteric orifices), until sufficient tissue has been removed to allow for direct visualisation of the ureteric orifices. In the occasion that a coagulating current has been applied to the ureteric orifice, a ureteric stent should be inserted to allow healing with stricture. (Welliver et al. 2017).

13.8.7.5 Haemorrhage

Significant haemorrhage requiring blood transfusion has an incidence of approximately 2.0% in both monopolar and bipolar TURP (Ahyai et al. 2010). The incidence of transfusion dependant haemorrhage is 10.0% in open prostatectomy (Rao et al. 2013). Other studies have found an incidence of 27% of patients who underwent open prostatectomy required blood transfusion. Tubaro and de Nunzio 2006).

There is negligible risk of transfusion dependent haemorrhage in HoLEP and KTP.

Complications related to bleeding are largely exclusively due to TURP and prostatectomy.

During TURP, arterial bleeding is managed by slow resection over bleeding areas, where coagulation may stop the bleeding. Complete control of arterial bleeding often requires resection till the prostatic capsule is reached. Excessive use of coagulation can cause delayed haemorrhage due to sloughing of heat-damaged tissue. (Welliver et al. 2017).

Venous bleeding during TURP is common and may be slowed by filling of the bladder. Breach of the prostatic capsule can lead to significant venous bleeding, due to exposure and damage to the venous sinusoids. Haemostasis should be confirmed at the end of the case. Ongoing venous bleeding may be conservatively managed with catheter traction. (Welliver et al. 2017).

Protracted haematuria after TURP may lead to clot retention and may require re-operation and diathermy to control. The incidence of re-operation to manage bleeding after monopolar TURP is 1.0%, and 0% in bipolar TURP (Ahyai et al. 2010).

13.8.7.6 Perforation of the Prostatic Capsule/Bladder

13.8.7.6.1 Monopolar/Bipolar TURP

Over-resection or overdistension of a thinned area of the prostatic capsule can cause perforation and can lead to profuse bleeding from the prostatic venous sinuses. Bladder perforation can also occur though this is almost always an extraperitoneal perforation in the context of TURP. If intraperitoneal rupture of the bladder is suspected, cystography should be utilised to rule this out—as intraperitoneal rupture of the bladder requires conversion to open laparotomy and repair of the bladder defect. Extraperitoneal bladder rupture can be managed with indwelling catheter drainage for a period of at least two weeks, until the defect has healed. (Partin et al. 2021).

13.8.7.6.2 Laser Procedures

Mucosal bleeding obstructing the surgeon's vision can lead to over-application of the laser and perforation of the prostatic capsule in KTP and HoLEP. This is an exceedingly rare event, as the coagulative effect of the lasers are excellent. In the case that prostatic capsular perforation occurs, and bleeding is not able to be controlled with laser, the procedure should be converted to a TURP to achieve haemostasis.

13.8.7.7   Late Complications of Operative Management of BPH

13.8.7.7.1   Bladder Neck Stenosis

The incidence of bladder neck stenosis was 5% in KTP, 2% in monopolar TURP 1.2% in HoLEP and 0.5% in bipolar TURP (Ahyai et al. 2010). The incidence of bladder neck stenosis in open prostatectomy was 8% (Tubaro and de Nunzio 2006).

Bladder neck stenosis is thought to occur from overtreatment at the bladder neck. Patients may experience an improvement in voiding symptoms that slowly deteriorate over weeks or months. The average time to development of bladder neck stenosis is 6 months, and can be demonstrated clearly by worsening serial peak urinary flow rates on uroflowmetric testing.

Bladder neck stenosis may be conservatively managed with dilatation using sounds, however may require re-operation in the form of a bladder neck incision. (Partin et al. 2021).

13.8.7.7.2   Urethral Stricture

The incidence of urethral stricture was 6.3% in KTP, 4.4% in HoLEP, 4.2% in monopolar TURP and 2.4% in bipolar TURP (Ahyai et al. 2010).

Urethral stricture may be a result of resectoscope related trauma, indwelling catheter use or post-operative urinary tract infection. This can be prevented by liberal use of lubricant for resectoscope insertion and gentle dilatation of the urethra prior to resectoscope insertion. (Partin et al. 2021).

13.8.7.7.3   Urinary Incontinence

The incidence of stress urinary incontinence was 0.6% after monopolar TURP, 0.2% after bipolar TURP, 0.9% after HoLEP and 0.0% after KTP (Ahyai et al. 2010).

The internal urethral sphincter is resected during TURP, prostatectomy and laser procedures for BPH. It is of utmost importance that the external sphincter is preserved, or the patient will develop stress urinary incontinence. Resections and laser application that terminates proximal to the verumontanum are highly unlikely to damage the external urethral sphincter. Fibres of the external sphincter may extend as far as to the distal prostate, and great care should be taken when resecting near this region. (Partin et al. 2021).

13.8.7.7.4   Dysuria

Transient dysuria had an incidence of 0.8% after monopolar TURP, 0% after bipolar TURP, 1.2% after HoLEP and 8.5% after KTP (Ahyai et al. 2010).

Dysuria is likely secondary to mucosal irritation after laser prostate procedures and self-resolved within two months of the procedure (Choi et al. 2013).

13.8.7.7.5 Retrograde Ejaculation and Sexual Dysfunction

TURP, HoLEP, KTP and prostatectomy cause retrograde ejaculation in a large proportion of patients. The average incidence of ejaculatory dysfunction is between 40–60% in TURP, 33.6% in KTP and 75% in HoLEP(Frieben et al. 2010).

The HoLEP and KTP interventions for BPH have comparable effects to TURP with regards to erectile function. Patients may have either an decrease or an increase in their erectile function after these procedures, though no change is observed for the majority (Frieben et al. 2010). No statistically significant difference in the incidence of retrograde ejaculation was found when comparing TURP to HoLEP, (Briganti et al. 2006).

It is important to discuss these complications with all patients prior to undertaking operative management of BPH.

13.8.8 Comparative Efficacy of Surgical Options for BPH

Two clear competitors stand out with regards to optimal surgical management of BPH;HoLEP and TURP (Table 13.3).

13.9 Urethral Stricture

13.9.1 Epidemiology

Urethral stricture disease has an overall prevalence of 0.6%, and sharply rises in incidence in males over the age of 55, and frequently cause bothersome lower urinary tract symptoms in those affected (Santucci et al. 2007). Urethral stricture is the commonest cause of lower urinary tract obstruction in younger male patients between 20 and 40 years of age.

Patients with urethral stricture disease have high rates of urinary tract infection (41%) and urinary incontinence (11%) (Santucci et al. 2007).

13.9.2 Anatomy

The urethra is divided into seven subsections as follows:

- Urethral Meatus: The vertical slit-like opening of the urethra, located on the ventral aspect of the glans penis

Table 13.3 Comparison between Holep and TURP

	TURP	HoLEP
Prostate volume	Limited to a maximum of 100–150 cc	No limit, can be used safely as an alternative to open prostatectomy
Re-operation rate at 7 years (Gilling et al. 2012)	18%	0%
Incidence of Urethral Stricture as a complication (Humphreys et al. 2008)	7.4%	2.2%
Decreased erectile function as a complication (Frieben et al. 2010)	7.7%	7.5%
INCREASED erectile function (Frieben et al. 2010; Michalak et al. 2015)	6.2%	7.1%
Retrograde Ejaculation as a complication (Frieben et al. 2010)	50–86%	50–96%
Improvement in IPSS Score at 12 months post-op (Ahyai et al. 2010; Michalak et al. 2015)	–	HoLEP superior, $P < 0.01$
Improvement in peak urinary flow rate at 12 months post-op, (Ahyai et al. 2010; Michalak et al. 2015)	–	HoLEP superior, $P < 0.0001$
Duration of hospital stay (Michalak et al. 2015)	–	HoLEP superior, $P = 0.001$
Pre-operative anticoagulation/ anti-platelet management (Rivera et al. 2017)	Antiplatelet and anticoagulant medications contra-indicated (must be ceased prior to surgery)	Safe in patients on antiplatelet or anticoagulant medication
Total Intraoperative Complication Rate (Ahyai et al. 2010)	3.2%	3.5%
Overall Morbidity (Ahyai et al. 2010)	No statistically significant difference between TURP and HoLEP ($P = 0.545$)	

- Fossa Navicularis: A slight 'hourglass' shaped widening of the urethra located immediately proximal to the meatus. This portion is located within the spongy erectile tissue of the glans penis. This portion of the urethra is shaped to strategically create 'spin' of the flow of urine, allowing males to direct or 'aim' their flow of urine as required. (Similar to creating 'spin' when throwing a football, to improve aim)
- Penile Urethra: Beginning distal to the investment of the ischiocavernosus muscle, the penile urethra, contained in the corpus spongiosum maintains a constant luminal diameter throughout its course.
- Bulbar Urethra: The proximal continuation of the penile urethra, it is covered by the ischiocavernosus muscle proximally and invested by the bulbospongiosus of the corpus spongiosum. The proximal portion of the bulbar urethra (where it is covered by these muscles) assists in ejaculation. With regards to surface anatomy, the bulbar urethra is found in the perineum (under the scrotum), and hence

is susceptible to injury (particularly straddle trauma during sports such as bike riding).

- Membranous Urethra: Proximal to the bulbar urethra, the membranous urethra passes through the pelvic floor and traverses the deep perineal pouch. It is surrounded by the external urethral sphincter which provides voluntary control of micturition. This is the narrowest portion of the urethra.
- Prostatic Urethra: Proximal to the membranous urethra, the proximal urethra is surrounded by glandular tissue of the prostate and is continuous with the bladder neck. It receives the openings of the ejaculatory ducts (from the seminal vesicles) and the prostatic ducts. It is the widest and most dilatable portion of urethra.
- Bladder neck: represents the commencement of the urethra.

The 'anterior urethra' comprises of the urethral meatus, navicular fossa, penile urethra and bulbous urethra.

The 'posterior urethra' comprises of the bulbar urethra, membranous urethra and bladder neck.

13.9.3   Pathophysiology

Urethral stricture refers to an abnormal narrowing of the lumen of a portion of the urethra. Strictures occurring in the anterior urethra have a different underlying pathology compared to posterior urethral disease.

A stricture involving a portion of urethra that is surrounded by corpus spongiosum is called an anterior urethral stricture. The characteristic pathology found in anterior urethral stricture disease is 'ischaemic spongiofibrosis', whereby the scar tissue of the surrounding spongy erectile tissue called corpus spongiosum is what manifests as anterior urethral stricture (Chapple et al. 2014; Mangera et al. 2016).

Of anterior urethral strictures, the most commonly affected region is the bulbar urethra (52%) (Fenton et al. 2005). The average length of anterior strictures is 4.1 cm (Fenton et al. 2005).

A number of causes have been implicated in the formation of anterior urethral strictures (Fenton et al. 2005):

- Idiopathic (33%): There may be the delayed result of unrecognised childhood injury, typically straddle injuries or pelvic fractures. Urethral injuries resulting in stricture disease may not manifest any symptoms until adulthood (Baskin and McAninch 1993).
- Iatrogenic (33%): Urethral instrumentation such as diagnostic cystoscopy or indwelling catheter insertion are causes of anterior urethral stricture. Transurethral procedures are the commonest cause of anterior urethral stricture disease. Mucosal ischaemia resulting from prolonged passage of a large resectoscope is thought to contribute to the high incidence of stricture disease in transurethral surgeries such as TURP. (Fenton et al. 2005)

- Post-traumatic (19%): Post-traumatic urethral stricture disease found in adults were found to be shorter in length compared to strictures of other aetiologies and were almost exclusively located in the bulbar urethra (given its vulnerability in straddle injuries).
- Children are more vulnerable to urethral stricture disease because the pre-pubertal prostate, puboprostatic ligament and bladder are not stable like in adulthood. Therefore, during a traumatic deceleration force, children may suffer urethral injury at the membranous urethra, through the prostatic urethra or at the bladder neck (Baskin and McAninch 1993).
- In contrast, due to the stabilising forces of the prostate and puboprostatic ligament in adults, trauma to the urethra in adulthood almost exclusively occurs at the membranous urethra (the only portion of the male urethra that is not supported by investing structures). (Baskin and McAninch 1993)
- Inflammatory (15%): Strictures caused by inflammatory or infectious pathology is commonly found in the distal urethra (fossa navicularis strictures), due to their proximity to infectious/inflammatory conditions that may affect the external meatus, glans or prepuce.

- Lichen sclerosis is a common cause of inflammatory strictures. It begins as an infection of the glans penis, but progresses to involve the meatus, causing meatal stenosis. This in turn causes high pressure micturition, resulting in intravasation of urine into the mucous-secreting periurethral glands.

- The glands become inflamed and generate an inflammatory response leading to fibrotic changes of the adjacent corpus spongiosum—resulting in the characteristic pathology seen in anterior stricture disease; spongiofibrosis. (Partin et al. 2021)

- Other infectious causes include chlamydial or gonococcal urethritis, though these are rarer since the advent of effective antibiotic treatment (Partin et al. 2021).

In contrast, the equivalent of a 'stricture' occurring in the posterior urethra is more appropriately called 'posterior urethral injury' or 'posterior urethral stenosis'. The underlying pathology in posterior urethral stenosis differs in that it involves reactive fibrotic changes to the urethra itself, in response to trauma, instrumentation or surgery.

13.9.4 Clinical Presentation and Diagnosis

13.9.4.1 History

Patients with urethral strictures describe a long history of worsening obstructive voiding symptoms or frequent urinary tract infections, including prostatitis or epididymitis.

Symptoms may include straining to void, weak or slow stream, the feeling of incomplete emptying, or taking a longer time to empty their bladder than their peers.

Specific questions that should be asked during history taking include:

- The presence of lower urinary tract symptoms
- History of urinary tract infections
- Any history of trauma to the pelvis or perineum, even in childhood
- A detailed sexual history, including any previously diagnosed sexually transmitted infections
- Any previous instrumentation of the urethra, including urological surgery or urethral catheterisation

If the patient cannot void at the time of presentation, the ideal temporising measure is for placement of an indwelling catheter under cystoscopic guidance, or placement of a suprapubic catheter (Partin et al. 2021).

13.9.4.2 Examination

The external genitalia should be examined, looking for meatal stenosis, evidence of lichen sclerosis, urethritis and urethral discharge. Any signs of infection or urethritis should prompt screening for sexually transmitted infection.

Note that the diameter of an unobstructed male urethra is approximately 10 French, and there is usually no significant effect on the flow of urine until the stricture narrows to smaller than this size, (Partin et al. 2021).

13.9.5 Investigations

13.9.5.1 Uroflowmetry

This is a simple, noninvasive method of evaluating voiding function in patients with lower urinary tract symptoms. In patients with urethral structure disease, the average urinary flow rate and peak urinary flow rate are likely to be reduced. The mean peak urinary flow rate in a study of 189 males with stricture disease was found to be 9.44 mL/second (Tam et al. 2016).

13.9.5.2 Post-Void Residual Volume

Long-standing stricture disease may lead to compensatory changes in the bladder, resulting in increased bladder capacity and less efficient bladder emptying. This can manifest as increased post-void residual volumes, which are easily measured by a simple and non-invasive bedside bladder ultrasound scan. The average post-void residual volume in a study of 189 males with urethral stricture disease was found to be 162 mL (Tam et al. 2016).

13.9.5.3   Retrograde Urethrogram

A urethral catheter is inserted into the distal urethra and used to instill 30 mL of water-soluble contrast into the urethra. A plain xray or continuous fluoroscopy is then performed. The contrast highlights areas of narrowing in the urethra which are likely to represent stricture disease. This method of imaging is used to evaluate the anterior urethra. The sensitivity of retrograde urethogram in diagnosing stricture disease is between 75–100%. The specificity is between 72–97% (Maciejewski and Rourke 2015).

13.9.5.4   Voiding Cystourethrogram

A urinary catheter is advanced into the bladder and used to instil contrast into the bladder until it is filled and dilated. Once the patients feel the urge to urinate, the catheter is removed and fluoroscopy is performed as the patient urinates to visualise the passage of contrast through the posterior urethra.

13.9.5.5   Cystoscopy/Urethroscopy

Endoscopic evaluation of the urethra is the gold standard in diagnosis of urethral stricture disease and in planning operative intervention for strictures. The limitation of standard cystoscopy, however, is that it may not be feasible to pass the cystoscopy through the stricture without trauma. This limitation can be overcome in centres where there is access to urethroscopes (narrow 6-French diameter), paediatric cystoscopes or flexible ureteroscope. (Maciejewski and Rourke 2015).

Flexible cystoscopy/urethroscopy will reveal a white/grey coloured urethra—indicating ischaemic changes. Healthy and vascularised urethra is pink in colour.

13.9.5.6   Ultrasound

Ultrasound is a newly emerging investigation for urethral stricture disease. The benefit of ultrasound over the traditional urethrogram methods are that it has better visualisation of soft tissue structures, allowing for identification of fibrotic changes in the corpus spongiosum that often extend beyond the luminal narrowing seen on urethrogram (Mangera et al. 2016).

This is clinically relevant as longer strictures may necessitate a different operative approach (such as urethroplasty rather than simple dilatation). In a series of 232 males who underwent ultrasound as part of investigation of stricture disease, the operative approach changed in 19% of their patients, where the true length of the stricture would have been underestimated if only urethrogram was used (Buckley et al. 2012).

The standard of care is to use imaging which allows for visualisation of both the anterior and posterior portions of urethra.

13.9.5.7 Complications of Urethral Stricture

13.9.5.7.1 Recurrent Urinary Tract Infections

41% of males with urethral stricture disease were found to have a history of recurrent urinary tract infections. As the bladder compensates for the chronic obstructive voiding difficulties, the post-void residual volume increases and the bladder does not empty efficiently. The static urine remaining in the bladder can host infection and this can be debilitating for the patient (Partin et al. 2021).

13.9.5.8 Urinary Incontinence

11% of male patients with urethral stricture disease were also diagnosed with urinary incontinence at the same time (Anger et al. 2010). This is thought to be represented by the portion of urethral strictures that occur as a complication of TURP, prostatectomy or bladder neck incision, where there may have been concomitant damage to the urethral sphincter as a result of surgery (Anger et al. 2010).

13.9.6 Management of Urethral Stricture

Stricture location, size, aetiology and patient preferences are all taken into consideration when choosing an operative approach to manage stricture disease. Where many patients consider the ideal procedure to be minimally invasive and do not mind returning for repeat procedures, others may prefer a definitive treatment, even if it is more involved.

The most commonly performed intervention is urethral dilatation and urethrotomy (Mangera et al. 2016).

Poor prognosis for surgical management should urge the Urologist towards performing primary urethroplasty rather than urethrotomy/dilatation that are likely to fail. Risk factors for poor prognosis are: stricture length > 2 cm, stricture location in the penile urethra, stricture aetiology being inflammatory/iatrogenic/lichen sclerosis and previous stricture recurrence (Mangera et al. 2016). The presence of two or more of these risk factors is more successfully treated by urethroplasty.

13.9.7 Conservative Management

13.9.7.1 Urethral Dilatation

Urethral dilatation is the commonest intervention performed for stricture disease and can be done with local anaesthesia in an outpatient setting. The goal of urethral dilatation is to gently *stretch* the scar, without causing damage that may induce reactive scarring. If bleeding is seen during the procedure, this generally indicates that the stricture has been torn rather than stretched (Partin et al. 2021).

Urethral dilatation may be performed by insertion of progressively larger sized urethral sounds, or ideally, by the insertion of a urethral balloon-dilating catheter.

Though it is less invasive than other operative techniques, the comparative success rate is low. In a small cohort of patients with short and uncomplicated strictures, urethral dilatation has the potential to be curative (Partin et al. 2021).

13.9.7.2 Internal Urethrotomy

Internal urethrotomy uses a transurethral approach to reach the urethral stricture and cut through the scar tissue from the inside. This procedure can be done under local anaesthesia in the outpatient setting. The aim is to extend the incision made from the strictured area to an area of healthy and vascularised urethral tissue, to allow for the release of scar contracture which hopefully results in enlargement of the urethral lumen. Internal urethrotomy capitalises on the body's capacity to heal by secondary intention, where epithelialization gradually progresses from the new wound edges (Partin et al. 2021).

No differences were found in the efficacy of urethral dilatation when compared to internal urethrotomy. Strictures shorter than 2 cm have the highest success rate, of 60% at 12 months. Strictures greater than 4 cm in length had a 20% success rate at 12 months (Steenkamp et al. 1997).

Urethral dilatation and internal urethrotomy are reserved as initial treatment for uncomplicated short strictures. Where there has been recurrence of the stricture after a second treatment of urethral dilatation/internal urethrotomy, alternative approaches should be considered, as there has been no evidence of benefit from a third treatment of this approach (Heyns et al. 1998).

13.9.8 Surgical Management

13.9.8.1 Anastomotic Urethroplasty

This technique, best used for uncomplicated bulbar strictures shorter than 3 cm in length, involves total excision of the strictured portion of urethra followed by primary anastomotic repair of the cut ends of urethra. It is performed under general anaesthesia.

The corpus cavernosum, spongiosum and ventral urethra are exposed using positioning sutures. The affected length of urethra is marked. The urethra is slowly dissected off the cavernosum, and a retractor is placed dorsal to the urethra to protect surrounding structures. The strictured segment of urethra is resected. A urethral catheter is inserted which allows easier approximation of the cut ends of urethra. This catheter generally remains in place for 10–20 days at least, until the anastomotic connection has healed.

Optimal results are achieved when the entirety of the fibrosed area is resected and the newly created anastomosis is free of tension (Partin et al. 2021). The success rates of this technique are between 91–98% over an average follow up time of 53 months (Barbagli et al. 2008; Chapple et al. 2014).

This technique, although effective, is unsuitable for complicated or long strictures.

13.9.8.2 Augmentation/Substitution Urethral Reconstruction

Autograft tissue is used to replace the length of strictured urethra that is excised and may be done in one or two staged procedures.

Where performed as a two-staged procedure, the first stage involves placement of the selected graft over the dartos fascia, creating a 'urethral plate'. The second surgery is performed once the graft has matured and involves the use of a tube-like instrument to 'inroll' the matured graft, creating the lumen of the new urethra (Chapple et al. 2014).

In the single-staged procedure, the entire fibrosed area of the urethral stricture is excised and primary urethral reconstruction with graft on lay may be performed using bladder epithelium, rectal mucosa, full thickness skin grafts or oral mucosal grafts (Partin et al. 2021). Oral mucosal grafts have the highest success rates of all the grafts (Chapple et al. 2014).

13.9.8.3 Perineal Urethrostomy

This involves the creation of a new urethral opening in the perineum and is considered to be a drastic intervention utilised where all other interventions have been unsuccessful in managing complex anterior urethral stricture disease.

The patient is placed in the lithotomy position. An inverted U incision is made in the perineum and dissection is performed until the bulbocavernosus is identified and divided, exposing the bulbar urethra. A longitudinal incision is made in the ventral aspect of the bulbar urethra. This incision is extended proximally until healthy urethra is encountered. The perineal skin is sutured to the edges of the proximal urethrotomy—which is now the new urethral opening. An indwelling catheter is placed into the new urethral opening, to remain for 4–7 days, (Myers and McAninch 2011).

Perineal urethrostomy is suitable for severe, recurrent anterior urethral stricture disease and for older patients who may not be suitable for major surgery.

13.9.8.4   Operative Complications

13.9.8.4.1   Alteration in Penile Appearance

Loss of urethral length may cause curvature or shortening of the penis, that is more noticeable during erection. This risk can be mitigated by inducing artificial erection intra-operatively during the reconstruction (Chapple et al. 2014). 10–39% of patients who underwent urethral reconstruction demonstrated contraction secondary to scarring of the graft, which points to the need to identify areas of improvement with these techniques (Barbagli et al. 2006).

13.9.8.4.2   Sexual Dysfunction

Though there is a risk of erectile dysfunction that occurs in the first few months post-operatively, the majority of men have a full recovery by 7 months post-operatively without any intervention (Chapple et al. 2014). 18.3% of patients reported decreased glans sensitivity after urethroplasty and 11.6% reported a glans that no longer swelled with erection, (Baradaran et al. 2018).

13.9.8.5   Spraying of Urine

10–13% of patients who undergo surgery for urethral stricture disease will experience spraying of the urine (rather than a stream). A large proportion of patients may have to sit down to urinate due to the unpredictability or weakness of their urinary stream. Patients also report a weakness in the force of ejaculation after urethral surgery. (Baradaran et al. 2018).

13.9.8.6   Donor Site Complications

In the case of oral mucosal grafts, buccal mucosa inferior to the parotid duct is excised for use as graft material. Haemorrhage at the donor site is usually easily controlled with pressure. Complications related to this form of graft include pain, infection, loss of sensation within the check or lower lip and damage to the parotid duct. A palpable scar may develop where the graft was harvested. Despite this, in a survey of 295 patients who underwent buccal mucosal graft harvestation, 98.4% were satisfied with the result of the urethral construction and would undergo the procedure again, (Chapple et al. 2014).

13.9.8.7   Prolonged Indwelling Catheter

Most surgical management techniques for treatment of urethral stricture disease require the patient to keep an indwelling catheter in situ for several weeks, to prevent recurrence of the stricture disease.

13.9.8.8   Stricture Recurrence

Recurrence rates are highest with urethral dilatation and urethrotomy and particularly when the stricture length is >4 cm. Patients should be advised that these treatment methods are more likely to be a temporizing measure than a cure.

Recurrence rates are lower with urethroplasty and augmentation reconstruction. The success rate of urethroplasty was found to be approximately 85%, where 15% of patients had to undergo a further surgical procedure or urethral dilatation due to stricture recurrence (Meeks et al. 2009).

The dorsal onlay technique for urethral augmentation reconstruction was found to have an 83% success rate during the average follow up period of 42 months (Chapple et al. 2014).

References

Abe T, Shinohara N, Harabayashi T, Sazawa A, Suzuki S, Kawarada Y, Nonomura K. Postoperative inguinal hernia after radical prostatectomy for prostate cancer. Urology. 2007;69:326–9.

Ahn SJ, Woo TY, Lee DW, Lew DH, Song SY. Nipple-areolar complex ischemia and necrosis in nipple-sparing mastectomy. Eur J Surg Oncol. 2018;44:1170–6.

Ahyai SA, Gilling P, Kaplan SA, Kuntz RM, Madersbacher S, Montorsi F, Speakman MJ, Stief CG. Meta-analysis of functional outcomes and complications following transurethral procedures for lower urinary tract symptoms resulting from benign prostatic enlargement. Eur Urol. 2010;58:384–97.

Ali Abulmeaty MM, Almajwal AM, Elsadek MF, Berika MY, Razak S. Metabolic effects of testosterone hormone therapy in normal and orchiectomized male rats: from indirect calorimetry to Lipolytic enzymes. Int J Endocrinol. 2019;2019:7546385.

Al-Said S, Al-Naimi A, Al-Ansari A, Younis N, Shamsodini A, Shokeir AA. Varicocelectomy for male infertility: a comparative study of open, laparoscopic and microsurgical approaches. J Urol. 2008;180:266–70.

Ameli M, Boroumand-Noughabi S, Gholami-Mahtaj L. A 14-year-old boy with torsion of the Epididymal cyst. Case Rep Urol. 2015;2015:731987.

Anger JT, Santucci R, Grossberg AL, Saigal CS. The morbidity of urethral stricture disease among male medicare beneficiaries. BMC Urol. 2010;10:3.

Asala S, Chaudhary SC, Masumbuko-Kahamba N, Bidmos M. Anatomical variations in the human testicular blood vessels. Ann Anat. 2001;183:545–9.

Asimakopoulos AD, De Nunzio C, Kocjancic E, Tubaro A, Rosier PF, Finazzi-AGRÒ E. Measurement of post-void residual urine. Neurourol Urodyn. 2016;35:55–7.

Autorino R, Perdonà S, D'armiento M, De Sio M, Damiano R, Cosentino L, Di Lorenzo G. Gynecomastia in patients with prostate cancer: update on treatment options. Prostate Cancer Prostatic Dis. 2006;9:109–14.

Bachman E, Travison TG, Basaria S, Davda MN, Guo W, Li M, Connor Westfall J, Bae H, Gordeuk V, Bhasin S. Testosterone induces erythrocytosis via increased erythropoietin and suppressed hepcidin: evidence for a new erythropoietin/hemoglobin set point. J Gerontol A Biol Sci Med Sci. 2014;69:725–35.

Bachmann A, Muir GH, Collins EJ, Choi BB, Tabatabaei S, Reich OM, Gómez-Sancha F, Woo HH. 180-W XPS GreenLight laser therapy for benign prostate hyperplasia: early safety, efficacy, and perioperative outcome after 201 procedures. Eur Urol. 2012;61:600–7.

Baradaran N, Hampson LA, Edwards TC, Voelzke BB, Breyer BN. Patient-reported outcome measures in urethral reconstruction. Curr Urol Rep. 2018;19:48.

Barbagli G, De Angelis M, Palminteri E, Lazzeri M. Failed hypospadias repair presenting in adults. Eur Urol. 2006;49:887–95.

Barbagli G, Guazzoni G, Lazzeri M. One-stage bulbar urethroplasty: retrospective analysis of the results in 375 patients. Eur Urol. 2008;53:828–33.

Baskin LS, Mcaninch JW. Childhood urethral injuries: perspectives on outcome and treatment. Br J Urol. 1993;72:241–6.

Berry SJ, Coffey DS, Walsh PC, Ewing LL. The development of human benign prostatic hyperplasia with age. J Urol. 1984;132:474–9.

Blumer CG, Restelli AE, Giudice PTD, Soler TB, Fraietta R, Nichi M, Bertolla RP, Cedenho AP. Effect of varicocele on sperm function and semen oxidative stress. BJU Int. 2012;109:259–65.

Boccardo F, Rubagotti A, Battaglia M, Di Tonno P, Selvaggi F, Conti G, Comeri G, Bertaccini A, Martorana G, Galassi P. Evaluation of tamoxifen and anastrozole in the prevention of gynecomastia and breast pain induced by bicalutamide monotherapy of prostate cancer. J Clin Oncol. 2005;23:808–15.

Bordoni D, Falco G, Cadenelli P, Ornelli M, Patriti A, Tessone A, Serafini M, Magalotti C. How to avoid nipple–areola complex complications in high-grade gynecomastia patients treated by mastectomy: surgical pearls. Cham: Springer International Publishing; 2018.

Bouchot O, Branchereau J, Perrouin-Verbe MA. Influence of inguinal hernia repair on male fertility. J Visc Surg. 2018;155:S37–40.

Bozeman CB, Carver BS, Caldito G, Venable DD, Eastham JA. Prostate cancer in patients with an abnormal digital rectal examination and serum prostate-specific antigen less than 4.0 ng/mL. Urology. 2005;66:803–7.

Braunstein GD. Gynecomastia. N Engl J Med. 2007;357:1229–37.

Briganti A, Naspro R, Gallina A, Salonia A, Vavassori I, Hurle R, Scattoni E, Rigatti P, Montorsi F. Impact on sexual function of holmium laser enucleation versus transurethral resection of the prostate: results of a prospective, 2-center, randomized trial. J Urol. 2006;175:1817–21.

Buckley JC, Wu AK, Mcaninch JW. Impact of urethral ultrasonography on decision-making in anterior urethroplasty. BJU Int. 2012;109:438–42.

Calogero AE, Burgio G, Condorelli RA, Cannarella R, La Vignera S. Epidemiology and risk factors of lower urinary tract symptoms/benign prostatic hyperplasia and erectile dysfunction. Aging Male. 2019;22:12–9.

Celigoj FAMD, Costabile RAMD. Surgery of the scrotum and seminal vesicles. In: Campbell MF, Retik AB, Walsh PC, editors. Campbell-Walsh urology. 11th ed. Philadelphia: Saunders; 2016.

Chan P. Management options of varicoceles. Indian J Urol. 2011;27:65–73.

Chancellor MB, Blaivas JG, Kaplan SA, Axelrod S. Bladder outlet obstruction versus impaired detrusor contractility: the role of outflow. J Urol. 1991;145:810–2.

Chapple C, Andrich D, Atala A, Barbagli G, Cavalcanti A, Kulkarni S, Mangera A, Nakajima Y. SIU/ICUD consultation on urethral strictures: the management of anterior urethral stricture disease using substitution urethroplasty. Urology. 2014;83:S31–47.

Cheng CY, Mruk DD. The blood-testis barrier and its implications for male contraception. Pharmacol Rev. 2012;64:16.

Chiba K, Ramasamy R, Lamb DJ, Lipshultz LI. The varicocele: diagnostic dilemmas, therapeutic challenges and future perspectives. Asian J Androl. 2016;18:276–81.

Choi YS, Bae WJ, Kim SJ, Kim KS, Cho HJ, Hong S-H, Lee JY, Hwang T-K, Kim SW. Efficacy and safety of 120-W GreenLight high-performance system laser photo vaporization of the prostate: 3-year results with specific considerations. Prostate Int. 2013;1:169–76.

Clavijo RI, Carrasquillo R, Ramasamy R. Varicoceles: prevalence and pathogenesis in adult men. Fertil Steril. 2017;108:364–9.

Coogan CL. Testicular mass. In: Myers JA, Millikan KW, Saclarides TJ, editors. Common surgical diseases: an algorithmic approach to problem solving. Springer New York: New York, NY; 2008.

Cooper TG, Raczek S, Yeung CH, Schwab E, Schulze H, Hertle L. Composition of fluids obtained from human epididymal cysts. Urol Res. 1992;20:275–80.

Cordova A, Tripoli M. Gynecomastia. Cham: Springer International Publishing; 2020.

Corona G, Giagulli VA, Maseroli E, Vignozzi L, Aversa A, Zitzmann M, Saad F, Mannucci E, Maggi M. Testosterone supplementation and body composition: results from a meta-analysis of observational studies. J Endocrinol Investig. 2016;39:967–81.

Costanzo PR, Pacenza NA, Aszpis SM, Suárez SM, Pragier UM, Usher JGS, Vásquez Cayoja M, Iturrieta S, Gottlieb SE, Rey RA, Knoblovits P. Clinical and etiological aspects of gynecomastia in adult males: a multicenter study. Biomed Res Int. 2018;2018:8364824.

Coward RM, Rajanahally S, Kovac JR, Smith RP, Pastuszak AW, Lipshultz LI. Anabolic steroid induced hypogonadism in young men. J Urol. 2013;190:2200–5.

Cross SS, Underwood JCE. Underwood's pathology. Churchill: Livingstone; 2013.

Dagur G, Gandhi J, Suh Y, Weissbart S, Sheynkin YR, Smith NL, Joshi G, Khan SA. Classifying hydroceles of the pelvis and groin: an overview of etiology, secondary complications, evaluation, and management. Curr Urol. 2017;10:1–14.

Dandapat MC, Padhi NC, Patra AP. Effect of hydrocele on testis and spermatogenesis. BJS. 1990;77:1293–4.

Dickson G. Gynecomastia. Am Fam Physician. 2012;85:716–22.

Diegidio P, Jhaveri JK, Ghannam S, Pinkhasov R, Shabsigh R, Fisch H. Review of current varicocelectomy techniques and their outcomes. BJU Int. 2011;108:1157–72.

Dipaola RS, Kumar P, Hait WN, Weiss RE. State-of-the-art prostate cancer treatment and research. A report from the Cancer Institute of new Jersey. N J Med. 2001;98:23–33.

Diraimondo CV, Roach AC, Meador CK. Gynecomastia from exposure to vaginal estrogen cream. N Engl J Med. 1980;302:1089–90.

Dong Z, Wang Z, Yang K, Liu Y, Gao W, Chen W. Tamsulosin versus terazosin for benign prostatic hyperplasia: a systematic review. Syst Biol Reprod Med. 2009;55:129–36.

Drake MJ, Chapple C, Sokol R, Oelke M, Traudtner K, Klaver M, Drogendijk T, Van Kerrebroeck P. Long-term safety and efficacy of single-tablet combinations of solifenacin and tamsulosin oral controlled absorption system in men with storage and voiding lower urinary tract symptoms: results from the NEPTUNE study and NEPTUNE II open-label extension. Eur Urol. 2015;67:262–70.

Dubin L, Amelar RD. Varicocelectomy: 986 cases in a twelve-year study. Urology. 1977;10:446–9.

Ergün S, Bruns T, Soyka A, Tauber R. Angioarchitecture of the human spermatic cord. Cell Tissue Res. 1997;288:391–8.

Esposito C, Valla JS, Najmaldin A, Shier F, Mattioli G, Savanelli A, Castagnetti M, Mckinley G, Stayaert H, Settimi A, Jasonni V, Guys JM. Incidence and management of hydrocele following varicocele surgery in children. J Urol. 2004;171:1271–3.

Fagerlund A, Lewin R, Rufolo G, Elander A, Santanelli di Pompeo F, Selvaggi G. Gynecomastia: a systematic review. J Plast Surg Hand Surg. 2015;49:311–8.

Fenton AS, Morey AF, Aviles R, Garcia CR. Anterior urethral strictures: etiology and characteristics. Urology. 2005;65:1055–8.

Fitzgibbons RJ, Forse RA. Groin hernias in adults. N Engl J Med. 2015;372:756–63.

Foley SJ, Soloman LZ, Wedderburn AW, Kashif KM, Summerton D, Basketter V, Holmes SA. A prospective study of the natural history of hematuria associated with benign prostatic hyperplasia and the effect of finasteride. J Urol. 2000;163:496–8.

Foo KT. What is a disease? What is the disease clinical benign prostatic hyperplasia (BPH)? World J Urol. 2019;37:1293–6.

Fourie N, Banieghbal B. Pediatric hydrocele: a comprehensive review. Clin Surg. 2017;2:1.

França LR, Hess RA, Dufour JM, Hofmann MC, Griswold MD. The Sertoli cell: one hundred fifty years of beauty and plasticity. Andrology. 2016;4:189–212.

Frieben RW, Lin HC, Hinh PP, Berardinelli F, Canfield SE, Wang R. The impact of minimally invasive surgeries for the treatment of symptomatic benign prostatic hyperplasia on male sexual function: a systematic review. Asian J Androl. 2010;12:500–8.

Fruhstorfer BH, Malata CM. A systematic approach to the surgical treatment of gynaecomastia. Br J Plast Surg. 2003;56:237–46.

Fukuta F, Masumori N, Mori M, Tsukamoto T. Natural history of lower urinary tract symptoms in Japanese men from a 15-year longitudinal community-based study. BJU Int. 2012;110:1023–9.

Gerstenberg TC, Andersen JT, Klarskov P, Ramirez D, Hald T. High flow infravesical obstruction in men: symptomatology, urodynamics and the results of surgery. J Urol. 1982;127:943–5.

Giagulli VA, Triggiani V, Carbone MD, Corona G, Tafaro E, Licchelli B, Guastamacchia E. The role of long-acting parenteral testosterone undecanoate compound in the induction of secondary sexual characteristics in males with hypogonadotropic hypogonadism. J Sex Med. 2011;8:3471–8.

Gilling PJ, Wilson LC, King CJ, Westenberg AM, Frampton CM, Fraundorfer MR. Long-term results of a randomized trial comparing holmium laser enucleation of the prostate and transurethral resection of the prostate: results at 7 years. BJU Int. 2012;109:408–11.

Goldenberg L, So A, Fleshner N, Rendon R, Drachenberg D, Elhilali M. The role of 5-alpha reductase inhibitors in prostate pathophysiology: is there an additional advantage to inhibition of type 1 isoenzyme? Can Urol Assoc J. 2009;3:S109–14.

Goluboff ET, Chang DT, Kirsch AJ, Fisch H. Incidence of external spermatic veins in patients undergoing inguinal varicocelectomy. Urology. 1994;44:893–6.

Goonewardene SS, Pietrzak P, Albala D. Basic Urological Management. Cham: Springer International Publishing AG; 2018.

Grosse H. Frequency, localization and associated disorders in urinary calculi. Analysis of 1671 autopsies in urolithiasis. Z Urol Nephrol. 1990;83:469–74.

Hagberg KW, Divan HA, Fang SC, Nickel JC, Jick SS. Risk of gynecomastia and breast cancer associated with the use of 5-alpha reductase inhibitors for benign prostatic hyperplasia. Clin Epidemiol. 2017;9:83–91.

Hägglöf C, Bergh A. The stroma-a key regulator in prostate function and malignancy. Cancers. 2012;4:531–48.

Hahn RG. Fluid absorption in endoscopic surgery. Br J Anaesth. 2006;96:8–20.

Harrington JM, Stein GF, Rivera RO, De Morales AV. The occupational hazards of formulating oral contraceptives--a survey of plant employees. Arch Environ Health. 1978;33:12–5.

Harrison RM, Lewis RW, Roberts JA. Testicular blood flow and fluid dynamics in monkeys with surgically induced varicoceles. J Androl. 1983;4:256–60.

Hattori M, Tonooka A, Zaitsu M, Mikami K, Suzue-Yanagisawa A, Uekusa T, Takeuchi T. Overexpression of aquaporin 1 in the tunica vaginalis may contribute to adult-onset primary hydrocele testis. Adv Urol. 2014;2014:202434.

Henriksen NA, Yadete DH, Sorensen LT, Agren MS, Jorgensen LN. Connective tissue alteration in abdominal wall hernia. Br J Surg. 2011;98:210–9.

Herniasurge Group, T. International guidelines for groin hernia management. Hernia. 2018;22:1–165.

Heyns CF, Steenkamp JW, De Kock ML, Whitaker P. Treatment of male urethral strictures: is repeated dilation or internal urethrotomy useful? J Urol. 1998;160:356–8.

Hikosaka A, Iwase Y. Spermatocele presenting as acute scrotum. Urol J. 2008;5:206–8.

Holtgrewe HL, Mebust WK, Dowd JB, Cockett AT, Peters PC, Proctor C. Transurethral prostatectomy: practice aspects of the dominant operation in American urology. J Urol. 1989;141: 248–53.

Hong SK, Lee ST, Jeong SJ, Byun SS, Hong YK, Park DS, Hong JY, Son JH, Kim C, Jang SH, Lee SE. Chronic kidney disease among men with lower urinary tract symptoms due to benign prostatic hyperplasia. BJU Int. 2010;105:1424–8.

Hou Y, Zhang Y, Li G, Wang W, Li H. Microsurgical epididymal cystectomy does not impact upon sperm count, motility or morphology and is a safe and effective treatment for epididymal cystic lesions (ECLs) in young men with fertility requirements. Urology. 2018;122:97–103.

Humphreys MR, Miller NL, Handa SE, Terry C, Munch LC, Lingeman JE. Holmium laser enucleation of the prostate--outcomes independent of prostate size? J Urol. 2008;180:2431–5. discussion 2435

Iglesias P, Carrero JJ, Díez JJ. Gonadal dysfunction in men with chronic kidney disease: clinical features, prognostic implications and therapeutic options. J Nephrol. 2012;25:31–42.

Irwin DE, Milsom I, Hunskaar S, Reilly K, Kopp Z, Herschorn S, Coyne K, Kelleher C, Hampel C, Artibani W, Abrams P. Population-based survey of urinary incontinence, overactive bladder, and other lower urinary tract symptoms in five countries: results of the EPIC study. Eur Urol. 2006;50:1306–14. discussion 1314-5

Isaacs JT. Prostate stem cells and benign prostatic hyperplasia. Prostate. 2008;68:1025–34.

Isidori AM, Giannetta E, Gianfrilli D, Greco EA, Bonifacio V, Aversa A, Isidori A, Fabbri A, Lenzi A. Effects of testosterone on sexual function in men: results of a meta-analysis. Clin Endocrinol. 2005;63:381–94.

Itoh M, Li XQ, Miyamoto K, Takeuchi Y. Degeneration of the seminiferous epithelium with ageing is a cause of spermatoceles? Int J Androl. 1999;22:91–6.

Jacobsen SJ. Risk factors for benign prostatic hyperplasia. Curr Urol Rep. 2007;8:281–8.

Jahnson S, Sandblom D, Holmäng S. A randomized trial comparing 2 doses of polidocanol sclerotherapy for hydrocele or spermatocele. J Urol. 2011;186:1319–23.

Jenkins JT, O'dwyer PJ. Inguinal hernias. BMJ (Clinical Research ed). 2008;336:269–72.

Jones. To store or mature spermatozoa? The primary role of the epididymis. Int J Androl. 1999;22:57–67.

Kaplan Steven A, Wein Alan J, Staskin David R, Roehrborn Claus G, Steers William D. Urinary retention and post-void residual urine in men: separating truth from tradition. J Urol. 2008;180:47–54.

Kasielska A, Antoszewski B. Surgical management of gynecomastia: an outcome analysis. Ann Plast Surg. 2013;71:471–5.

Kauffman EC, Kim HH, Tanrikut C, Goldstein M. Microsurgical Spermatocelectomy: technique and outcomes of a novel surgical approach. J Urol. 2011;185:238–42.

Kiddoo DA, Wollin TA, Mador DR. A population based assessment of complications following outpatient hydrocelectomy and spermatocelectomy. J Urol. 2004;171:746–8.

Kordzadeh A, Liu MO, Jayanthi NV. Male infertility following inguinal hernia repair: a systematic review and pooled analysis. Hernia. 2017;21:1–7.

Kumar P, Kumar N, Thakur DS, Patidar A. Male hypogonadism: symptoms and treatment. J Adv Pharm Technol Res. 2010;1:297–301.

Kuo RL, Paterson RF, Kim SC, Siqueira TM, Elhilali MM, Lingeman JE. Holmium laser enucleation of the prostate (HoLEP): a technical update. World J Surg Oncol. 2003;1:6.

Kupis Ł, Dobroński PA, Radziszewski P. Varicocele as a source of male infertility - current treatment techniques. Cent Eur J Urol. 2015;68:365–70.

Kurobe M, Harada A, Sugihara T, Baba Y, Hiramatsu T, Ohashi S, Otsuka M. The outcomes of conservative management and the natural history of asymptomatic hydroceles in children. Pediatr Surg Int. 2020;36:1189–95.

Lake MG, Krook LS, Cruz SV. Pituitary adenomas: an overview. Am Fam Physician. 2013;88:319–27.

Langeveld HR, Van't Riet M, Weidema WF, Stassen LPS, Steyerberg EW, Lange J, Bonjer HJ, Jeekel J. Total extraperitoneal inguinal hernia repair compared with Lichtenstein (the level-trial): a randomized controlled trial. Ann Surg. 2010;251:819–24.

Lapid O, Jolink F, Meijer SL. Pathological findings in gynecomastia: analysis of 5113 breasts. Ann Plast Surg. 2015;74:163–6.

Lashen AMAH, Nafea MA-A, Elsayed MAM. Hydrocelectomy through inguinal approach in adults. Egypt J Hosp Med. 2019;76:3801–6.

Lei J, Luo C, Zhang Y, Guo Y, Su X, Wang X. A comparison of a novel endoscopic "Su-Wang technique" with the open "Jaboulay's procedure" for the surgical treatment of adult primary vaginal hydrocele. Sci Rep. 2019;9:9152.

Leung ML, Gooding GA, Williams RD. High-resolution sonography of scrotal contents in asymptomatic subjects. AJR Am J Roentgenol. 1984;143:161–4.

Lewis DS, Grimm LJ, Kim CY. Left renal vein compression as cause for varicocele: prevalence and associated findings on contrast-enhanced CT. Abdom Imaging. 2015;40:3147–51.

Li C-C, Fu J-P, Chang S-C, Chen T-M, Chen S-G. Surgical treatment of gynecomastia: complications and outcomes. Ann Plast Surg. 2012;69:510–5.

Light D, Ratnasingham K, Banerjee A, Cadwallader R, Uzzaman MM, Gopinath B. The role of ultrasound scan in the diagnosis of occult inguinal hernias. Int J Surg. 2011;9:169–72.

Lizcano F, Guzmán G. Estrogen deficiency and the origin of obesity during menopause. Biomed Res Int. 2014;2014:757461.

Lockhart K, Dunn D, Teo S, Ng JY, Dhillon M, Teo E, Van Driel ML. Mesh versus non-mesh for inguinal and femoral hernia repair. Cochrane Database Syst Rev. 2018;9(2):16.

Lomboy JR, Coward RM. The varicocele: clinical presentation, evaluation, and surgical management. Semin Interv Radiol. 2016;33:163–9.

Lord PH. A bloodless operation for the radical cure of idiopathic hydrocele. BJS. 1964;51:914–6.

Low LS, Nair SM, Davies AJW, Akapita T, Holmes MA. Aspiration and sclerotherapy of hydroceles and spermatoceles/epididymal cysts with 100% alcohol. ANZ J Surg. 2020;90:57–61.

Lundström KJ, Söderström L, Jernow H, Stattin P, Nordin P. Epidemiology of hydrocele and spermatocele; incidence, treatment and complications. Scand J Urol. 2019;53:134–8.

Maciejewski C, Rourke K. Imaging of urethral stricture disease. Transl Androl Urol. 2015;4:2–9.

Mangera A, Osman N, Chapple C. Evaluation and management of anterior urethral stricture disease. F1000Res. 2016;5:153.

Marra AR, Puig-Asensio M, Edmond MB, Schweizer ML, Nepple KG. Infectious complications of conventional laparoscopic vs robotic laparoscopic prostatectomy: a systematic literature review and meta-analysis. J Endourol. 2019;33:179–88.

Matsuda T, Muguruma K, Horii Y, Ogura K, Yoshida O. Serum antisperm antibodies in men with vas deferens obstruction caused by childhood inguinal herniorrhaphy. Fertil Steril. 1993;59:1095–7.

McConnell JD, Barry MJ, Bruskewitz RC. Benign prostatic hyperplasia: diagnosis and treatment. Agency for Health Care Policy and Research. Clin Pract Guidel Quick Ref Guide Clin. 1994;1:1–17.

Meeks Joshua J, Maschino Alexandra C, Mcvary Kevin T, Sandhu Jaspreet S. Clinically significant prostate cancer is rarely missed by ablative procedures of the prostate in men with prostate specific antigen less than 4 ng/ml. J Urol. 2013;189:111–5.

Meeks JJ, Erickson BA, Granieri MA, Gonzalez CM. Stricture recurrence after Urethroplasty: a systematic review. J Urol. 2009;182:1266–70.

Michalak J, Tzou D, Funk J. HoLEP: the gold standard for the surgical management of BPH in the 21(st) century. Am J Clin Exp Urol. 2015;3:36–42.

Mihai R. Diagnosis, treatment and outcome of adrenocortical cancer. Br J Surg. 2015;102:291–306.

Mirone V, Imbimbo C, Longo N, Fusco F. The detrusor muscle: an innocent victim of bladder outlet obstruction. Eur Urol. 2007;51:57–66.

Mohamad NV, Soelaiman IN, Chin KY. A concise review of testosterone and bone health. Clin Interv Aging. 2016;11:1317–24.

Musa O, Roy A, Ansari NA, Sharan J. Evaluation of the role of sodium tetradecyl sulfate as a sclerosant in the treatment of primary hydrocele. Indian J Surg. 2015;77:432–7.

Myers JB, Mcaninch JW. Perineal urethrostomy. BJU Int. 2011;107:856–65.

Narula HS, Carlson HE. Gynaecomastia--pathophysiology, diagnosis and treatment. Nat Rev Endocrinol. 2014;10:684–98.

Nilsson H, Stranne J, Stattin P, Nordin P. Incidence of groin hernia repair after radical prostatectomy: a population-based nationwide study. Ann Surg. 2014;259:1223–7.

Nitti VW. Pressure flow urodynamic studies: the gold standard for diagnosing bladder outlet obstruction. Rev Urol. 2005;7(Suppl 6):S14–21.

Nitti VW, Khullar V, Van Kerrebroeck P, Herschorn S, Cambronero J, Angulo JC, Blauwet MB, Dorrepaal C, Siddiqui E, Martin NE. Mirabegron for the treatment of overactive bladder: a prespecified pooled efficacy analysis and pooled safety analysis of three randomised, double-blind, placebo-controlled, phase III studies. Int J Clin Pract. 2013;67:619–32.

Norões J, Dreyer G. A mechanism for chronic filarial hydrocele with implications for its surgical repair. PLoS Negl Trop Dis. 2010;4:e695.

Ntali G, Capatina C, Grossman A, Karavitaki N. Functioning Gonadotroph adenomas. J Clin Endocrinol Metabol. 2014;99:4423–33.

O'donnell AM, Foo ITH. Anaesthesia for transurethral resection of the prostate. Continuing Education in Anaesthesia Critical Care & Pain. 2009;9:92–6.

Olsson H, Bladstrom A, ALM P. Male gynecomastia and risk for malignant tumours--a cohort study. BMC Cancer. 2002;2:26.

Osifo OD, Osaigbovo EO. Congenital hydrocele: prevalence and outcome among male children who underwent neonatal circumcision in Benin City, Nigeria. J Pediatr Urol. 2008;4: 178–82.

Özalp B, Berköz Ö, Aydinol M. Is the transposition of the nipple-areolar complex necessary in Simon grade 2b gynecomastia operations using suction-assisted liposuction? J Plast Surg Hand Surg. 2018;52:7–13.

Palmer JR, Herbst AL, Noller KL, Boggs DA, Troisi R, Titus-Ernstoff L, Hatch EE, Wise LA, Strohsnitter WC, Hoover RN. Urogenital abnormalities in men exposed to diethylstilbestrol in utero: a cohort study. Environ Health. 2009;8:37.

Partin AW, Kavoussi LR, Peters CA, editors. Campbell-Walsh-Wein urology. 12th ed. Philadelphia, PA: Elsevier; 2021.

Partrick DA, Bensard DD, Karrer FM, Ruyle SZ. Is routine pathological evaluation of pediatric hernia sacs justified? J Pediatr Surg. 1998;33:1090–2. discussion 1093-4

Perdonà S, Autorino R, De Placido S, D'armiento M, Gallo A, Damiano R, Pingitore D, Gallo L, De Sio M, Bianco AR, Di Lorenzo G. Efficacy of tamoxifen and radiotherapy for prevention and treatment of gynaecomastia and breast pain caused by bicalutamide in prostate cancer: a randomised controlled trial. Lancet Oncol. 2005;6:295–300.

Petty PM, Solomon M, Buchel EW, Tran NV. Gynecomastia: evolving paradigm of management and comparison of techniques. Plast Reconstr Surg. 2010;125:1301–8.

Pfitzenmaier J, Gilfrich C, Pritsch M, Herrmann D, Buse S, Haferkamp A, Djakovic N, Pahernik S, Hohenfellner M. Vaporization of prostates of > or =80 mL using a potassium-titanyl-phosphate laser: midterm-results and comparison with prostates of <80 mL. BJU Int. 2008;102:322–7.

Poelman MM, Van Den Heuvel B, Deelder JD, Abis GS, Beudeker N, Bittner RR, Campanelli G, Van Dam D, Dwars BJ, Eker HH, Fingerhut A, Khatkov I, Koeckerling F, Kukleta JF, Miserez M, Montgomery A, Munoz Brands RM, Morales Conde S, Muysoms FE, Soltes M, Tromp W, Yavuz Y, Bonjer HJ. EAES consensus development conference on endoscopic repair of groin hernias. Surg Endosc. 2013;27:3505–19.

Presicce F, De Nunzio C, Tubaro A. Clinical implications for the early treatment of benign prostatic enlargement (BPE): a systematic review. Curr Urol Rep. 2018;19:70.

Rahnema CD, Lipshultz LI, Crosnoe LE, Kovac JR, Kim ED. Anabolic steroid induced hypogonadism: diagnosis and treatment. Fertil Steril. 2014;101:1271–9.

Rao J-M, Yang J-R, Ren Y-X, He J, Ding P, Yang J-H. Plasmakinetic enucleation of the prostate versus Transvesical open prostatectomy for benign prostatic hyperplasia 80 mL: 12-month follow-up results of a randomized clinical trial. Urology. 2013;82:176–81.

Ridha H, Colville RJ, Vesely MJ. How happy are patients with their gynaecomastia reduction surgery? J Plast Reconstr Aesthet Surg. 2009;62:1473–8.

Rioja J, Sánchez-Margallo FM, USÓN J, RIOJA LA. Adult hydrocele and spermatocele. BJU Int. 2011;107:1852–64.

Rivera M, Krambeck A, Lingeman J. Holmium laser enucleation of the prostate in patients requiring anticoagulation. Curr Urol Rep. 2017;18:77.

Robinson A, Light D, Nice C. Meta-analysis of sonography in the diagnosis of inguinal hernias. J Ultrasound Med. 2013;32:339–46.

Roehrborn CG. Alfuzosin 10 mg once daily prevents overall clinical progression of benign prostatic hyperplasia but not acute urinary retention: results of a 2-year placebo-controlled study. BJU Int. 2006;97:734–41.

Roehrborn CG, Dolte KS, Ross KS, Girman CJ. Incidence and risk reduction of long-term outcomes: a comparison of benign prostatic hyperplasia with several other disease areas. Urology. 2000;56:9–18.

Roosen JU, Larsen T, Iversen E, Berg JBS. A comparison of aspiration, Antazoline sclerotherapy and surgery in the treatment of hydrocele. Br J Urol. 1991;68:404–6.

Roy JR, Chakraborty S, Chakraborty TR. Estrogen-like endocrine disrupting chemicals affecting puberty in humans--a review. Med Sci Monit. 2009;15:Ra137–45.

Russo GI, Morgia G. Lower urinary tract symptoms and benign prostatic hyperplasia: from research to bedside. Elsevier Science & Technology: Saint Louis; 2018.

Saber A. Minimally access versus conventional hydrocelectomy: a randomized trial. Int Braz J Urol. 2015;41:750–6.

Salama N, Blgozah S. Immediate development of post-varicocelectomy hydrocele: a case report and review of the literature. J Med Case Rep. 2014;8:70.

Sanda MG, Beaty TH, Stutzman RE, Childs B, Walsh PC. Genetic susceptibility of benign prostatic hyperplasia. J Urol. 1994;152:115–9.

Santucci RA, Joyce GF, Wise M. Male urethral stricture disease. J Urol. 2007;177:1667–74.

Schmedt CG, Sauerland S, Bittner R. Comparison of endoscopic procedures vs Lichtenstein and other open mesh techniques for inguinal hernia repair: a meta-analysis of randomized controlled trials. Surg Endosc. 2005;19:188–99.

Schröder L, Rudlowski C, Walgenbach-Brünagel G, Leutner C, Kuhn W, Walgenbach KJ. Surgical strategies in the treatment of gynecomastia grade I-II: the combination of liposuction and subcutaneous mastectomy provides excellent patient outcome and satisfaction. Breast Care. 2015;10:184–8.

Senger JL, Chandran G, Kanthan R. Is routine pathological evaluation of tissue from gynecomastia necessary? A 15-year retrospective pathological and literature review. Plast Surg (Oakv). 2014;22:112–6.

Shan CJ, Lucon AM, Pagani R, Srougi M. Sclerotherapy of hydroceles and spermatoceles with alcohol: results and effects on the semen analysis. Int Braz J Urol. 2011;37:307–12. discussion 312-33

Shapiro E, Becich MJ, Hartanto V, Lepor H. The relative proportion of stromal and epithelial hyperplasia is related to the development of symptomatic benign prostate hyperplasia. J Urol. 1992;147:1293–7.

Shigemura K, Tanaka K, Haraguchi T, Yamamichi F, Muramaki M, Miyake H, Fujisawa M. Postoperative infectious complications in our early experience with holmium laser enucleation of the prostate for benign prostatic hyperplasia. Korean J Urol. 2013;54:189–93.

Simon BE, Hoffman S, Kahn S. Classification and surgical correction of gynecomastia. Plast Reconstr Surg. 1973;51:48–52.

Simons MP, Aufenacker T, Bay-Nielsen M, Bouillot JL, Campanelli G, Conze J, De Lange D, Fortelny R, Heikkinen T, Kingsnorth A, Kukleta J, Morales-Conde S, Nordin P, Schumpelick V, Smedberg S, Smietanski M, Weber G, Miserez M. European hernia society guidelines on the treatment of inguinal hernia in adult patients. Hernia. 2009;13:343–403.

Singano V, Amberbir A, Garone D, Kandionamaso C, Msonko J, Van Lettow M, Kalima K, Mataka Y, Kawalazira G, Mateyu G, Kwekwesa A, Matengeni A, Van Oosterhout JJ. The burden of gynecomastia among men on antiretroviral therapy in Zomba, Malawi. PLoS One. 2017;12:e0188379.

Sollie M. Management of gynecomastia-changes in psychological aspects after surgery-a systematic review. Gland Surg. 2018;7:S70–s76.

Song YN, Wang YB, Huang R, He XG, Zhang JF, Zhang GQ, Ren YL, Pang JH, Pang D. Surgical treatment of gynecomastia: mastectomy compared to liposuction technique. Ann Plast Surg. 2014;73:275–8.

Ssentongo AE, Kwon EG, Zhou S, Ssentongo P, Soybel DI. Pain and dysfunction with sexual activity after inguinal hernia repair: systematic review and meta-analysis. J Am Coll Surg. 2020;230:237–250.e7.

Steenkamp JW, Heyns CF, De Kock ML. Internal urethrotomy versus dilation as treatment for male urethral strictures: a prospective, randomized comparison. J Urol. 1997;157:98–101.

Steigman CK, Sotelo-Avila C, Weber TR. The incidence of spermatic cord structures in inguinal hernia sacs from male children. Am J Surg Pathol. 1999;23:880–5.

Stimson JB, Fihn SD. Benign prostatic hyperplasia and its treatment. J Gen Intern Med. 1990;5:153–65.

Stranne J, Johansson E, Nilsson A, Bill-Axelson A, Carlsson S, Holmberg L, Johansson J-E, Nyberg T, Ruutu M, Wiklund NP, Steineck G. Inguinal hernia after radical prostatectomy for prostate cancer: results from a randomized setting and a nonrandomized setting. Eur Urol. 2010;58:719–26.

Suba Z. Interplay between insulin resistance and estrogen deficiency as co- activators in carcinogenesis. Pathol Oncol Res. 2012;18:123–33.

Tam CA, Voelzke BB, Elliott SP, Myers JB, Mcclung CD, Vanni AJ, Breyer BN, Erickson BA, Trauma & Urologic Reconstruction Network of, S. Critical analysis of the use of Uroflowmetry for urethral stricture disease surveillance. Urology. 2016;91:197–202.

Telegrafo M, Introna T, Coi L, Cornacchia I, Rella L, Stabile Ianora AA, Angelelli G, Moschetta M. Breast US as primary imaging modality for diagnosing gynecomastia. G Chir. 2016;37:118–22.

Tombal B, De Visccher L, Cosyns JP, Lorge F, Opsomer R, Wese FX, Van Cangh PJ. Assessing the risk of unsuspected prostate cancer in patients with benign prostatic hypertrophy: a 13-year retrospective study of the incidence and natural history of T1a-T1b prostate cancers. BJU Int. 1999;84:1015–20.

Traish AM, Mulgaonkar A, Giordano N. The dark side of 5α-reductase inhibitors' therapy: sexual dysfunction, high Gleason grade prostate cancer and depression. Korean J Urol. 2014;55:367–79.

Tsai L, Milburn PA, Cecil CLT, Lowry PS, Hermans MR. Comparison of recurrence and postoperative complications between 3 different techniques for surgical repair of idiopathic hydrocele. Urology. 2019;125:239–42.

Tubaro A, de Nunzio C. The current role of open surgery in BPH. EAU-EBU Update Series. 2006;4:191–201.

Tyrrell CJ, Payne H, Tammela TLJ, Bakke A, Lodding P, Goedhals L, Van Erps P, Boon T, Van De Beek C, Andersson S-O, Morris T, Carroll K. Prophylactic breast irradiation with a single dose of electron beam radiotherapy (10 Gy) significantly reduces the incidence of bicalutamide-induced gynecomastia. Int J Radiat Oncol Biol Phys. 2004;60:476–83.

Untergasser G, Madersbacher S, Berger P. Benign prostatic hyperplasia: age-related tissue-remodeling. Exp Gerontol. 2005;40:121–8.

Vahlensieck W, Hesse A. Calculi in hydrocele: incidence and results of infrared spectroscopy analysis. Korean J Urol. 2010;51:362–4.

Van Poppel H, Tyrrell CJ, Haustermans K, Cangh PV, Keuppens F, Colombeau P, Morris T, Garside L. Efficacy and tolerability of radiotherapy as treatment for Bicalutamide-induced Gynaecomastia and breast pain in prostate cancer. Eur Urol. 2005;47:587–92.

Vickman RE, Franco OE, Moline DC, Vander Griend DJ, Thumbikat P, Hayward SW. The role of the androgen receptor in prostate development and benign prostatic hyperplasia: a review. Asian J Urol. 2020;7:191–202.

Vivien A, Lazard T, Rauss A, Laisné MJ, Bonnet F. Infection after transurethral resection of the prostate: variation among centers and correlation with a long-lasting surgical procedure. Eur Urol. 1998;33:365–9.

Walden PD, Gerardi C, Lepor H. Localization and expression of the alpha1A-1, alpha1B and alpha1D-adrenoceptors in hyperplastic and non-hyperplastic human prostate. J Urol. 1999;161:635–40.

Walsh TJ, Seeger KT, Turek PJ. Spermatoceles in adults: when does size matter? Arch Androl. 2007;53:345–8.

Wang H, Sun Y, Wang L, Xu C, Yang Q, Liu B, Liu Z. Hypoxia-induced apoptosis in the bilateral testes of rats with left-sided varicocele: a new way to think about the varicocele. J Androl. 2010;31:299–305.

Wantz GE. Testicular atrophy as a sequela of inguinal hernioplasty. Int Surg. 1986;71:159–63.

Wasson JH, Reda DJ, Bruskewitz RC, Elinson J, Keller AM, Henderson WG. A comparison of transurethral surgery with watchful waiting for moderate symptoms of benign prostatic hyperplasia. N Engl J Med. 1995;332:75–9.

Weatherly D, Wise PG, Mendoca S, Loeb A, Cheng Y, Chen JJ, Steinhardt G. Epididymal cysts: are they associated with infertility? Am J Mens Health. 2018;12:612–6.

Weingertner N, Bellocq J-P, Foschini MP, Morandi L, Weingertner N, Bellocq J-P, Vreuls C, Van Diest PJ. Gynecomastia. Cham: Springer International Publishing; 2019.

Welliver C, Helo S, Mcvary KT. Technique considerations and complication management in transurethral resection of the prostate and photoselective vaporization of the prostate. Transl Androl Urol. 2017;6:695–703.

Wessells H, Roy J, Bannow J, Grayhack J, Matsumoto AM, Tenover L, Herlihy R, Fitch W, Labasky R, Auerbach S, Parra R, Rajfer J, Culbertson J, Lee M, Bach MA, Waldstreicher J. Incidence and severity of sexual adverse experiences in finasteride and placebo-treated men with benign prostatic hyperplasia. Urology. 2003;61:579–84.

Williams MJ. Gynecomastia: its incidence, recognition and host characterization in 447 autopsy cases. Am J Med. 1963;34:103–12.

Wilson JM, Aaronson DS, Schrader R, Baskin LS. Hydrocele in the pediatric patient: inguinal or scrotal approach? J Urol. 2008;180:1724–7. discussion 1727-8

Zahalsky MP, Berman AJ, Nagler HM. Evaluating the risk of epididymal injury during hydrocelectomy and spermatocelectomy. J Urol. 2004;171:2291–2.

Zhou T, Hu ZY, Zhang HP, Zhao K, Zhang Y, Li Y, Wei JJ, Yuan HF. Effects of testosterone supplementation on body composition in HIV patients: a meta-analysis of double-blinded randomized controlled trials. Curr Med Sci. 2018;38:191–8.

Chapter 14
Benign Gastrointestinal Conditions

Wei Mou Lim, Nathan Lawrentschuk, and Alexander G. Heriot

14.1 Oesophagus

14.1.1 Gastroesophageal Reflux Disease

14.1.1.1 Background

Reflux of gastric contents into the distal oesophagus is a physiological phenomenon. When this causes symptoms, erosions or metaplastic changes to the epithelium lining, it is termed gastroesophageal reflux disease (GERD) (Vakil et al. 2006). Reflux disease can be categorised into 3 main subgroups: nonerosive GERD, erosive GERD and Barrett's oesophagus. Prevalence is 10–30% in Western societies but this is probably underreported as endoscopic evidence of GERD has a poor correlation to symptomatology (El-Serag et al. 2014). Impaired oesophageal peristalsis, poor lower oesophageal sphincter tone and delayed gastric emptying are involved in the pathogenesis of this common disease (Kahrilas et al. 1986; Buckles et al. 2004).

14.1.1.2 Symptoms and Signs

Classic symptoms include heartburn, retrosternal chest pain, regurgitation, waterbrash, dysphagia or odynophagia. Patients can present atypically with cough, wheezing, hoarseness, otitis media and dental erosions when effluents from the reflux affects the upper and lower respiratory tracts.

W. M. Lim · N. Lawrentschuk · A. G. Heriot (✉)
Department of Surgical Oncology, Peter MacCallum Cancer Centre,
Melbourne, VIC, Australia
e-mail: Alexander.Heriot@petermac.org

© The Author(s), under exclusive license to Springer Nature
Switzerland AG 2022
S. S. Goonewardene et al. (eds.), *Men's Health and Wellbeing*,
https://doi.org/10.1007/978-3-030-84752-4_14

14.1.1.3 Investigations

Most individuals are managed based on clinical symptoms alone. Patients with persistent symptoms and red flags such as weight loss, anaemia or gastrointestinal bleeding warrant further investigation. Gastroscopy can detect erosive oesophagitis, Barrett's oesophagus and hiatus hernia which are commonly associated with this disease. Oesophageal manometry and pH studies can be used to confirm reflux if there is lack of endoscopic or histological confirmation of oesophagitis and metaplastic changes. This is pertinent before any anti-reflux operations. Contrast studies such as barium swallow can demonstrate reflux but are not widely used clinically. Cross sectional imaging can detect incidental hiatal hernia.

14.1.1.4 Management

Non-pharmacotherapy management includes weight loss and exercise. Furthermore, a modified eating habit such as avoiding large meals, late meals and precipitants such as alcohol and caffeinated beverages, which can relax the lower oesophageal sphincter mechanism. First line therapy with histamine 2 receptor antagonist or proton pump inhibitor (PPI) is commonly used to reduce gastric acid secretion. Adjuncts like prokinetics and antacids may be prescribed to improve symptoms (Hunt et al. 2017). Patients on maximal medical therapy with persistent reflux can be referred for laparoscopic fundoplication (Katz et al. 2013b). Anti-reflux surgery has been shown to have good efficacy at managing symptoms (Grant et al. 2013).

14.1.1.5 Relevance to Men's Health

Reflux is common in both men and women and have no sexual predilection. However, men have a higher incidence of non-erosive GERD, erosive oesophagitis and Barrett's oesophagus (Cook et al. 2005; Asanuma et al. 2016) Barrett's oesophagus is a pre-malignant disease that is common amongst Caucasian men. This condition is correlated with a 30–125 times increased risk of progression to developing oesophageal adenocarcinoma (Stein and Siewert 1993). Typical appearance at endoscopy is a salmon pink coloured mucosa at the gastroesophageal junction and high-risk features include ulceration, stricture and nodularity. Increased surveillance programs for Barrett's and the utility of upper endoscopy as first line investigation for dyspepsia has increased the incidence of lower third oesophageal malignancy, particularly in the Western world (Pohl and Welch 2005). Men with Barrett's progressed to cancer more quickly than their female counterparts and pooled estimates from a large meta-analysis showed that cancer incidence was two times higher in men (10.2/1000 person-years) compared to women (4.5/1000 person-years) (Yousef et al. 2008).

Erosive oesophagitis is another condition prevalent amongst men. It is characterised by ulceration of the gastroesophageal junction mucosa and graded according to the Los Angeles Classification by the extent of mucosal breaks in a longitudinal and circumferential manner (Armstrong et al. 1996). Male gender and associated Metabolic X syndrome including obesity, hypertension, diabetes, alcohol and smoking use are predictors of erosive oesophagitis (Labenz et al. 2004; Kim et al. 2011; Park et al. 2008). Duration and severity of reflux symptoms have both been associated with increased risk of developing oesophageal malignancy (Rubenstein and Taylor 2010).

14.2 Stomach

14.2.1 Peptic Ulcer Disease

14.2.1.1 Background

Peptic ulcers typically involve the stomach and proximal duodenum. They are caused by increased gastric acid secretion and impaired mucosal defence mechanisms leading to ulcer formation which extend beyond the muscularis mucosae. Major aetiology includes helicobacter pylori infection, non-steroidal anti-inflammatory drugs (NSAIDS), aspirin and smoking.

The prevalence of peptic ulcer disease is highly variable worldwide as it coincides with the presence of helicobacter co-infection, which is more common in developing countries (Li et al. 2010; Hooi et al. 2017; Azhari et al. 2018). Helicobacter pylori is a gram-negative spirochete that is transmitted via saliva, water and food via contamination. This virulent bacteria resides in the stomach, produces proinflammatory cytokines, urease, proteases leading to upregulation of gastrin and acid secretion (Blaser and Atherton 2004).

On the contrary, increased use of NSAIDS and antithrombotic have increased due to the ageing population for treatment of osteoarthritis and cardiovascular disease. This is a significant contributor to peptic disease in developed countries (Lanas et al. 2006).

14.2.1.2 Symptoms and Signs

Epigastric pain is the most common symptom. Other associated symptoms include bloating, early satiety, weight loss, nausea and vomiting. Occult bleeding can present with iron deficiency anaemia. Acute complications such as haemorrhage and perforation warrant urgent intervention. Late complications can develop with stricture formation and gastric outlet obstruction.

14.2.1.3 Investigations

In evaluating patients with peptic ulcer disease, an upper gastrointestinal endoscopy is the investigation of choice. It enables direct visualisation of the mucosa, facilitate biopsies and rapid urease test to confirm helicobacter pylori, which is the most common cause for ulcers. Other methods include faecal antigen and urea breath tests which are used in primary care setting. Patients who present with complicated peptic ulcer disease with obstruction, perforation and bleeding may need urgent computer tomography (CT) or angiography.

14.2.1.4 Management

Treatment is dependent on the aetiology of the ulcers. For uncomplicated peptic ulcer disease, cessation of NSAIDS, aspirin and smoking are recommended. Helicobacter co-infection should be investigated and eradicated if found positive. Triple therapy consists of combination penicillin, macrolide-based antibiotics and a PPI. Resistant cases may need an addition of another antimicrobial or bismuth.

Acute upper gastrointestinal bleeding with haematemesis or melaena should undergo endoscopy within 24 h and high dose PPI according to international consensus (Barkun et al. 2019). The ulcers can be treated with a combination of adrenaline injection, thermocoagulation or clipping. Re-bleeding should be managed with a repeat endoscopy and failing that, urgent angiographic embolization or surgery is indicated (Abe et al. 2010). Peptic perforation with generalised peritonism requires a patch repair. Resection is indicated if the ulcer is larger than 2 cm and not amenable to a Graham patch.

14.2.1.5 Relevance to Men's Health

According to historical epidemiological data, men are twice as likely to have peptic ulcer disease as compared to women (Kurata et al. 1985). Despite a decreased in overall pooled incidence rates in the last century, complicated disease associated with bleeding still remains at a 2 to 1 ratio compared to women (Lin et al. 2011). Peptic perforation is likewise more common in men compared to their partners, particularly in the duodenum (Bardhan and Royston 2008; Wysocki et al. 2011). Ulcers are also the most common cause for occult gastrointestinal bleeding resulting in iron deficiency in young men (Carter et al. 2013).

14.3 Gallbladder

14.3.1 Gallstone Disease

14.3.1.1 Background

Cholelithiasis is due to an imbalance of cholesterol, bile salts, and lecithin leading to crystallisation and stone formation. This is exacerbated by abnormal gallbladder contraction and stasis. Stones are predominantly cholesterol based, pigment or mixed. Gallstone disease is very common in Western populations (Everhart et al. 1999) and risk factors include female gender, a positive family history, pregnancy, age, obesity and diabetes. Gallstones can also develop post bariatric surgery, parenteral nutrition and from lithogenic drugs such as somatostatin analogues, oral contraceptive pill and hormone replacement therapy.

14.3.1.2 Symptoms and Signs

Biliary colic is characterised by upper abdominal pain typically in the epigastrium or lower chest with radiation to the right upper quadrant, shoulder and back. Patients can appear ill with diaphoresis, nausea, vomiting and often misdiagnosed as cardiac chest pain.

When an impacted stone causes cystic duct obstruction, this causes inflammation and infection of the bile leading to acute cholecystitis. Fever and tachycardia indicate sepsis and the presence of jaundice signifies choledocholithiasis and/or associated cholangitis. Murphy's sign is secondary to parietal irritation from a distended gallbladder that is aggravated on contact with the examining fingers during deep inspiration. Complications from gallstones include pancreatitis, obstruction from gallstone ileus and Mirizzi Syndrome.

14.3.1.3 Investigations

White cell count with leucocytosis, raised bilirubin levels and cholestatic liver enzyme profile should raise suspicion of complicated gallstone disease. Raised serum amylase or lipase indicate concurrent pancreatitis. Transabdominal ultrasound is routinely used to evaluate gallstones. It is readily available, cheap and a non-invasive imaging modality with a sensitivity of 88% for calculous cholecystitis (Shea et al. 1994). Gallstones have a characteristic echogenic appearance with post acoustic shadowing. Other associated features on ultrasound such as hyperaemia, sonographic probe tenderness, gallbladder wall thickening greater than 4mm, oedema, pericholecystic fluid and common bile duct diameter can assist clinicians in differentiating uncomplicated biliary disease from cholecystitis. In cases of diagnostic uncertainty, cholescintigraphy with hepatobiliary iminodiacetic acid (HIDA) is 96% sensitive (Kiewiet et al. 2012). However, this requires an injection of intravenous radiotracer technetium Tc 99m, which is more time consuming and costly.

Suspected choledocholithiasis or cholangitic patients should undergo evaluation of intrahepatic and extrahepatic bile ducts with magnetic resonance cholangiopancreatography (MRCP) or endoscopic retrograde cholangiopancreatography (ERCP). Endoscopic ultrasound (EUS) or percutaneous transhepatic cholangiography (PTC) are alternative options, however these investigations are reserved as they are more invasive.

14.3.1.4 Management

Most individuals are asymptomatic and only 2–4% progress to biliary colic (Tazuma et al. 2017). Laparoscopic cholecystectomy is the preferred operation for symptomatic disease with low complication rate, reduced hospital stay and quicker recovery compared to open techniques (Keus et al. 2006; Zha et al. 2010). Non-surgical therapy with ursodeoxycholic acid or extracorporeal shock wave lithotripsy have poor success, with higher recurrence rate and generally not recommended (May et al. 1993; Rabenstein et al. 2005).

Acute cholangitis secondary to bile duct stones is a surgical emergency. Resuscitation and broad-spectrum penicillin or cephalosporin-based antibiotics should be instituted (Gomi et al. 2018). Endoscopic transpapillary drainage with a stent via ERCP is the most effective and minimal invasive method. Sphincterotomy and stone extraction can be facilitated in the same setting to remove the obstruction. Failing this, PTC and EUS guided biliary drainage and stenting are salvage options in difficult cases especially in altered anatomy from previous gastrointestinal surgery or failure in cannulating the ampulla during ERCP (Mukai et al. 2017).

14.3.1.5 Relevance to Men's Health

With the advent of minimally invasive surgery, laparoscopic cholecystectomy is considered the gold standard treatment for gallstones. However, an open operation is a fallback option in the setting of severe inflammation, unclear anatomy, bile duct injury and bleeding (Bingener-Casey et al. 2002). Overall, conversion to open cholecystectomy is more common in men (Russell et al. 1998). Reasons for conversion are delayed presentation, severe disease with inflammatory adhesions, longer operating times and difficult anatomy compared to women (Zisman et al. 1996; Gharaibeh et al. 2002; Thesbjerg et al. 2010).

Men have a higher rate of complicated gallstone disease, especially gangrenous cholecystitis. Full thickness necrosis of the gallbladder occurs in the setting of sustained obstruction of the cystic duct leading to epithelial injury and ischaemia. Severe cases can result in gallbladder perforation (Stefanidis et al. 2006). The incidence of gangrenous cholecystitis ranges from 2% to 30% in the literature with the majority of case series describing a male predominance (Nikfarjam et al. 2011). Associated risk factors include cardiovascular disease and diabetes, which are common co-morbidities in men (Hunt and Chu 2000; Fagan et al. 2003). Men with gangrenous cholecystitis who undergo laparoscopic cholecystectomy have a higher risk of conversion to open procedure and mortality (Merriam et al. 1999; Bourikian et al. 2015).

Iatrogenic bile duct injury from laparoscopic cholecystectomy is a rare but serious complication. Male gender is a risk factor for major bile duct injury necessitating bile duct reconstruction due to more severe symptomatic gallstone disease (Fletcher et al. 1999; Waage and Nilsson 2006; Giger et al. 2011; El-Dhuwaib et al. 2016). This is also predictive of all-cause mortality (Halbert et al. 2016).

14.4 Pancreas

14.4.1 Pancreatitis

14.4.1.1 Background

Pancreatitis is an inflammatory disease with significant morbidity and mortality due to its local and systemic complications, with a rising incidence in the last decade (Lowenfels et al. 2005). Inflammation of the gland is caused by inappropriate activation of acinar cells leading to autodigestion. Gallstone and alcohol induced pancreatitis constitute more than 80% of cases. Other less common risk factors include anatomical anomalies such as pancreatic divisum, strictures, malignancy and causes like hypertriglyceridemia, trauma, ERCP, medications (azathioprine/thiazides) and autoimmune IgG4 disease (Whitcomb 2006; Yang et al. 2008).

Approximately 10–20% of acute pancreatitis will pursue a severe course with multi organ dysfunction. Progression to chronic pancreatitis occurs in 10–30% of patients (Sankaran et al. 2015).

14.4.1.2 Symptoms and Signs

The pancreas is a retroperitoneal organ that has both exocrine and endocrine function. When it is inflamed in the setting of pancreatitis, the pain typically starts in the epigastrium radiating to the back due to his embryological location. This is often accompanied by nausea and vomiting. Tachycardia, hypotension, oliguria, reduced conscious state and respiratory distress are signs of organ hypoperfusion.

14.4.1.3 Investigations

Full blood count often demonstrates leucocytosis and increased haematocrit due to intravascular volume depletion from third space losses. Amylase or lipase level greater than 3 times the reference range is diagnostic of pancreatitis. A high C reactive protein (CRP), urea, creatinine and hypocalcaemia are indicative of severe disease. Abdominal ultrasonography is paramount to confirm or exclude gallstones. CT can provide useful prognostic information regarding the severity of pancreatitis (Balthazar 2002).

Acute pancreatitis can be confidently diagnosed when 2 of 3 criteria based on clinical, laboratory and imaging modalities are fulfilled.

14.4.1.4 Management

This disease can be broadly divided into two categories: interstitial oedematous and necrotising acute pancreatitis according to the Revised Atlanta Classification. Local complications are based on time interval of 4 weeks. Acute peripancreatic or necrotic collections are early complications and pseudocyst or walled off necrosis a late sequelae (Banks et al. 2013).

Treatment of acute pancreatitis is primarily supportive. Analgesia, aggressive fluid resuscitation and early nutrition are important in the early stages. Antibiotic use is shown to be ineffective in the prophylactic setting and not indicated unless in the presence of sepsis and infected collections (Isenmann et al. 2004). Patients with multiorgan dysfunction should be resuscitated in intensive care units. A step-up approach is preferred in the management of peripancreatic collections and necrosis with less morbidity and complications than open surgery. This involves percutaneous catheter insertion, followed by endoscopic transluminal drainage or minimally invasive retroperitoneal necrosectomy (van Santvoort et al. 2010; van Brunschot et al. 2018). Evidence supports the use of ERCP in scenarios when there is presence of cholangitis and ductal obstruction (Working Group IAP/APA Acute Pancreatitis Guidelines 2013).

Cholecystectomy is safe for mild or moderate gallstone pancreatitis and advocated in the index admission to prevent recurrence of pancreatitis (Hernandez et al. 2004). On the other hand, surgery for advanced disease should be delayed for more than 6 weeks in patients with persistent peripancreatic collections and necrosis, as this is associated with organ failure and hospitalisation (Kwong and Vege 2017).

14.4.1.5 Relevance to Men's Health

Alcohol induced pancreatitis is second most common aetiology, constituting about 20–30% of acute presentations, causing a significant health burden (Yang et al. 2008). It affects men more than women due to higher prevalence of liquor intake (Yen et al. 1982). This observation has been validated in a dose response manner (Corrao et al. 1999). Large population cohort studies have shown an association with alcohol use across all disease severity including mild, severe and chronic pancreatitis (Jaakkola and Nordback 1993; Kristiansen et al. 2008). In the acute setting, pancreatic related complications such as haemorrhagic, necrotic pancreatic collections and extrapancreatic sequalae like gastrointestinal bleeding, sepsis and mortality are higher in males (Akhtar and Shaheen 2004; Yen et al. 1982).

Chronic pancreatitis is a late disorder due to progressive loss of pancreatic function secondary to recurrent episodes of acute inflammation and fibrosis. It can manifest with recurrent chronic abdominal pain, cachexia, metabolic dysfunction with malabsorption, steatorrhea and diabetes (Lankisch et al. 1993). Features on imaging include pancreatic gland atrophy, ductal stricture or stones, parenchymal calcifications, pancreatic pseudocysts, fistulas and ascites (Büchler et al. 2009). Akin to alcohol pancreatitis, chronic pancreatitis have a higher predilection for men. In a large multicentre analysis of prospective patients with chronic pancreatitis in North America, there is a higher incidence of chronic pancreatitis in men. This is associated with a likewise higher alcohol and tobacco use (Romagnuolo et al. 2016). Smoking is a well-recognised cofactor potentiating the toxic effects of alcohol on the pancreas (Andriulli et al. 2010). It is also an independent, dose dependent risk factor in the pathogenesis of chronic pancreatitis in men (Maisonneuve et al. 2005).

14.5 Small Bowel

14.5.1 Small Bowel Obstruction

14.5.1.1 Background

The small bowel is the longest viscera suspended by a mesentery in the abdominal cavity. It has a large surface area lined by mucosa with microvilli projections for absorption of micro/macronutrients, trace elements and fluid (Volk and Lacy 2017).

Due to its long course, the small bowel is susceptible to mechanical obstruction. The most common cause is adhesions from previous abdominal surgery, accounting for more than half presentations for small bowel obstruction (Miller et al. 2000). Abdominal wall/groin/internal hernias, volvulus, stricture from tumours, radiation or Crohn's, Meckel's intussusception and foreign bodies are other causes (Tong et al. 2020).

14.5.1.2 Symptoms and Signs

Colicky abdominal pain associated with bilious vomiting and abdominal distension are hallmark features of acute small bowel obstruction. Reduced bowel actions and flatus are less specific symptoms.

Patients may present with hypovolaemia with dry mucous membranes, foetor and reduced urine output. Tachycardia, hypotension, fevers and peritonism indicate a more serious course like small bowel ischaemia or perforation. It is important to assess abdominal hernial orifices in all patients to exclude incarcerated or strangulated hernia as a cause for the obstruction.

14.5.1.3 Investigations

Electrolyte and renal function evaluation are of utmost importance during the workup for these patients as the small bowel is involved in 7–8 l of fluid secretion and absorption capabilities. Elevated urea, creatinine, hyponatraemia and hypokalaemia are usual laboratory findings. Raised white cell count, leucocytosis, CRP, acidosis and serum lactate may indicate bowel ischaemia but lack specificity (Lange and Jäckel 1994).

An abdominal radiography will often show dilated loops of small bowel with air fluid levels. CT scan is utilised to identify the site of obstruction termed the "transition point", particularly useful in patients who have a virgin abdomen or minimal past abdominal surgeries. This subset of patients may require surgical intervention. Abdominal free fluid, mesenteric stranding and pneumatosis are concerning findings on CT suggestive of bowel ischaemia.

14.5.1.4 Management

The initial management of mechanical small bowel obstruction is aimed at fluid resuscitation, electrolyte replacement and decompression of the gastrointestinal tract with a nasogastric tube. Stable patients are managed conservatively with fasting and diet upgraded once bowel function resumes. Water soluble contrast agent, such as gastrografin, has been shown to reduce surgery, length of hospitalisation and improve resolution of adhesive small bowel obstruction (Ceresoli et al. 2016). Nevertheless, about 10–30% of patients will require operative intervention (Ten Broek et al. 2018). Surgery is reserved for individuals who fail non operative treatment or presence of strangulation, ischaemiaor peritonitis.

Laparotomy and adhesiolysis has been the standard approach to patients who fail to progress with conservative management. This involves a large midline incision to access the peritoneal cavity, divide the adhesions and resect any obstructing mass or devitalised bowel. There is some evidence of minimally invasive methods using laparoscopic adhesiolysis with reduced medical, surgical complications, mortality and hospitalisation (Sajid et al. 2016). However, laparoscopic adhesiolysis involves advanced surgical skills and must be used with caution in select patients with minimal abdominal surgeries to minimise risk of unrecognised enterotomies (Farinella et al. 2009).

14.5.1.5 Relevance to Men's Health

Although the overall incidence of acute small bowel obstruction between men and women are comparable (Drożdż and Budzyński 2012), men have a higher frequency of groin, epigastric hernias and congenital anomalies such as paraduodenal hernias, congenital peritoneal bands and Meckel's diverticulum, which are alternative causes of bowel obstruction.

Inguinal hernias are about 10 times more common in men than women (Primatesta and Goldacre 1996). Though the overall risk of strangulation and bowel obstruction is low for inguinal hernias (Gallegos et al. 1991), they are still overrepresented in male compared to female in an acute hernia in all age groups (Rai et al. 1998; Kurt et al. 2003; Chen et al. 2020). Similarly, epigastric hernia is more common in men (Ponten et al. 2012). Repair with mesh is safe in select patients with obstruction in clean or clean-contaminated fields (Nieuwenhuizen et al. 2011; Henriksen et al. 2020).

Paraduodenal hernias, although infrequent, is the most common form of congenital internal hernias. It is due to malrotation of the small bowel leading to mesenteric fusion defects lateral to the duodenojejunal flexure or first part of the jejunal mesentery causing small bowel entrapment. It happens more commonly in males than females in a 3:1 ratio (Shadhu et al. 2018). Similarly, congenital band adhesions are a rare entity and cause for mechanical small bowel obstruction, more common in men (Sozen et al. 2012; Skoglar et al. 2018). A Meckel's diverticulum is a remnant of the vitellointestinal duct in the mid small bowel and may form fibrous bands making it susceptible to bowel obstruction. It is reported to be 4 times more common in males. 46.7% of paediatric and 35.6% of adults with Meckel's present with bowel obstruction (Hansen and Søreide 2018).

14.5.2 Inflammatory Bowel Disease: Crohn's Disease/ Ulcerative Colitis

14.5.2.1 Background

Inflammatory bowel disease (IBD) can be divided into two separate disease process: Crohn's and ulcerative colitis with distinct clinical, pathological and surgical presentations. Crohn's disease involves full thickness inflammation of the gastrointestinal tract and can affect any segment from the oropharynx to the anus. It typically involves the terminal ileum and can have "skip lesions". On the other hand, inflammation in ulcerative colitis is limited to the colonic mucosa and extends from distal to proximal colon in a contiguous fashion.

IBD is a disease in developed countries of temperate climate. However, in recent decades, there has been a rising incidence in developing countries in Asia and South America (Khalili et al. 2012; Ng et al. 2018). The pathogenesis of IBD is complex and involves genetic predisposition, environmental factors such as smoking, hygiene, and impaired bowel mucosal barrier. This causes a disproportionate host immune response leading to chronic and relapsing inflammation of the gastrointestinal tract (Abegunde et al. 2016).

14.5.2.2 Symptoms and Signs

General symptoms include abdominal cramps, altered bowel habits with haematochezia or mucus, weight loss and fatigue. Extraintestinal organ systems can be affected including skin (erythema nodosum, pyoderma gangrenous), joints (arthralgia, sacroiliitis, ankylosing spondylitis), ocular (iritis/uveitis), liver (primary sclerosing cholangitis/cholangiocarcinoma), metabolic (gallstones/renal stones/ osteoporosis) and haematopoiesis (anaemia) (Vavricka et al. 2011). Due to the transmural inflammation and multifocality of Crohn's disease, specific complications such as bowel strictures/obstruction/perforation, mural/mesenteric abscess, perianal sepsis and fistulas can develop.

14.5.2.3 Investigations

As the symptoms are non-specific, patients who was suspected of having IBD should undergo routine laboratory testing and nutritional evaluation. This involves full blood count, renal function, liver function tests, iron studies, B12, folate and vitamin D levels. An upper and lower endoscopy is mandated to enable mucosal evaluation and biopsies. Faecal calprotectin is a useful screening marker for diagnosis of IBD, surveillance of disease activity and remission (D'Haens et al. 2012).

In diagnostic uncertainty for mid small bowel Crohn's disease, adjuncts such as capsule endoscopy, CT enteroclysis and MRI enterogram can help investigate occult Crohn's segment and differentiate inflammatory from fibrotic strictures.

14.5.2.4 Management

A multidisciplinary approach should be adopted for patients with IBD. Patients should be comanaged by gastroenterologist, colorectal surgeon, nurse and dietician. Principles of treatment are to induce and maintain remission, minimise use of steroids, prevent complications and screen for colorectal cancer.

A step-up approach is generally adopted for mild disease. 5-aminosalicylate oral medications and rectal enemas are recommended as first line therapy for ulcerative colitis. This can be escalated to immunomodulators such as thiopurines, methotrexate, biologics like anti-tumour necrosis factor (anti-TNF) inhibitors(infliximab, adalimumab), anti-integrin antibodies (vedolizumab) or anti-interleukin antibodies (ustekinumab) (Rubin et al. 2019). Refractory disease and complications such as colorectal cancer, toxic megacolon, fulminant colitis and perforations are referred for surgical intervention. Colonic resection in ulcerative colitis is considered "curative" as it removes the diseased colon. This may involve segmental resection, a total colectomy with end ileostomy or a staged proctocolectomy with ileal pouch anal anastomosis (IPAA) depending on disease status and patient stability (Gallo et al. 2018).

Unlike ulcerative colitis, Crohn's disease is generally managed in a "bowel conserving" manner. This is due to high disease recurrence and risk of skip lesions. Medical therapy should be optimised prior to embarking on surgery. Short segment strictures can be managed with endoscopic dilatation (Navaneethan et al. 2016). Multifocal disease can undergo strictureplasty to preserve bowel length and minimise short bowel syndrome (Bellolio et al. 2012). Ileocaecal resection is indicated for distal ileal disease. Management of perianal Crohn's disease is mainly medical therapy with anti-TNF inhibitors (Sands et al. 2004). Surgery is reserved to drain perianal sepsis and preserve continence. Long term draining setons, faecal diversion or proctectomy may be needed in complex perianal Crohn's (Gu et al. 2015).

14.5.2.5 Relevance to Men's Health

Diagnosis of ulcerative colitis is slightly more common in men (Nørgård et al. 2014; Shivashankar et al. 2017). Failure of medical optimisation is an indication for surgery. Male gender is associated with a trend in developing anti-TNF inhibitor antibodies (Brandse et al. 2016) and increased drug clearance (Ternant et al. 2008; Fasanmade et al. 2009) leading to poor response to biologic therapy. In a recent large population based surveillance study, men progressed faster in the disease course needing a colectomy (Al-Darmaki et al. 2017). This has been validated in a meta-analysis with multiple observational and retrospective studies (Dias et al. 2014).

An IPAA is a reconstruction of a "neorectum" to restore gastrointestinal tract continuity after proctocolectomy. Failure rates is reported to be in the order of 10–20% over 5–20 years (Tekkis et al. 2010; Mark-Christensen et al. 2018). Male gender is associated with an increased risk of pouch failure with leakage, fistulas, pelvic sepsis, acute or chronic pouchitis, ultimately leading to pouch excision (Tiainen and Matikainen 2000; Joelsson et al. 2006; Helavirta et al. 2020; Kayal et al. 2020). Histological assessment of pouch biopsies also indicate a more severe disease course in men (Setti Carraro et al. 1994). IPAA in a male pelvis poses inherent surgical challenges due to the narrow anatomical pelvic inlet and visceral obesity (Wu et al. 2016).

14.6 Large Bowel

14.6.1 Appendicitis

14.6.1.1 Background

The vermiform appendix is a vestigial structure situated in the right lower quadrant of the peritoneal cavity. Inflammation of the appendix occurs as a result of luminal obstruction from lymphoid follicles, appendicolith and rarely parasites or tumour. A cycle of congestion and arterial ischaemia can lead to perforation and abscess formation (Bhangu et al. 2015).

Appendicitis is a disease of the young, with a peak incidence during early childhood and late adolescence. The estimated incidence is 100 to 150/100,000 in Western nations and as high as 200/100,000 in Asia (Ferris et al. 2017). It is one of the most common acute abdominal pain presentations to general practice and the emergency department (Viniol et al. 2014; Mahajan et al. 2020).

14.6.1.2 Symptoms and Signs

Due to its midgut origin, appendicitis can present with early symptoms of non-localised vague central abdominal pain. Localisation to the right iliac fossa happens when parietal peritoneum is secondarily inflamed. Anorexia, nausea, vomiting and fevers are predominant features. Irritation of the rectosigmoid from an inflamed appendix may manifest as diarrhoea.

At beside review, patients typically have a flushed appearance with a dry fetor due to emesis and dehydration. Tachycardia and pyrexia of 38 degrees or above indicate signs of sepsis. Abdominal examination often reveals localised pain and rebound over McBurney's point. Associated signs may include cough, movement or percussion tenderness.

14.6.1.3   Investigations

Diagnosis is based on a combination of clinical acumen, laboratory tests and imaging. Whilst there are many scoring systems developed to help diagnose appendicitis, such as the Paediatric Appendicitis Score (PAS) or Alvarado Score, none have been proven to have high sensitivity and specificity (Ebell and Shinholser 2014).

Elevated white cell count with a raised neutrophil differential and CRP should raise suspicion of an infectious process. Urinalysis and beta human chorionic gonadotrophin (HCG) are essential for women of child bearing age to rule out other causes of right lower quadrant pain such as urinary tract infection or ectopic pregnancy.

In situations of diagnostic dilemma, imaging can be utilised to reduce the rate of negative appendicectomy. Ultrasound is the imaging modality of choice in children and young women due to accessibility and non-radiation technique. Features suggestive of appendicitis on ultrasound include identification of a non-compressible tubular structure, diameter greater than 6 mm, wall thickness more than 3 mm, probe tenderness, increased blood flow, peri-appendiceal fat echogenicity or free fluid.

Visualisation of the appendix using sonography is operator dependent, and when the appendix is not visualised, the sensitivity for appendicitis is poor (D'Souza et al. 2015) CT or MRI have higher sensitivity and specificity, with a high diagnostic accuracy where the clinical presentation is atypical (Zhang et al. 2017).

14.6.1.4   Management

Appendicitis is the one of the most common surgical emergencies in the world (Disease, Injury, and Prevalence 2016). Once the diagnosis is made, patients should receive analgesia, intravenous fluids, and broad-spectrum antibiotics. Historically, an open appendicectomy has been performed at McBurney's point through a gridiron muscle splitting incision. Laparoscopic appendicectomy has superseded open surgery, with many advantages such as reduced surgical site infection, length of stay, less post-operative pain and quicker recovery for patients described in a recent Cochrane review (Jaschinski et al. 2018). It also reduces the risk of adhesion formation and subsequent post-operative bowel obstruction in the short and long term (Markar et al. 2014). Patients needing conversion to open appendicectomy have a higher risk of re-laparotomy and incisional hernias in the future (Swank et al. 2011).

Non-operative management of uncomplicated appendicitis has been under intense debate for the last decade with multiple studies concluding that this subset of patients can be managed with an antibiotic only approach instead of surgery. However, there is a failure rate of 8% at the index admission and 20% risk of subsequent hospitalisations for repeated attacks (Podda et al. 2019). Late recurrent appendicitis has been quoted to be as high as 39.1% at 5 years (Salminen et al. 2018).

On the contrary, in complicated disease with phlegmon and abscess formation, conservative management with analgesia, intravenous fluids, antibiotics and percutaneous drainage are advocated in stable patients (Di Saverio et al. 2020). Non operation is associated with less overall complications, wound infection and re-operation rates (Simillis et al. 2010). Although laparoscopic appendicectomy by experienced surgeons have been shown to reduce re-admission and interventions for recurrent appendicitis, there is still a 10% risk of bowel resection and 13% risk of incomplete appendicectomy (Mentula et al. 2015).

14.6.1.5 Relevance to Men's Health

Men have a higher rate of appendicitis to women at all age groups, with an overall ratio of 1.4 to 1 (Al-Omran et al. 2003; Stein et al. 2012). Lifetime risk of appendicitis is 8.6% for males and 6.7% for females (ADDISS et al. 1990). Anatomically, male appendices have been found to have higher proportion of lymphoid tissue making it more susceptible to obstruction, a common aetiology for appendicitis (Hwang and Krumbhaar 1940). Pathological evaluation also revealed more appendicitis and peritonitis in men than women (Lee 1962; Chang 1981).

Overall rate of perforated appendix is reported to be in the order of 20–30%. Extremes of age and men are particularly vulnerable, with a higher perforation rate (Andersson et al. 1992; Guss and Richards 2000). They are likely to be older with comorbidities, with an Odds Ratio of 1.5 and 2 for male gender and age above 80 years old respectively (Young et al. 2007; Drake et al. 2014).

14.6.2 Diverticular Disease

14.6.2.1 Background

Diverticular disease is a common presentation to primary care physicians and hospital admissions. This is particularly significant in western and developed countries where the prevalence is high. It is estimated that 50–60% of individuals age 60 and above will have diverticulosis. This trend increases with every decade of life (Peery et al. 2020).

Majority of diverticulosis are situated in the sigmoid colon (Peery et al. 2016). On the contrary, right sided diverticulosis is more common in the Asian population (Fong et al. 2011; Wang et al. 2015). This disease is acquired over time, with a complex interplay of genetic predisposition such as age (Commane et al. 2009), ethnicity (Golder et al. 2011), connective tissue disorders (Reichert et al. 2018) and environmental factors including increased weight (Rosemar et al. 2008), diet (Crowe et al. 2011), smoking (Aune et al. 2017), inflammation (Tursi and Elisei 2019) and alteredcolonicmicrobiome (Schieffer et al. 2017).

14.6.2.2 Symptoms and Signs

A diverticulum forms when an outpouching of colonic mucosa and submucosa develops through the wall of the colon where the vasa recta penetrates (Hughes 1969). This unique anatomical anomaly explains the symptoms that patients present with. Most individuals with diverticulosis are asymptomatic. It is the most common reported finding during routine colonoscopy (Everhart and Ruhl 2009). Others present with acute symptoms such as fevers, anorexia, left lower quadrant abdominal pain, bleeding and altered bowel habits. Pneumaturia, faecaluria and per vaginal stool raises suspicion of fistulating diverticular disease.

Diverticulitis is caused by micro or macro perforation of diverticula as a result offaecolith impaction, bacterial overgrowth and ischaemia. Patients can present with localised or generalised peritonism, intra-abdominal abscess, phlegmon, large bowel obstruction, colo-enteric, colo-vesical or colo-cutaneous fistula. Contrary to earlier studies, the incidence of diverticulitis is less than previously reported. Initial publications have quoted a rate of 10–25% (Parks 1975; Boles Jr. and Jordan 1958), however, contemporary data have shown this to be much lower, between 1.7 and 4% (Salem et al. 2007; Shahedi et al. 2013).

Diverticular bleeding is caused by rupture of the vasa recta in the diverticulum, which most commonly occur in the ascending colon. Patients can present with rapid transit rectal bleeding, clots and haemodynamic instability.

14.6.2.3 Investigations

Asymptomatic individuals are diagnosed incidentally during colonoscopy screening for polyps or colorectal cancer. Some are found during imaging. With the advent of CT, this has superseded fluoroscopic enema as the investigation of choice for patients who present with symptomatic diverticular disease. Features of diverticulitis on CT include colonic wall thickening, mesenteric stranding, pelvic abscess, localised extraluminal gas or free intraperitoneal gas suggestive of gross perforation.

Patients should undergo routine laboratory testing including full blood count to assess for leucocytosis and CRP. An elevated white cell count and CRP are predictive of complicated diverticular disease (Bolkenstein et al. 2018).

14.6.2.4 Management

Treatment of diverticular disease is related to symptomatology and is highly variable. Asymptomatic diverticulosis is managed conservatively with lifestyle modifications including exercise, weight loss, cessation of smoking and diet.

Antibiotics, analgesia and bowel rest are routine for uncomplicated diverticulitis.

Complicated diverticulitis warrants radiological drainage or surgery. Localised diverticular abscess smaller than 3–4 cm can be managed with antibiotics alone, failing that percutaneous drainage or surgery (Brandt et al. 2006; Elagili et al. 2015). Patients who have recovered from this index admission should be counselled to undergo sigmoid resection due to 2 times risk of recurrent diverticulitis and 4 times risk of recurrent abscess (Rose et al. 2015).

Unstable patients with diffuse peritonitis or obstruction will require emergent sigmoid colectomy. Options include a Hartmann end colostomy or a primary colorectal anastomosis with or without a diverting stoma. A recent large meta-analysis found no significant difference between the 2 techniques in randomised controlled trials, albeit a slight superiority in mortality in the primary anastomosis group in observational studies (Halim et al. 2019).

14.6.2.5   Relevance to Men's Health

Men have a higher incidence of diverticulitis before the age of 50 (Wheat and Strate 2016). The disease course in males tends to be more aggressive with perforation, (Schauer et al. 1992) bleeding (McConnell et al. 2003) and complications including obstruction, fistula, peritonitis, with poorer survival (Bharucha et al. 2015). With the obesity epidemic, the increase disease burden in younger individuals is a cause for concern, particularly in young men (Broad et al. 2019). In a large prospective study of men, increased body mass index has been linked with a 1.7 fold to 3 fold increase in diverticulitis and diverticular bleeding (Strate et al. 2009). Tobacco use is also more prevalent in men and is a known risk factor for diverticulitis and its complications (Humes et al. 2016).

Recurrent diverticulitis is common in young males (Katz et al. 2013a). Young patients tend to pursue surgery for their disease following their admission for diverticular disease. This is reflected in time trend analysis with higher percentage rise in operations in men in a study conducted in England (Kang et al. 2003) This is a significant loss of function and economic burden to this working class of adults.

14.7   Anorectum

14.7.1   Perianal Sepsis and Fistulas

14.7.1.1   Background

Perianal sepsis is one of the most common problems encountered in the emergency room and surgical practice. It arises due to obstruction of the anal crypt gland leading to bacterial overgrowth and abscess formation (Parks 1961). Location of the abscess is relative to the anatomy of the anorectum and the sphincters. Most common locations are perianal, followed up ischiorectal, intersphincteric, supralevator and submucosal (Ramanujam et al. 1984) 90% of abscesses are cryptoglandular in nature. Other unusual causes include Crohn disease, tuberculosis, malignancy, trauma and pelvic sepsis (Sahnan et al. 2017). As high as 30–70% of patients have associated fistula in ano in the index infection, while another 30% will recur with a fistula after initial treatment of their abscess.

14.7.1.2 Symptoms and Signs

Patients often present with severe anorectal pain with a localised perianal mass and erythema. Ischiorectal abscess can present with generalised induration of the ischiorectal fossa with no obvious mass lesion and this should raise suspicion of a deeper collection. Less common abscesses in the intersphincteric, supralevator and submucosal planes lack physical signs and are usually felt during a digital rectal examination (DRE). Systemic signs of sepsis including fevers, tachycardia and lethargy may be present.

At times, patients can present with an indolent course with intermittent swelling, purulent or mucus discharge and bleeding suggestive of recurrent abscesses and fistula in ano. An external opening may be visible with a palpable cord extending towards the anus. Occasionally an internal opening can be palpated during a digital exam of the anal canal.

14.7.1.3 Investigations

Superficial abscesses and simple fistulas do not require further investigation. Deep, recurrent abscesses or complex fistulas should undergo cross sectional imaging to delineate the anatomy. MRI is preferable to CT with better soft tissue resolution and high accuracy. An alternative modality is endoanal ultrasound (Buchanan et al. 2004).

14.7.1.4 Management

Once diagnosed, perianal sepsis should undergo prompt incision and drainage. Initial management should be aimed at draining the sepsis and preserving continence. A draining seton may be needed if a fistula is found during the index operation. Although a primary fistulotomy or fistulectomy can be performed, it has an increased risk of incontinence (Malik et al. 2010).

Recurrent abscesses are frequently associated with fistula in ano. Simple fistulas are low intersphincteric or transsphincteric involving less than 30% of the sphincter complex. They are amenable to fistulotomy in select patients with good preoperative continence and high success rate (Abramowitz et al. 2016).

Complex fistulas involving more than 30% of sphincter, suprasphincteric, extrasphincteric locations, anterior fistulas in women, fistulas associated with Crohn disease, radiotherapy or malignancy warrants specialist input. Sphincter sparing operations such as ligation of intersphincteric tract (LIFT) procedure and mucosal advancement flaps can be considered with moderate success rates (Madbouly et al. 2014). Less effective methods such as fistula plugs and fibrin glue are other options but have variable effectiveness (Vogel et al. 2016).

14.7.1.5   Relevance to Men's Health

Males have a higher incidence of perianal sepsis with a ratio of 2:1 to females (Read and Abcarian 1979; Sainio 1984). This is even higher in paediatric populations below the age of 1 (Festen and van Harten 1998). Obesity, diabetes, smoking and alcohol history, which are conditions prevalent in men, have been linked to an increased risk of fistula formation (Wang et al. 2014). Men have been found to have a higher risk of high transsphincteric fistula and complex posterior based fistula tracks, leading to higher recurrence rates (Emile et al. 2018).

There are several hypotheses in the literature to explain this gender difference. Anatomical studies have found a higher frequency of anal glands in the male anal canal compared to females (Lilius 1968). Another theory is the role of higher levels of circulating androgen hormone in men and protective effects of oestrogen in women (Takatsuki 1976; Lunniss et al. 1995). Oestrogen has anti-inflammatory properties and testosterone induces inflammatory cytokines like TNF alpha and IL-6 (El-Tawil 2012). Other authors postulate the stronger sphincter complex in men leading to anal gland obstruction and sepsis (Hamadani et al. 2009).

14.7.2   Anal Fissures

14.7.2.1   Background

Anal fissure is characterised by linear break in the mucosa and skin of the anoderm. It is common in young women and middle aged men (Mapel et al. 2014). More than 90% of anal fissures are single and occur in the posterior midline. Anterior placed fissures are more common in women. Lateral and multiple fissures reflect other pathologies such as inflammatory bowel disease, trauma, malignancy and sexually transmitted infections.

A fissure forms when the anal mucosa is traumatised after passage of a hard stool. This causes a tear in the epithelium exposing underlying internal sphincter muscle and reflex spasm. The posterior midline is known to have poor blood supply (Klosterhalfen et al. 1989). This relative ischaemia is exacerbated in the presence of a fissure when pain causes contraction of the sphincter complex leading to poor perfusion to the region. Symptomatic patients have higher anal resting tone and hypertrophy of the internal anal sphincter causing a vicious cycle of repetitive trauma, ischaemia and persistence of the fissure (Schouten et al. 1994; Farouk et al. 1994).

14.7.2.2 Symptoms and Signs

As the mucosal break is distal to the dentate line which is innervated by somatic pain fibres, the pain experienced is typically sharp, burning and well localised in nature. A classical history described by patients is severe pain during defaecation lasting minutes to hours. Associated symptoms include bleeding and tenesmus.

On perianal examination, an acute fissure has a typical characteristic fresh pink appearance to the mucosa with hypertonia of the surrounding sphincter complex. Chronic fissures that are more than 6 weeks duration have other defining features such as a well-defined mucosa edge, sentinel anal skin tag, hypertrophied anal papilla and whitish base lined by horizontal fibres of the internal sphincter muscle.

14.7.2.3 Investigations

Patients who cannot tolerate a beside assessment warrants an examination under anaesthesia and proctoscopy. Atypical fissures will need further investigations including biopsy and colonoscopy to exclude perianal Crohn's, cancer and communicable diseases such as human immunodeficiency virus, syphilis or chlamydia.

14.7.2.4 Management

Conservative and medical therapy are first line treatment options for this disease. Patients are educated to avoid trauma and constipation that can exacerbate the fissure. Adequate fluid intake, fibre and stool softeners can be utilised to relieve constipation. Topical nitrate or calcium channel blocker ointment work by relaxing the sphincter complex, improving the blood supply to the fissures to promote healing (Ezri and Susmallian 2003). Common side effects of these ointments are due to systemic absorption causing headaches, hypotension and dizziness.

With failure of medical therapy, examination under anaesthesia and surgery may be needed. Botulinum toxin can be injected into the internal sphincter to induce a chemical sphincterotomy (Berkel et al. 2014). More invasive lateral sphincterotomy can be done with a high healing rate (Gandomkar et al. 2015). Anal advancement flap can be utilised for resistant cases to cover the sphincter, minimise spasm and avoid risk of incontinence (Patel et al. 2011).

14.7.2.5 Relevance to Men's Health

There are distinct differences between the anal canal between a man and woman. Men have a longer and thicker internal/external anal sphincter mechanism in all dimensions (Regadas et al. 2007). This is reflected with higher resting and maximal squeeze anal manometry pressures at all age groups (McHugh and Diamant 1987). Striking a balance between sphincterotomy to relieve persistent symptoms and risk of incontinence is important during the decision making for patients. Female patients have a higher incidence of seepage and loss of control to flatus and bowels (Garg et al. 2013). Due to the shorter length of the anal canal, females have a risk of inadvertent excessive muscle division (Sultan et al. 1994). Conversely, men have a higher risk of recurrent fissures post lateral sphincterotomy (Acar et al. 2019).

14.7.3 Haemorrhoids

14.7.3.1 Background

The anal canal is approximately 3–4 cm in length lined by fibrovascular cushions containing a network of arteriovenous anastomoses that drain into the superior, middle and inferior haemorrhoidal arteries and veins. When these anal cushions enlarge and prolapse causing symptoms, they are termed haemorrhoids. The true prevalence of this disease is not well reported and probably underquoted (Johanson and Sonnenberg 1990; Riss et al. 2011). Haemorrhoids are commonly develop between the age of 40–60 with a slight preponderance for women (Xia et al. 2021). The pathophysiology is multifactorial. It is theorised that the anal cushions descent with disruption of the Treitz muscle, connective tissue and suspensory ligaments (Thomson 1975; Margetis 2019). This is exacerbated by ageing, (Haas et al. 1984) impaired venous return leading to engorgement of the sinusoids and high anal resting tone (Sun et al. 1990). Increased abdominal pressure such as constipation, straining, obesity and pregnancy are other risk factors.

14.7.3.2 Symptoms and Signs

Haemorrhoids can be divided into three main anatomical groups: internal, external or mixed. This is in relation to their positions to the dentate line and embryological origin. Haemorrhoids that arise from the endoderm above the dentate line are typically situated at the 3, 7 and 11 o'clock positions. Pain from these are visceral in nature. On the other hand, external haemorrhoids are distal to the dentate line lined by squamous epithelium and confer somatic sensation.

Rectal bleeding is the most common symptom. The bleeding is typically fresh, bright red in colour in the toilet bowl or paper. Other symptoms include anorectal pain, prolapsing mass, redundant skin, soiling and pruritus ani. Acute complications can develop when internal haemorrhoids prolapse, strangulate triggering sphincter spasm and severe anismus. Similarly, an external haemorrhoid can acutely thrombose with an underlying haematoma causing intense pain from the richly innervated perianal skin derived from ectoderm origin.

14.7.3.3 Investigations

A thorough history and physical examination should be performed to exclude other causes of rectal bleeding such as cancer, anal fissure, abscess or fistula that can masquerade as haemorrhoids. A visual inspection of the perianal area combined with a DRE should be performed on all patients. A bedside anoscopy or proctosigmoidoscopy can be easily performed to evaluate the anal canal and rectum. Select patients who are above 40–50 years of age with family history of colorectal cancer or adenomas should be evaluated with a flexible sigmoidoscopy or colonoscopy (Davis et al. 2018).

14.7.3.4 Management

Initial treatment of haemorrhoids is focused on lifestyle modifications such as high fibre diet, increasing fluid intake, stool softeners and reduced straining (Alonso-Coello et al. 2005). Topical therapy with analgesia, steroids, antispasmodic and venoactive agents can been prescribed with some relief of symptoms (Tjandra et al. 2007; Perera et al. 2012). Patients who are refractory to these measures present to the surgeon for further treatments. Rubber band ligation is the most effective modality with reported high success at controlling bleeding in the outpatient setting (Cocorullo et al. 2017) Compared to sclerotherapy and infrared coagulation, banding has less reintervention rates (MacRae and McLeod 1995).

Larger internal haemorrhoids, particularly those associated with an external component should be referred for surgery. Excisional haemorrhoidectomy can be performed in the open or closed technique (Bhatti et al. 2016). Internal haemorrhoidal columns are commonly excised at the left lateral, right posterior and right anterior positions with suture ligation of the pedicle. Adequate mucocutaneous bridges should be left behind to minimise risk of anal stenosis. Excisional haemorrhoidectomy is overall most effective at managing haemorrhoids with a low failure rate (Similis et al. 2015). Other alternatives such as stapled haemorrhoidopexy and doppler guided transanal haemorrhoidal artery ligation rectoanal repair (HAL-RAR) have less pain but higher recurrence.

14.7.3.5 Relevance to Men's Health

Acute thrombosed external haemorrhoid is a debilitating condition that often presents to medical attention in a delayed setting. This commonly occurs in young men secondary to constipation and trauma to the anoderm (Barrios and Khubchandani 1979; Oh 1989; Greenspon et al. 2004). Management can vary from conservative management with sitz bath, analgesia, aperients, topical nifedipine/nitroglycerin, to incision and evacuation of the clot, to haemorrhoidectomy. In a prospective randomised trial comparing 3 modalities of treatment, excision has the lowest recurrence rate, better symptom resolution and less residual anal skin tags compared to less invasive management (Cavcić et al. 2001). Prolapsed strangulated internal haemorrhoids is likewise another agonising condition more common in men (Eu et al. 1994; Ceulemans et al. 2000). Haemorrhoidectomy is proven efficacious but with a risk of sphincter injury (Allan et al. 2006).

Clinicians need to be mindful that in treating both acute and chronic haemorrhoidal disease, perianal sepsis is a rare but serious adverse event. Presentation can range from perianal abscess to Fournier's gangrene, retroperitoneal/pelvic/liver abscess and distant sites such as septic pulmonary emboli and pneumomediastinum. The male gender is a predominant feature in these complications (McCloud et al. 2006).

References

Abe N, Takeuchi H, Yanagida O, Sugiyama M, Atomi Y. Surgical indications and procedures for bleeding peptic ulcer. Dig Endosc. 2010;22(Suppl 1):S35–7.

Abegunde AT, Muhammad BH, Bhatti O, Ali T. Environmental risk factors for inflammatory bowel diseases: evidence based literature review. World J Gastroenterol. 2016;22:6296–317.

Abramowitz L, Soudan D, Souffran M, Bouchard D, Castinel A, Suduca JM, Staumont G, Devulder F, Pigot F, Ganansia R, Varastet M, for the Groupe de Recherche en Proctologie de la Société Nationale Française de Colo-Proctologie, and the Club de Réflexion des Cabinets et Groupe d'Hépato-Gastroentérologie. The outcome of fistulotomy for anal fistula at 1 year: a prospective multicentre French study. Color Dis. 2016;18:279–85.

Acar T, Acar N, Güngör F, Kamer E, Güngör H, Candan MS, Bağ H, Tarcan E, Dilek ON, Haciyanli M. Treatment of chronic anal fissure: Is open lateral internal sphincterotomy (LIS) a safe and adequate option? Asian J Surg. 2019;42:628–33.

Addiss DG, Shaffer N, Fowler BS, Tauxe RV. The epidemiology of appendicitis and appendectomy in the United States. Am J Epidemiol. 1990;132:910–25.

Akhtar AJ, Shaheen M. Extrapancreatic manifestations of acute pancreatitis in African-American and hispanic patients. Pancreas. 2004;29:291–7.

Al-Darmaki A, Hubbard J, Seow CH, Leung Y, Novak K, Shaheen AA, Panaccione R, Kaplan GG. Clinical predictors of the risk of early colectomy in ulcerative colitis: a population-based study. Inflamm Bowel Dis. 2017;23:1272–7.

Allan A, Samad AJ, Mellon A, Marshall T. Prospective randomised study of urgent haemorrhoidectomy compared with non-operative treatment in the management of prolapsed thrombosed internal haemorrhoids. Color Dis. 2006;8:41–5.

Al-Omran M, Mamdani M, McLeod RS. Epidemiologic features of acute appendicitis in Ontario, Canada. Can J Surg. 2003;46:263–8.

Alonso-Coello P, Guyatt G, Heels-Ansdell D, Johanson JF, Lopez-Yarto M, Mills E, Zhou Q. Laxatives for the treatment of hemorrhoids. Cochrane Database Syst Rev. 2005:Cd004649.

Andersson RE, Hugander A, Thulin AJ. Diagnostic accuracy and perforation rate in appendicitis: association with age and sex of the patient and with appendicectomy rate. Eur J Surg. 1992;158:37–41.

Andriulli A, Botteri E, Almasio PL, Vantini I, Uomo G, Maisonneuve P. Smoking as a cofactor for causation of chronic pancreatitis: a meta-analysis. Pancreas. 2010;39:1205–10.

Armstrong D, Bennett JR, Blum AL, Dent J, De Dombal FT, Galmiche JP, Lundell L, Margulies M, Richter JE, Spechler SJ, Tytgat GN, Wallin L. The endoscopic assessment of esophagitis: a progress report on observer agreement. Gastroenterology. 1996;111:85–92.

Asanuma K, Iijima K, Shimosegawa T. Gender difference in gastro-esophageal reflux diseases. World J Gastroenterol. 2016;22:1800–10.

Aune D, Sen A, Leitzmann MF, Tonstad S, Norat T, Vatten LJ. Tobacco smoking and the risk of diverticular disease – a systematic review and meta-analysis of prospective studies. Color Dis. 2017;19:621–33.

Azhari H, Underwood F, King J, Coward S, Shah S, Chan C, Ho G, Ng S, Kaplan G. The global incidence of peptic ulcer disease and its complications at the turn of the 21st century: a systematic review: 1199. Am Coll Gastroenterol. 2018;113:S684–S85.

Balthazar EJ. Staging of acute pancreatitis. Radiol Clin N Am. 2002;40:1199–209.

Banks PA, Bollen TL, Dervenis C, Gooszen HG, Johnson CD, Sarr MG, Tsiotos GG, Vege SS. Classification of acute pancreatitis--2012: revision of the Atlanta classification and definitions by international consensus. Gut. 2013;62:102–11.

Bardhan KD, Royston C. Time, change and peptic ulcer disease in Rotherham, UK. Dig Liver Dis. 2008;40:540–6.

Barkun AN, et al. Management of nonvariceal upper gastrointestinal bleeding: guideline recommendations from the International Consensus Group. Ann Int Med. 2019;171:805–22.

Barrios G, Khubchandani M. Urgent hemorrhoidectomy for hemorrhoidal thrombosis. Dis Colon Rectum. 1979;22:159–61.

Bellolio F, Cohen Z, MacRae HM, O'Connor BI, Victor JC, Huang H, McLeod RS. Strictureplasty in selected Crohn's disease patients results in acceptable long-term outcome. Dis Colon Rectum. 2012;55:864–9.

Berkel AE, Rosman C, Koop R, van Duijvendijk P, van der Palen J, Klaase JM. Isosorbide dinitrate ointment vs botulinum toxin A (Dysport) as the primary treatment for chronic anal fissure: a randomized multicentre study. Color Dis. 2014;16:O360–6.

Bhangu A, Søreide K, Di Saverio S, Assarsson JH, Drake FT. Acute appendicitis: modern understanding of pathogenesis, diagnosis, and management. Lancet. 2015;386:1278–87.

Bharucha AE, Parthasarathy G, Ditah I, Fletcher JG, Ewelukwa O, Pendlimari R, Yawn BP, Melton LJ, Schleck C, Zinsmeister AR. Temporal trends in the incidence and natural history of diverticulitis: a population-based study. Am J Gastroenterol. 2015;110:1589–96.

Bhatti MI, Sajid MS, Baig MK. Milligan-Morgan (Open) versus ferguson haemorrhoidectomy (closed): a systematic review and meta-analysis of published randomized, controlled trials. World J Surg. 2016;40:1509–19.

Bingener-Casey J, Richards ML, Strodel WE, Schwesinger WH, Sirinek KR. Reasons for conversion from laparoscopic to open cholecystectomy: a 10-year review. J Gastrointest Surg. 2002;6:800–5.

Blaser MJ, Atherton JC. Helicobacter pylori persistence: biology and disease. J Clin Invest. 2004;113:321–33.

Boles RS Jr, Jordan SM. The clinical significance of diverticulosis. Gastroenterology. 1958;35:579–82.

Bolkenstein HE, van de Wall BJ, Consten EC, van der Palen J, Broeders IA, Draaisma WA. Development and validation of a diagnostic prediction model distinguishing complicated from uncomplicated diverticulitis. Scand J Gastroenterol. 2018;53:1291–7.

Bourikian S, Anand RJ, Aboutanos M, Wolfe LG, Ferrada P. Risk factors for acute gangrenous cholecystitis in emergency general surgery patients. Am J Surg. 2015;210:730–3.

Brandse JF, Mathôt RA, van der Kleij D, Rispens T, Ashruf Y, Jansen JM, Rietdijk S, Löwenberg M, Ponsioen CY, Singh S, van den Brink GR, D'Haens GR. Pharmacokinetic features and presence of antidrug antibodies associate with response to infliximab induction therapy in patients with moderate to severe ulcerative colitis. Clin Gastroenterol Hepatol. 2016;14:251–8.e1–2.

Brandt D, Gervaz P, Durmishi Y, Platon A, Morel P, Poletti PA. Percutaneous CT scan-guided drainage vs. antibiotherapy alone for Hinchey II diverticulitis: a case-control study. Dis Colon Rectum. 2006;49:1533–8.

Broad JB, Wu Z, Xie S, Bissett IP, Connolly MJ. Diverticular disease epidemiology: acute hospitalisations are growing fastest in young men. Tech Coloproctol. 2019;23:713–21.

Buchanan GN, Halligan S, Bartram CI, Williams AB, Tarroni D, Richard C, Cohen G. Clinical examination, endosonography, and MR imaging in preoperative assessment of fistula in ano: comparison with outcome-based reference standard. Radiology. 2004;233:674–81.

Büchler MW, Martignoni ME, Friess H, Malfertheiner P. A proposal for a new clinical classification of chronic pancreatitis. BMC Gastroenterol. 2009;9:93.

Buckles DC, Sarosiek I, McCallum RW, McMillin C. Delayed gastric emptying in gastroesophageal reflux disease: reassessment with new methods and symptomatic correlations. Am J Med Sci. 2004;327:1–4.

Carter D, Levi G, Tzur D, Novis B, Avidan B. Prevalence and predictive factors for gastrointestinal pathology in young men evaluated for iron deficiency anemia. Dig Dis Sci. 2013;58:1299–305.

Cavcić J, Turcić J, Martinac P, Mestrović T, Mladina R, Pezerović-Panijan R. Comparison of topically applied 0.2% glyceryl trinitrate ointment, incision and excision in the treatment of perianal thrombosis. Dig Liver Dis. 2001;33:335–40.

Ceresoli M, Coccolini F, Catena F, Montori G, Di Saverio S, Sartelli M, Ansaloni L. Water-soluble contrast agent in adhesive small bowel obstruction: a systematic review and meta-analysis of diagnostic and therapeutic value. Am J Surg. 2016;211:1114–25.

Ceulemans R, Creve U, Van Hee R, Martens C, Wuyts FL. Benefit of emergency haemorrhoidectomy: a comparison with results after elective operations. Eur J Surg. 2000;166:808–12. discussion 13

Chang AR. An analysis of the pathology of 3003 appendices. Aust N Z J Surg. 1981;51:169–78.

Chen P, Yang W, Zhang J, Wang C, Yu Y, Wang Y, Yang L, Zhou Z. Analysis of risk factors associated bowel resection in patients with incarcerated groin hernia. Medicine. 2020;99:e20629–e29.

Cocorullo G, Tutino R, Falco N, Licari L, Orlando G, Fontana T, Raspanti C, Salamone G, Scerrino G, Gallo G, Trompetto M, Gulotta G. The non-surgical management for hemorrhoidal disease. A systematic review. G Chir. 2017;38:5–14.

Commane DM, Arasaradnam RP, Mills S, Mathers JC, Bradburn M. Diet, ageing and genetic factors in the pathogenesis of diverticular disease. World J Gastroenterol. 2009;15:2479–88.

Cook MB, Wild CP, Forman D. A systematic review and meta-analysis of the sex ratio for Barrett's esophagus, erosive reflux disease, and nonerosive reflux disease. Am J Epidemiol. 2005;162:1050–61.

Corrao G, Bagnardi V, Zambon A, Arico S. Exploring the dose-response relationship between alcohol consumption and the risk of several alcohol-related conditions: a meta-analysis. Addiction. 1999;94:1551–73.

Crowe FL, Appleby PN, Allen NE, Key TJ. Diet and risk of diverticular disease in Oxford cohort of European Prospective Investigation into Cancer and Nutrition (EPIC): prospective study of British vegetarians and non-vegetarians. BMJ (Clinical Research ed). 2011;343:d4131–d31.

D'Haens G, Ferrante M, Vermeire S, Baert F, Noman M, Moortgat L, Geens P, Iwens D, Aerden I, Van Assche G, Van Olmen G, Rutgeerts P. Fecal calprotectin is a surrogate marker for endoscopic lesions in inflammatory bowel disease. Inflamm Bowel Dis. 2012;18:2218–24.

Davis BR, Lee-Kong SA, Migaly J, Feingold DL, Steele SR. The American society of colon and rectal surgeons clinical practice guidelines for the management of hemorrhoids. Dis Colon Rectum. 2018;61:284–92.

Dias CC, Rodrigues PP, da Costa-Pereira A, Magro F. Clinical predictors of colectomy in patients with ulcerative colitis: systematic review and meta-analysis of cohort studies. J Crohn's Colitis. 2014;9:156–63.

Disease GBD, Incidence Injury, and Collaborators Prevalence. Global, regional, and national incidence, prevalence, and years lived with disability for 310 diseases and injuries, 1990–2015: a systematic analysis for the Global Burden of Disease Study 2015. Lancet (London, England). 2016;388:1545–602.

Drake FT, Mottey NE, Farrokhi ET, Florence MG, Johnson MG, Mock C, Steele SR, Thirlby RC, Flum DR. Time to appendectomy and risk of perforation in acute appendicitis. JAMA Surg. 2014;149:837–44.

Drożdż W, Budzyński P. Change in mechanical bowel obstruction demographic and etiological patterns during the past century: observations from one health care institution. Arch Surg. 2012;147:175–80.

D'Souza N, D'Souza C, Grant D, Royston E, Farouk M. The value of ultrasonography in the diagnosis of appendicitis. Int J Surg. 2015;13:165–9.

Ebell MH, Shinholser J. What are the most clinically useful cutoffs for the alvarado and pediatric appendicitis scores? A systematic review. Ann Emerg Med. 2014;64:365–72.e2.

Elagili F, Stocchi L, Ozuner G, Kiran RP. Antibiotics alone instead of percutaneous drainage as initial treatment of large diverticular abscess. Tech Coloproctol. 2015;19:97–103.

El-Dhuwaib Y, Slavin J, Corless DJ, Begaj I, Durkin D, Deakin M. Bile duct reconstruction following laparoscopic cholecystectomy in England. Surg Endosc. 2016;30:3516–25.

El-Serag HB, Sweet S, Winchester CC, Dent J. Update on the epidemiology of gastro-oesophageal reflux disease: a systematic review. Gut. 2014;63:871–80.

El-Tawil AM. Mechanism of non-specific-fistula-in-ano: hormonal aspects-review. Pathophysiology. 2012;19:55–9.

Emile SH, Elgendy H, Sakr A, Youssef M, Thabet W, Omar W, Khafagy W, Farid M. Gender-based analysis of the characteristics and outcomes of surgery for anal fistula: analysis of more than 560 cases. J Coloproctol. 2018;38:199–206.

Eu KW, Seow-Choen F, Goh HS. Comparison of emergency and elective haemorrhoidectomy. Br J Surg. 1994;81:308–10.

Everhart JE, Ruhl CE. Burden of digestive diseases in the United States part II: lower gastrointestinal diseases. Gastroenterology. 2009;136:741–54.

Everhart JE, Khare M, Hill M, Maurer KR. Prevalence and ethnic differences in gallbladder disease in the United States. Gastroenterology. 1999;117:632–9.

Ezri T, Susmallian S. Topical nifedipine vs. topical glyceryl trinitrate for treatment of chronic anal fissure. Dis Colon Rectum. 2003;46:805–8.

Fagan SP, Awad SS, Rahwan K, Hira K, Aoki N, Itani KMF, Berger DH. Prognostic factors for the development of gangrenous cholecystitis. Am J Surg. 2003;186:481–5.

Farinella E, Cirocchi R, La Mura F, Morelli U, Cattorini L, Delmonaco P, Migliaccio C, De Sol AA, Cozzaglio L, Sciannameo F. Feasibility of laparoscopy for small bowel obstruction. World J Emerg Surg. 2009;4:3.

Farouk R, Duthie GS, MacGregor AB, Bartolo DCC. Sustained internal sphincter hypertonia in patients with chronic anal fissure. Dis Colon Rectum. 1994;37:424–9.

Fasanmade AA, Adedokun OJ, Ford J, Hernandez D, Johanns J, Hu C, Davis HM, Zhou H. Population pharmacokinetic analysis of infliximab in patients with ulcerative colitis. Eur J Clin Pharmacol. 2009;65:1211–28.

Ferris M, Quan S, Kaplan BS, Molodecky N, Ball CG, Chernoff GW, Bhala N, Ghosh S, Dixon E, Ng S, Kaplan GG. The global incidence of appendicitis: a systematic review of population-based studies. Ann Surg. 2017;266:237–41.

Festen C, van Harten H. Perianal abscess and fistula-in-ano in infants. J Pediatr Surg. 1998;33:711–3.

Fletcher DR, Hobbs MS, Tan P, Valinsky LJ, Hockey RL, Pikora TJ, Knuiman MW, Sheiner HJ, Edis A. Complications of cholecystectomy: risks of the laparoscopic approach and protective effects of operative cholangiography: a population-based study. Ann Surg. 1999;229:449–57.

Fong SS, Tan EY, Foo A, Sim R, Cheong DM. The changing trend of diverticular disease in a developing nation. Color Dis. 2011;13:312–6.

Gallegos NC, Dawson J, Jarvis M, Hobsley M. Risk of strangulation in groin hernias. Br J Surg. 1991;78:1171–3.

Gallo G, Kotze PG, Spinelli A. Surgery in ulcerative colitis: When? How? Best Pract Res Clin Gastroenterol. 2018;32-33:71–8.

Gandomkar H, Zeinoddini A, Heidari R, Amoli HA. Partial lateral internal sphincterotomy versus combined botulinum toxin A injection and topical diltiazem in the treatment of chronic anal fissure: a randomized clinical trial. Dis Colon Rectum. 2015;58:228–34.

Garg P, Garg M, Menon GR. Long-term continence disturbance after lateral internal sphincterotomy for chronic anal fissure: a systematic review and meta-analysis. Color Dis. 2013;15:e104–17.

Gharaibeh KI, Qasaimeh GR, Al-Heiss H, Ammari F, Bani-Hani K, Al-Jaberi TM, Al-Natour S. Effect of timing of surgery, type of inflammation, and sex on outcome of laparoscopic cholecystectomy for acute cholecystitis. J Laparoendosc Adv Surg Tech Part A. 2002;12:193–8.

Giger U, Ouaissi M, Schmitz SF, Krähenbühl S, Krähenbühl L. Bile duct injury and use of cholangiography during laparoscopic cholecystectomy. Br J Surg. 2011;98:391–6.

Golder M, Ster IC, Babu P, Sharma A, Bayat M, Farah A. Demographic determinants of risk, colon distribution and density scores of diverticular disease. World J Gastroenterol. 2011;17:1009–17.

Gomi H, Solomkin JS, Schlossberg D, Okamoto K, Takada T, Strasberg SM, Ukai T, Endo I, Iwashita Y, Hibi T, Pitt HA, Matsunaga N, Takamori Y, Umezawa A, Asai K, Suzuki K, Han H-S, Hwang T-L, Mori Y, Yoon Y-S, Huang WS-W, Belli G, Dervenis C, Yokoe M, Kiriyama S, Itoi T, Palepu J, James Garden O, Miura F, de Santibañes E, Shikata S, Noguchi Y, Wada K, Honda G, Supe AN, Yoshida M, Mayumi T, Gouma DJ, Deziel DJ, Liau K-H, Chen M-F, Liu K-H, Cheng-Hsi S, Chan ACW, Yoon D-S, Choi I-S, Jonas E, Chen X-P, Fan ST, Ker C-G, Giménez ME, Kitano S, Inomata M, Mukai S, Higuchi R, Hirata K, Inui K, Sumiyama Y, Yamamoto M. Tokyo guidelines 2018: antimicrobial therapy for acute cholangitis and cholecystitis. J Hepatobiliary Pancreat Sci. 2018;25:3–16.

Grant AM, Cotton SC, Boachie C, Ramsay CR, Krukowski ZH, Heading RC, Campbell MK. Minimal access surgery compared with medical management for gastro-oesophageal reflux disease: five year follow-up of a randomised controlled trial (REFLUX). BMJ. 2013;346:f1908.

Greenspon J, Williams SB, Young HA, Orkin BA. Thrombosed external hemorrhoids: outcome after conservative or surgical management. Dis Colon Rectum. 2004;47:1493–8.

Gu J, Valente MA, Remzi FH, Stocchi L. Factors affecting the fate of faecal diversion in patients with perianal Crohn's disease. Color Dis. 2015;17:66–72.

Guss DA, Richards C. Comparison of men and women presenting to an ED with acute appendicitis. Am J Emerg Med. 2000;18:372–5.

Haas PA, Fox TA Jr, Haas GP. The pathogenesis of hemorrhoids. Dis Colon Rectum. 1984;27:442–50.

Halbert C, Altieri MS, Yang J, Meng Z, Chen H, Talamini M, Pryor A, Parikh P, Telem DA. Long-term outcomes of patients with common bile duct injury following laparoscopic cholecystectomy. Surg Endosc. 2016;30:4294–9.

Halim H, Askari A, Nunn R, Hollingshead J. Primary resection anastomosis versus Hartmann's procedure in Hinchey III and IV diverticulitis. World J Emerg Surg. 2019;14:32.

Hamadani A, Haigh PI, Liu I-LA, Abbas MA. Who is at risk for developing chronic anal fistula or recurrent anal sepsis after initial perianal abscess? Dis Colon Rectum. 2009;52:217–21.

Hansen C-C, Søreide K. Systematic review of epidemiology, presentation, and management of Meckel's diverticulum in the 21st century. Medicine. 2018;97:e12154.

Helavirta I, Lehto K, Huhtala H, Hyöty M, Collin P, Aitola P. Pouch failures following restorative proctocolectomy in ulcerative colitis. Int J Color Dis. 2020;35:2027–33.

Henriksen NA, Montgomery A, Kaufmann R, Berrevoet F, East B, Fischer J, Hope W, Klassen D, Lorenz R, Renard Y, Garcia Urena MA, Simons MP. Guidelines for treatment of umbilical and epigastric hernias from the European Hernia Society and Americas Hernia Society. Br J Surg. 2020;107:171–90.

Hernandez V, Pascual I, Almela P, Añon R, Herreros B, Sanchiz V, Minguez M, Benages A. Recurrence of acute gallstone pancreatitis and relationship with cholecystectomy or endoscopic sphincterotomy. Am J Gastroenterol. 2004;99:2417–23.

Hooi JKY, Lai WY, Ng WK, Suen MMY, Underwood FE, Tanyingoh D, Malfertheiner P, Graham DY, Wong VWS, Wu JCY, Chan FKL, Sung JJY, Kaplan GG, Ng SC. Global prevalence of helicobacter pylori infection: systematic review and meta-analysis. Gastroenterology. 2017;153:420–9.

Hughes LE. Postmortem survey of diverticular disease of the colon. I. Diverticulosis and diverticulitis. Gut. 1969;10:336–44.

Humes DJ, Ludvigsson JF, Jarvholm B. Smoking and the risk of hospitalization for symptomatic diverticular disease: a population-based Cohort Study from Sweden. Dis Colon Rectum. 2016;59:110–4.

Hunt DRH, Chu FCK. Gangrenous Cholecystitis in the Laparoscopic Era. Aust N Z J Surg. 2000;70:428–30.

Hunt R, Armstrong D, Katelaris P, Afihene M, Bane A, Bhatia S, Chen M-H, Choi MG, Melo AC, Fock KM, Ford A, Hongo M, Khan A, Lazebnik L, Lindberg G, Lizarzabal M, Myint T, Moraes-Filho JP, Salis G, Lin JT, Vaidya R, Abdo A, LeMair A, Review Team. World gastroenterology organisation global guidelines: GERD global perspective on gastroesophageal reflux disease. J Clin Gastroenterol. 2017;51:467–78.

Hwang JM, Krumbhaar KB. The amount of lymphoid tissue of the human appendix and its weight at different age periods. Am J Med Sci. 1940;199:75–83.

Isenmann R, Rünzi M, Kron M, Kahl S, Kraus D, Jung N, Maier L, Malfertheiner P, Goebell H, Beger HG. Prophylactic antibiotic treatment in patients with predicted severe acute pancreatitis: a placebo-controlled, double-blind trial. Gastroenterology. 2004;126:997–1004.

Jaakkola M, Nordback I. Pancreatitis in Finland between 1970 and 1989. Gut. 1993;34:1255–60.

Jaschinski T, Mosch CG, Eikermann M, Neugebauer EA, Sauerland S. Laparoscopic versus open surgery for suspected appendicitis. Cochrane Database Syst Rev. 2018;11:Cd001546.

Joelsson M, Benoni C, Oresland T. Does smoking influence the risk of pouchitis following ileal pouch anal anastomosis for ulcerative colitis? Scand J Gastroenterol. 2006;41:929–33.

Johanson JF, Sonnenberg A. The prevalence of hemorrhoids and chronic constipation. An epidemiologic study. Gastroenterology. 1990;98:380–6.

Kahrilas PJ, Dodds WJ, Hogan WJ, Kern M, Arndorfer RC, Reece A. Esophageal peristaltic dysfunction in peptic esophagitis. Gastroenterology. 1986;91:897–904.

Kang JY, Hoare J, Tinto A, Subramanian S, Ellis C, Majeed A, Melville D, Maxwell JD. Diverticular disease of the colon--on the rise: a study of hospital admissions in England between 1989/1990 and 1999/2000. Aliment Pharmacol Ther. 2003;17:1189–95.

Katz LH, Guy DD, Lahat A, Gafter-Gvili A, Bar-Meir S. Diverticulitis in the young is not more aggressive than in the elderly, but it tends to recur more often: systematic review and meta-analysis. J Gastroenterol Hepatol. 2013a;28:1274–81.

Katz PO, Gerson LB, Vela MF. Guidelines for the diagnosis and management of gastroesophageal reflux disease. Official J Am Coll Gastroenterol. 2013b;108:308–28.

Kayal M, Plietz M, Rizvi A, Radcliffe M, Riggs A, Yzet C, Tixier E, Trivedi P, Ungaro RC, Khaitov S, Sylla P, Greenstein A, Frederic Colombel J, Dubinsky MC. Inflammatory pouch conditions are common after ileal pouch anal anastomosis in ulcerative colitis patients. Inflamm Bowel Dis. 2020;26:1079–86.

Keus F, de Jong J, Gooszen HG, Laarhoven CJHM. Laparoscopic versus open cholecystectomy for patients with symptomatic cholecystolithiasis. Cochrane Database Syst Rev. 2006;(4):CD006231.

Khalili H, Huang ES, Ananthakrishnan AN, Higuchi L, Richter JM, Fuchs CS, Chan AT. Geographical variation and incidence of inflammatory bowel disease among US women. Gut. 2012;61:1686–92.

Kiewiet JJ, Leeuwenburgh MM, Bipat S, Bossuyt PM, Stoker J, Boermeester MA. A systematic review and meta-analysis of diagnostic performance of imaging in acute cholecystitis. Radiology. 2012;264:708–20.

Kim BJ, Cheon WS, Hyoung-Chul O, Kim JW, Park JD, Kim JG. Prevalence and risk factor of erosive esophagitis observed in Korean National Cancer Screening Program. J Korean Med Sci. 2011;26:642–6.

Klosterhalfen B, Vogel P, Rixen H, Mittermayer C. Topography of the inferior rectal artery: a possible cause of chronic, primary anal fissure. Dis Colon Rectum. 1989;32:43–52.

Kristiansen L, Grønbæk M, Becker U, Tolstrup JS. Risk of pancreatitis according to alcohol drinking habits: a population-based Cohort Study. Am J Epidemiol. 2008;168:932–7.

Kurata JH, Haile BM, Elashoff JD. Sex differences in peptic ulcer disease. Gastroenterology. 1985;88:96–100.

Kurt N, Oncel M, Ozkan Z, Bingul S. Risk and outcome of bowel resection in patients with incarcerated groin hernias: retrospective study. World J Surg. 2003;27:741–3.

Kwong WT, Vege SS. Unrecognized necrosis at same admission cholecystectomy for pancreatitis increases organ failure and infected necrosis. Pancreatology. 2017;17:41–4.

Labenz J, Jaspersen D, Kulig M, Leodolter A, Lind T, Meyer-Sabellek W, Stolte M, Vieth M, Willich S, Malfertheiner P. Risk factors for erosive esophagitis: a multivariate analysis based on the ProGERD study initiative. Am J Gastroenterol. 2004;99:1652–6.

Lanas A, García-Rodríguez LA, Arroyo MT, Gomollón F, Feu F, González-Pérez A, Zapata E, Bástida G, Rodrigo L, Santolaria S, Güell M, de Argila CM, Quintero E, Borda F, Piqué JM. Risk of upper gastrointestinal ulcer bleeding associated with selective cyclo-oxygenase-2 inhibitors, traditional non-aspirin non-steroidal anti-inflammatory drugs, aspirin and combinations. Gut. 2006;55:1731–8.

Lange H, Jäckel R. Usefulness of plasma lactate concentration in the diagnosis of acute abdominal disease. Eur J Surg. 1994;160:381–4.

Lankisch PG, Löhr-Happe A, Otto J, Creutzfeldt W. Natural course in chronic pancreatitis. Pain, exocrine and endocrine pancreatic insufficiency and prognosis of the disease. Digestion. 1993;54:148–55.

Lee JA. The influence of sex and age on appendicitis in children and young adults. Gut. 1962; 3:80–4.

Li Z, Zou D, Ma X, Chen J, Shi X, Gong Y, Man X, Gao L, Zhao Y, Wang R, Yan X, Dent J, Sung JJ, Wernersson B, Johansson S, Liu W, He J. Epidemiology of peptic ulcer disease: endoscopic results of the systematic investigation of gastrointestinal disease in China. Am J Gastroenterol. 2010;105:2570–7.

Lilius HG. Fistula-in-ano, an investigation of human foetal anal ducts and intramuscular glands and a clinical study of 150 patients. Acta Chir Scand Suppl. 1968;383:7–88.

Lin KJ, García Rodríguez LA, Hernández-Díaz S. Systematic review of peptic ulcer disease incidence rates: do studies without validation provide reliable estimates? Pharmacoepidemiol Drug Saf. 2011;20:718–28.

Lowenfels AB, Sullivan T, Fiorianti J, Maisonneuve P. The epidemiology and impact of pancreatic diseases in the United States. Curr Gastroenterol Rep. 2005;7:90–5.

Lunniss PJ, Jenkins PJ, Besser GM, Perry LA, Phillips RKS. Gender differences in incidence of idiopathic fistula-in-ano are not explained by circulating sex hormones. Int J Color Dis. 1995;10:25–8.

MacRae HM, McLeod RS. Comparison of hemorrhoidal treatment modalities. A meta-analysis. Dis Colon Rectum. 1995;38:687–94.

Madbouly KM, El Shazly W, Abbas KS, Hussein AM. Ligation of intersphincteric fistula tract versus mucosal advancement flap in patients with high transsphincteric fistula-in-ano: a prospective randomized trial. Dis Colon Rectum. 2014;57:1202–8.

Mahajan P, Basu T, Pai C-W, Singh H, Nancy P, Fernanda Bellolio M, Gadepalli SK, Kamdar NS. Factors associated with potentially missed diagnosis of appendicitis in the Emergency Department. JAMA Netw Open. 2020;3:e200612–e12.

Maisonneuve P, Lowenfels AB, Müllhaupt B, Cavallini G, Lankisch PG, Andersen JR, Dimagno EP, Andrén-Sandberg A, Domellöf L, Frulloni L, Ammann RW. Cigarette smoking accelerates progression of alcoholic chronic pancreatitis. Gut. 2005;54:510–4.

Malik AI, Nelson RL, Tou S. Incision and drainage of perianal abscess with or without treatment of anal fistula. Cochrane Database Syst Rev. 2010:Cd006827.

Mapel DW, Schum M, Von Worley A. The epidemiology and treatment of anal fissures in a population-based cohort. BMC Gastroenterol. 2014;14:129.

Margetis N. Pathophysiology of internal hemorrhoids. Ann Gastroenterol. 2019;32:264–72.

Markar SR, Penna M, Harris A. Laparoscopic approach to appendectomy reduces the incidence of short- and long-term post-operative bowel obstruction: systematic review and pooled analysis. J Gastrointest Surg. 2014;18:1683–92.

Mark-Christensen A, Erichsen R, Brandsborg S, Pachler FR, Nørager CB, Johansen N, Pachler JH, Thorlacius-Ussing O, Kjaer MD, Qvist N, Preisler L, Hillingsø J, Rosenberg J, Laurberg S. Pouch failures following ileal pouch-anal anastomosis for ulcerative colitis. Color Dis. 2018;20:44–52.

May GR, Sutherland LR, Shaffer EA. Efficacy of bile acid therapy for gallstone dissolution: a meta-analysis of randomized trials. Aliment Pharmacol Ther. 1993;7:139–48.

McCloud JM, Jameson JS, Scott AND. Life-threatening sepsis following treatment for haemorrhoids: a systematic review. Color Dis. 2006;8:748–55.

McConnell EJ, Tessier DJ, Wolff BG. Population-based incidence of complicated diverticular disease of the sigmoid colon based on gender and age. Dis Colon Rectum. 2003;46:1110–4.

McHugh SM, Diamant NE. Effect of age, gender, and parity on anal canal pressures. Contribution of impaired anal sphincter function to fecal incontinence. Dig Dis Sci. 1987;32:726–36.

Mentula P, Sammalkorpi H, Leppäniemi A. Laparoscopic surgery or conservative treatment for appendiceal abscess in adults? A randomized controlled trial. Ann Surg. 2015;262:237–42.

Merriam LT, Kanaan SA, Dawes LG, Angelos P, Prystowsky JB, Rege RV, Joehl RJ. Gangrenous cholecystitis: analysis of risk factors and experience with laparoscopic cholecystectomy. Surgery. 1999;126:680–6.

Miller G, Boman J, Shrier I, Gordon PH. Etiology of small bowel obstruction. Am J Surg. 2000;180:33–6.

Mukai S, Itoi T, Baron TH, Takada T, Strasberg SM, Pitt HA, Ukai T, Shikata S, Teoh AYB, Kim M-H, Kiriyama S, Mori Y, Miura F, Chen M-F, Lau WY, Wada K, Supe AN, Giménez ME, Yoshida M, Mayumi T, Hirata K, Sumiyama Y, Inui K, Yamamoto M. Indications and techniques of biliary drainage for acute cholangitis in updated Tokyo Guidelines 2018. J Hepatobiliary Pancreat Sci. 2017;24:537–49.

Navaneethan U, Lourdusamy V, Njei B, Shen B. Endoscopic balloon dilation in the management of strictures in Crohn's disease: a systematic review and meta-analysis of non-randomized trials. Surg Endosc. 2016;30:5434–43.

Ng SC, Shi HY, Hamidi N, Underwood FE, Tang W, Benchimol EI, Panaccione R, Ghosh S, Wu JCY, Chan FKL, Sung JJY, Kaplan GG. Worldwide incidence and prevalence of inflammatory bowel disease in the 21st century: a systematic review of population-based studies. Lancet. 2018;390:2769–78.

Nieuwenhuizen J, van Ramshorst GH, ten Brinke JG, de Wit T, van der Harst E, Hop WC, Jeekel J, Lange JF. The use of mesh in acute hernia: frequency and outcome in 99 cases. Hernia. 2011;15:297–300.

Nikfarjam M, Niumsawatt V, Sethu A, Fink MA, Muralidharan V, Starkey G, Jones RM, Christophi C. Outcomes of contemporary management of gangrenous and non-gangrenous acute cholecystitis. HPB. 2011;13:551–8.

Nørgård BM, Nielsen J, Fonager K, Kjeldsen J, Jacobsen BA, Qvist N. The incidence of ulcerative colitis (1995-2011) and Crohn's disease (1995-2012) - based on nationwide Danish registry data. J Crohns Colitis. 2014;8:1274–80.

Oh C. Acute thrombosed external hemorrhoids. Mt Sinai J Med. 1989;56:30–2.

Park J-H, Park D-I, Kim H-J, Cho Y-K, Sohn C-I, Jeon W-K, Kim B-I. Metabolic syndrome is associated with erosive esophagitis. World J Gastroenterol. 2008;14:5442–7.

Parks AG. Pathogenesis and treatment of fistuila-in-ano. Br Med J. 1961;1:463–9.

Parks TG. Natural history of diverticular disease of the colon. Clin Gastroenterol. 1975;4:53–69.

Patel SD, Oxenham T, Praveen BV. Medium-term results of anal advancement flap compared with lateral sphincterotomy for the treatment of anal fissure. Int J Color Dis. 2011;26:1211–4.

Peery AF, Keku TO, Martin CF, Eluri S, Runge T, Galanko JA, Sandler RS. Distribution and characteristics of colonic diverticula in a united states screening population. Clin Gastroenterol Hepatol. 2016;14:980–85 e1.

Peery AF, Keku TO, Galanko JA, Sandler RS. Sex and race disparities in diverticulosis prevalence. Clin Gastroenterol Hepatol. 2020;18:1980–6.

Perera N, Liolitsa D, Iype S, Croxford A, Yassin M, Lang P, Ukaegbu O, van Issum C. Phlebotonics for haemorrhoids. Cochrane Database Syst Rev. 2012:Cd004322

Podda M, Gerardi C, Cillara N, Fearnhead N, Gomes CA, Birindelli A, Mulliri A, Davies RJ, Di Saverio S. Antibiotic treatment and appendectomy for uncomplicated acute appendicitis in adults and children: a systematic review and meta-analysis. Ann Surg. 2019;270:1028–40.

Pohl H, Welch HG. The role of overdiagnosis and reclassification in the marked increase of esophageal adenocarcinoma incidence. J Natl Cancer Inst. 2005;97:142–6.

Ponten JE, Somers KY, Nienhuijs SW. Pathogenesis of the epigastric hernia. Hernia. 2012;16:627–33.

Primatesta P, Goldacre MJ. Inguinal hernia repair: incidence of elective and emergency surgery, readmission and mortality. Int J Epidemiol. 1996;25:835–9.

Rabenstein T, Radespiel-Tröger M, Höpfner L, Benninger J, Farnbacher M, Greess H, Lenz M, Hahn EG, Schneider HT. Ten years experience with piezoelectric extracorporeal shockwave lithotripsy of gallbladder stones. Eur J Gastroenterol Hepatol. 2005;17:629–39.

Rai S, Chandra SS, Smile SR. A study of the risk of strangulation and obstruction in groin hernias. Aust NZ J Surg. 1998;68:650–4.

Ramanujam PS, Prasad ML, Abcarian H, Tan AB. Perianal abscesses and fistulas. A study of 1023 patients. Dis Colon Rectum. 1984;27:593–7.

Read DR, Abcarian H. A prospective survey of 474 patients with anorectal abscess. Dis Colon Rectum. 1979;22:566–8.

Regadas FS, Murad-Regadas SM, Lima DM, Silva FR, Barreto RG, Souza MH, Regadas Filho FS. Anal canal anatomy showed by three-dimensional anorectal ultrasonography. Surg Endosc. 2007;21:2207–11.

Reichert MC, Kupcinskas J, Krawczyk M, Jungst C, Casper M, Grunhage F, Appenrodt B, Zimmer V, Weber SN, Tamelis A, Lukosiene JI, Pauziene N, Kiudelis G, Jonaitis L, Schramm C, Goeser T, Schulz A, Malinowski M, Glanemann M, Kupcinskas L, Lammert F. A variant of COL3A1 (rs3134646) is associated with risk of developing diverticulosis in white men. Dis Colon Rectum. 2018;61:604–11.

Riss S, Weiser FA, Schwameis K, Mittlbock M, Stift A. Haemorrhoids, constipation and faecal incontinence: Is there any relationship? Color Dis. 2011;13:e227–33.

Romagnuolo J, Talluri J, Kennard E, Sandhu BS, Sherman S, Cote GA, Al-Kaade S, Gardner TB, Gelrud A, Lewis MD, Forsmark CE, Guda NM, Conwell DL, Banks PA, Muniraj T, Wisniewski SR, Ye T, Mel Wilcox C, Anderson MA, Brand RE, Slivka A, Whitcomb DC, Yadav D. Clinical profile, etiology, and treatment of chronic pancreatitis in North American women: analysis of a large multicenter cohort. Pancreas. 2016;45:934–40.

Rose J, Parina RP, Faiz O, Chang DC, Talamini MA. Long-term outcomes after initial presentation of diverticulitis. Ann Surg. 2015;262:1046–53.

Rosemar A, Angeras U, Rosengren A. Body mass index and diverticular disease: a 28-year follow-up study in men. Dis Colon Rectum. 2008;51:450–5.

Rubenstein JH, Taylor JB. Meta-analysis: the association of oesophageal adenocarcinoma with symptoms of gastro-oesophageal reflux. Aliment Pharmacol Ther. 2010;32:1222–7.

Rubin DT, Ananthakrishnan AN, Siegel CA, Sauer BG, Long MD. ACG clinical guideline: ulcerative colitis in adults. Am J Gastroenterol. 2019;114:384–413.

Russell JC, Walsh SJ, Reed-Fourquet L, Mattie A, Lynch J. Symptomatic cholelithiasis: a different disease in men? Connecticut laparoscopic cholesystectomy registry. Ann Surg. 1998;227:195–200.

Sahnan K, Adegbola SO, Tozer PJ, Watfah J, Phillips RKS. Perianal abscess. BMJ. 2017;356:j475.

Sainio P. Fistula-in-ano in a defined population. Incidence and epidemiological aspects. Ann Chir Gynaecol. 1984;73:219–24.

Sajid MS, Khawaja AH, Sains P, Singh KK, Baig MK. A systematic review comparing laparoscopic vs open adhesiolysis in patients with adhesional small bowel obstruction. Am J Surg. 2016;212:138–50.

Salem TA, Molloy RG, O'Dwyer PJ. Prospective, five-year follow-up study of patients with symptomatic uncomplicated diverticular disease. Dis Colon Rectum. 2007;50:1460–4.

Salminen P, Tuominen R, Paajanen H, Rautio T, Nordström P, Aarnio M, Rantanen T, Hurme S, Mecklin JP, Sand J, Virtanen J, Jartti A, Grönroos JM. Five-year follow-up of antibiotic therapy for uncomplicated acute appendicitis in the APPAC Randomized Clinical Trial. JAMA. 2018;320:1259–65.

Sands BE, Anderson FH, Bernstein CN, Chey WY, Feagan BG, Fedorak RN, Kamm MA, Korzenik JR, Lashner BA, Onken JE, Rachmilewitz D, Rutgeerts P, Wild G, Wolf DC, Marsters PA, Travers SB, Blank MA, van Deventer SJ. Infliximab maintenance therapy for fistulizing Crohn's disease. N Engl J Med. 2004;350:876–85.

Sankaran SJ, Xiao AY, Wu LM, Windsor JA, Forsmark CE, Petrov MS. Frequency of progression from acute to chronic pancreatitis and risk factors: a meta-analysis. Gastroenterology. 2015;149:1490–500.e1.

Saverio D, Salomone MP, De Simone B, Ceresoli M, Augustin G, Gori A, Boermeester M, Sartelli M, Coccolini F, Tarasconi A, de'Angelis N, Weber DG, Tolonen M, Birindelli A, Biffl W, Moore EE, Kelly M, Soreide K, Kashuk J, Ten Broek R, Gomes CA, Sugrue M, Davies RJ, Damaskos D, Leppäniemi A, Kirkpatrick A, Peitzman AB, Fraga GP, Maier RV, Coimbra R, Chiarugi M, Sganga G, Pisanu A, de'Angelis GL, Tan E, Van Goor H, Pata F, Di Carlo I, Chiara O, Litvin A, Campanile FC, Sakakushev B, Tomadze G, Demetrashvili Z, Latifi R, Abu-Zidan F, Romeo O, Segovia-Lohse H, Baiocchi G, Costa D, Rizoli S, Balogh ZJ, Bendinelli C, Scalea T, Ivatury R, Velmahos G, Andersson R, Kluger Y, Ansaloni L, Catena F. Diagnosis and treatment of acute appendicitis: 2020 update of the WSES Jerusalem guidelines. World J Emerg Surg. 2020;15:27.

Schauer PR, Ramos R, Ghiatas AA, Sirinek KR. Virulent diverticular disease in young obese men. Am J Surg. 1992;164:443–6. discussion 46–8

Schieffer KM, Sabey K, Wright JR, Toole DR, Drucker R, Tokarev V, Harris LR, Deiling S, Eshelman MA, Hegarty JP, Yochum GS, Koltun WA, Lamendella R, Stewart DB Sr. The microbial ecosystem distinguishes chronically diseased tissue from adjacent tissue in the sigmoid colon of chronic, recurrent diverticulitis patients. Sci Rep. 2017;7:8467.

Schouten WR, Briel JW, Auwerda JJ. Relationship between anal pressure and anodermal blood flow. The vascular pathogenesis of anal fissures. Dis Colon Rectum. 1994;37:664–9.

Setti Carraro P, Talbot IC, Nicholls RJ. Longterm appraisal of the histological appearances of the ileal reservoir mucosa after restorative proctocolectomy for ulcerative colitis. Gut. 1994;35:1721–7.

Shadhu K, Ramlagun D, Ping X. Para-duodenal hernia: a report of five cases and review of literature. BMC Surg. 2018;18:32.

Shahedi K, Fuller G, Bolus R, Cohen E, Michelle V, Shah R, Agarwal N, Kaneshiro M, Atia M, Sheen V, Kurzbard N, van Oijen MGH, Yen L, Hodgkins P, Erder MH, Spiegel B. Long-term risk of acute diverticulitis among patients with incidental diverticulosis found during colonoscopy. Clin Gastroenterol Hepatol. 2013;11:1609–13.

Shea JA, Berlin JA, Escarce JJ, Clarke JR, Kinosian BP, Cabana MD, Tsai WW, Horangic N, Malet PF, Schwartz JS, et al. Revised estimates of diagnostic test sensitivity and specificity in suspected biliary tract disease. Arch Intern Med. 1994;154:2573–81.

Shivashankar R, Tremaine WJ, Harmsen WS, Loftus EV Jr. Incidence and prevalence of Crohn's disease and ulcerative colitis in Olmsted County, Minnesota from 1970 through 2010. Clin Gastroenterol Hepatol. 2017;15:857–63.

Simillis C, Symeonides P, Shorthouse AJ, Tekkis PP. A meta-analysis comparing conservative treatment versus acute appendectomy for complicated appendicitis (abscess or phlegmon). Surgery. 2010;147:818–29.

Simillis C, Thoukididou SN, Slesser AA, Rasheed S, Tan E, Tekkis PP. Systematic review and network meta-analysis comparing clinical outcomes and effectiveness of surgical treatments for haemorrhoids. Br J Surg. 2015;102:1603–18.

Skoglar A, Gunnarsson U, Falk P. Band adhesions not related to previous abdominal surgery – a retrospective cohort analysis of risk factors. Ann Med Surg. 2018;36:185–90.

Sozen S, Emir S, Yazar FM, Altinsoy HK, Topuz O, Vurdem UE, Cetinkunar S, Ozkan Z, Guzel K. Small bowel obstruction due to anomalous congenital peritoneal bands - case series in adults. Bratislavske Lekarske Listy. 2012;113:186–9.

Stefanidis D, Sirinek KR, Bingener J. Gallbladder perforation: risk factors and outcome. J Surg Res. 2006;131:204–8.

Stein HJ, Siewert JR. Barrett's esophagus: pathogenesis, epidemiology, functional abnormalities, malignant degeneration, and surgical management. Dysphagia. 1993;8:276–88.

Stein GY, Rath-Wolfson L, Zeidman A, Atar E, Marcus O, Joubran S, Ram E. Sex differences in the epidemiology, seasonal variation, and trends in the management of patients with acute appendicitis. Langenbeck's Arch Surg. 2012;397:1087–92.

Strate LL, Liu YL, Aldoori WH, Syngal S, Giovannucci EL. Obesity increases the risks of diverticulitis and diverticular bleeding. Gastroenterology, 22.e1. 2009;136:115.

Sultan AH, Kamm MA, Nicholls RJ, Bartram CI. Prospective study of the extent of internal anal sphincter division during lateral sphincterotomy. Dis Colon Rectum. 1994;37:1031–3.

Sun WM, Read NW, Shorthouse AJ. Hypertensive anal cushions as a cause of the high anal canal pressures in patients with haemorrhoids. Br J Surg. 1990;77:458–62.

Swank HA, Eshuis EJ, van Berge Henegouwen MI, Bemelman WA. Short- and long-term results of open versus laparoscopic appendectomy. World J Surg. 2011;35:1221–8.

Takatsuki S. An etiology of anal fistula in infants. Keio J Med. 1976;25:1–4.

Tazuma S, Unno M, Igarashi Y, Inui K, Uchiyama K, Kai M, Tsuyuguchi T, Maguchi H, Mori T, Yamaguchi K, Ryozawa S, Nimura Y, Fujita N, Kubota K, Shoda J, Tabata M, Mine T, Sugano K, Watanabe M, Shimosegawa T. Evidence-based clinical practice guidelines for cholelithiasis 2016. J Gastroenterol. 2017;52:276–300.

Tekkis PP, Lovegrove RE, Tilney HS, Smith JJ, Sagar PM, Shorthouse AJ, Mortensen NJ, Nicholls RJ. Long-term failure and function after restorative proctocolectomy - a multi-centre study of patients from the UK National Ileal Pouch Registry. Color Dis. 2010;12:433–41.

Ten Broek RPG, Krielen P, Di Saverio S, Coccolini F, Biffl WL, Ansaloni L, Velmahos GC, Sartelli M, Fraga GP, Kelly MD, Moore FA, Peitzman AB, Leppaniemi A, Moore EE, Jeekel J, Kluger Y, Sugrue M, Balogh ZJ, Bendinelli C, Civil I, Coimbra R, De Moya M, Ferrada P, Inaba K, Ivatury R, Latifi R, Kashuk JL, Kirkpatrick AW, Maier R, Rizoli S, Sakakushev B, Scalea T, Søreide K, Weber D, Wani I, Abu-Zidan FM, De'Angelis N, Piscioneri F, Galante JM, Catena F, van Goor H. Bologna guidelines for diagnosis and management of adhesive small bowel obstruction (ASBO): 2017 update of the evidence-based guidelines from the world society of emergency surgery ASBO working group. World J Emerg Surg. 2018;13:24.

Ternant D, Aubourg A, Magdelaine-Beuzelin C, Degenne D, Watier H, Picon L, Paintaud G. Infliximab pharmacokinetics in inflammatory bowel disease patients. Ther Drug Monit. 2008;30:523–9.

Thesbjerg SE, Harboe KM, Bardram L, Rosenberg J. Sex differences in laparoscopic cholecystectomy. Surg Endosc. 2010;24:3068–72.

Thomson WH. The nature of haemorrhoids. Br J Surg. 1975;62:542–52.

Tiainen J, Matikainen M. Long-term clinical outcome and anemia after restorative proctocolectomy for ulcerative colitis. Scand J Gastroenterol. 2000;35:1170–3.

Tjandra JJ, Tan JJ, Lim JF, Murray-Green C, Kennedy ML, Lubowski DZ. Rectogesic (glyceryl trinitrate 0.2%) ointment relieves symptoms of haemorrhoids associated with high resting anal canal pressures. Color Dis. 2007;9:457–63.

Tong JWV, Lingam P, Shelat VG. Adhesive small bowel obstruction – an update. Acute Med Surg. 2020;7:e587.

Tursi A, Elisei W. Role of inflammation in the pathogenesis of diverticular disease. Mediat Inflamm. 2019;2019:8328490.

Vakil N, van Zanten SV, Kahrilas P, Dent J, Jones R. The Montreal definition and classification of gastroesophageal reflux disease: a global evidence-based consensus. Am J Gastroenterol. 2006;101:1900–20. quiz 43

van Brunschot S, van Grinsven J, van Santvoort HC, Bakker OJ, Besselink MG, Boermeester MA, Bollen TL, Bosscha K, Bouwense SA, Bruno MJ, Cappendijk VC, Consten EC, Dejong CH, van Eijck CH, Erkelens WG, van Goor H, van Grevenstein WMU, Haveman JW, Hofker SH, Jansen JM, Laméris JS, van Lienden KP, Meijssen MA, Mulder CJ, Nieuwenhuijs VB, Poley JW, Quispel R, de Ridder RJ, Römkens TE, Scheepers JJ, Schepers NJ, Schwartz MP, Seerden T, Spanier BWM, Straathof JWA, Strijker M, Timmer R, Venneman NG, Vleggaar FP, Voermans RP, Witteman BJ, Gooszen HG, Dijkgraaf MG, Fockens P. Endoscopic or surgical step-up approach for infected necrotising pancreatitis: a multicentre randomised trial. Lancet. 2018;391:51–8.

van Santvoort HC, Besselink MG, Bakker OJ, Hofker HS, Boermeester MA, Dejong CH, van Goor H, Schaapherder AF, van Eijck CH, Bollen TL, van Ramshorst B, Nieuwenhuijs VB, Timmer R, Laméris JS, Kruyt PM, Manusama ER, van der Harst E, van der Schelling GP,

Karsten T, Hesselink EJ, van Laarhoven CJ, Rosman C, Bosscha K, de Wit RJ, Houdijk AP, van Leeuwen MS, Buskens E, Gooszen HG. A step-up approach or open necrosectomy for necrotizing pancreatitis. N Engl J Med. 2010;362:1491–502.

Vavricka SR, Brun L, Ballabeni P, Pittet V, Prinz Vavricka BM, Zeitz J, Rogler G, Schoepfer AM. Frequency and risk factors for extraintestinal manifestations in the Swiss inflammatory bowel disease cohort. Am J Gastroenterol. 2011;106:110–9.

Viniol A, Keunecke C, Biroga T, Stadje R, Dornieden K, Bösner S, Donner-Banzhoff N, Haasenritter J, Becker A. Studies of the symptom abdominal pain--a systematic review and meta-analysis. Fam Pract. 2014;31:517–29.

Vogel JD, Johnson EK, Morris AM, Paquette IM, Saclarides TJ, Feingold DL, Steele SR. Clinical practice guideline for the management of anorectal abscess, fistula-in-ano, and rectovaginal fistula. Dis Colon Rectum. 2016;59:1117–33.

Volk N, Lacy B. Anatomy and physiology of the small bowel. Gastrointest Endosc Clin. 2017;27:1–13.

Waage A, Nilsson M. Iatrogenic bile duct injury: a population-based study of 152 776 cholecystectomies in the Swedish Inpatient Registry. Arch Surg. 2006;141:1207–13.

Wang D, Yang G, Qiu J, Song Y, Wang L, Gao J, Wang C. Risk factors for anal fistula: a case-control study. Tech Coloproctol. 2014;18:635–9.

Wang FW, Chuang HY, Tu MS, King TM, Wang JH, Hsu CW, Hsu PI, Chen WC. Prevalence and risk factors of asymptomatic colorectal diverticulosis in Taiwan. BMC Gastroenterol. 2015;15:40.

Wheat CL, Strate LL. Trends in hospitalization for diverticulitis and diverticular bleeding in the United States from 2000 to 2010. Clin Gastroenterol Hepatol. 2016;14:96–103 e1.

Whitcomb DC. Clinical practice. Acute pancreatitis. N Engl J Med. 2006;354:2142–50.

Working Group IAP/APA Acute Pancreatitis Guidelines. IAP/APA evidence-based guidelines for the management of acute pancreatitis. Pancreatology. 2013;13:e1–15.

Wu XR, Ashburn J, Remzi FH, Li Y, Fass H, Shen B. Male gender is associated with a high risk for chronic antibiotic-refractory pouchitis and ileal pouch anastomotic sinus. J Gastrointest Surg. 2016;20:631–9.

Wysocki A, Budzyński P, Kulawik J, Drożdż W. Changes in the Localization of perforated peptic ulcer and its relation to gender and age of the patients throughout the last 45 years. World J Surg. 2011;35:811–6.

Xia W, Barazanchi AWH, Coomarasamy C, Jin J, Maccormick AD, Sammour T, Hill AG. Epidemiology of haemorrhoids and publicly funded excisional haemorrhoidectomies in New Zealand (2007–2016): a population-based cross-sectional study. Colorectal Dis. 2021;23(1):265–73.

Yang AL, Vadhavkar S, Singh G, Omary MB. Epidemiology of alcohol-related liver and pancreatic disease in the United States. Arch Intern Med. 2008;168:649–56.

Yen S, Hsieh CC, MacMahon B. Consumption of alcohol and tobacco and other risk factors for pancreatitis. Am J Epidemiol. 1982;116:407–14.

Young YR, Chiu TF, Chen JC, Tung MS, Chang MW, Chen JH, Sheu BF. Acute appendicitis in the octogenarians and beyond: a comparison with younger geriatric patients. Am J Med Sci. 2007;334:255–9.

Yousef F, Cardwell C, Cantwell MM, Galway K, Johnston BT, Murray L. The incidence of esophageal cancer and high-grade dysplasia in Barrett's esophagus: a systematic review and meta-analysis. Am J Epidemiol. 2008;168:237–49.

Zha Y, Chen XR, Luo D, Jin Y. The prevention of major bile duct injures in laparoscopic cholecystectomy: the experience with 13,000 patients in a single center. Surg Laparosc Endosc Percutan Tech. 2010;20:378–83.

Zhang H, Liao M, Chen J, Zhu D, Byanju S. Ultrasound, computed tomography or magnetic resonance imaging - which is preferred for acute appendicitis in children? A meta-analysis. Pediatr Radiol. 2017;47:186–96.

Zisman A, Gold-Deutch R, Zisman E, Negri M, Halpern Z, Lin G, Halevy A. Is male gender a risk factor for conversion of laparoscopic into open cholecystectomy? Surg Endosc. 1996;10:892–4.

Chapter 15
Malignant Gastrointestinal Conditions

Michael P. Flood, Nathan Lawrentschuk, and Alexander G. Heriot

15.1 Oesophageal Cancer

Oesophageal cancer is predominantly made up of two histological subtypes, adenocarcinoma and squamous cell carcinoma. Adenocarcinomas tend to arise in the distal oesophagus, whereas squamous cell carcinomas can occur throughout. In recent decades, the incidence of adenocarcinoma has risen and is now more prevalent compared to squamous cell carcinoma in western countries (Kirby and Rice 1994). In a worldwide context, squamous cell carcinoma remains the predominant histology, particularly so in the Orient (Bray et al. 2018).

Oesophageal cancer has a male preponderance, with a male to female ratio of 3–4:1 and an occurrence most commonly in the sixth and seventh decades of life. The American Cancer Society estimates that over 18,000 new cases of oesophageal cancer will be diagnosed in the United States in 2020, with almost 14,000 of these in the male population (American Cancer Society 2020). Modifiable risk factors include smoking, alcohol, obesity and gastro-oesophageal reflux disease (GORD). GORD is the most common predisposing factor for adenocarcinoma of the oesophagus. The sequelae of chronic mucosal irritation by acid reflux includes the sequence of Barrett's intestinal metaplasia, low-grade dysplasia, high-grade dysplasia and ultimately adenocarcinoma. The risk of adenocarcinoma in patients with Barrett's metaplasia has been estimated to be 30–60 times that of the general population (Solaymani-Dodaran et al. 2004). An elderly male with a short history of progressive (solids initially, then liquids) dysphagia is the most common clinical presentation. Unintentional weight loss, retrosternal discomfort and anorexia are symptoms that also heavily feature.

M. P. Flood · N. Lawrentschuk · A. G. Heriot (✉)
Department of Surgical Oncology, Peter MacCallum Cancer Centre,
Melbourne, VIC, Australia
e-mail: Alexander.Heriot@petermac.org

© The Author(s), under exclusive license to Springer Nature Switzerland AG 2022
S. S. Goonewardene et al. (eds.), *Men's Health and Wellbeing*,
https://doi.org/10.1007/978-3-030-84752-4_15

Oesophagoscopy is the investigation of choice, allowing direct visualisation and biopsy of oesophageal mucosa. Endoscopic ultrasound (EUS), computed tomography (CT) and positron emission tomography (PET) provide additional staging information. Bronchoscopy and laparoscopy play a role in assessing more advanced disease.

Despite optimally treated early stage disease the survival rates remain poor, with a 5-year overall survival (OS) of only 47.1% (National Cancer Instutute 2020). Neoadjuvant chemoradiotherapy followed by surgery has become the standard of care since the inception of the CROSS regimen in 2015 (Shapiro et al. 2015). Definitive chemoradiotherapy is indicated for squamous cell carcinoma of the upper third of the oesophagus, given the morbidity of surgery and non-inferiority in survival (Bedenne et al. 2007). In the palliative setting, chemotherapy or radiotherapy, stenting and laser are used to restore swallowing.

Resectable patients are offered either a transhiatal or transthoracic oesophagectomy, depending on tumour location and surgeon preference. Transhiataloesophagectomy involves a midline laparotomy and left sided neck incision, with resection of the oesophagus in between. The mobilised stomach is anastomosed with the cervical oesophagus at the neck. Transthoracic oesophagectomy is done either as a two-stage (Ivor-Lewis) or three-stage (McKeown) operation. The key difference to the transthoracic approach being access to the right chest, which is thought to provide a superior oncological operation given the direct visualisation of the mid oesophagus and its lymphatic drainage. Minimally invasive oesophagectomy has recently come to the fore, incorporating laparoscopy, thoracoscopy and robotics into the aforementioned procedures, resulting in improved perioperative morbidity.

15.2 Gastric Cancer

Once the second most common cancer worldwide, gastric cancer has fallen to sixth in males and ninth in females (Bray et al. 2018). Akin to oesophageal cancer, sex plays a role, with a 2:1 male to female ratio. There is a steady increase in risk with age, peaking in the eighth decade (National Cancer Instutute 2020). Tremendous geographic variation exists in incidence, with the highest rates found in Japan (Bray et al. 2018). This variation and recent dramatic falls in incidence across the developed world, suggest that environmental factors play a key role in gastric carcinogenesis.

Adenocarcinoma accounts for over 90% of cases, with Helicobacter pylori (H. pylori) infection of the stomach the most important risk factor. The high prevalence of H. pylori infection in the developing world has been attributed to the increased incidence of gastric cancer in these countries. Other risk factors include diets high in salt and processed meat, smoking, alcohol consumption, obesity and the autoimmune disorder, pernicious anaemia (Fenoglio-Preiser et al. 1996; Scheiman and Cutler 1999).

Gastric cancer lacks specific symptomatology early in its course. Vague complaints of dyspepsia and epigastric discomfort are often mistaken for GORD and are ignored by patients. This can lead to quite advanced disease at presentation. Important characteristics that should raise clinical suspicion include unremitting pain that is not relieved by food or antacid medication. More advanced disease can additionally present with weight loss, early satiety, anorexia and vomiting.

Gastroscopy is the diagnostic modality of choice. Size, site, orientation and morphology of the tumour can be noted and biopsied. In the case of malignant-appearing ulceration, several ulcer-edge biopsies are taken to avoid non-diagnostic ulcerative central tissue. Upper endoscopy is complimented by CT and EUS to complete tumour staging. EUS can assess the extent of gastric wall invasion and nodal status. It is particularly useful in the sampling of equivocal lymph nodes. Laparoscopy and peritoneal fluid cytology sampling is routinely used in patients with advanced locoregional disease that appear resectable on imaging. About one third of these patients will have peritoneal metastases (Muntean et al. 2009). PET is not routinely used in most centres.

Early-stage or local gastric cancer only accounts for about 20% of all cases diagnosed (Hundahl et al. 2000). With such a high proportion of patients presenting with locally advanced or metastatic disease, 5-year OS rates are at a dismal 31% (National Cancer Institute 2020). Perioperative chemotherapy has become mainstay inoperable gastric cancer. Initially this was implemented using the ECF regimen (epirubicin, cisplatin and 5-fluorouracil) from the 2006 MAGIC trial (Cunningham et al. 2006) and was recently replaced in most institutions with FLOT (5-fluorouracil, leucovorin, oxaliplatin and docetaxel) following the 2019 FLOT4 trial (Al-Batran et al. 2019).

Surgical management of gastric cancer is largely dependent on tumour location. Proximal gastric cancers are generally managed by a total gastrectomy and roux en Y reconstruction. Subtotal gastrectomy is the operation of choice for tumours of the distal stomach, which account for about a third of cases. Given the extensive lymphatic network around the stomach and proclivity for these tumours to extend microscopically, a 5 cm resection margin is employed.

Palliation for irresectable or widespread metastatic disease can involve radiotherapy or endoscopic management for bleeding, stenting or gastrojejunal bypass for obstruction and enteral feeding for malnutrition.

15.3   Small Bowel Cancer & Neuroendocrine Neoplasms

Small bowel tumours are rare entities, accounting for less than 5% of all gastrointestinal malignancies. They are histologically diverse, with adenocarcinoma, lymphoma, gastrointestinal stromal tumour (GIST) and neuroendocrine tumour accounting for the majority of subtypes. A quarter of these patients will have associated cancers in the colon, prostate, breast and other sites (Neugut et al. 1998). Hereditary conditions such as familial adenomatous polyposis (FAP), Gardner's syndrome, coeliac disease and Peutz-Jeghers syndrome account for these associations. As is the trend with other gastrointestinal cancers, diet, smoking and alcohol are risk factors.

Clinical presentation of small bowel malignancy is highly variable, with abdominal pain, weight loss, bowel obstruction/intussusception, and gastrointestinal bleeding making up the majority of symptoms and signs.

Given the diagnostic dilemma that these tumours present, a number of imaging and endoscopy modalities exist. From a radiological perspective, the traditional use of fluoroscopy with small bowel follow through has largely been replaced by magnetic resonance imaging (MRI) small bowel and CT enteroclysis. Push enteroscopy, double balloon enteroscopy and capsule endoscopy also provide an array of diagnostic approaches.

Primary surgical resection is mainstay in the majority of cases. Adjuvant chemotherapy for resectable adenocarcinoma is controversial, but its use has become more frequent in recent times (Singhal and Singhal 2007; Ecker et al. 2016). Tyrosine kinase inhibitors are routinely used in the adjuvant setting for selected resectable GISTs. Primary chemotherapy is generally mainstay for intestinal lymphoma, though surgical resection may often be required to obtain a histological diagnosis or to manage local complications such as perforation.

Gastrointestinal neuroendocrine neoplasms (NEN) can arise anywhere along the gastrointestinal tract and account for about 0.5% of all newly diagnosed malignancies (Taal and Visser 2004). Nomenclature has changed in recent years, to primarily divide these tumours according to their proliferation, making the traditional term, "carcinoid", obsolete (Nagtegaal et al. 2020). Neuroendocrine tumours have low proliferation, while neuroendocrine carcinomas have high proliferation. The most common sites of these neoplasms of the GI tract are the small bowel, rectum, colon and appendix (Tsikitis et al. 2012; Halperin et al. 2017). NEN follow a more benign natural history than most other malignancies, with 5-year survival at 94%. (National Cancer Institute 2020).

Diagnosis is usually incidental or late in the clinical course of the disease, when the lesion manifests as a mechanical complication (bleeding or obstruction) or as a result of significant catecholamine production (carcinoid syndrome), which occurs in about 20% of cases (Halperin et al. 2017). Carcinoid syndrome is typified by diarrhoea, flushing, tachycardia, colicky abdominal pain and wheezing, and is most common in the setting of disseminated disease, in particular liver metastases (Modlin et al. 2005).

Biochemical markers such as chromogranin A and 5-hydroxyindoleacetic acid are useful during investigation and surveillance of disease. Radionucleotide scans such as somatostatin receptor scintigraphy (OctreoScan) and PET that utilises somatostatin receptor tracers such as 68-Ga DOTATATE (GaTate) can aid in tumour localisation. Surgical resection is the standard therapeutic modality, with chemotherapeutic regimens employed in the metastatic setting.

15.4 Colorectal and Appendiceal Cancer

Colorectal cancer is the third most common cancer and the second leading cause of cancer-related mortality worldwide (Bray et al. 2018). An estimated 1.8 million cases occurred globally in 2018, with the highest rates found in the Western world-USA, Canada, Japan, parts of Europe, New Zealand and Australia (World Health Organization 2020). The incidence of colorectal malignancy is marginally higher in males than in females, with a male to female ratio of 1.3:1 Incidence starts to increase after the age of 35 and rises rapidly after age 50, with 88% of cases occurring during that period (Siegel et al. 2020).

Adenocarcinoma accounts for over 90% of lesions, with neuroendocrine neoplasm, lymphoma and sarcoma fulfilling the remainder of histological types. In terms of anatomical site, 70% originate in the colon and 30% in the rectum, with a synchronous lesion encountered in 4–5% of cases (Siegel et al. 2020). Risk factors other than increasing age include obesity, a diet high in fat and low in fibre, smoking, a personal or family history of colorectal cancer, hereditary conditions such as Lynch syndrome and FAP, and inflammatory bowel disease.

Alike the majority of cancer, colorectal cancer has traditionally been diagnosed after the onset of symptoms. With the advent of screening colonoscopy and national faecal immunochemical testing programs, asymptomatic early-stage malignancy is being diagnose more frequently, translating to a decrease in cancer-related mortality rates (Yang et al. 2014). Typical signs and symptoms associated with colorectal cancer include change in bowel habit (74%), rectal bleeding (71%), tenesmus (24.5%), abdominal mass (12.5%) and abdominal pain (3.8%) (Thompson et al. 2017).

Colonoscopy accompanied by digital rectal examination is the most accurate and versatile test for colorectal cancer, not only localising and biopsying lesions, but detecting synchronous malignancy and removing polyps. CT colonography is utilised in instances of incomplete colonoscopy (primarily due to a tortuous colon or poor bowel preparation), providing a similarly sensitive and less invasive alternative (Atkin et al. 2013). Once a diagnosis has been established, staging is completed by CT thorax abdomen pelvis and MRI pelvis, in the case of rectal cancer.

The approximate 5-year survival rate for all stages of colorectal cancer in the United States is 64.4%. More specifically, 5-year survival rates range from 14% for disseminated disease to 90% for localised disease (SEER 2020a, b, c). Colon and rectal cancer have very different treatment algorithms. Tumours originating from the caecum to the sigmoid colon are generally treated with a surgery first approach followed by adjuvant chemotherapy for high-risk disease. Clinical trials such as FOXTROT (Morton 2019) and NICHE (Chalabi et al. 2020) have begun to raise the question regarding potential benefits of neoadjuvant chemo- and immunotherapy for colon cancers. Locally advanced rectal cancer is treated with neoadjuvant chemoradiotherapy followed by surgery and adjuvant chemotherapy. Surgical options include anterior resection, abdominoperineal resection or pelvic exenteration, depending on tumour location, involvement of local structures and response to therapy. Transanal minimally invasive surgery (TAMIS) is a recently devised technique used to locally excise early rectal cancers, saving the morbidity of formal resection. Low patient compliance with adjuvant chemotherapy in rectal cancer has led to the investigation of total neoadjuvant therapy (TNT). Promising results from the RAPIDO trial will look to change clinical practice in this regard (Hospers et al. 2020).

Appendix malignancies account for only 3% of large bowel neoplasms (Siegel et al. 2020), with an incidence of 0.97 per 100,000 population (Marmor et al. 2015). They are genetically distinct from gastrointestinal cancers and comprise several histological types, that can be broadly characterised into epithelial neoplasms (e.g. adenocarcinoma, low-grade and high-grade mucinous neoplasm) and neuroendocrine neoplasms. Patients usually present with acute appendicitis or as an incidental finding on CT. Right hemicolectomy with/without adjuvant chemotherapy (in the presence of node positive disease) is the standard treatment for adenocarcinoma.

Pseudomyxoma peritonei (PMP) is a rare clinical condition manifesting from a ruptured mucin-filled tumour. It is characterised by diffuse collections of mucinous peritoneal deposits and gelatinous ascites. The primary tumour is most commonly derived from the appendix, though colorectal, ovarian and gastric origins are possible (Mittal et al. 2017). Treatment has historically been with debulking surgery. As abdominopelvic mucin accumulates, obstructive-type symptoms such as nausea, vomiting, abdominal bloating and pain, and eventually obstipation occur. In the last 30 years, cytoreductive surgery (CRS) and hyperthermic intraperitoneal chemotherapy (HIPEC) has become the standard of care for PMP (Sugarbaker 1995).

15.5 Anal Cancer

Although anal cancer accounts for only 4% of large bowel malignancies, the incidence is increasing throughout the world (Islami et al. 2017). Squamous cell carcinoma (SCC) accounts for 90% of cases, with adenocarcinoma, melanoma and neuroendocrine neoplasms making up the remainder. Current annual incidence of anal SCC is 1.6 per 100,000, with a female preponderance (Deshmukh et al. 2020). Anal SCC develops at the anal squamocolumnarjunction and arises from anal intraepithelial neoplasia (AIN), a precancerous lesion. Human immunodeficiencyvirus (HIV), anoreceptive intercourse and human papillomavirus (HPV), particularly strain 16, and smoking are associated risk factors.

Rectal bleeding is the most common symptom, occurring in approximately 45% of patients. Pain and pruritus are also common. Notably, 20% of patients will be asymptomatic at diagnosis (Schneider and Schulte 1981; Singh et al. 1981). Patients with anal cancer often delay seeking medical attention. To compound matters, anal cancer is often misdiagnosed, or the diagnosis is delayed significantly, during which time a patient may receive treatment for a perceived benign disorder.

Lower GI endoscopy with biopsy is suitable for lesions proximal to the dentate line. However, given the tumour's proclivity to manifest below the dentate line and in the perianal skin, an examination under anaesthesia is often required to adequately assess and biopsy. Also depending on tumour position, lymph node metastases may spread to the inguinal and external iliac nodes if distal to the dentate line or to the mesorectal and internal iliac nodes if proximal. This is important during clinical examination and staging evaluation. MRI of the pelvis/transrectal ultrasound is imperative to accurate local staging. CT and PET can assess for regional and distant disease.

Anal SCC is a radiosensitive tumour. Five-year survival for all stages is 69.7%, improving to 81.9% in those presenting with localised disease (SEER 2020, Anal Cancer). For the majority of patients, treatment will consist of definitive radiation therapy to the tumour and inguinal lymph nodes (often performed prophylactically), combined with 5-FU and mitomycin/cisplatin chemotherapies. This treatment methodology has remained largely unchanged since its advent as the Nigro protocol in 1974 (Nigro et al. 1974).

Surgery plays a limited role in the treatment of anal cancer. A temporary defunctioning stoma can be performed for those with poor sphincter function or at risk of radiotherapy complications e.g. rectovaginal fistula. Surgery, in the form of abdominoperineal resection, is reserved for patients who have residual or recurrent disease after chemoradiation. Local excision is an option for small, early and well differentiated lesions.

15.6 Metastatic Gastrointestinal Cancer

Cancers of the gastrointestinal tract metastasise most commonly to the liver, peritoneum and lung. Systemic treatment such as chemotherapy and, more recently, immunotherapy have been shown to prolonging patient survival. However, in recent years surgery has begun to play a major role in select patients presenting with stage four disease in a synchronous (the colorectal primary is in situ) or metachronous (the primary has been resected at an earlier stage) fashion. This change in strategy is particularly evident in colorectal metastases to the liver and the peritoneum.

One in four patients will experience liver metastases from colorectal cancer at some point during their cancer journey (Engstrand et al. 2018). Although two-thirds of affected patients will have extrahepatic disease, some have disease that is isolated to the liver and can benefit from surgical management as an alternative to or as a combination with systemic chemotherapy. Five-year survival is up to 46% in patients with completely resected disease of the liver (Dupre et al. 2018). Staging PET, to out rule any distant metastases not visible on CT, is accompanied by a liver MRI, which fully characterises the extent of liver disease. The management of patients with colorectal liver metastases is a complex process, where cases are routinely discussed at specialist multidisciplinary team meetings. There is currently no standard approach to patients presenting with potentially resectable disease. The integration of systemic chemotherapy with surgery varies between institutions. For patients that present with synchronous hepatic metastases, a simultaneous or staged resection of the liver and colorectal primary is possible. A simultaneous resection is obviously preferable from a patient perspective but a staged resection is often required if the primary is symptomatic or there is extensive disease of the liver. For patients with borderline or unresectable liver metastases, upfront systemic chemotherapy is an appropriate option. However, down staging from unresectable to resectable disease occurs in only up to 47% of patients when employing modern systemic chemotherapy with novel drug-administration techniques (D'Angelica et al. 2015).

Colorectal peritoneal metastases (CRPM) present in 8% of patients, in either a synchronous or a metachronous fashion (Segelman et al. 2012) and confer the worst OS amongst all sites of colorectal metastases. After rigorous diagnostic workup, that includes CT, PET and diagnostic laparoscopy, select patients with low volume, isolated peritoneal disease are offered CRS and HIPEC in specialised centres. Similar to colorectal liver metastases, there is no consensus on the use of perioperative chemotherapy, though patients generally receive some form of systemic chemotherapy in a neoadjuvant or adjuvant fashion. Routine use of CRS and HIPEC as treatment option for CRPM remains controversial as evidence to conclude efficacy is lacking. Despite this, many centres have published promising results, with 5-year survival rates of up to 52% (Baratti et al. 2018).

References

Al-Batran SE, Homann N, Pauligk C, Goetze TO, Meiler J, Kasper S, Kopp HG, Mayer F, Haag GM, Luley K, Lindig U, Schmiegel W, Pohl M, Stoehlmacher J, Folprecht G, Probst S, Prasnikar N, Fischbach W, Mahlberg R, Trojan J, Koenigsmann M, Martens UM, Thuss-Patience P, Egger M, Block A, Heinemann V, Illerhaus G, Moehler M, Schenk M, Kullmann F, Behringer DM, Heike M, Pink D, Teschendorf C, Lohr C, Bernhard H, Schuch G, Rethwisch V, von Weikersthal LF, Hartmann JT, Kneba M, Daum S, Schulmann K, Weniger J, Belle S, Gaiser T, Oduncu FS, Guntner M, Hozaeel W, Reichart A, Jager E, Kraus T, Monig S, Bechstein WO, Schuler M, Schmalenberg H, Hofheinz RD, Flot Aio Investigators. Perioperative chemotherapy with fluorouracil plus leucovorin, oxaliplatin, and docetaxel versus fluorouracil or capecitabine plus cisplatin and epirubicin for locally advanced, resectable gastric or gastro-oesophageal junction adenocarcinoma (FLOT4): a randomised, phase 2/3 trial. Lancet. 2019;393:1948–57.
American Cancer Society. Cancer facts & figures 2020. 2020. Available at https://www.cancer.org/content/dam/cancer-org/research/cancer-facts-and-statistics/annual-cancer-facts-and-figures/2020/cancer-facts-and-figures-2020.pdf. Accessed 12 Nov 2020.
Atkin W, Dadswell E, Wooldrage K, Kralj-Hans I, von Wagner C, Edwards R, Yao G, Kay C, Burling D, Faiz O, Teare J, Lilford RJ, Morton D, Wardle J, Halligan S, Siggar Investigators. Computed tomographic colonography versus colonoscopy for investigation of patients with symptoms suggestive of colorectal cancer (SIGGAR): a multicentre randomised trial. Lancet. 2013;381:1194–202.
Baratti D, Kusamura S, Iusco D, Cotsoglou C, Guaglio M, Battaglia L, Virzi S, Mazzaferro V, Deraco M. Should a history of extraperitoneal disease be a contraindication to cytoreductive surgery and hyperthermic intraperitoneal chemotherapy for colorectal cancer peritoneal metastases? Dis Colon Rectum. 2018;61:1026–34.

Bedenne L, Michel P, Bouche O, Milan C, Mariette C, Conroy T, Pezet D, Roullet B, Seitz JF, Herr JP, Paillot B, Arveux P, Bonnetain F, Binquet C. Chemoradiation followed by surgery compared with chemoradiation alone in squamous cancer of the esophagus: FFCD 9102. J Clin Oncol. 2007;25:1160–8.

Bray F, Ferlay J, Soerjomataram I, Siegel RL, Torre LA, Jemal A. Global cancer statistics 2018: GLOBOCAN estimates of incidence and mortality worldwide for 36 cancers in 185 countries. CA Cancer J Clin. 2018;68:394–424.

Chalabi M, Fanchi LF, Dijkstra KK, Van den Berg JG, Aalbers AG, Sikorska K, Lopez-Yurda M, Grootscholten C, Beets GL, Snaebjornsson P, Maas M, Mertz M, Veninga V, Bounova G, Broeks A, Beets-Tan RG, de Wijkerslooth TR, van Lent AU, Marsman HA, Nuijten E, Kok NF, Kuiper M, Verbeek WH, Kok M, Van Leerdam ME, Schumacher TN, Voest EE, Haanen JB. Neoadjuvant immunotherapy leads to pathological responses in MMR-proficient and MMR-deficient early-stage colon cancers. Nat Med. 2020;26:566–76.

Cunningham D, Allum WH, Stenning SP, Thompson JN, Van de Velde CJ, Nicolson M, Scarffe JH, Lofts FJ, Falk SJ, Iveson TJ, Smith DB, Langley RE, Verma M, Weeden S, Chua YJ, Magic Trial Participants. Perioperative chemotherapy versus surgery alone for resectable gastroesophageal cancer. N Engl J Med. 2006;355:11–20.

D'Angelica MI, Correa-Gallego C, Paty PB, Cercek A, Gewirtz AN, Chou JF, Capanu M, Kingham TP, Fong Y, DeMatteo RP, Allen PJ, Jarnagin WR, Kemeny N. Phase II trial of hepatic artery infusional and systemic chemotherapy for patients with unresectable hepatic metastases from colorectal cancer: conversion to resection and long-term outcomes. Ann Surg. 2015;261:353–60.

Deshmukh AA, Suk R, Shiels MS, Sonawane K, Nyitray AG, Liu Y, Gaisa MM, Palefsky JM, Sigel K. Recent trends in squamous cell carcinoma of the anus incidence and mortality in the United States, 2001-2015. J Natl Cancer Inst. 2020;112:829–38.

Dupre A, Malik HZ, Jones RP, Diaz-Nieto R, Fenwick SW, Poston GJ. Influence of the primary tumour location in patients undergoing surgery for colorectal liver metastases. Eur J Surg Oncol. 2018;44:80–6.

Ecker BL, McMillan MT, Datta J, Mamtani R, Giantonio BJ, Dempsey DT, Fraker DL, Drebin JA, Karakousis GC, Roses RE. Efficacy of adjuvant chemotherapy for small bowel adenocarcinoma: a propensity score-matched analysis. Cancer. 2016;122:693–701.

Engstrand J, Nilsson H, Stromberg C, Jonas E, Freedman J. Colorectal cancer liver metastases – a population-based study on incidence, management and survival. BMC Cancer. 2018;18:78.

Fenoglio-Preiser CM, Noffsinger AE, Belli J, Stemmermann GN. Pathologic and phenotypic features of gastric cancer. Semin Oncol. 1996;23:292–306.

Halperin DM, Shen C, Dasari A, Xu Y, Chu Y, Zhou S, Shih Y-CT, Yao JC. Frequency of carcinoid syndrome at neuroendocrine tumour diagnosis: a population-based study. Lancet Oncol. 2017;18:525–34.

Hospers G, Bahadoer RR, Dijkstra EA, van Etten B, Marijnen C, Putter H, Meershoek E, Kranenbarg K, Roodvoets AG, Nagtegaal ID, Beets-Tan RGH, Blomqvist LK, Fokstuen T, ten Tije AJ, Capdevila J, Hendriks MP, Edhemovic I, Cervantes A, Nilsson PJ, Glimelius B, Van De Velde CJH. Short-course radiotherapy followed by chemotherapy before TME in locally advanced rectal cancer: the randomized RAPIDO trial. J Clin Oncol. 2020;38:4006.

Hundahl SA, Phillips JL, Menck HR. The National Cancer Data Base Report on poor survival of U.S. gastric carcinoma patients treated with gastrectomy: fifth edition American Joint Committee on Cancer staging, proximal disease, and the "different disease" hypothesis. Cancer. 2000;88:921–32.

Islami F, Ferlay J, Lortet-Tieulent J, Bray F, Jemal A. International trends in anal cancer incidence rates. Int J Epidemiol. 2017;46:924–38.

Kirby TJ, Rice TW. The epidemiology of esophageal carcinoma. The changing face of a disease. Chest Surg Clin N Am. 1994;4:217–25.

Marmor S, Portschy PR, Tuttle TM, Virnig BA. The rise in appendiceal cancer incidence: 2000-2009. J Gastrointest Surg. 2015;19:743–50.

Mittal R, Chandramohan A, Moran B. Pseudomyxoma peritonei: natural history and treatment. Int J Hyperth. 2017;33:511–9.

Modlin IM, Kidd M, Latich I, Zikusoka MN, Shapiro MD. Current status of gastrointestinal carcinoids. Gastroenterology. 2005;128:1717–51.

Morton D. FOxTROT: an international randomised controlled trial in 1053 patients evaluating neoadjuvant chemotherapy (NAC) for colon cancer. On behalf of the FOxTROT Collaborative Group. Ann Oncol. 2019;30:v198.

Muntean V, Mihailov A, Iancu C, Toganel R, Fabian O, Domsa I, Muntean MV. Staging laparoscopy in gastric cancer. Accuracy and impact on therapy. J Gastrointestin Liver Dis. 2009;18:189–95.

Nagtegaal ID, Odze RD, Klimstra D, Paradis V, Rugge M, Schirmacher P, Washington KM, Carneiro F, Cree IA, W. H. O. Classification of Tumours Editorial Board. The 2019 WHO classification of tumours of the digestive system. Histopathology. 2020;76:182–8.

National Cancer Institute. Cancer stat facts: esophageal cancer. 2020. Available at https://seer.cancer.gov/statfacts/html/esoph.html. Accessed 12 Nov 2020.

Neugut AI, Jacobson JS, Suh S, Mukherjee R, Arber N. The epidemiology of cancer of the small bowel. Cancer Epidemiol Biomark Prev. 1998;7:243–51.

Nigro ND, Vaitkevicius VK, Considine B Jr. Combined therapy for cancer of the anal canal: a preliminary report. Dis Colon Rectum. 1974;17:354–6.

Scheiman JM, Cutler AF. Helicobacter pylori and gastric cancer. Am J Med. 1999;106:222–6.

Schneider TC, Schulte WJ. Management of carcinoma of anal canal. Surgery. 1981;90:729–34.

SEER Stat Facts Sheets: Anal Cance. NIH surveillance, epidemiology and end results program. 2020. Available at https://seer.cancer.gov/statfacts/html/anus.html. Accessed 19 Nov 2020.

Segelman J, Granath F, Holm T, Machado M, Mahteme H, Martling A. Incidence, prevalence and risk factors for peritoneal carcinomatosis from colorectal cancer. Br J Surg. 2012;99:699–705.

Shapiro J, van Lanschot JJB, Hulshof M, van Hagen P, van Berge Henegouwen MI, Wijnhoven BPL, van Laarhoven HWM, Nieuwenhuijzen GAP, Hospers GAP, Bonenkamp JJ, Cuesta MA, Blaisse RJB, Busch ORC, Ten Kate FJW, Creemers GM, Punt CJA, Plukker JTM, Verheul HMW, Bilgen EJS, van Dekken H, van der Sangen MJC, Rozema T, Biermann K, Beukema JC, Piet AHM, van Rij CM, Reinders JG, Tilanus HW, Steyerberg EW, van der Gaast A, Cross Study Group. Neoadjuvant chemoradiotherapy plus surgery versus surgery alone for oesophageal or junctional cancer (CROSS): long-term results of a randomised controlled trial. Lancet Oncol. 2015;16:1090–8.

Siegel RL, Miller KD, Goding Sauer A, Fedewa SA, Butterly LF, Anderson JC, Cercek A, Smith RA, Jemal A. Colorectal cancer statistics, 2020. CA Cancer J Clin. 2020;70:145–64.

Singh R, Nime F, Mittelman A. Malignant epithelial tumors of the anal canal. Cancer. 1981;48:411–5.

Singhal N, Singhal D. Adjuvant chemotherapy for small intestine adenocarcinoma. Cochrane Database Syst Rev. 2007;3:CD005202.

Solaymani-Dodaran M, Logan RF, West J, Card T, Coupland C. Risk of oesophageal cancer in Barrett's oesophagus and gastro-oesophageal reflux. Gut. 2004;53:1070–4.

Sugarbaker PH. Peritonectomy procedures. Ann Surg. 1995;221:29–42.

Surveillance, Epidemiology and End Results Program. SEER Stat Fact Sheet: colon and rectum cancer. National Cancer Institute. 2020a. Available at https://seer.cancer.gov/statfacts/html/colorect.html. Accessed 19 Nov 2020.

Surveillance, Epidemiology and End Results Program. SEER Stat Facts Sheet: stomach cancer. National Cancer Institute. 2020b. Available at https://seer.cancer.gov/statfacts/html/stomach.html. Accessed 12 Nov 2020.

Surveillance, Epidemiology and End Results Program. SEER Stat Facts Sheets: colorectal cancer. National Cancer Institute. 2020c. Available at https://seer.cancer.gov/statfacts/html/colorect.html. Accessed 20 Nov 2020.

Taal BG, Visser O. Epidemiology of neuroendocrine tumours. Neuroendocrinology. 2004;80(Suppl 1):3–7.

Thompson MR, O'Leary DP, Flashman K, Asiimwe A, Ellis BG, Senapati A. Clinical assessment to determine the risk of bowel cancer using symptoms, age, mass and iron deficiency anaemia (SAMI). Br J Surg. 2017;104:1393–404.

Tsikitis VL, Wertheim BC, Guerrero MA. Trends of incidence and survival of gastrointestinal neuroendocrine tumors in the United States: a seer analysis. J Cancer. 2012;3:292–302.

World Health Organization. Colorectal cancer. WHO. 2020. Available at https://gco.iarc.fr/today/data/factsheets/cancers/10_8_9-Colorectum-fact-sheet.pdf. Accesssed 19 Nov 2020.

Yang DX, Gross CP, Soulos PR, Yu JB. Estimating the magnitude of colorectal cancers prevented during the era of screening: 1976 to 2009. Cancer. 2014;120:2893–901.

Chapter 16
Lung Cancer and Smoking

Nima Yaftian, William Hoang, and Phillip Antippa

16.1 Background

Tobacco has been grown and smoked in the Americas since 6000 BC. It was introduced to Europe by Christopher Columbus in the fifteenth century following his voyages to the Americas. The tobacco plant was named Nicotiana tabacum after Jean Nicot of Villemain, the French ambassador to Portugal who sent tobacco in the form of snuff to the Queen Mother of France Catherine de Medicis as a treatment for her migraines (de Micheli 2015). Tobacco consumption in Europe was primarily limited to snuff consumption until the eighteenth century when smoking tobacco gained popularity across mainland Europe.

While tobacco use gained popularity there were some early concerns raised about the negative health impacts of smoking. King James I, the first monarch from the House of Stuart was an early critic of tobacco. In his treatise—a counterblaste to tobacco he speaks against the purported health benefits of tobacco and talks of the health and social impacts of tobacco in the context of medical knowledge of the era (1604). Even at the time of writing the treatise the banning of tobacco was not viable due to its widespread use and popularity, however, in 1604 the English Parliament passed an effective tax of 4000% on tobacco. This was thought to be one of the first public health measures taken against tobacco.

Tobacco use in society has a significant economic impact. There are estimates that smoking attributable diseases have cost the global economy over US$420 billion in direct healthcare related expenditure and US$1.4 trillion in combined direct expenditure and productivity loss in the year 2012 which equates to 1.8% of global

N. Yaftian · W. Hoang · P. Antippa (✉)
Department of Cardiothoracic Surgery, Royal Melbourne Hospital, Parkville, Australia
e-mail: Nima.Yaftian@mh.org.au; william.hoang@mh.org.au; Phillip.Antippa@mh.org.au

© The Author(s), under exclusive license to Springer Nature
Switzerland AG 2022
S. S. Goonewardene et al. (eds.), *Men's Health and Wellbeing*,
https://doi.org/10.1007/978-3-030-84752-4_16

GDP for that year (Goodchild et al. 2018). The burden of smoking on the developing world is concerning with almost 40% of these costs being incurred in developing countries.

Despite the early evidence tobacco use has remained prevalent late into the twentieth century and into the twenty-first century. In the year 2000, 33.3% of all people and 50% of all men aged over 15 were consumers of some form of tobacco. Encouragingly these numbers are reducing with 22.8% of all people and 37.5% of men aged over 15 consuming tobacco in the year 2020 (WHO 2019). There is a significant association between smoking and socio-economic status. People with a university education are less likely to smoke than people without a tertiary education. Between both groups of smokers those with a tertiary education are likely to be lighter smokers when compared to those without (Coleman et al. 2019). This disparity shows the challenges that are faced when trying to reduce the uptake and use of tobacco.

Lung cancer was an uncommon disease prior to the twentieth century and the link between smoking tobacco and lung cancer was not established until halfway through the century (Proctor 2012). Early pathological studies of cadaveric lungs with neoplasms suggested a link between lung cancer and smoking (Adler 1913). Some of the earliest and strongest evidence to emerge was from Doll and Hill who in 1954 performed a prospective cohort analysis of smoking habits among medical practitioners and incidence of lung cancer. They found that there was a proportional increase in incidence of lung cancer and the amount of tobacco smoked daily (Doll and Hill 1954). In the US the 1964 Surgeon General's report on smoking and health was crucial in raising the awareness of the links between smoking and cancers of the lung and larynx (US Department of Health 1964).

16.2   Epidemiology

Throughout its emergence from the twentieth century to now, lung cancer has steadily become a significant contributor to global morbidity and mortality. Globally lung cancer has the highest incidence of diagnosis and is the leading cause of cancer death. While smoking and the incidence of lung cancer have reduced throughout the developing world the overall incidence of lung cancer has increased due to the increased rates of smoking and lung cancer in the developed world. Lung cancer is the second most common cancer among men in developed countries while it is the most common cancer in developing countries. Smoking plays a significant contribution with approximately 80% of the lung cancers among males being attributed to smoking (Jemal et al. 2011).

A diagnosis of lung cancer has traditionally been associated with very poor outcomes. While there have been new developments in treatments the overall 5-year survival for lung cancer remains between 9 and 15% (Dela Cruz et al. 2011). A significant contributor to the poor outcomes associated with lung cancer is the late stage of presentation. Approximately 65% of newly diagnosed lung cancers will be stage III or IV (Morgensztern et al. 2010). Later stages of presentation will significantly limit the number of interventions that can be offered and may take a patient from curative intent treatment to palliative.

Lung cancer can be roughly subdivided into two main subtypes—small cell lung cancer (SCLC) and non-small cell lung cancer (NSCLC). SCLC is seen almost exclusively in smokers and accounts for approximately 15% of all lung cancers (Dela Cruz et al. 2011). Patients with SCLC may present with either limited stage or extensive stage disease. Median survival varies from 8 to 20 months following diagnosis depending on the extent of the disease. However, despite new advances in therapy the prognosis for SCLC remains poor with 5-year survival of 10% for limited stage disease and 2% for extensive stage disease (Sculier et al. 2008).

The majority of lung cancers that are diagnosed are NSCLC (making up approximately 85% of lung cancers). Of the NSCLC the most common histological subtypes are adenocarcinoma in approximately 40%, squamous cell carcinoma in approximately 25% and large cell carcinoma in approximately 5%. The other 30% is comprised of various other less common subtypes (Morgensztern et al. 2010).

16.3 Assessment for Suspected Lung Cancer

The two pillars of lung cancer evaluation are a timely diagnosis and accurate staging that allow for prompt appropriate treatment and improved survival (Buccheri and Ferrigno 2004). Assessment of a patient with suspected lung cancer should include a thorough history and examination, evaluation of risk factors, and appropriate investigations with an aim to prioritise timely referrals, imaging, tissue biopsy, and ultimately treatment. Guidelines suggest that patient distress is minimised if results of their investigations are shared with them within 1 week, and specialist appointments should take place within 2 weeks of the referral (Cancer Australia 2020).

Each patient's journey throughout the evaluation, treatment and subsequent surveillance of lung cancer must be individualised not just for their symptoms, but also, for their own preferences, values, psychosocial state and personal circumstances (Ost et al. 2013). Early establishment and frequent revisitation of patient preferences help guide decision-making that is mutual (between clinician and patient) and ultimately in the patient's best wishes.

The pathways to arriving at a suspicion or diagnosis of lung cancer are numerous, but generally present in two forms: the symptomatic patient, and those in whom a lesion is detected incidentally. Currently, there are multiple lung cancer screening trials being rolled out globally as evidence has suggested that screening high-risk patients via low dose CT will reduce lung cancer related and all cause mortality (Usman Ali et al. 2016).

16.4   Signs and Symptoms

Patients who present with clinical signs or symptoms due to lung cancer tend to have advanced disease and are more likely to have a poor prognosis. The most common symptoms among patients with lung cancer include cough, whether new or changed in character from a chronic cough (45% of lung cancer patients); haemoptysis (27%); dyspnoea, wheeze, or stridor (37%); and chest pain (27%), typically dull and aching (Chute et al. 1985). Other concerning symptoms and signs include persistent or recurrent chest infections, hoarseness of voice, cervical and/or supraclavicular lymphadenopathy, digital clubbing, deep vein thrombosis, and systemic symptoms such as fevers, chills, night sweats, anorexia, and loss of weight. The presence of unexplained symptoms should prompt an urgent chest radiograph. Obstruction of the superior vena cava occurs in around 12% of SCLC (Urban et al. 1993). Symptoms include a sense of fullness in the head, which can be reflected on examination by facial plethora or oedema, and congested neck veins.

The Pancoast syndrome is a rare manifestation of apical lung cancers of the upper lobe. The pathognomonic signs and symptoms of a Pancoast tumour are due to local compression of the brachial plexus and sympathetic chain by direct, local tumour invasion. The syndrome comprises ipsilateral shoulder pain, radicular pain and eventual muscular atrophy in the C8-T2 (ulnar) distribution, and Horner's syndrome (meiosis, ptosis, and anhidrosis).

Complications arising from local invasion or metastasis also need to be considered. Common sites of metastasis in lung cancer include brain (28.4%), bone (34.3%), adrenal glands (26.7%) and liver (13.4%) (Popper 2016). Intracerebral metastases can present with headache, nausea and vomiting, or a first seizure. Likewise, bone pain and pathological fractures, or signs of liver obstruction without explanation should all prompt consideration of a primary lung malignancy.

16.5   Past Medical History

A thorough evaluation of past medical history is key to elicit risk factors for lung cancer. A history of infective and inflammatory chronic lung pathologies is associated with a higher incidence of lung malignancy. These include chronic obstructive pulmonary disease, chronic bronchitis, emphysema, pneumonia, and tuberculosis. Many of these have independent associations with smoking, and there is some debate regarding their causation or correlation with lung cancer. However, for non-smokers, a past history of pneumonia, tuberculosis, pulmonary interstitial fibrosis and scleroderma have also been correlated with lung malignancy (Abu-Shakra et al. 1993; Brenner et al. 2011; Hubbard et al. 2000).

16.6 Risk Factors

Recognition of modifiable and fixed risk factors for lung cancer is important in approaching the assessment of patients at risk of lung cancer. As described above, the largest modifiable risk factor for development of lung cancer is tobacco smoking. Smoking cessation or reduction lowers the risk of developing a lung malignancy, as well as the mortality in patients with existing diagnoses of lung cancer (US Department of Health 1964). A thorough history of a patient's smoking behaviours is essential with important information to elicit including:

- Primary smoking habit: Onset and duration of habit, pack year history, year of cessation if relevant, type or method of tobacco smoked—e.g., cigarettes, self-rolled (if so, whether other substances are included, such as marijuana), cigars, e-cigarettes or other vaping devices.
- Second-hand or environmental tobacco smoke—e.g., marriage to a smoking partner; other smokers in the household [National Research Council Committee on Passive Smoking (Oberg et al. 2011; Asomaning et al. 2008)].

Further enquiry should be made about any history of exposure, especially those that are occupational, to other carcinogens including radon, asbestos, diesel exhaust, and air pollution. Asbestos and silica dust exposure are particularly noted as independent risks for lung cancer, however, its combination with smoking has been shown to be multiplicative for lung cancer risk and mortality compared to exposure amongst non-smokers (Markowitz et al. 2013; Brown 2009).

16.7 Investigation

An initial chest radiograph should be requested for all patients in whom a lung cancer is suspected. Comparison with previous radiography will help to identify chronicity and growth patterns of lung pathology. Concerning features on chest radiography include: intraparenchymal lesions, pleural effusion, pleural thickening/nodularity or atelectasis localised to a lobe or anatomical segment of lung. Further evaluation with computed tomography (CT) with contrast enhancement if possible should be considered in patients with concerning features on chest radiography, or if there is a higher suspicion of lung malignancy with notable thoracic symptoms despite an apparently normal chest radiograph.

A full blood examination can detect anaemia, leucocytosis, and thrombocytosis, which are common findings in patients with lung malignancy (Hamilton et al. 2005; Kasuga et al. 2001; Kosmidis and Krzakowski 2005).

Deranged liver function tests can indicate liver metastases, while increased serum alkaline phosphatase and hypercalcaemia can be seen with bone metastases.

Serum biochemistry is useful in the diagnosis of paraneoplastic syndromes. Paraneoplastic syndromes are rare but important to recognise, due to the risk of further metabolic derangement and symptom burden for the patient. They are more commonly associated with SCLC. These include hypercalcaemia, hyperparathyroidism, elevated calcitriol and osteoclast activating factors.

The syndrome of inappropriate anti-diuretic hormone (SIADH) is seen in 11% of patients with SCLC, and manifests as euvolaemic hyponatraemia (List et al. 1986).

16.8 Smoking Cessation

Ultimately, reducing smoking is the key to reducing the risk of lung malignancies. It is important for clinicians to routinely and sensitively address patients' smoking behaviour, their willingness to cease or reduce, and to support them through the process.

Enquiry should be made to understand the patient's level of psychological and physical dependence, including time and effort directly or indirectly dedicated to accessing tobacco. Validated tools consisting of a few questions exist which can be readily applied during the clinical encounter; examples include the Fagerstrom test for nicotine dependence (Heatherton et al. 1991).

The clinician should seek to understand the role that smoking serves for the patient and their current will to reduce or quit. The Stages of Changes model provides a useful framework and terminology for considering readiness to change. An understanding of the patient's circumstances, such as the financial strain caused by a smoking habit, or guilt when contemplating the impact of smoking on one's children, can be advantageous in applying motivational interviewing techniques to explore the patient's preparedness to progress along the Stages of Change.

Where relevant, understanding should also be sought about the patient's previous attempts to reduce or quit smoking, their relative success, and what, if any, behavioural, psychological, and pharmacological strategies the patient found useful. With this in mind, the clinician should be proactive in offering referral to individual or group counselling programs, and prescribing nicotine replacement or other pharmacological therapies for nicotine dependence.

16.9 Psychosocial Considerations

At every stage of the process through detection, diagnosis, and treatment of lung cancer, care must be taken to acknowledge the patient's needs, preferences, and values. It is essential to identify common times of increased vulnerability during this period. It is important for primary care physicians to acknowledge these moments and take a holistic approach to the care they provide. Primary care physicians are the best equipped to help manage the significant psychological impact a diagnosis of lung cancer has on patients and their loved ones.

It is also important to recognise that there is a social stigma that has been associated with lung cancer and its association with smoking. This stigma continues to persist when it comes to funding and research for lung cancer. Although lung cancer is the leading cause of cancer death worldwide, there is a disproportionate reduction in funding for research when compared to other cancers (Carter et al. 2015). The stigma associated with lung cancer has the potential to delay medical help seeking. It is clear that a non-judgemental approach from both clinicians and society in general is in the patient's best interest.

16.10   Conclusion

Smoking is the greatest modifiable risk factor and an amplifier of other risk factors for developing lung cancers. Although smoking rates in the developed world are reducing they are increasing in the developing world and rates of smoking among men remain worryingly high. While there are developments in the treatment of lung cancer, the prognosis still remains poor. The early cessations of smoking and early diagnosis of lung cancer remain the two pillars in successfully reducing the disease burden.

References

Abu-Shakra M, Guillemin F, Lee P. Cancer in systemic sclerosis. Arthritis Rheum. 1993;36:460–4.
Adler I. Primary malignant growths of the lungs and bronchi—A pathological and clinical study. By I. Adler, A. M., M. D. pp. 325, with 16 plates. price 16s net. London: Longmans, Green & Co., 1912. The Laryngoscope. 1913;23:80.
Asomaning K, Miller DP, Liu G, Wain JC, Lynch TJ, Su L, Christiani DC. Second hand smoke, age of exposure and lung cancer risk. Lung Cancer. 2008;61:13–20.
Brenner DR, Mclaughlin JR, Hung RJ. Previous lung diseases and lung cancer risk: a systematic review and meta-analysis. PLoS One. 2011;6:e17479.
Brown T. Silica exposure, smoking, silicosis and lung cancer—complex interactions. Occupational Medicine. 2009;59:89–95.
Buccheri G, Ferrigno D. Lung cancer: clinical presentation and specialist referral time. European Respiratory Journal. 2004;24:898–904.
Cancer Australia. Evidence report for Investigating symptoms of lung cancer: a guide for all health professionals. Surry Hills, NSW: Cancer Australia; 2020.
Carter AJ, Delarosa B, Hur H. An analysis of discrepancies between United Kingdom cancer research funding and societal burden and a comparison to previous and United States values. Health Res Policy Syst. 2015;13:62.
Chute CG, Greenberg ER, Baron J, Korson R, Baker J, Yates J. Presenting conditions of 1539 population-based lung cancer patients by cell type and stage in New Hampshire and Vermont. Cancer. 1985;56:2107–11.
Coleman SRM, Gaalema DE, Nighbor TD, Kurti AA, Bunn JY, Higgins ST. Current cigarette smoking among U.S. college graduates. Prev Med. 2019;128:105853.
De Micheli A. The tobacco in the light of history and medicine. Arch Cardiol Mex. 2015;85:318–22.

Dela Cruz CS, Tanoue LT, Matthay RA. Lung cancer: epidemiology, etiology, and prevention. Clin Chest Med. 2011;32:605–44.

Doll R, Hill AB. The mortality of doctors in relation to their smoking habits. Br Med J. 1954;1:1451–5.

Goodchild M, Nargis N, Tursan D'espaignet E. Global economic cost of smoking-attributable diseases. Tobacco Control. 2018;27:58–64.

Hamilton W, Peters TJ, Round A, Sharp D. What are the clinical features of lung cancer before the diagnosis is made? A population based case-control study. Thorax. 2005;60:1059–65.

Heatherton TF, Kozlowski LT, Frecker RC, Fagerström KO. The Fagerström test for nicotine dependence: a revision of the Fagerström tolerance questionnaire. Br J Addict. 1991;86:1119–27.

Hubbard R, Venn A, Lewis S, Britton J. Lung cancer and cryptogenic fibrosing alveolitis. A population-based cohort study. Am J Respir Crit Care Med. 2000;161:5–8.

Jemal A, Bray F, Center MM, Ferlay J, Ward E, Forman D. Global cancer statistics. CA: A Cancer Journal for Clinicians. 2011;61:69–90.

Kasuga I, Makino S, Kiyokawa H, Katoh H, Ebihara Y, Ohyashiki K. Tumor-related leukocytosis is linked with poor prognosis in patients with lung carcinoma. Cancer. 2001;92:2399–405.

King James I. A counterblaste to tobacco [Online]. Document Bank of Virginia. 1604. https://edu. lva.virginia.gov/dbva/items/show/124. Accessed 13 Oct 2020.

Kosmidis P, Krzakowski M. Anemia profiles in patients with lung cancer: what have we learned from the European Cancer Anaemia Survey (ECAS)? Lung Cancer. 2005;50:401–12.

List AF, Hainsworth JD, Davis BW, Hande KR, Greco FA, Johnson DH. The syndrome of inappropriate secretion of antidiuretic hormone (SIADH) in small-cell lung cancer. J Clin Oncol. 1986;4:1191–8.

Markowitz SB, Levin SM, Miller A, Morabia A. Asbestos, asbestosis, smoking, and lung cancer. New findings from the North American insulator cohort. Am J Respir Crit Care Med. 2013;188:90–6.

Morgensztern D, Ng SH, Gao F, Govindan R. Trends in stage distribution for patients with non-small cell lung cancer: a National Cancer Database survey. J Thorac Oncol. 2010;5:29–33.

Oberg M, Jaakkola MS, Woodward A, Peruga A, Prüss-Ustün A. Worldwide burden of disease from exposure to second-hand smoke: a retrospective analysis of data from 192 countries. Lancet. 2011;377:139–46.

Ost DE, Yeung S-CJ, Tanoue LT, Gould MK. Clinical and organizational factors in the initial evaluation of patients with lung cancer: diagnosis and management of lung cancer: American College of Chest Physicians evidence-based clinical practice guidelines. Chest. 2013;143:e121S–41S.

Popper HH. Progression and metastasis of lung cancer. Cancer Metastasis Rev. 2016;35:75–91.

Proctor RN. The history of the discovery of the cigarette–lung cancer link: evidentiary traditions, corporate denial, global toll: Table 1. Tobacco Control. 2012;21:87–91.

Sculier JP, Chansky K, Crowley JJ, van Meerbeeck J, Goldstraw P. The impact of additional prognostic factors on survival and their relationship with the anatomical extent of disease expressed by the 6th Edition of the TNM Classification of Malignant Tumors and the proposals for the 7th Edition. J Thorac Oncol. 2008;3:457–66.

Urban T, Lebeau B, Chastang C, Leclerc P, Botto MJ, Sauvaget J. Superior vena cava syndrome in small-cell lung cancer. Arch Intern Med. 1993;153:384–7.

US Department of Health, E., and Welfare. Smoking and Health Report of the Advisory Committee to the Surgeon General of the Public Health Service. 1964.

Usman Ali M, Miller J, Peirson L, Fitzpatrick-Lewis D, Kenny M, Sherifali D, Raina P. Screening for lung cancer: a systematic review and meta-analysis. Prev Med. 2016;89:301–14.

WHO. WHO global report on trends in prevalence of tobacco smoking 2000–2025. 3rd ed. Geneva: World Health Organisation; 2019.

Chapter 17
Testicular Malignancy

Herney Andres García-Perdomo, Carlos Toribio-Vázquez, Álvaro Yebes, Diego M. Carrion, and Juan Gómez Rivas

17.1 Introduction

Testicular cancer (TC) comprises different types of neoplasms, depending on the cell of origin and the age at presentation. Still, germ cell tumours (GCT) constitute the vast majority of cases.

Germ cell tumours can be diagnosed in every age group, but this malignancy typically affects young men (90%) (Rajpert-De Meyts et al. 2018), and is also the most common solid tumour in young men. It usually shows as a clinical-stage I and is potentially curable with a single surgical procedure. Most cases represent sporadic occurrences. In more advanced stages, TC might also be curable with multimodal therapy. The diagnosis and risk stratification would be based on the pathological specimen; however, multiple images are necessary for decision-making. There are no screening strategies for early detection of GCT, although for gonadal, the regular self-assessment might be acceptable (Stevenson and Lowrance 2015). This chapter describes a summary of the general epidemiology, aetiology, classification, diagnosis, and treatment modalities.

H. A. García-Perdomo
Division of Urology, Department of Surgery, School of Medicine, Universidad del Valle, Cali, Colombia

C. Toribio-Vázquez · Á. Yebes
Urology Department, La Paz University Hospital, Madrid, Spain

Autonomous University of Madrid, Madrid, Spain

D. M. Carrion · J. G. Rivas (✉)
Department of Urology, Hospital Clinico San Carlos, Madrid, Spain

Autonomous University of Madrid, Madrid, Spain

La Paz University Hospital Institute for Health Research (IdiPaz), Madrid, Spain

17.2   Epidemiology

The annual incidence of TC is 5–14 cases per 100,000 men (Stevenson and Lowrance 2015; Laguna et al. 2020). Also, its incidence rate peaks at the third decade of life for non-seminoma and mixed GCT patients and the fourth decade for seminoma (Gurney et al. 2019; Nigam et al. 2015). This neoplasm represents 1% of all male tumours and 5% of all urological tumours (Park et al. 2018). Predominantly, 90–95% of patients have GCT, and 1–2% are bilateral at diagnosis (Park et al. 2018).

GCTs are categorised into two histologic subtypes: seminoma and non-seminoma. The first one is more frequent, but the non-seminoma is more aggressive and usually include multiple cell types (Gilligan et al. 2020; Vasdev et al. 2013). Most non-seminomas are mixed tumours of four subtypes: embryonal carcinoma, choriocarcinoma, yolk sac tumour, and teratoma (Vasdev et al. 2013).

Eighty percent of patients with seminoma and about 60% of non-seminomatous germ cell tumour (NSGCT) patients have stage I disease at diagnosis (Laguna et al. 2020; Verhoeven et al. 2014; Klepp et al. 1990).

17.3   Aetiology

There are multiple risk factors for developing TC (Fig. 17.1). All these are components of a syndrome called testicular dysgenesis, which encompasses cryptorchidism, hypospadias, decreased spermatogenesis, familial history of TC, and the presence of a contralateral tumour or germ cell neoplasia in situ (GCNIS) (Laguna et al. 2020; Peng et al. 2009; Jørgensen et al. 2010; Greene et al. 2010; Holzik et al. 2004; Kharazmi et al. 2015).

Testicular microcalcifications might be associated with TC, as these ultrasound findings are more frequent in young patients with TC than in patients without TC (Stevenson and Lowrance 2015). Having an ultrasound with testicular microcalcifications increases the risk 12 times (RR 12 95%CI 8.1–17) (Wang et al. 2015).

Undescended testis increases the risk of TC in about 2–14 times over the general population. This finding might be associated with the change in the scrotal temperature, based on the location. However, the literature describes that there are multiple environmental and genetic factors influencing this transformation (Richiardi et al. 2007). If early surgical correction is performed prior to puberty, the risk lowers to twofold (Stevenson and Lowrance 2015; Kurz and Tasian 2016).

There are multiple genetic factors associated with TC. There are polymorphisms of P53 in 66% of cases of in situ tumours (Kuczyk et al. 1996). Also, there is an association between polymorphisms in the phosphatase and tensin homolog (PTEN) gen (Andreassen et al. 2013). An isochromosome of the short arm of chromosome 12 (i12p) might be present in almost all the adult patients with GCTs and in situ neoplasia (Laguna et al. 2020; Bosl and Motzer 1997).

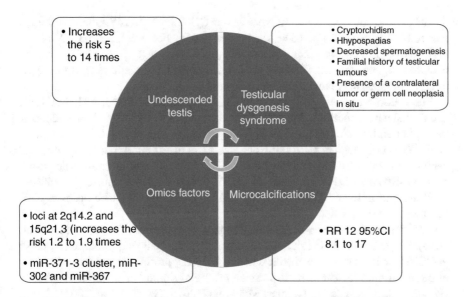

Fig. 17.1 Risk factors and aetiology of testicular cancer. *RR* relative risk, *CI* confidence interval (adapted from Laguna et al. 2020; Peng et al. 2009; Jørgensen et al. 2010; Greene et al. 2010; Holzik et al. 2004; Kharazmi et al. 2015)

Table 17.1 Genes and polymorphisms associated with testis cancer

Polymorphism	Gene	Localization	OR (95%CI)
Rs2713206	TFCP2L1	2q14.2	1.26 (1.16–1.36)
rs3755605	GPR160	3q26.2	1.19 (1.12–1.26)
rs55873183	DMRT1	9p24.3	1.89 (1.67–2.14)
rs61408740	LHPP	10q26.13	1.65 (1.38–1.96)
rs12912292	PRTG	15q21.3	1.22 (1.15–1.29)
rs60180747	MAP2K1, TIPIN	15q22.31	1.23 (1.16–1.32)
rs34601376	ZNF726	19p12a	1.29 (1.2–1.39)

There are multiple independent polymorphisms in established loci associated with TC (Wang et al. 2017) (Table 17.1). Also, there is a special interest for the loci at 2q14.2 and 15q21.3 (Loveday et al. 2018).

17.4 Pathology

GCTs are incredibly heterogeneous tumours. They have pluripotency features, and 50% of them might differentiate in any somatic tissue, forming teratomas (Rajpert-De Meyts et al. 2018; Solter 2006).

GCTs originate from a common precursor; GCNIS, previously called carcinoma in situ of the testis (Rajpert-De Meyts et al. 2018; Fosså et al. 2003). This one derived from arrested immature germ cells (gonocytes) that persisted outside perinatal life (Sonne et al. 2009). They are located at the basal membrane of the seminiferous tubules; they have irregular chromatin in the nuclei.

The following genes: KIT, P53, OCT4, AP2-gamma/TFAP2C, KIT, NANOG, and LIN28 have important biological functions such as germ cell survival and maintenance of embryonic stem cell pluripotency (Rajpert-Meyts et al. 2015; Jørgensen et al. 1995; Looijenga et al. 2003; Hoei-Hansen et al. 2004). Accordingly, they are useful in the histopathological diagnosis of TC. Also, TC displays a specific profile of micro-RNAs (miRNA) in overt all GCTs, except teratoma: miR-371-3 cluster, miR-302 and miR-367 (Novotny et al. 2012). The latter might be necessary for diagnosis and prognosis.

Both seminoma and non-seminoma tumours originate from GCNIS (Moch et al. 2016). Seminoma resembles a mass of immature germ cells; morphologically close to GCNIS cells. In this case, the cells homogeneously proliferate and retain features of germinal lineage. The gene expression profile of seminoma is similar to that of GCNIS and fetalgonocytes (Rajpert-Meyts et al. 2015; Gashaw et al. 2007).

Nonseminomatous tumours might have different histological forms and contain undifferentiated embryonal carcinoma and somatic components (Rajpert-De Meyts et al. 2016). These tumours also contain extra-embryonic tissue components (yolk sac tumour and choriocarcinoma). The mixed tumours contain both, but they are classified and treated as non-seminoma; which is usually more aggressive than seminoma (Rajpert-De Meyts et al. 2016).

Besides, the recommended pathological classification is based on the 2016 update of the World Health Organization (WHO) pathological classification (Moch et al. 2016):

1. Germ cell tumours

• Germ cell neoplasia in situ (GCNIS)

2. Derived from germ cell neoplasia in situ (GCNIS)

• Seminoma
• Embryonal carcinoma
• Yolk sac tumour, post-pubertal type
• Trophoblastic tumours
• Teratoma, post-pubertal type
• Teratoma with somatic-type malignancies
• Mixed germ cell tumours

3. Germ cell tumours unrelated to GCNIS

• Spermatocytic tumour
• Yolk sac tumour, pre-pubertal type
• Mixed germ cell tumour, pre-pubertal type

4. Sex cord/stromal tumours

- Leydig cell tumour

  – Malignant Leydig cell tumour

- Sertoli cell tumour

  – Malignant Sertoli cell tumour
  – Large cell calcifying Sertoli cell tumour
  – Intratubular large cell hyalinising Sertoli cell neoplasia

- Granulosa cell tumour

  – Adult type
  – Juvenile type

- Thecoma/fibroma group of tumours
- Other sex cord/gonadal stromal tumours

  – Mixed
  – Unclassified

- Tumours containing both germ cell and sex cord/gonadal stromal

  – Gonadoblastoma

5. Miscellaneous non-specific stromal tumours

- Ovarian epithelial tumours
- Tumours of the collecting ducts and rete testis

  – Adenoma
  – Carcinoma

- Tumours of paratesticular structures

  – Adenomatoid tumour
  – Mesothelioma (epithelioid, biphasic)
  – Epididymal tumours

- Cystadenoma of the epididymis
- Papillary cystadenoma
- Adenocarcinoma of the epididymis
- Mesenchymal tumours of the spermatic cord and testicular adnexae

17.5 Staging and Prognosis Classification Systems

The International Union Against Cancer (UICC) described the classification for the anatomical extend of the disease in 2016—Tumour, Node, and Metastasis (TNM) (Table 17.2) (Laguna et al. 2020; Brierley et al. 2016).

Table 17.2 TNM classification

pT—Primary Tumour 1	
pTX	Primary tumour cannot be assessed[a]
pT0	No evidence of primary tumour (e.g. histological scar in testis) pTis Intratubular germ cell neoplasia (carcinoma in situ)[b]
pT1	Tumour limited to testis and epididymis without vascular/lymphatic invasion; tumour may invade tunica albuginea but not tunica vaginalis[c]
pT2	Tumour limited to testis and epididymis with vascular/lymphatic invasion, or tumour extending through tunica albuginea with involvement of tunica vaginalis[d]
pT3	Tumour invades spermatic cord with or without vascular/lymphatic invasion[d]
pT4	Tumour invades scrotum with or without vascular/lymphatic invasion
N—Regional lymph nodes: clinical	
NX	Regional lymph nodes cannot be assessed
N0	No regional lymph node metastasis
N1	Metastasis with a lymph node mass 2 cm or less in greatest dimension or multiple lymph nodes, none more than 2 cm in greatest dimension
N2	Metastasis with a lymph node mass more than 2 cm but not more than 5 cm in greatest dimension; or more than five nodes positive, none more than 5 cm; or greatest dimension; or more than five nodes positive, none more than 5 cm; or evidence of extranodal extension of tumour
N3	Metastasis with a lymph node mass more than 5 cm in greatest dimension
Pn—Regional Lymph Nodes: Pathological	
pNX	Regional lymph nodes cannot be assessed
pN0	No regional lymph node metastasis
pN1	Metastasis with a lymph node mass 2 cm or less in greatest dimension and five or fewer positive nodes, none more than 2 cm in greatest dimension
pN2	Metastasis with a lymph node mass more than 2 cm but not more than 5 cm in greatest dimension; or more than five nodes positive, none more than 5 cm; or greatest dimension; or more than five nodes positive, none more than 5 cm; or evidence or extranodal extension of tumour
pN3	Metastasis with a lymph node mass more than 5 cm in greatest dimension
M—Distant metastasis	
MX	Distant metastasis cannot be assessed
M0	No distant metastasis
M1	Distant metastasis[d]
M1a	Non-regional lymph node(s) or lung metastasis
M1b	Distant metastasis other than non-regional lymph nodes and lung
S—Serum Tumour Markers (Pre chemotherapy)	
SX	Serum marker studies not available or not performed
S0	Serum marker study levels within normal limits

	LDH (U/I)	hCG (mIU/mL)	AFP (ng/mL)
S1	$<1.5 \times N$ and	<5000 and	<1000

Table 17.2 (continued)

S2	1.5–10 × N or	5000–50,000 or	1000–10,000
S3	>10 × N or	>50,000 or	>10,000

N indicates the upper limit of normal

LDH lactate dehydrogenase, *hCG* human chorionic gonadotrophin, *AFP* alpha-fetoprotein

[a]Except for pTis and pT4, where radical orchidectomy is not always necessary for classification purposes, the extent of the primary tumour is classified after radical orchidectomy; see pT. In other circumstances, TX is used if no radical orchidectomy has been performed

[b]The current "Carcinoma in situ" GCNIS replaces nomenclature

[c]AJCC eighth edition subdivides T1 Pure Seminoma by T1a and T1b depending on size no greater than 3 cm or greater than 3 cm in greatest dimension

[d]AJCC eighth edition considers the hilar soft tissue invasion as pT2, while the discontinuous involvement of the spermatic cord is considered as pM1

Table 17.3 Testicular tumour stages

Stage	Tumour	Node	Metastases	Serum markers
Stage 0	pTis	N0	M0	S0
Stage I	pT1-4	N0	M0	SX
Stage IA	pT1	N0	M0	S0
Stage IB	pT2-4	N0	M0	S0
Stage IS	Any pT/X	N0	M0	S1-3
Stage II	Any pT/X	N1-3	M0	SX
Stage IIA	Any pT/X	N1	M0	S0-1
Stage IIB	Any pT/X	N2	M0	S0-1
Stage IIC	Any pT/X	N3	M0	S0-1
Stage III	Any pT/X	Any N	M1a	SX
Stage IIIA	Any pT/X	Any N	M1a	S0-1
Stage IIIB	Any pT/X	N1-N3	M0	S2
Stage IIIB	Any pT/X	Any N	M1a	S2
Stage IIIC	Any pT/X	N1-N3	M0	S3
Stage IIIC	Any pT/X	Any N	M1a	S3
Stage IIIC	Any pT/X	Any N	M1b	Any S

17.5.1 UICC Prognostic Groups

According to the previous classification, there are prognostic groups defined as in Table 17.3 (Laguna et al. 2020; Brierley et al. 2016). Regarding the Clinical Stage (CS) I disease: A is for tumours limited to the testis and epididymis with no evidence of metastasis; B is for more locally invasive tumour, but no evidence of metastasis; and S is for patients with persistent evidence of elevated serum tumour markers after orchidectomy.

Additionally, there is a critical prognosis classification for metastatic TC based on clinically independent adverse factors. This one uses histology, location of the primary tumour and metastasis, and pre-chemotherapy tumour markers. It classifies the patient in one of three groups: good, intermediate or poor prognosis (Table 17.4) (Mead 1997).

Table 17.4 The prognostic-based staging system for metastatic germ cell cancer (International Germ Cell Cancer Collaborative Group—IGCCCG)

Good-prognosis group		
Non-seminoma (56% of cases) 5-year PFS 5-year survival	All of the following criteria: • Testis/retro-peritoneal primary • No non-pulmonary visceral metastases • AFP < 1000 ng/mL • hCG < 5000 IU/L (1000 ng/mL) • LDH < 1.5 × ULN	
	5-year PFS	89%
	5-year survival	92%
Seminoma (90% of cases) 5-year PFS 5-year survival	All of the following criteria: • Any primary site • No non-pulmonary visceral metastases • Normal AFP • Any hCG • Any LDH	
	5-year PFS	82%
	5-year survival	86%
Intermediate-prognosis group		
Non-seminoma (28% of cases) 5-year PFS 5-year survival	Any of the following criteria: • Testis/retro-peritoneal primary • No non-pulmonary visceral metastases • AFP 1000–10,000 ng/mL or • hCG 5000–50,000 IU/L or • LDH 1.5–10 × ULN	
	5-year PFS	75%
	5-year survival	80%
Seminoma (10% of cases) 5-year PFS 5-year survival	All of the following criteria: • Any primary site • Non-pulmonary visceral metastases • Normal AFP • Any hCG • Any LDH	
	5-year PFS	67%
	5-year survival	72%
Poor-prognosis group		
Non-seminoma (16% of cases) 5-year PFS 5-year survival	Any of the following criteria: • Mediastinal primary • Non-pulmonary visceral metastases • AFP > 10,000 ng/mL or • hCG > 50,000 IU/L (10,000 ng/mL) or • LDH > 10 × ULN	
	5-year PFS	41%
	5-year survival	48%
Seminoma	No patients classified as poor prognosis	

17.6 Diagnostic Assessment

TC usually presents as an incidental finding in ultrasound (US) or as a scrotal mass detected by the patient. Pain is only present in 25% of cases. When suspecting a TC, the physical exploration must include scrotal, abdominal and supraclavicular exploration (Laguna et al. 2020; Germà-Lluch et al. 2002; Moul 2007) (Fig. 17.2).

17.7 Imaging

17.7.1 Scrotal Imaging

First of all, the physician needs to confirm the lesion found in a physical examination with a high-frequency (>10 MHz) ultrasound. With ultrasound, we can determine the volume and anatomical location, whether it is intra or extra-testicular and examine the contralateral testicle to exclude other lesions (Moul 2007; Shaw 2008).

Ultrasound is also recommended for men with retroperitoneal or visceral masses in the absence of palpable testicular mass. There is no need to have elevated tumour markers to suspect TC (Laguna et al. 2020).

Magnetic resonance imaging (MRI) has higher sensitivity and specificity than ultrasound when considering the diagnosis of TC. However, it has a higher cost; therefore, its use is recommended for unclear scenarios (Kim et al. 2007; Cassidy et al. 2010).

Fig. 17.2 Diagnostic tools for testicular cancer (adapted from Laguna et al. 2020; Germà-Lluch et al. 2002; Moul 2007)

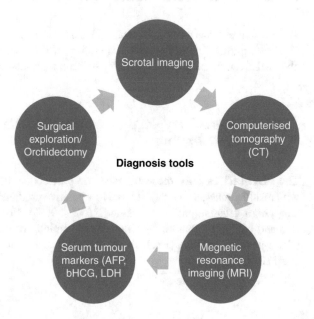

17.7.2 Computerised Tomography (CT)

For staging, contrast-enhanced computerised tomography (CT) is the most sensitive way to assess the abdomen, pelvis and thorax. It is recommended for all patients before a surgical procedure. Nonetheless, it can be performed after the histopathological confirmation.

Pierorazio et al. described a sensitivity, specificity, positive predictive value (PPV), negative predictive value (NPV) and accuracy of 66.7% (range 37–100%), 95.2% (range 58–100%), 87.4% (60–100%), 73.4% (67–100%) and 83% (range 71–100%), respectively for abdominal contrast-enhanced CT (Pierorazio et al. 2020). Moreover, specificity increases and sensitivity decreases when lymph node size increases (Leibovitch et al. 1995).

Pierorazio et al. also found a median sensitivity, specificity, PPV, NPV and accuracy of 100% (range 95–100%), 92.7% (range 89–97%), 67.7% (range 25–84%), 100% (range 99–100%) and 93% (range 91–97%), respectively for thoracic CT. This tool is even more sensitive for thoracic staging (Pierorazio et al. 2020).

When findings are not conclusive, for example, lymph nodes less than 2 cm in the retroperitoneum, restaging after 6–8 weeks might be suggested. On the other hand, in patients with NCGCT, multiple lung metastases and poor-prognosis risk group (>5000 UI/L hCG values) or clinical symptoms, a brain CT should be done (Feldman et al. 2016).

17.7.3 Magnetic Resonance Imaging (MRI)

This critical tool has a role in detecting brain metastasis. MRI is more sensitive than CT, but it requires more expertise. Therefore, if CT is contraindicated and if available, MRI should be used to characterise brain metastasis (Pope 2018).

17.7.4 Fluorodeoxyglucose-Positron Emission Tomography (FDG-PET)

The FDG-PET is only recommended for post-chemotherapy seminoma patients with residual masses (>3 cm). It has a low positive predictive value (PPV) (23–69%) and a high negative predictive value (NPV) (>90%). Necrosis, fibrosis, inflammation and benign tumours might increase the activity of fluorodeoxyglucose (FDG) (False-positive results) (Bachner et al. 2012; De Santis et al. 2014; Cathomas et al. 2018).

17.8   Serum Tumour Markers

Alpha-fetoprotein (AFP), beta subunit of human chorionic gonadotropin (b-hCG) and lactate dehydrogenase (LDH) should be measured before and after the orchidectomy. To establish the S in the TNM classification, the serum markers should be measure before the surgical procedure. On the other side, for the TC prognosis group classification, these markers should be measure after orchidectomy. For metastatic TC, risk stratification is based on serum markers before chemotherapy.

AFP and b-hCG are increased in around 60% of patients with NSGCT. Pure seminomas do not have AFP elevation, but as much as 20% may increase b-hCG. AFP rarely elevates in pure choriocarcinoma. Teratoma rarely elevates AFP and hCG (Stevenson and Lowrance 2015; Barlow et al. 2010; Gilligan et al. 2010) (Table 17.5).

The serum half-life of AFP is 5–7 days and for b-HCG is 1–3 days. Therefore, the physician needs to measure them post-surgery after 3–5 half-lives; until normalisation expected to occur. This assessment is used for prognosis, although normalisation after orchiectomy does not exclude metastatic disease. LDH is the less specific marker; it suggests tumour volume (Laguna et al. 2020; Mead 1997).

17.9   Surgical Exploration

Surgical exploration becomes mandatory in patients with suspected TC. This approach will help to determine the histopathology of the tumour and local staging. Radical orchidectomy (up to the internal inguinal ring) is the standard of care in these patients (Laguna et al. 2020).

Testis sparing surgery might be indicated if solitary testis to preserve fertility and hormone function. It should be performed with frozen section examination (FSE). Data is limited for this approach; therefore, the patient should be clear that he may need a radical orchidectomy after (Laguna et al. 2020; Matei et al. 2017; Bienek et al. 2018; Scandura et al. 2018).

Contralateral two-site surgical biopsy is advised to rule out GCNIS. It is not indicated for all patients, only for those at high risk such as testicular volume <12 mL, and/or history of undescended testis (Laguna et al. 2020; Ruf et al. 2015; Heidenreich and Moul 2002; Dieckmann et al. 2007).

Table 17.5 Description of tumour markers and spread pattern according to the type of tumour

Histologic subtype	AFP	hCG	LDH	Lymphatic spread	Hematogenous spread
Seminoma	Never	Possible	Possible	Common	Rare
Choriocarcinoma	Rare	Usually	Possible	Common	Possible
Teratoma	Rare	Rare	Possible	Common	Rare
Embryonal	Possible	Possible	Possible	Common	Rare
Yolk sac	Possible	Possible	Possible	Common	Possible

17.10   Treatment

17.10.1   Germ Cell Neoplasia in Situ (GCNIS)

GCNIS is a precursor lesion for GCT often diagnosed by orchiectomy or testicular biopsy for a cause different from malignant disease suspicion. The risk of developing subsequent invasive GCT is 50% within 5 years (KP et al. 2003). Radical orchiectomy is the standard treatment, although low-dose radiotherapy ($\leq$20 Gy) is also considered due to the advantage of preserving testicular endocrine function, as Leydig cells are relatively radio-resistant compared with germinal epithelium (Montironi 2002).

17.11   Clinical Stage I Seminoma Germ Cell Tumours

Treatment in this stage has been debated in the last decade, with active surveillance, chemotherapy and radiotherapy being accepted. These treatments provide long-term control of the disease in almost 100% of cases (Stephenson and Gilligan 2020). Compared with other types of GCT, seminomas have a relatively favourable natural history with increased sensitivity to radiation therapy and platin-based chemotherapy (Powles et al. 2005).

Active surveillance includes schedules with reviews every 2–4 months during the first 3 years, then every 6 months until completing 7 years, and therefore annually. As serum tumour markers are less useful for evaluating relapses in seminomas compared to NSGCT, follow-up with other additional tests such as chest radiography and abdominopelvic CT, in addition to physical examination, is necessary. This treatment modality should be used in patients with a low risk of relapse, which according to some studies, corresponds to tumours <4 cm and without invasion of the rete testis (Warde et al. 2002).

The mean time to relapse in CS I seminomas is usually longer compared to NSGCT (14 months vs. 4–8 months), in addition to a larger percentage of patients who relapse after more than 5 years (Chung et al. 2015). This implies the need for a longer-term follow-up than in other types of GCT, with special attention to retroperitoneal involvement, since approximately 15% of CS I seminomas present subclinical metastatic disease at that level at diagnosis, although it usually remits after orchiectomy (Chung et al. 2015; Tandstad et al. 2016; Cohn-Cedermark et al. 2015; Kollmannsberger et al. 2015).

Adjuvant chemotherapy with carboplatin agent using a dosage of one or two courses of area under the curve (AUC) 7. The use of carboplatin derives from the excellent response obtained in advanced seminomas (complete response rates of 65–90%) and its lower toxicity compared to cisplatin (Horwich et al. 2000). The mean time to relapse is around 19 months, with a 15% relapse at 3 years after treatment, which is a longer time if compared with active surveillance. The majority of patients relapsing after adjuvant chemotherapy with carboplatin can be successfully treated with a standard cisplatin-based chemotherapy regimen appropriate to their disease stage (Fischer et al. 2017).

Radiation therapy is also considered for seminomas due to its highly radiosensitivity; adjuvant radiation therapy with a total dose of 20–24 Gy reduces recurrence rates by up to 1–3% (Melchior et al. 2001). Radiation therapy regimens include the para-aortic (PA) regions, and occasionally also the ipsilateral inguinal region (dog-leg configuration). This last configuration entails a lower pelvic recurrence rate at the expense of a longer time for spermatogenesis recovery (Brenner and Hall 2007). A scrotal shield should always be used in the contralateral testis to prevent toxicity (Bieri et al. 1999).

17.12   Clinical Stage I Non-seminoma Germ Cell Tumours

NSGCT can present subclinical occult metastases in the retroperitoneum or at distant sites in 20–30% of patients (Albers et al. 2003). This condition makes the decision between the different treatment options controversial. Surveillance, primary retroperitoneal lymph node dissection (RPLND) and primary chemotherapy are accepted treatment options with similar long-term survival rates approaching 100%.

Some risk factors have been studied to predict the presence of hidden metastases at diagnosis in order to individualize the optimal treatment for each patient. The most commonly identified risk factors are lympho-vascular invasion (LVI) and a predominant component of embryonal carcinoma (EC). In the absence of these two factors, the risk of occult metastases is less than 20%, so surveillance could be a reasonable recommended approach for these type of patients (Vergouwe et al. 2003).

Active surveillance within a rigorous protocol offers the potential of reducing the treatment toxicity in selected patients with non-risk NSGCT. Approximately 70–80% of patients with CS I NSGCT are cured by orchiectomy alone and 14–48% of patients undergoing surveillance recur within 2 years of surgery (Kollmannsberger et al. 2015; Groll et al. 2007). The location of relapses is usually retroperitoneal in 60% of cases, with 11% presenting a large-volume metastatic recurrent disease. Serum tumour markers can be normal in 35% of patients at the time of relapse, so imaging and clinical evaluation should be performed regularly during follow-up (Hamilton et al. 2019). The standard treatment for patients who relapse under surveillance is induction chemotherapy based on the IGCCCG risk (Mead 1997).

For men undergoing primary chemotherapy, bleomycin etoposide and cisplatin (BEP) ×1 is recommended with a reported relapse rate of 2.3% and a mortality rate of 0.2%. These results are not significantly different from the rates seen with BEP×2 (Gilligan et al. 2020; Huddart and Reid 2018). The disadvantages of primary chemotherapy are that it is not useful for the treatment of retroperitoneal teratomas, favouring chemo-resistance and late relapses, the need for long-term follow-up with CT imaging, and its related toxicity (particularly cardiovascular). Primary chemotherapy is thus associated with the lowest risk of relapse, but these relapses are frequently chemo-resistant, especially if they have received a regimen other than BEP (Stephenson and Gilligan 2020).

The role of RPLND has diminished due to the high cancer-specific survival (CSS) obtained with chemotherapy and salvage treatment in relapse cases. It is recommended to perform a full, bilateral template RPLND, employing a nerve-sparing technique to avoid ejaculatory disfunction, as it should not compromise the oncologic efficacy (Eggener et al. 2007; Subramanian et al. 2010) (Fig. 17.3). RPLND should only be performed by experienced surgeons in high volume centres. In cases of node involvement (pN1), RPLND is curative in 60–90% of patients, although the risk of relapse in patients with pN2-3 disease is greater than 50% (Sheinfeld 2003). In these cases, adjuvant chemotherapy reduces the rates of relapse to less than 1% (Kondagunta et al. 2004). Rates of relapse after RPLND for retroperitoneal teratoma are low and adjuvant chemotherapy is not recommended (Liu et al. 2015).

CS IS involves the presence of elevated post-orchiectomy serum markers regardless of clinical and radiological findings of metastatic disease (Fig. 17.3). These patients should be treated similar to the ones with CS IIC-III, with a CSS greater than 90% after induction chemotherapy.

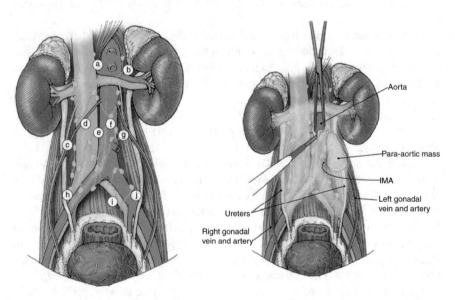

Fig. 17.3 Lymph node chains involved in RPLND: Right (a) and left suprahiliar (b), paracaval (c), precaval (d), inter-aortocaval (e), pre-aortic (f), para-aortic (g), right iliac (h), inter-iliac (i) and left iliac (j)

17.13   Clinical Stage IIA-B Seminoma Germ Cell Tumours

In patients with CS IIA-B Seminoma, radiotherapy provides recurrence rates of 9–24%, being one of the most standardized therapies. The employed doses are between 30 and 36 Gy over a field that includes the para-aortic and ipsilateral iliac lymph nodes. Relapses are treated in almost all cases with chemotherapy, reaching a CSS around 100% (Classen et al. 2003).

Chemotherapy is an alternative to radiotherapy: BEP×3 or etoposide plus cisplatin (EP) ×4, which can be used in case of contraindications to bleomycin and in older patients. Induction chemotherapy is preferable in CS IIB and patients with multiple or bulky (≥3 cm) masses, due to the lower risk of relapse in comparison with radiotherapy (Garcia-del-Muro et al. 2008).

Currently, there are no randomised studies assessing the difference between both therapies. However, some clinical trials are evaluating the role of treatment alternatives such as RPLND or the combination of both radiotherapy and a single course of carboplatin, with a potentially lower toxicity.

17.14   Clinical Stage IIA-B Non-seminoma Germ Cell Tumour

Induction chemotherapy (± post-chemotherapy RPLND) and RPLND (± adjuvant chemotherapy) are accepted treatments with survival rates over 95%. Induction chemotherapy should be the initial treatment in most of the cases, as 60–78% of patients achieve a complete response, avoiding a subsequent RPLND, with a CSS around 96–100% (Stephenson et al. 2007; Weissbach et al. 2000).

There is consensus that patients presenting elevated tumour markers (AFP and b-HCG) or bulky lymph nodes (≥3 cm) should receive induction chemotherapy because of the high risk of occult metastasis disease (Stephenson et al. 2005).

Arguments in favour of RPLND are that 30% of patients with CS IIA-B NSGCT have retroperitoneal teratoma that is resistant to chemotherapy (Stephenson et al. 2005), 13–35% of patients have pathologically negative lymph nodes (Weissbach et al. 2000) and 10–35% of patients avoid any chemotherapy (Donohue et al. 1995).

Patients with indeterminate lymph nodes on staging abdominal-pelvic CT imaging may be observed initially during a 6 weeks period to clarify which should be the first treatment option, given the low-risk of metastatic disease (Stephenson et al. 2007).

17.15   Clinical Stage IIC and III Seminoma Germ Cell Tumours

In these cases, the preferred treatment is first-line cisplatin-based chemotherapy (Bokemeyer et al. 2004). Depending on the IGCCCG prognosis risk group, the regimens and number of cycles may vary. Standard treatment in good-prognosis seminoma should be BEP×3 or EP×4, in the case of contraindications to bleomycin (De Wit 2007). Single-agent carboplatin in advanced seminoma is associated with inferior survival compared to cisplatin-based regimen (Bokemeyer et al. 2004). In intermediate-prognosis seminoma BEP×4 or etoposide, cisplatin and ifosfamide (VIP) ×4 when contraindications to bleomycin are recommended options (Fizazi et al. 2014).

After first-line chemotherapy, 58–80% of patients have residual masses on control imaging, of which 50–66% show spontaneous resolution (30–50% for masses >3 cm). After histological analysis, viable malignancy only appears in 10% of cases (Bachner et al. 2012, A F et al. 2002). The size of residual masses is an important predictor of viable malignancy: 0-4% for masses <3 cm, compared with 13-55% for masses >3 cm (AF et al. 2002; Herr et al. 1997). FDG-PET has a high NPV in patients with residual masses after chemotherapy, in order to select patients for post-chemotherapy surgery if information on disease viability is provided (Bachner et al. 2012; De Santis et al. 2004). There is consensus that patients with discrete residual masses >3 cm should be evaluated with FDG-PET at least 6–8 weeks after completion of chemotherapy. Those who are FDG-PET negative or those with masses smaller than 3 cm should be observed.

Regarding the appropriate management of patients with FDG-PET positive residual masses larger than 3 cm, there is controversy given the reported rates of false-positive imaging (De Santis et al. 2004). Post-chemotherapy RPLND should be considered in those cases, particularly if residual masses are observed after second-line chemotherapy (Rice et al. 2014).

17.16   Clinical Stage IIC and III Non-seminoma Germ Cell Tumour

As in Seminomas, CS IIC and III NSGCT should be managed with cisplatin-based chemotherapy. The regimens and chemotherapy cycles also would vary depending on the IGCCCG risk criteria.

For good-prognosis tumours, the treatments of choice are BEP×3 or EP×4. The 5-year overall survival (OS) with this regimen is 91–94%. Two randomized trials have shown that BEP×3 is no inferior to BEP×4 in good-risk patients (De Wit et al. 2001; Saxman et al. 1998). Nevertheless, BEP×3 is superior to cisplatin, vinblastine and bleomycin (PVB) in with advanced disease (De Wit et al. 1997, 2001).

For patients with an intermediate or poor-prognosis NSGCT as defined by the IGCCCG, standard treatment consists of BEP×4, with a 5-year survival rate of 79% and 48%, respectively (Mead 1997). A recent randomized clinical trial (RCT) showed no significant improvement in overall survival (OS) with BEP×4 plus paclitaxel (T-BEP) compared to BEP×4 alone (De Wit et al. 2012). Also, cisplatin, etoposide and ifosfamide (PEI) ×4 has similar efficacy in exchange for higher myelotoxicity (Nichols et al. 1998).

After completion of first-line BEP chemotherapy, between 38 and 68% of patients have residual masses larger than 1 cm, and there is clear consensus that they should undergo post-chemotherapy surgery (Honecker et al. 2018). After post-chemotherapy surgery, multiple series have reported persistent GCT on histological analysis in ≥50% of the resected specimens. Histology will demonstrate necrosis, teratoma and viable malignancy in 40%, 45% and 15% of cases, respectively (Stephenson and Gilligan 2020). Thus, RPLND should be performed because of the high probability of residual disease in the retroperitoneum. RPLND is a strong predictor of histology at other anatomical sites (chest and supraclavicular fossa are the most common). Observation of small residual masses at other sites different from retroperitoneum is a reasonable option of the histology of RPLND is necrosis (Masterson et al. 2012).

Brain metastases are associated with choriocarcinoma, presenting highly vascular tumours that tend to bleed, causing haemorrhage during chemotherapy, with death rates of 4–10% (Nonomura et al. 2009). Close monitoring of intracranial haemorrhage during chemotherapy should be carried out. After BEP×4 chemotherapy regimen, resection of residual masses is recommended too (Kollmannsberger et al. 2000).

17.17 Non-germ Cell Tumours

Sex cord-stromal tumours include Leydig cell, Sertoli cell, granulosa cell tumours and gonadoblastoma. Some of them can be suspected preoperatively as they may initially be seen with gynecomastia, impotence, decreased libido, infertility and painless testicular masses. Testis-sparing surgery may be performed in the case of tumours smaller than 3 cm with intraoperative histologic confirmation of clear surgical margins. Approximately 90% of these tumours are benign (Carmignani et al. 2007; Banerji et al. 2016). Orchiectomy should be completed if malignant features or GCT appear in final histology. Malignant features include size larger than 5 cm, infiltrative margins, extra-testicular extension, high mitotic index, necrosis or vascular invasion, nuclear atypia, DNA ploidy and increased MIB-1 expression (Cheville et al. 1998). Patients with ≥2 malignant features or metastatic disease should undergo RPLND, as metastatic Leydig cell tumours are resistant to radiotherapy and chemotherapy, with a poor prognosis (Nicolai et al. 2015). In case of gonadoblastoma, bilateral orchiectomy is required because of the risk of bilateral tumours (40%) and subsequent development of GCT in half of the cases (Ulbright 2005; Scully 1970).

17.18 Differences Between the European Urological Association (EAU) and National Comprehensive Cancer Network (NCCN) Testicular Cancer Guidelines

In order to better address the differences between the *National Comprehensive Cancer Network* (NCCN) and European Association of Urology (EAU) guidelines each area will be evaluated individually (Laguna et al. 2020; Gilligan et al. 2020). A summary of the main differences between the two guidelines can also be seen in Table 17.6.

17.18.1 Diagnosis and Staging

With regards to diagnosis and staging recommendations there are no differences between the two classifications; as an extra note, the EAU guidelines recommend family members with a family history of TC to perform regular self-exploration (Laguna et al. 2020).

Table 17.6 Main differences between National Comprehensive Cancer Network (NCCN) and European Association of Urology (EAU) guidelines

		NCCN guidelines	EAU guidelines
Diagnosis		Family members with a family history of testicular cancer should perform Auto-exploration.	
Treatment	CS I Seminoma with risk factors	Surveillance, chemotherapy or RPLND	Surveillance only for patients that do not accept or cannot undergo chemotherapy. RPLND for select cases that cannot undergo chemotherapy or surveillance
	CS IIIC Non seminoma	BEP ×4 or VIP ×4	One cycle of BEP/PEI[a] ->3 weeks tumour marker analysis. When decreased: BEP or PEI maximum of four cycles. When markers progress: treatment intensification is recommended.
Surveillance		Protocol by stage and histology. Tumour marker optional for CS I, IIA and IIB non-bulky seminoma.	Generalised protocol. Tumour marker surveillance recommended for all stages.

EAU European Urological Association, *NCCN* National Comprehensive Cancer Network, *CS* clinical stage, *RPLND* retroperitoneal lymph node dissection, *BEP* bleomycin, etoposide and platinum, *PEI* cisplatin, etoposide and ifosfamide

[a]Decreased lung function

17.18.2 Treatment

Treatment options in seminoma vs non-seminomatous tumour (Table 17.7).

17.18.2.1 Seminoma: Clinical Stage I

In the NCCN classification, surveillance is recommended for pT1-pT3 testicular tumours. On the other hand, the EAU guidelines suggest doing so when health services and patient compliance is available.

With regards to carboplatin chemotherapy the NCCN guidelines recommend performing 1–2 cycle treatment whereas the EAU guidelines recommend only 1.

In the case of RT, the NCCN guidelines recommend treatment with 20 Gy/10 fractions, whereas the EAU advises against it. They also recommend limiting the use of adjuvant treatments in patients with no risk factors (Laguna et al. 2020; Gilligan et al. 2020).

17.18.2.2 Seminoma: Clinical Stage II

The NCCN guidelines offer RT at a dose of 36 Gy as a treatment option only for patients with non-bulky (<3 cm) CS IIB (Gilligan et al. 2020).

Table 17.7 Treatment options in patients with seminoma and non-seminoma germ cell tumours

Clinical stage	Seminoma	Non-seminoma
CS I	– *Active surveillance* (preferable in patients with low-risk of relapse) – *Adjuvant chemotherapy* [single-agent carboplatin] (AUC = 7 ×1–2 cycles) – *Adjuvant radiotherapy* [20–24 Gy] (PA and ipsilateral iliac lymph nodes)	– *Active surveillance* (preferable in patients with low-risk of relapse) – *Adjuvant chemotherapy* [BEP ×1 cycle] – *RPLND*
CS IIA-B	– *Radiotherapy* [30–36 Gy] (PA and ipsilateral iliac lymph nodes) – *Chemotherapy* [BEP ×3 cycles or EP ×4 cycles]	– *Chemotherapy* [BEP ×3 cycles or EP ×4 cycles] – *RPLND*
CS IS, IIC, III	– *Chemotherapy* o Good-prognosis [BEP ×3 cycles or EP ×4 cycles] o Intermediate-prognosis [BEP ×4 cycles or VIP ×4 cycles]	– *Chemotherapy* o Good-prognosis [BEP ×3 cycles or EP ×4 cycles] o Intermediate-prognosis [BEP ×4 cycles] o Poor-prognosis [BEP ×4 cycles]
	• Brain metastases [Chemotherapy ± Radiotherapy ± Surgery (if clinically indicated)]	

17.18.2.3   Non-seminoma: Clinical Stage I

Both guidelines recommend surveillance for CS I non-seminoma without risk factors; when surveillance is not available: NCCN recommends RPLND or one course of BEP, which is also the suggested option by the EAU guidelines (Laguna et al. 2020; Gilligan et al. 2020).

The EAU guidelines also state that patients with marker-positive recurrence and/or progressing tumours should undergo three or four course BEP followed by RPLDN when necessary (Laguna et al. 2020).

In CS I with risk factors (pT2-pT4), the NCCN guidelines suggest surveillance, chemotherapy, or RPLND as possible treatments. The EAU guidelines only recommend surveillance for patients that do not accept or cannot undergo chemotherapy and reserve RPLND for select cases that cannot undergo chemotherapy, nor surveillance.

17.18.2.4   Non-seminoma: Clinical Stage II and III

- IIA/IIB (Negative markers): the EAU guidelines recommend excluding negative marker embryonal carcinoma by histological examination either from RPNLD or biopsy and 6-week surveillance before restaging for patients in which histological examination is not available (Laguna et al. 2020). Meanwhile, the NCCN guidelines recommend nerve-sparing RPLDN and primary chemotherapy with BEP ×3 or EP ×4 (Gilligan et al. 2020).
- IIIC: The NCCN guidelines concur on BEP ×4 or VIP ×4 for selected cases (Gilligan et al. 2020). The EAU guidelines recommend one cycle of BEP or cisplatin plus etoposide plus ifosfamide (PEI) for patients with decreased lung function. After 3 weeks, a new serum tumour marker analysis should be performed, and, if markers have declined, BEP or PEI treatments should be maintained for a maximum of four cycles. When markers progress, treatment intensification is recommended. Residual mass excision should be performed after chemotherapy and marker levels are normalising/normal (Laguna et al. 2020; Gilligan et al. 2020).

17.18.3   Surveillance

The NCCN guidelines recommend serum tumour marker analysis follow-up for CS I, IIA and IIB non-bulky seminoma tumours as optional, whereas the EAU recommends their analysis at all stages. Another difference is the NCCN guidelines have made specific recommendations for each stage, whereas the EAU guidelines have joined all tumours into three risk groups (Laguna et al. 2020; Gilligan et al. 2020).

17.19 Tumour Recurrence

17.19.1 Seminoma Recurrence

CS I seminoma tumours have a very good prognosis with lower recurrence rates compared to NSGCT independent of the treatment option (Groll et al. 2007). Patients with CSI seminoma present a risk of approximately 15% of relapse due to occult metastasis, generally in the retroperitoneum (Fischer et al. 2017). Specifically, the para-aortic lymph nodes are the most common site of recurrence, presenting asymptomatically in the majority of cases (Boujelbene et al. 2011). For tumours of the right testicle, inter-aortocaval lymph nodes are mostly affected, whereas tumours on the left testicle disseminate to para-aortic lymph nodes (Hale et al. 2018). Other less common areas of recurrence are pelvic lymph nodes and rarely inguinal lymph nodes and lung. Use of abdominopelvic CT allows for detection of possible recurrences. The eighth edition of American Joint Committee on Cancer (AJCC) presents imaging characteristics to facilitate differentiation between normal and pathologic findings. They recommend considering lymph nodes greater than 2 cm in size as positive for tumoural dissemination (Hale et al. 2018).

Previous studies have found an overall recurrence rate for seminoma tumours of 16.8% with a specific survival rate close to 100% (Laguna et al. 2020; Groll et al. 2007). Most recurrences appear within the first 2 years, with median recurrence times tending to be longer in cases of seminoma. With regards to the site of recurrence, the most commonly affected area was the retroperitoneal lymph nodes (Nayan et al. 2017). Cases with no rete testis invasion or with a size <4 cm are considered to have low-risk features, in such cases, disease recurrence is present in around 6% of patients (Nayan et al. 2017). Other studies have presented a 12.2% conditional recurrence rate for tumours <3 cm (Nayan et al. 2017).

Recurrence generally presents in patients with CS I on active surveillance and CS I-IIB that have undergone primary radiotherapy. In cases of CSI recurrence after active surveillance, treatment with primary radiotherapy offers curative rates of 70–90% (Stephenson and Gilligan 2020). On the other hand, patients with bulky (>3 cm) retroperitoneal masses and patients with systemic dissemination should undergo first-line chemotherapy. In these cases, response rates can reach 100% (Stephenson and Gilligan 2020). Treatment with first-line chemotherapy for patients who have previously undergone radiotherapy allows for disease eradication in nearly all cases with extra retroperitoneal affection (Stephenson and Gilligan 2020).

For patients that have previously undergone chemotherapy, a distinction must be made between those that presented an early relapse (<2 years after completion of primary treatment) and those with a late relapse (>2 years). Early relapse in post-chemotherapy seminoma patients presents in 15–20% of cases after induction chemotherapy, and 10% of these will even do so after presenting a complete initial response (Mead 1997). In both of these cases, prognosis tends to be poor, with long-term survival rates that range between 20 and 50% (Stephenson and Gilligan

2020). Studies evaluating effectiveness of second-line chemotherapy are limited due to the low incidence of cases. Most patients are treated following recommendations obtained from management of recurrent NSGCT. Some studies have evaluated the use of vinblastine, ifosfamide and cisplatin (VeIP)×4 as a second-line chemotherapy. Treatment with high-dose chemotherapy have also been evaluated in a small series of patients presenting a 50% complete response rate. Patients with tumour recurrence after first-line chemotherapy, especially cases with normal tumour markers, should undergo histological analysis to evaluate the possibility of teratoma.

17.19.2   Late Recurrences

Of all cases of late relapses, seminoma tumours are responsible for less than 8% of recurrences (Coward et al. 2009; Baniel et al. 1995).This percentage may in fact be higher, as other series such as Dieckmann et al., presenting 41% of pure seminoma late recurrences in their 122-patient series (Albers et al. 2008). Of these 122 cases, only 6% had received first-line chemotherapy, with most patients having undergone single-agent treatment with carboplatin or RT. From their series, persistent disease control was achieved in 88% of cases. Extrapolating from these findings, some authors have concluded that late relapse seminoma may, in fact, be a positive prognosis factor, most of all, for patients that have not received previous treatments with cisplatin (Stephenson and Gilligan 2020).

17.19.3   NSGCT Recurrence

Treatment of NSGCT recurrence depends on the initial treatment received and the site of relapse. Cases without prior chemotherapy present a better prognosis compared to those that have (Stephenson and Gilligan 2020).

Chemotherapy-naïve NSGCT recurrence occurs in patients with CSI on active surveillance or CSI-IIB cases that underwent primary RPLDN as single treatment. Elevation of tumour markers may be present in 60–75% of cases of CSI NSGCT on active surveillance (Alexandre et al. 2001). For these cases, chemotherapy following the International Germ Cell Cancer Collaborative Group (IGCCCG) risk regime and duration recommendations allow for curative rates of around 95% (Stephenson et al. 2007). Some select cases with non-bulky (<3 cm) retroperitoneal relapse with negative tumour markers on surveillance may be candidates for RPLDN or induction chemotherapy. Specifically, cases with teratoma in the initial histology show the best response rates in these scenarios (Stephenson et al. 2007). RPLDN long term curative rates are close to 100% and allow for a reduction of patient exposure to chemotherapy. Patients with CS I-IIB that present tumour recurrence after RPLDN are most commonly affected in lungs or mediastinum (Stephenson and Gilligan 2020). Recurrence generally occurs within the first 2 years, and most cases present a good response to first-line chemotherapy (Stephenson and Gilligan 2020; Luz et al. 2010; Zuniga et al. 2009). It is recommended that for cases of late recurrence (>2 years) after orchiectomy or RPLDN, especially in cases with normal levels of tumour markers, biopsy or tumour resection is performed in order to omit the possibility of teratoma (Michael et al. 2000; Oldenburg et al. 2006). Survival of chemotherapy-naïve patients appears similar between early and late recurrences.

Most of NSGCT recurrences appear in the first 2 years from treatment and are known as early recurrences (<2 years) (Michael et al. 2000; Nichols et al. 1997). A subset of early relapse with worse prognosis are those that do so within the first 6 months of complete response after first-line chemotherapy. These are known as incomplete responders (Fosså et al. 1999).

Lorch et al. performed a multicentre pooled analysis evaluating 1984 patients that underwent second-line chemotherapy that recurred after first-line. They observed a progression-free average time of 10 months, with an overall survival of 41 months. Factors associated with progression after second-line chemotherapy were: incomplete responders (HR 1.4–1.9), visceral non-pulmonary metastasis (HR 1.3), primary mediastinal involvement (HR 3), serum AFP elevation (HR 1.3–2.0) and HCG (HR 1.5) (Fizazi et al. 2014; Mead et al. 2005).

Currently, four main second-line chemotherapy combinations are accepted. These treatments are based around three-agent therapies based on cisplatin and ifosfamide combined with a third agent (i.e. paclitaxel (TIP), etoposide (PEI/VIP) or gemcitabine (GIP) (Fizazi et al. 2014; Mead et al. 2005). Due to the haematological toxicity associated to these treatments, the use of granulocyte colony-stimulating factor should be associated, and close control by an experienced oncologist is warranted. To this day, randomised controlled trials are still required to compare the efficacy between these regimes. Another proposed treatment option as second-line chemotherapy is high dose chemotherapy (HDCT). However, three randomised trials studies evaluating survival improvement when compared to standard-dose suggested it should not be considered a general approach. Current recommendations reserve these treatments for centres with larger experience (Beyer et al. 2002). HDCT treatment options include combinations of: gemcitabine, paclitaxel, oxaliplatin, and irinotecan (Bokemeyer et al. 2008; Nicolai et al. 2009; Bachner et al. 2012; Veenstra and Vaughn 2011).

It is important to evaluate tumour markers during follow-up as patients that present decreasing or normal levels after first-line chemotherapy, followed by an increase in tumour growth, may be due to cystic lesions. This rare case is known as growing teratoma syndrome. Management of such cases involves chemotherapy interruption, followed by radical resection. Patients that achieve resection present a favourable long-term prognosis (André et al. 2000).

In some cases, residual mass after salvage chemotherapy may occur. Surgical resection of these masses is necessary when a complete serological response is achieved after second-line chemotherapy. This is known as post-salvage chemotherapy surgical resection (PSCS) (Eggener et al. 2007; Cary et al. 2015; Debono et al. 1997). Viable malignant tumour is present in 38–45% of PSCS histology with rates of necrosis and teratoma averaging 26% and 21–24%, respectively. The presence of teratoma and necrosis is generally lower when compared to residual masses of patients after first-line chemotherapy. Patients with PSCS present poor 5-year survival rates ranging from 44 to 61%, most of all, for patients with viable malignant tissue (Hartmann et al. 1997).

Some patients will experience disease progression despite first- and second-line chemotherapy, presenting with a very limited prognosis. For cases that present elevating tumour markers with a single retroperitoneum site, surgical resection should be attempted when possible. This is known as "desperation surgery". Series evaluating these cases are still limited; however, some authors have reported postoperative tumour marker decline in 47–60% of cases with long-term survival ranging from 33 to 57% independent of postoperative chemotherapy association (Albers et al. 2003; Beck et al. 2005; Eastham et al. 1994). Nearly half of these patient's present necrosis or teratoma in the resected mass.

17.19.4   Late Recurrence

Late recurrence following chemotherapy occurs in 3% of NSGCT cases (Oldenburg et al. 2006). In such cases, histological analysis is recommended, especially in those with normal tumour markers. Upon histological analysis, three main possibilities may be seen: viable malignancy (54–88%) of which yolk sac tumour is the most common, teratoma (12–28%) or malignant alteration (10–20%) being adenocarcinoma the most common differentiation (Masterson et al. 2012; Baniel et al. 1995). The most common area of late recurrence is the retroperitoneum (50–72%), followed by the lungs (17%) mediastinum (9%), neck (7%) and the pelvis (4%) (Masterson et al. 2012; Albers et al. 2008; Oldenburg et al. 2006). Late recurrence is commonly resistant to chemotherapy; therefore, oncological control is largely centred on surgical resection. It is recommended that these surgeries are performed in experienced centres in order to reduce complications.

17.20   Surveillance

Surveillance during the first 5 years after diagnosis is centred on the detection of recurrent disease. Early detection allows for selection of the most effective treatment with the lowest risk of complications. Effective follow-up depends on correct previous histological analysis, staging, primary treatment, and close control of post-surgical baseline CT scan changes. Many authors recommend establishing follow up regimes with adequate with patients, physicians and health care capabilities (Laguna et al. 2020; Gilligan et al. 2020).

Surveillance for all stages should be centred on clinical history, physical examination and tumour marker analysis (CH+PE+TM). The time between each follow-up visit should correlate with risk of tumour recurrence for each case (Laguna et al. 2020; Gilligan et al. 2020). Surveillance is done primarily through computed tomography (CT), reserving ultrasound and other imaging techniques for specific situations. Recent recommendations suggest the need to reduce patient exposure to ionizing rays by decreasing the number of surveillance CT scans (Laguna et al. 2020; Gilligan et al. 2020). Consequent to these recommendations, magnetic resonance imaging (MRI) has been put forward as an equally valid follow-up imaging technique. Chest imagining through chest X-ray should be performed when required; if symptoms or suspicion of disease dissemination, a chest CT should be performed instead.

Tumour surveillance after 5 years is generally discouraged as late relapse (>5 years) is uncommon, presenting in less than 0.5% of cases. After the 5 years follow up, consultations should focus on detecting adverse events from treatments. Cases of late relapse generally present symptomatically, with half of the cases presenting with an elevation in tumour markers, both in seminoma and in NSGCTs. It is important to educate patients on possible symptoms of tumour recurrence to ensure early detection and treatment.

According to initial tumour histology, CS and choice of treatment, three surveillance groups have been proposed by the EAU (Tables 17.8, 17.9 and 17.10):

The following points summarise minimal surveillance recommendations made by the EAU consensus group members for testicular cancer for each stage:

– Seminoma CS I on active surveillance or after adjuvant treatment:

    a. First year: CH+PE+TM and abdominopelvic CT/MRI twice a year.
    b. Second year: CH+PE+TM and abdominopelvic CT/MRI twice a year.
    c. Third year: CH+PE+TM twice a year and abdominopelvic CT/MRI at 36 months.
    d. Fourth and fifth year: CH+PE+TM once a year and abdominopelvic CT/MRI at 60 months

Table 17.8 Surveillance groups proposed by EAU

1. Patients who have received chemotherapy (adjuvant or curative) for good and intermediate prognosis metastasis disease (Following IGCCCG criteria) achieving a complete remission with or without surgery (In cases of seminoma, this includes <3 cm lesions or >3 cm with a negative PET scan).
2. Patients with seminoma CS I
3. Patients with non-seminoma CS I on active surveillance

Patients with poor prognosis or that have not achieved a complete remission should perform individual surveillance regimes that best correlate to the clinical situation of the patient

Table 17.9 Non-seminoma CS I on active surveillance

First year: CH+PE+TM four[a] times a year, chest X-ray and abdominopelvic CT/MRI twice a year.
Second year: CH+PE+TM four times a year, chest X-ray twice a year and abdominopelvic CT/MRI at 24 months[b].
Third year: CH+PE+TM twice a year, chest X-ray once a year for high-risk cases (LVI+) and abdominopelvic CT/MRI at 36 months[c].
Fourth and fifth year: CH+PE+TM twice a year, chest X-ray at 60 months for high-risk cases (LVI+) and abdominopelvic CT/MRI at 60 months[c].

[a]In high-risk cases (LVI+) 6 CH+PE a year are recommended
[b]In high-risk cases (LVI+) a majority of the EAU consensus group recommends an additional CT scan at 18 months
[c]50% of the EAU consensus group recommends this

Table 17.10 After adjuvant treatment or complete remission for advanced disease (does not include poor prognosis or no remission)

First year: CH+PE+TM four times a year, chest X-ray and abdominopelvic CT/MRI once or twice a year.
Second year: CH+PE+TM four times a year, chest X-ray once a year and abdominopelvic CT/MRI at 24 months.
Third year: CH+PE+TM twice a year, chest X-ray once a year and abdominopelvic CT/MRI at 36 months.
Fourth and fifth year: CH+PE+TM twice a year, chest X-ray once a year and abdominopelvic CT/MRI at 60 months.

17.21 Conclusion

Studies are yet to show which is the optimal treatment for CSI seminoma. The two main accepted regimes are RT and primary chemotherapy. Both of these treatments have presented survival rates close to 100%. These tumours have a low rate of occult metastasis leading to most clinicians advocating for surveillance as the standard treatment for CS I seminoma tumours. Cases that present recurrence after surveillance may undergo RT.

For cases of CSIIA-B, RT and first-line chemotherapy are accepted treatments for non-bulky cases. CS IIC and III Seminoma should undergo first-line chemotherapy following recommendations of the IGCCCG risk group. Residual masses larger than 3 cm should undergo FDG-PET imaging to evaluate functionality. For positive cases, surgical resection is recommended, whereas negative masses and those under 3 cm should be followed closely.

Similar to seminoma, clinical management of CSI NSGCT is still under investigation, with the ideal treatment still to be identified. The three main treatment options: surveillance, primary RPLDN, or chemotherapy have presented survival rates close to 100%. It is recommended that patients with LVI or EC should be managed according to a risk-adapted approach undergoing active treatment, reserving surveillance for patients without these.

Tumour recurrence after surveillance should be treated with induction chemotherapy following based on the IGCCCG risk group. Cases of non-bulky (<3 cm) tumour with normal tumour markers may also undergo RPLDN. CS IS, IIC, IIC and III NSGCT should be managed with chemotherapy. Following risk-appropriate chemotherapy and surgery for necessary cases, the survival of good-, intermediate- and poor-risk is 89–94%, 75–83% and 41–71%, respectively.

References

Albers P, Siener R, Kliesch S, Weissbach L, Krege S, Sparwasser C, et al. Risk factors for relapse in clinical stage I nonseminomatous testicular germ cell tumors: results of the German Testicular Cancer Study Group Trial. J Clin Oncol [Internet]. 2003;21(8):1505–12. https://pubmed.ncbi. nlm.nih.gov/12697874/. Accessed 22 Oct 2020.

Albers P, Siener R, Krege S, Schmelz HU, Dieckmann KP, Heidenreich A, et al. Randomized phase III trial comparing retroperitoneal lymph node dissection with one course of bleomycin and etoposide plus cisplatin chemotherapy in the adjuvant treatment of clinical stage I nonseminomatous testicular germ cell tumors: AUO trial AH 01/94 by the German testicular cancer study group. J Clin Oncol [Internet]. 2008;26(18):2966–72. https://pubmed.ncbi.nlm. nih.gov/18458040/. Accessed 22 Oct 2020.

Alexandre J, Fizazi K, Mahé C, Culine S, Droz JP, Théodore C, et al. Stage I non-seminomatous germ-cell tumours of the testis: identification of a subgroup of patients with a very low risk of relapse. Eur J Cancer [Internet]. 2001;37(5):576–82. https://pubmed.ncbi.nlm.nih. gov/11290432/. Accessed 22 Oct 2020.

André F, Fizazi K, Culine S, Droz JP, Taupin P, Lhommé C, et al. The growing teratoma syndrome: results of therapy and long-term follow-up of 33 patients. Eur J Cancer [Internet]. 2000;36(11):1389–94. https://pubmed.ncbi.nlm.nih.gov/10899652/. Accessed 22 Oct 2020.

Andreassen KE, Kristiansen W, Karlsson R, Aschim EL, Dahl O, Fosså SD, et al. Genetic variation in AKT1, PTEN and the 8q24 locus, and the risk of testicular germ cell tumor. Hum Reprod. 2013;28(7):1995–2002.

Bachner M, Loriot Y, Gross-goupil M, Zucali PA, Horwich A, Germa-lluch JR, et al. 2-18fluoro-deoxy-D-glucose positron emission tomography (FDG-PET) for postchemotherapy seminoma residual lesions: a retrospective validation of the sempet trial. Ann Oncol. 2012;23(1):59–64.

Banerji JS, Odem-Davis K, Wolff EM, Nichols CR, Porter CR. Patterns of care and survival outcomes for malignant sex cord stromal testicular cancer: results from the National Cancer Data Base. J Urol [Internet]. 2016;196(4):1117–22. https://pubmed.ncbi.nlm.nih.gov/27036305/. Accessed 22 Oct 2020.

Baniel J, Foster RS, Gonin R, Messemer JE, Donohue JP, Einhorn LH. Late relapse of testicular cancer. J Clin Oncol [Internet]. 1995;13(5):1170–6. https://indiana.pure.elsevier.com/en/publications/late-relapse-of-testicular-cancer. Accessed 23 Oct 2020.

Barlow LMJ, Badalato GM, McKiernan JM. Serum tumor markers in the evaluation of male germ cell tumors. Nat Rev Urol. 2010;7:610–7.

Beck SDW, Foster RS, Bihrle R, Einhorn LH, Donohue JP. Outcome analysis for patients with elevated serum tumor markers at postchemotherapy retroperitoneal lymph node dissection. J Clin Oncol [Internet]. 2005;23(25):6149–56. https://pubmed.ncbi.nlm.nih.gov/16135481/. Accessed 23 Oct 2020.

Beyer J, Stenning S, Gerl A, Fossa S, Siegert W. High-dose versus conventional-dose chemotherapy as first-salvage treatment in patients with non-seminomatous germ-cell tumors: a matched-pair analysis. Ann Oncol [Internet]. 2002;13(4):599–605. https://pubmed.ncbi.nlm. nih.gov/12056711/. Accessed 23 Oct 2020.

Bieniek JM, Juvet T, Margolis M, Grober ED, Lo KC, Jarvi KA. Prevalence and management of incidental small testicular masses discovered on ultrasonographic evaluation of male infertility. J Urol. 2018;199(2):481–6.

Bieri S, Rouzaud M, Miralbell R. Seminoma of the testis: is scrotal shielding necessary when radiotherapy is limited to the para-aortic nodes? Radiother Oncol [Internet]. 1999;50(3):349–53. https://pubmed.ncbi.nlm.nih.gov/10392822/. Accessed 22 Oct 2020.

Bokemeyer C, Kollmannsberger C, Stenning S, Hartmann JT, Horwich A, Clemm C, et al. Metastatic seminoma treated with either single agent carboplatin or cisplatin-based combination chemotherapy: a pooled analysis of two randomised trials. Br J Cancer [Internet]. 2004;91(4):683–7. https://www.ncbi.nlm.nih.gov/pmc/articles/PMC2364800/. Accessed 22 Oct 2020.

Bokemeyer C, Oechsle K, Honecker F, Mayer F, Hartmann JT, Waller CF, et al. Combination chemotherapy with gemcitabine, oxaliplatin, and paclitaxel in patients with cisplatin-refractory or multiply relapsed germ-cell tumors: a study of the German Testicular Cancer Study Group. Ann Oncol [Internet]. 2008;19(3):448–53. https://pubmed.ncbi.nlm.nih.gov/18006893/. Accessed 23 Oct 2020.

Bosl GJ, Motzer RJ. Testicular germ-cell cancer. N Engl J Med. 1997;337:242.

Boujelbene N, Cosinschi A, Boujelbene N, Khanfir K, Bhagwati S, Herrmann E, et al. Pure seminoma: a review and update [Internet]. Radiat Oncol. 2011;6. https://pubmed.ncbi.nlm.nih.gov/21819630/. Accessed 23 Oct 2020.

Brenner DJ, Hall EJ. Computed tomography – An increasing source of radiation exposure [Internet]. N Engl J Med. 2007;357:2277–84. http://www.nejm.org/doi/abs/10.1056/NEJMra072149. Accessed 22 Oct 2020.

Brierley J, Gospodarowicz M, Wittekind C. The TNM Classification of malignant tumours. 8th ed. United Kingdom: Wiley-Blackwell; 2016.

Carmignani L, Colombo R, Gadda F, Galasso G, Lania A, Palou J, et al. Conservative surgical therapy for leydig cell tumor. J Urol [Internet]. 2007;178(2):507–11. https://moh-it.pure.elsevier.com/en/publications/conservative-surgical-therapy-for-leydig-cell-tumor. Accessed 22 Oct 2020.

Cary C, Pedrosa JA, Jacob J, Beck SDW, Rice KR, Einhorn LH, et al. Outcomes of postchemotherapy retroperitoneal lymph node dissection following high-dose chemotherapy with stem cell transplantation. Cancer [Internet]. 2015;121(24):4369–75. https://pubmed.ncbi.nlm.nih.gov/26371446/. Accessed 23 Oct 2020.

Cassidy FH, Ishioka KM, McMahon CJ, Chu P, Sakamoto K, Lee KS, et al. MR imaging of scrotal tumors and pseudotumors. Radiographics. 2010;30:665–83.

Cathomas R, Klingbiel D, Bernard B, Lorch A, Garcia del Muro X, Morelli F, et al. Questioning the value of fluorodeoxyglucose positron emission tomography for residual lesions after chemotherapy for metastatic seminoma: results of an International Global Germ Cell Cancer Group Registry. J Clin Oncol. 2018;36(34):3381–7.

Cheville JC, Sebo TJ, Lager DJ, Bostwick DG, Farrow GM. Leydig cell tumor of the testis: a clinicopathologic, DNA content, and MIB-1 comparison of nonmetastasizing and metastasizing tumors. Am J Surg Pathol [Internet]. 1998;22(11):1361–7. https://mayoclinic.pure.elsevier.com/en/publications/leydig-cell-tumor-of-the-testis-a-clinicopathologic-dna-content-a. Accessed 22 Oct 2020.

Chung P, Daugaard G, Tyldesley S, Atenafu EG, Panzarella T, Kollmannsberger C, et al. Evaluation of a prognostic model for risk of relapse in stage I seminoma surveillance. Cancer Med [Internet]. 2015;4(1):155–60. /pmc/articles/PMC4312129/?report=abstract. Accessed 22 Oct 2020.

Classen J, Schmidberger H, Meisner C, Souchon R, Sautter-Bihl ML, Sauer R, et al. Radiotherapy for stages IIA/B testicular seminoma: final report of a prospective multicenter clinical trial. J Clin Oncol [Internet]. 2003;21(6):1101–6. https://pubmed.ncbi.nlm.nih.gov/12637477/. Accessed 22 Oct 2020.

Cohn-Cedermark G, Stahl O, Tandstad T. Surveillance vs. adjuvant therapy of clinical stage I testicular tumors – a review and the SWENOTECA experience. Andrology [Internet]. 2015;3(1):102–10. https://pubmed.ncbi.nlm.nih.gov/25270123/. Accessed 22 Oct 2020.

Coward RM, Simhan J, Carson CC. Prostate-specific antigen changes and prostate cancer in hypogonadal men treated with testosterone replacement therapy. BJU Int [Internet]. 2009;103(9):1179–83. https://pubmed.ncbi.nlm.nih.gov/19154450/. Accessed 23 Oct 2020.

De Santis M, Becherer A, Bokemeyer C, Stoiber F, Oechsle K, Sellner F, et al. 2-18fluoro-deoxy-D-glucose positron emission tomography is a reliable predictor for viable tumor in postchemotherapy seminoma: an update of the prospective multicentric SEMPET trial. J Clin Oncol [Internet]. 2004;22(6):1034–9. https://pubmed.ncbi.nlm.nih.gov/15020605/. Accessed 22 Oct 2020.

De Santis M, Bachner M, Cerveny M, Kametriser G, Steininger T, Königsberg R, et al. Combined chemoradiotherapy with gemcitabine in patients with locally advanced inoperable transitional cell carcinoma of the urinary bladder and/or in patients ineligible for surgery: a phase I trial. Ann Oncol. 2014;25(9):1789–94.

De Wit R. Refining the optimal chemotherapy regimen in good prognosis germ cell cancer: interpretation of the current body of knowledge. J Clin Oncol. 2007;25:4346–9.

de Wit R, Stoter G, Kaye SB, Sleijfer DT, Jones WG, ten Bokkel Huinink WW, et al. Importance of bleomycin in combination chemotherapy for good-prognosis testicular non-seminoma: a randomized study of the European Organization for Research and Treatment of Cancer Genitourinary Tract Cancer Cooperative Group. J Clin Oncol [Internet]. 1997;15(5):1837–43. https://pubmed.ncbi.nlm.nih.gov/9164193/. Accessed 22 Oct 2020.

De Wit R, Roberts JT, Wilkinson PM, De Mulder PHM, Mead GM, Fosså SD, et al. Equivalence of three or four cycles of bleomycin, etoposide, and cisplatin chemotherapy and of a 3- or 5-day schedule in good-prognosis germ cell cancer: a randomized study of the European Organization for Research and Treatment of Cancer Genitourinary Tract Cancer Cooperative Group and the Medical Research Council. J Clin Oncol [Internet]. 2001;19(6):1629–40. https://pubmed.ncbi.nlm.nih.gov/11250991/. Accessed 22 Oct 2020.

De Wit R, Skoneczna I, Daugaard G, De Santis M, Garin A, Aass N, et al. Randomized phase III study comparing paclitaxel-bleomycin, etoposide, and cisplatin (BEP) to standard BEP in intermediate-prognosis germ-cell cancer: intergroup study EORTC 30983. J Clin Oncol [Internet]. 2012;30(8):792–9. https://pubmed.ncbi.nlm.nih.gov/22271474/. Accessed 22 Oct 2020.

Debono DJ, Heilman DK, Einhorn LH, Donohue JP. Decision analysis for avoiding postchemotherapy surgery in patients with disseminated nonseminomatous germ cell tumors. J Clin Oncol. 1997;15(4):1455–64.

Dieckmann KP, Classen J, Loy V. Diagnosis and management of testicular intraepithelial neoplasia (carcinoma in situ)–surgical aspects. APMIS [Internet]. 2003;111(1). https://pubmed.ncbi.nlm.nih.gov/12752236/. Accessed 22 Oct 2020.

Dieckmann KP, Kulejewski M, Pichlmeier U, Loy V. Diagnosis of contralateral testicular intraepithelial neoplasia (TIN) in patients with testicular germ cell cancer: systematic two-site biopsies are more sensitive than a single random biopsy. Eur Urol. 2007;51(1):175–85.

Donohue JP, Thornhill JA, Foster RS, Rowland RG, Bihrle R. Clinical stage B non-seminomatous germ cell testis cancer: the Indiana University experience (1965-1989) using routine primary retroperitoneal lymph node dissection. Eur J Cancer. 1995;31(10):1599–604.

Eastham JA, Wilson TG, Russell C, Ahlering TE, Skinner DG. Surgical resection in patients with nonseminomatous germ cell tumor who fail to normalize serum tumor markers after chemotherapy. Urology. 1994;43(1):74–80.

Eggener SE, Carver BS, Sharp DS, Motzer RJ, Bosl GJ, Sheinfeld J. Incidence of disease outside modified retroperitoneal lymph node dissection templates in clinical stage I or IIA nonseminomatous germ cell testicular cancer. J Urol [Internet]. 2007;177(3):937–43. https://pubmed.ncbi.nlm.nih.gov/17296380/. Accessed 22 Oct 2020.

Feldman DR, Lorch A, Kramar A, Albany C, Einhorn LH, Giannatempo P, et al. Brain metastases in patients with germ cell tumors: prognostic factors and treatment options – An analysis from the Global Germ Cell Cancer Group. J Clin Oncol. 2016;34(4):345–51.

Fischer S, Tandstad T, Wheater M, Porfiri E, Fléchon A, Aparicio J, et al. Outcome of men with relapse after adjuvant carboplatin for clinical stage I seminoma. J Clin Oncol [Internet]. 2017;35(2):194–200. https://pubmed.ncbi.nlm.nih.gov/27893332/. Accessed 22 Oct 2020.

Fizazi K, Delva R, Caty A, Chevreau C, Kerbrat P, Rolland F, et al. A risk-adapted study of cisplatin and etoposide, with or without ifosfamide, in patients with metastatic seminoma: results of the GETUG S99 multicenter prospective study. Eur Urol [Internet]. 2014;65(2):381–6. https://pubmed.ncbi.nlm.nih.gov/24094847/. Accessed 22 Oct 2020.

Flechon A, Bompas E, Biron P, Droz JP. Management of post-chemotherapy residual masses in advanced seminoma. J Urol [Internet]. 2002;168(5). https://pubmed.ncbi.nlm.nih.gov/12394688/. Accessed 22 Oct 2020.

Fosså SD, Stenning SP, Gerl A, Horwich A, Clark PI, Wilkinson PM, et al. Prognostic factors in patients progressing after cisplatin-based chemotherapy for malignant non-seminomatous germ cell tumours. Br J Cancer [Internet]. 1999;80(9):1392–9. /pmc/articles/PMC2363071/?report=abstract. Accessed 23 Oct 2020.

Fosså SD, Aass N, Heilo A, Daugaard G, Skakkebæk NE, Stenwig AE, et al. Testicular carcinoma in situ in patients with extragonadal germ-cell tumours: the clinical role of pretreatment biopsy. Ann Oncol. 2003;14(9):1412–8.

Garcia-del-Muro X, Maroto P, Gumà J, Sastre J, Brea ML, Arranz JA, et al. Chemotherapy as an alternative to radiotherapy in the treatment of stage IIA and IIB testicular seminoma: a Spanish germ cell cancer group study. J Clin Oncol [Internet]. 2008;26(33):5416–21. https://pubmed.ncbi.nlm.nih.gov/18936476/. Accessed 22 Oct 2020.

Gashaw I, Dushaj O, Behr R, Biermann K, Brehm R, Rübben H, et al. Novel germ cell markers characterize testicular seminoma and fetal testis. Mol Hum Reprod. 2007;13(10):721–7.

Germà-Lluch JR, Garcia del Muro X, Maroto P, Paz-Ares L, Arranz JA, Gumà J, et al. Clinical pattern and therapeutic results achieved in 1490 patients with germ-cell tumours of the testis: the experience of the Spanish Germ-Cell Cancer Group (GG). Eur Urol. 2002;42(6):553–63.

Gilligan TD, Seidenfeld J, Basch EM, Einhorn LH, Fancher T, Smith DC, et al. American society of clinical oncology clinical practice guideline on uses of serum tumor markers in adult males with germ cell tumors. J Clin Oncol. 2010;28:3388–404.

Gilligan T, Lin D, Aggarwal R, Chism D, Cost N, Derweesh I, et al. NCCN Clinical Practice Guidelines in Oncology – Testicular Cancer. 2020.

Greene MH, Kratz CP, Mai PL, Mueller C, Peters JA, Bratslavsky G, et al. Familial testicular germ cell tumors in adults: 2010 summary of genetic risk factors and clinical phenotype. Endocr Relat Cancer. 2010;17.

Groll RJ, Warde P, Jewett MAS. A comprehensive systematic review of testicular germ cell tumor surveillance [Internet]. Crit Rev Oncol Hematol. 2007;64:182–97. https://pubmed.ncbi.nlm.nih.gov/17644403/. Accessed 22 Oct 2020.

Gurney JK, Florio AA, Znaor A, Ferlay J, Laversanne M, Sarfati D, et al. International trends in the incidence of testicular cancer: lessons from 35 years and 41 countries. Eur Urol. 2019;76(5):615–23.

Hale GR, Teplitsky S, Truong H, Gold SA, Bloom JB, Agarwal PK. Lymph node imaging in testicular cancer [Internet]. Transl Androl Urol. 2018;7:864–74. /pmc/articles/PMC6212624/?report=abstract. Accessed 23 Oct 2020.

Hamilton RJ, Nayan M, Anson-Cartwright L, Atenafu EG, Bedard PL, Hansen A, et al. Treatment of relapse of clinical stage I nonseminomatous germ cell tumors on surveillance [Internet]. J Clin Oncol. 2019;37:1919–26. https://pubmed.ncbi.nlm.nih.gov/30802156/. Accessed 22 Oct 2020.

Hartmann JT, Candelaria M, Kuczyk MA, Schmoll HJ, Bokemeyer C. Comparison of histological results from the resection of residual masses at different sites after chemotherapy for metastatic non-seminomatous germ cell tumours. Eur J Cancer Part A. 1997;33(6):843–7.

Heidenreich A, Moul JW. Contralateral testicular biopsy procedure in patients with unilateral testis cancer: is it indicated? Semin Urol Oncol. 2002;20:234–8.

Herr HJ, Sheinfeld J, Puc H, Heelan R, Bajorin D, Mencel P, et al. Surgery for a post-chemotherapy residual mass in seminoma. J Urol. 1997;157:860–2.

Hoei-Hansen CE, Nielsen JE, Almstrup K, Sonne SB, Graem N, Skakkebaek NE, et al. Transcription factor AP-2γ is a developmentally regulated marker of testicular carcinoma in situ and germ cell tumors. Clin Cancer Res. 2004;10(24):8521–30.

Holzik MFL, Rapley EA, Hoekstra HJ, Sleijfer DT, Nolte IM, Sijmons RH. Genetic predisposition to testicular germ-cell tumours. Lancet Oncol. 2004;5:363–71.

Honecker F, Aparicio J, Berney D, Beyer J, Bokemeyer C, Cathomas R, et al. ESMO consensus conference on testicular germ cell cancer: diagnosis, treatment and follow-up. Ann Oncol [Internet]. 2018:1658–86. https://pubmed.ncbi.nlm.nih.gov/30113631/. Accessed 22 Oct 2020.

Horwich A, Oliver RTD, Wilkinson PM, Mead GM, Harland SJ, Cullen MH, et al. A Medical Research Council randomized trial of single agent carboplatin versus etoposide and cisplatin for advanced metastatic seminoma. Br J Cancer [Internet]. 2000;83(12):1623–9. http://www.bjcancer.com. Accessed 22 Oct 2020.

Huddart RA, Reid AM. Adjuvant therapy for stage IB germ cell tumors: one versus two cycles of BEP [Internet]. Adv Urol. 2018;2018. https://pubmed.ncbi.nlm.nih.gov/29808086/. Accessed 22 Oct 2020.

Jørgensen N, Rajpert-De Meyts E, Graem N, Müller J, Giwercman A, Skakkebaek N. Expression of immunohistochemical markers for testicular carcinoma in situ by normal human fetal germ cells. Lab Invest. 1995;72(2):223–31.

Jørgensen N, De Meyts ER, Main KM, Skakkebæk NE. Testicular dysgenesis syndrome comprises some but not all cases of hypospadias and impaired spermatogenesis. Int J Androl. 2010;33:298–303.

Kharazmi E, Hemminki K, Pukkala E, Sundquist K, Tryggvadottir L, Tretli S, et al. Cancer risk in relatives of testicular cancer patients by histology type and age at diagnosis: a joint study from five Nordic countries. Eur Urol. 2015;68(2):283–9.

Kim W, Rosen MA, Langer JE, Banner MP, Siegelman ES, Ramchandani P. US-MR imaging correlation in pathologic conditions of the scrotum. Radiographics. 2007;27:1239–53.

Klepp O, Flodgren P, Maartman-moe H, Lindholm CE, Unsgaard B, Teigum H, et al. Early clinical stages (CS1, CS1MK+ and CS2A) of non-seminomatous testis cancer: value of pre- and post-orchiectomy serum tumor marker information in prediction of retroperitoneal lymph node metastases. Ann Oncol. 1990;1(4):281–8.

Kollmannsberger C, Nichols C, Bamberg M, Hartmann JT, Schleucher N, Beyer J, et al. First-line high-dose chemotherapy ± radiation therapy in patients with metastatic germ-cell cancer and brain metastases. Ann Oncol [Internet]. 2000;11(5):553–9. https://pubmed.ncbi.nlm.nih.gov/10907948/. Accessed 22 Oct 2020.

Kollmannsberger C, Tandstad T, Bedard PL, Cohn-Cedermark G, Chung PW, Jewett MA, et al. Patterns of relapse in patients with clinical stage I testicular cancer managed with active surveillance. J Clin Oncol [Internet]. 2015;33(1):51–7. https://pubmed.ncbi.nlm.nih.gov/25135991/. Accessed 22 Oct 2020.

Kondagunta GV, Sheinfeld J, Mazumdar M, Mariani TV, Bajorin D, Bacik J, et al. Relapse-free and overall survival in patients with pathologic stage II nonseminomatous germ cell cancer treated with etoposide and cisplatin adjuvant chemotherapy. J Clin Oncol [Internet]. 2004;22(3):464–7. https://pubmed.ncbi.nlm.nih.gov/14752068/. Accessed 22 Oct 2020.

Kuczyk MA, Serth J, Bokemeyer C, Jonassen J, Machtens S, Werner M, et al. Alterations of the p53 tumor suppressor gene in carcinoma in situ of the testis. Cancer. 1996;78(9):1958–66.

Kurz D, Tasian G. Current management of undescended testes. Curr Treat Options Pediatr. 2016;2(1):43–51.

Laguna M, Albers P, Albrecht W, Algaba F, Bokemeyer C, Boormans J, et al. EAU Guidelines on testicular cancer. 2020.

Leibovitch I, Foster RS, Kopecky KK, Donohue JP. Improved accuracy of computerized tomography based clinical staging in low stage nonseminomatous germ cell cancer using size criteria of retroperitoneal lymph nodes. J Urol. 1995;154(5):1759–63.

Liu NW, Cary C, Strine AC, Beck SDW, Masterson TA, Bihrle R, et al. Risk of recurrence for clinical stage I and II patients with teratoma only at primary retroperitoneal lymph node dissection. Urology [Internet]. 2015;86(5):981–4. https://pubmed.ncbi.nlm.nih.gov/26232690/. Accessed 22 Oct 2020.

Looijenga LHJ, Stoop H, De Leeuw HPJC, De Gouveia Brazao CA, Gillis AJM, Van Roozendaal KEP, et al. POU5F1 (OCT3/4) identifies cells with pluripotent potential in human germ cell tumors. Cancer Res. 2003;63(9):2244–50.

Loveday C, Litchfield K, Levy M, Holroyd A, Broderick P, Kote-Jarai Z, et al. Validation of loci at 2q14.2 and 15q21.3 as risk factors for testicular cancer. Oncotarget. 2018;9(16):12630–8.

Luz MA, Kotb AF, Aldousari S, Brimo F, Tanguay S, Kassouf W, et al. Retroperitoneal lymph node dissection for residual masses after chemotherapyin nonseminomatous germ cell testicular tumor. World J Surg Oncol [Internet]. 2010;8(1):97. http://wjso.biomedcentral.com/articles/10.1186/1477-7819-8-97. Accessed 23 Oct 2020.

Masterson TA, Shayegan B, Carver BS, Bajorin DF, Feldman DR, Motzer RJ, et al. Clinical impact of residual extraretroperitoneal masses in patients with advanced nonseminomatous germ cell testicular cancer. Urology [Internet]. 2012;79(1):156–9. https://pubmed.ncbi.nlm.nih.gov/22202548/. Accessed 22 Oct 2020.

Matei DV, Vartolomei MD, Renne G, Tringali VML, Russo A, Bianchi R, et al. Reliability of frozen section examination in a large cohort of testicular masses: what did we learn? Clin Genitourin Cancer. 2017;15(4):e689–96.

Mead GM. International germ cell consensus classification: a prognostic factor-based staging system for metastatic germ cell cancers. J Clin Oncol [Internet]. 1997;15(2):594–603. https://pubmed.ncbi.nlm.nih.gov/9053482/. Accessed 22 Oct 2020.

Mead GM, Cullen MH, Huddart R, Harper P, Rustin GJS, Cook PA, et al. A phase II trial of TIP (paclitaxel, ifosfamide and cisplatin) given as second-line (post-BEP) salvage chemotherapy for patients with metastatic germ cell cancer: a medical research council trial. Br J Cancer [Internet]. 2005;93(2):178–84. https://pubmed.ncbi.nlm.nih.gov/15999102/. Accessed 23 Oct 2020.

Melchior D, Hammer P, Fimmers R, Schüller H, Albers P. Long term results and morbidity of paraaortic compared with paraaortic and iliac adjuvant radiation in clinical stage I seminoma. Anticancer Res. 2001;21:2989.

Michael H, Lucia J, Foster RS, Ulbright TM. The pathology of late recurrence of testicular germ cell tumors. Am J Surg Pathol. 2000;24(2):257–73.

Moch H, Cubilla AL, Humphrey PA, Reuter VE, Ulbright TM. The 2016 WHO classification of tumours of the urinary system and male genital organs—Part A: Renal, penile, and testicular tumours. Eur Urol. 2016;70(1):93–105.

Montironi R. Intratubular germ cell neoplasia of the testis: testicular intraepithelial neoplasia. Eur Urol [Internet]. 2002;41(6):651–4. https://pubmed.ncbi.nlm.nih.gov/12074783/. Accessed 22 Oct 2020.

Moul JW. Timely diagnosis of testicular cancer. Urol Clin North Am. 2007;34:109–17.

Nayan M, Jewett MAS, Hosni A, Anson-Cartwright L, Bedard PL, Moore M, et al. Conditional risk of relapse in surveillance for clinical stage I testicular cancer. Eur Urol [Internet]. 2017;71(1):120–7. https://pubmed.ncbi.nlm.nih.gov/27527805/. Accessed 22 Oct 2020.

Nichols CR, Catalano PJ, Crawford ED, Vogelzang NJ, Einhorn LH, Loehrer PJ. Randomized comparison of cisplatin and etoposide and either bleomycin or ifosfamide in treatment of advanced disseminated germ cell tumors: an Eastern Cooperative Oncology Group, Southwest Oncology Group, and Cancer and Leukemia Group B Study. J Clin Oncol. 1997;16.

Nichols CR, Catalano PJ, Crawford ED, Vogelzang NJ, Einhorn LH, Loehrer PJ. Randomized comparison of cisplatin and etoposide and either bleomycin or ifosfamide in treatment of advanced disseminated germ cell tumors: an Eastern Cooperative Oncology Group, Southwest Oncology

Group, and cancer and leukemia group B study. J Clin Oncol [Internet]. 1998;16(4):1287–93. https://pubmed.ncbi.nlm.nih.gov/9552027/. Accessed 22 Oct 2020.

Nicolai N, Necchi A, Gianni L, Piva L, Biasoni D, Torelli T, et al. Long-term results of a combination of paclitaxel, cisplatin and gemcitabine for salvage therapy in male germ-cell tumours. BJU Int [Internet]. 2009;104(3):340–6. https://pubmed.ncbi.nlm.nih.gov/19239440/. Accessed 23 Oct 2020.

Nicolai N, Necchi A, Raggi D, Biasoni D, Catanzaro M, Piva L, et al. Clinical outcome in testicular sex cord stromal tumors: testis sparing vs radical orchiectomy and management of advanced disease. Urology [Internet]. 2015;85(2):402–6. https://pubmed.ncbi.nlm.nih.gov/25623702/. Accessed 22 Oct 2020.

Nigam M, Aschebrook-Kilfoy B, Shikanov S, Eggener S. Increasing incidence of testicular cancer in the United States and Europe between 1992 and 2009. World J Urol. 2015;33(5):623–31.

Nonomura N, Nagahara A, Oka D, Mukai M, Nakai Y, Nakayama M, et al. Brain metastases from testicular germ cell tumors: a retrospective analysis: Original Article: Clinical Investigation. Int J Urol [Internet]. 2009;16(11):887–93. https://pubmed.ncbi.nlm.nih.gov/19863625/. Accessed 22 Oct 2020.

Novotny GW, Belling KC, Bramsen JB, Nielsen JE, Bork-Jensen J, Almstrup K, et al. MicroRNA expression profiling of carcinoma in situ cells of the testis. Endocr Relat Cancer. 2012;19(3):365–79.

Oldenburg J, Alfsen GC, Wæhre H, Fosså SD. Late recurrences of germ cell malignancies: a population-based experience over three decades. Br J Cancer [Internet]. 2006;94(6):820–7. www.bjcancer.com. Accessed 23 Oct 2020.

Park JS, Kim J, Elghiaty A, Ham WS. Recent global trends in testicular cancer incidence and mortality. Med (United States). 2018;97(37)

Peng X, Zeng X, Peng S, Deng D, Zhang J. The association risk of male subfertility and testicular cancer: a systematic review. PLoS One. 2009;4

Pierorazio PM, Cheaib JG, Tema G, Patel HD, Gupta M, Sharma R, et al. Performance characteristics of clinical staging modalities for early stage testicular germ cell tumors: a systematic review. J Urol. 2020;203(5):894–901.

Pope WB. Brain metastases: neuroimaging. In: Handbook of clinical neurology. Elsevier B.V.; 2018. p. 89–112.

Powles TB, Bhardwa J, Shamash J, Mandalia S, Oliver T. The changing presentation of germ cell tumours of the testis between 1983 and 2002. BJU Int [Internet]. 2005;95(9):1197–200. https://pubmed.ncbi.nlm.nih.gov/15892800/. Accessed 22 Oct 2020.

Rajpert-De Meyts E, McGlynn KA, Okamoto K, Jewett MAS, Bokemeyer C. Testicular germ cell tumours. Lancet. 2016;387:1762–74.

Rajpert-De Meyts E, Skakkebaek NE, Toppari J. Testicular cancer pathogenesis, diagnosis and endocrine aspects. Endotext. MDText.com, Inc.; 2018.

Rajpert-Meyts E, Nielsen JE, Skakkebæk NE, Almstrup K. Diagnostic markers for germ cell neoplasms: from placental-like alkaline phosphatase to micro-RNAs. Folia Histochem Cytobiol. 2015;53(3):177–88.

Rice KR, Beck SDW, Bihrle R, Cary KC, Einhorn LH, Foster RS. Survival analysis of pure seminoma at post-chemotherapy retroperitoneal lymph node dissection. J Urol [Internet]. 2014;192(5):1397–402. https://pubmed.ncbi.nlm.nih.gov/24813309/. Accessed 22 Oct 2020.

Richiardi L, Pettersson A, Akre O. Genetic and environmental risk factors for testicular cancer. Int J Androl. 2007;30(4):230–41.

Ruf CG, Gnoss A, Hartmann M, Matthies C, Anheuser P, Loy V, et al. Contralateral biopsies in patients with testicular germ cell tumours: patterns of care in Germany and recent data regarding prevalence and treatment of testicular intra-epithelial neoplasia. Andrology. 2015;3(1):92–8.

Saxman SB, Finch D, Gonin R, Einhorn LH. Long-term follow-up of a phase III study of three versus four cycles of bleomycin, etoposide, and cisplatin in favorable-prognosis germ-cell tumors: The Indiana University experience. J Clin Oncol [Internet]. 1998;16(2):702–6. https://pubmed.ncbi.nlm.nih.gov/9469360/. Accessed 22 Oct 2020.

Scandura G, Verrill C, Protheroe A, Joseph J, Ansell W, Sahdev A, et al. Incidentally detected testicular lesions <10 mm in diameter: can orchidectomy be avoided? BJU Int. 2018;121(4):575–82.

Scully RE. Gonadoblastoma. A review of 74 cases. Cancer [Internet]. 1970;25(6):1340–56. https://pubmed.ncbi.nlm.nih.gov/4193741/. Accessed 22 Oct 2020.

Shaw J. Diagnosis and treatment of testicular cancer. Am Family Physician. 2008;77

Sheinfeld J. Risks of the uncontrolled retroperitoneum [Internet]. Ann Surg Oncol. 2003;10:100–1. https://link.springer.com/article/10.1245/ASO.2003.01.011. Accessed 22 Oct 2020.

Solter D. From teratocarcinomas to embryonic stem cells and beyond: a history of embryonic stem cell research. Nat Rev Genet. 2006;7:319–27.

Sonne SB, Almstrup K, Dalgaard M, Juncker AS, Edsgard D, Ruban L, et al. Analysis of gene expression profiles of microdissected cell populations indicates that testicular carcinoma in situ is an arrested gonocyte. Cancer Res. 2009;69(12):5241–50.

Stephenson A, Gilligan T. Neoplasms of the testis. In: Campbell-Walsh-Wein urology. 12th ed. Elsevier; 2020. p. 1680–710.

Stephenson AJ, Bosl GJ, Motzer RJ, Kattan MW, Stasi J, Bajorin DF, et al. Retroperitoneal lymph node dissection for nonseminomatous germ cell testicular cancer: impact of patient selection factors on outcome. J Clin Oncol [Internet]. 2005;23(12):2781–8. https://pubmed.ncbi.nlm. nih.gov/15837993/. Accessed 22 Oct 2020.

Stephenson AJ, Bosl GJ, Motzer RJ, Bajorin DF, Stasi JP, Sheinfeld J. Nonrandomized comparison of primary chemotherapy and retroperitoneal lymph node dissection for clinical stage IIA and IIB nonseminomatous germ cell testicular cancer. J Clin Oncol [Internet]. 2007;25(35):5597–602. https://pubmed.ncbi.nlm.nih.gov/18065732/. Accessed 22 Oct 2020.

Stevenson SM, Lowrance WT. Epidemiology and diagnosis of testis cancer. Urol Clin North Am. 2015;42:269–75.

Subramanian VS, Nguyen CT, Stephenson AJ, Klein EA. Complications of open primary and post-chemotherapy retroperitoneal lymph node dissection for testicular cancer. Urol Oncol Semin Orig Investig [Internet]. 2010;28(5):504–9. https://pubmed.ncbi.nlm.nih.gov/19097812/. Accessed 22 Oct 2020.

Tandstad T, Ståhl O, Dahl O, Haugnes HS, Håkansson U, Karlsdottir A, et al. Treatment of stage I seminoma, with one course of adjuvant carboplatin or surveillance, risk-adapted recommendations implementing patient autonomy: a report from the Swedish and Norwegian Testicular Cancer Group (SWENOTECA). Ann Oncol [Internet]. 2016;27(7):1299–304. https://pubmed. ncbi.nlm.nih.gov/27052649/. Accessed 22 Oct 2020.

Ulbright TM. Germ cell tumors of the gonads: a selective review emphasizing problems in differential diagnosis, newly appreciated, and controversial issues. Mod Pathol. 2005. https:// pubmed.ncbi.nlm.nih.gov/15761467/. Accessed 22 Oct 2020.

Vasdev N, Moon A, Thorpe AC. Classification, epidemiology and therapies for testicular germ cell tumours. Int J Dev Biol. 2013;57(2–4):133–9.

Veenstra CM, Vaughn DJ. Third-line chemotherapy and novel agents for metastatic germ cell tumors. Hematol Oncol Clin North Am [Internet]. 2011;25(3):577–91. https://linkinghub.elsevier.com/retrieve/pii/S0889858811000190. Accessed 23 Oct 2020.

Vergouwe Y, Steyerberg EW, Eijkemans MJC, Albers P, Habbema JDF. Predictors of occult metastasis in clinical stage I non-seminoma: a systematic review [Internet]. J Clin Oncol. 2003;21:4092–9. https://pubmed.ncbi.nlm.nih.gov/14559885/. Accessed 22 Oct 2020.

Verhoeven RHA, Karim-Kos HE, Coebergh JWW, Brink M, Horenblas S, De Wit R, et al. Markedly increased incidence and improved survival of testicular cancer in the Netherlands. Acta Oncol (Madr). 2014;53(3):342–50.

Wang T, Liu LH, Luo JT, Liu TS, Wei AY. A meta-analysis of the relationship between testicular microlithiasis and incidence of testicular cancer. Urol J. 2015;12(2):2057–64.

Wang Z, McGlynn KA, Rajpert-De Meyts E, Bishop DT, Chung CC, Dalgaard MD, et al. Meta-analysis of five genome-wide association studies identifies multiple new loci associated with testicular germ cell tumor. Nat Genet. 2017;49(7):1141–6.

Warde P, Specht L, Horwich A, Oliver T, Panzarella T, Gospodarowicz M, et al. Prognostic factors for relapse in stage I seminoma managed by surveillance: a pooled analysis. J Clin Oncol [Internet]. 2002;20(22):4448–52. https://pubmed.ncbi.nlm.nih.gov/12431967/. Accessed 22 Oct 2020.

Weissbach L, Bussar-Maatz R, Flechtner H, Pichlmeier U, Hartmann M, Keller L. RPLND or primary chemotherapy in clinical stage IIA/B nonseminomatous germ cell tumors? Results of a prospective multicenter trial including quality of life assessment. Eur Urol [Internet]. 2000;37(5):582–94. https://pubmed.ncbi.nlm.nih.gov/10765098/. Accessed 22 Oct 2020.

Zuniga A, Kakiashvili D, Jewett MAS. Surveillance in stage I nonseminomatous germ cell tumours of the testis. BJU Int [Internet]. 2009;104(9b):1351–6. http://doi.wiley.com/10.1111/j.1464-410X.2009.08858.x. Accessed 23 Oct 2020.

Chapter 18
Prostate Cancer Malignancy

Sanchia S. Goonewardene, Hanif Motiwala, Raj Persad, and Declan Cahill

18.1 Prostate Cancer Epidemiology

18.1.1 Factors Impacting Prostate Cancer Epidemiology

There have been a number of factors identified which can impact on prostate cancer. Klein determined the long-term effect of vitamin E and selenium on risk of prostate cancer (Klein et al. 2011). Compared with the placebo (referent group) in which 529 men developed prostate cancer, 620 men in the vitamin E group developed prostate cancer, as did 575 in the selenium group and 555 in the selenium plus vitamin E group (Klein et al. 2011). Compared with placebo, the absolute increase in risk of prostate cancer per 1000 person-years was 1.6 for vitamin E, 0.8 for selenium, and 0.4 for the combination (Klein et al. 2011). However, in contrast, the SELECT study reported that neither selenium nor vitamin E had any beneficial effect (Klein et al. 2011). At the moment, the overall belief, is that there is no association between selenium, vitamin E and Prostate Cancer.

S. S. Goonewardene (✉)
The Princess Alexandra Hospital, Harlow, UK

H. Motiwala
Southend University Hospital, Southend, UK

R. Persad
North Bristol NHS Foundation Trust, Bristol, UK
e-mail: rajpersad@bristolurology.com

D. Cahill
The Royal Marsden Hospital, London, UK

© The Author(s), under exclusive license to Springer Nature
Switzerland AG 2022
S. S. Goonewardene et al. (eds.), *Men's Health and Wellbeing*,
https://doi.org/10.1007/978-3-030-84752-4_18

Kristal examined inadequate vitamin D intake could increase prostate cancer risk (Kristal et al. 2014). What was demonstrated were vitamin D being protective among African American men (n = 250 cases), with a reduced risk of high grade cancer.

Regular aspirin use probably protects against some malignancies including prostate cancer (PC), but its impact on lethal PC is uncertain (Downer et al. 2017). Downer investigated the association between regular aspirin and the risk of lethal PC in a large prospective cohort and survival after PC diagnosis (Downer et al. 2017). There were 22,071 healthy male physicians randomised to aspirin, β-carotene, both, or placebo. Risk analysis revealed that 502 out of 22,071 developed lethal PC by 2009. Current and past regular aspirin was associated with a lower risk of lethal PC compared to never users (Downer et al. 2017). In the survival analysis, 407/3277 men diagnosed with nonlethal PC developed lethal disease by 2015 (Downer et al. 2017). Current post diagnostic aspirin was associated with lower risks of lethal PC and overall mortality (Downer et al. 2017). The evidence still suggests there is no link between aspirin and prostate cancer deaths. However, aspirin can provide a protective effect against CVS risk from ADT. Consideration should be given to starting appropriate therapies including lifestyle advice, antihypertensive and lipid-lowering agents, plus possibly aspirin in these patients. Having started ADT, the patients should have a regular (possibly annual) assessment of their cardiovascular risk factors.

Smith assessed the relationship between prevalent diabetes and mortality using data from Radiation Therapy Oncology Group Protocol 92-02, a large randomized trial of men (N = 1554) treated with radiation therapy and short-term versus long-term adjuvant goserelin for locally advanced prostate cancer (Smith et al. 2008). There were a total of 765 deaths; 210 (27%) were attributed to prostate cancer (Smith et al. 2008). In univariate analyses, prevalent diabetes was associated with greater all-cause mortality and non-prostate cancer mortality but not prostate cancer mortality (Smith et al. 2008). After controlling for other covariates, prevalent diabetes remained significantly associated with greater all-cause mortality and non-prostate cancer mortality (P < .0001) but not prostate cancer mortality (P = .34) (Smith et al. 2008). In contrast, weight was associated with greater prostate cancer mortality (P = .002) but not all-cause or non-prostate cancer mortality (Smith et al. 2008). However, Antunes demonstrated diabetic patients had significantly higher prostatic volumes, yet with no significant difference was found between the relapse risk classification or the ISUP classification in diabetic vs. non diabetic patients (Antunes et al. 2018). Diabetes is a risk factor for death, but not prostate cancer.

18.1.2   Prostate Cancer Prevention

Kristal examined nutritional risk factors for prostate cancer among 9559 participants in the Prostate Cancer Prevention Trial (United States and Canada, 1994–2003) (Kristal et al. 2011). Neither dietary nor supplemental intakes of nutrients often suggested for prostate cancer prevention, including lycopene, long-chain n-3 fatty acids, vitamin D, vitamin E, and selenium, were significantly associated with cancer risk (Kristal et al. 2011). High intake of n-6 fatty acids, through their effects on inflammation and oxidative stress, may increase prostate cancer risk (Kristal et al. 2011). There is some evidence that higher consumption of dairy products could increase the risk of prostate cancer, but the evidence is variable (Lopez-Plaza et al. 2019).

Meyer assessed whether a supplementation with low doses of antioxidant vitamins and minerals could reduce the occurrence of prostate cancer and influence biochemical markers (Meyer et al. 2005). Overall, there was a moderate nonsignificant reduction in prostate cancer rate associated with the supplementation (Meyer et al. 2005). However, the effect differed significantly between men with normal baseline PSA (<3 µg/L) and those with elevated PSA (P = 0.009) (Meyer et al. 2005). Among men with normal PSA, there was a marked statistically significant reduction in the rate of prostate cancer for men receiving the supplements (Meyer et al. 2005). In men with elevated PSA at baseline, the supplementation was associated with an increased incidence of prostate cancer of borderline statistical significance (Meyer et al. 2005). However, this could be an observation effect. The supplementation had no effect on PSA or IGF levels. In contrast, Zhang demonstrated sequence variants in inflammatory genes interact with plasma antioxidants to modulate prostate cancer risk, and confer a protective effect (Zhang et al. 2010). However this effect may be observational and not due to true change.

Wang studied the metabolism and bioactivity of green tea polyphenols in human prostate tissue, men with clinically localized prostate cancer consumed six cups of green tea (n = 8) daily or water (n = 9) for 3–6 weeks before undergoing radical prostatectomy (Wang et al. 2010). In the urine of men consuming green tea, 50–60% of both (−)-epigallocatechin and (−)-epicatechin were present in methylated form with 4′-O-MeEGC being the major methylated form of (−)-epigallocatechin (Wang et al. 2010). When incubated with EGCG, LNCaP prostate cancer cells were able to methylate EGCG to 4″-MeEGCG (Wang et al. 2010). The capacity of 4″-MeEGCG to inhibit proliferation and NF-kappaB activation and induce apoptosis in LNCaP cells was decreased significantly compared with EGCG (Wang et al. 2010). Yet in contrast, other studies have found conflicting evidence for use of green tea for the prevention of prostate cancer (Filippini et al. 2020).

Compound	Impact on epidemiology	Level of evidence	Outcome
Vit D	Men with higher serum 25(OH)D were less likely to die from their prostate cancer(Mondul 2016)	1	Higher serum 25(OH)D years prior to diagnosis was associated with longer prostate cancer survival
Calcium	Dietary calcium was modestly associated with increased risks for nonaggressive prostate cancer (Ahn et al. 2007)	1	Dietary calcium was unrelated to aggressive disease

Compound	Impact on epidemiology	Level of evidence	Outcome
Green tea	There was no difference in prostate cancer incidence between treated and placebo groups (Micalli 2017)	1	This study did not find any association between green tea and decrease in prostate cancer incidence
Lycopenes	Belby reported a null association between lycopenes and prostate cancer risk (Belby 2010)	1	There was no association between prostate cancer risk and intake of lycopenes
Antioxidants	There was no significant correlation between selenium. Levels and prostate cancer risk (Year mismatch please check et al. 2019)	1	There was no association between selenium and prevention of prostate cancer

18.1.3   Prostate Cancer Diagnostics

PSA is the commonly used marker in Prostate Cancer, but there are alternatives that can be used. The use of biomarkers for prostate cancer screening, diagnosis and prognosis has the potential to improve the clinical management of the patients (Saini 2016).

Zhao have investigated cell-free miRNAs in the urinary supernatant combined with urinary DNA methylation markers to form an integrative panel for prediction of AS patient reclassification (Zhao et al. 2019). Zhao identified a three-marker panel, that was significant for prediction of patient reclassification with a negative predictive value of 90.9% (Zhao et al. 2019). This three-marker panel also demonstrated additive value to PSA for prediction of patient reclassification which could be used as an adjunct to prostate cancer diagnostics (c-statistic = 0.717, ROC bootstrapped 1000 iteration P = 0.041) (Zhao et al. 2019).

18.1.4   TRUS vs. Transperineal Prostate Biopsies

The rationale for directing targeted biopsy towards the centre of lesions has been questioned in light of prostate cancer grade heterogeneity (El-Shater 2016). Ninety-four men were included, with 184 (94 ± 3.2%) lesions showed concordant hotspots and 11/47 (23 ± 12.1%) of heterogeneous lesions showed discordant hotspots (El-Shater 2016). Guiding one biopsy needle to the maximum cancer diameter would lead to correct Gleason grade attribution in 94% of all lesions and 79% of heterogeneous ones if a true hit was obtained (El-Shater 2016). This paper clearly highlights the more favourable staging with targeted biopsy (El-Shater 2016). However, it is important to note that targeted not systematic biopsies will miss some cancers. Although these may be insignificant cancers, they can be detected further down the line by ongoing PSA testing, and assessing screening/risk.

Very often clinicians will encounter situations such as a negative MRI, in a patient with a benign smooth prostate and elevated PSA. Drost reviewed this and demonstrated MRI has a favourable diagnostic accuracy in clinically significant prostate cancer detection (Drost et al. 2019). Compared to systematic biopsy, it increased the number of significant cancers detected while reducing the number of insignificant cancer diagnosed (Drost et al. 2019).

Kasivinathan examined MRI guided targeted biopsies vs. standard TRUS biopsies as part of an RCT (Kasivisvanathan et al. 2018). Clinically significant cancer was detected in 95 men (38%) in the MRI-targeted biopsy group, as compared with 64 of 248 (26%) in the standard-biopsy group (adjusted difference, 12 percentage points; 95% confidence interval [CI], 4–20; P = 0.005) (Kasivisvanathan et al. 2018). MRI, with or without targeted biopsy, was noninferior to standard biopsy (Kasivisvanathan et al. 2018). Fewer men in the MRI-targeted biopsy group than in the standard-biopsy group received a diagnosis of clinically insignificant cancer (adjusted difference, −13 percentage points; 95% CI, −19 to −7; P < 0.001) (Kasivisvanathan et al. 2018).

18.1.5 Prostate Cancer Imaging

Sonni determined the impact of ^{68}Ga-PSMA-11 PET/CT on initial and subsequent management decisions in a cohort of PCa patients (Sonni 2020). PSMA PET/CT changed the disease stage in 135 of 197 (69%) patients (upstaging in 38%, downstaging in 30%, and no change in stage in 32%) (Sonni 2020). Management was affected in 104 of 182 (57%) patients (Sonni 2020). Specifically, PSMA PET/CT impacted the management of patients who were restaged after radiation therapy without meeting the Phoenix criteria for BCR, after other definitive local treatments, and with advanced metastatic disease in 13 of 18 (72%), 8 of 12 (67%), and 59 of 96 (61%), respectively (Sonni 2020).

Fendler assessed 68Ga-PSMA-11 PET as part of a multicentre clinical trial (Fendler et al. 2019). A total of 635 men were enrolled (Fendler et al. 2019). The PPV was 0.84 (95% CI, 0.75–0.90) (Fendler et al. 2019). 68Ga-PSMA-11 PET localized recurrent prostate cancer in 475 of 635 (75%) patients; detection rates significantly increased with prostate-specific antigen (PSA): 38% for <0.5 ng/mL (n = 136), 57% for 0.5 to <1.0 ng/mL (n = 79), 84% for 1.0 to <2.0 ng/mL (n = 89), 86% for 2.0 to <5.0 ng/mL (n = 158), and 97% for ≥5.0 ng/mL (n = 173, P < .001) (Fendler et al. 2019). This quite clearly demonstrates a role for PSMA PET in diagnosis of recurrent disease.

18.1.6 Risk Stratification

It was D'Amico that stratified prostate cancer into low, intermediate and high risk disease (D'Amico 1998), which was then further developed by Epstein (2016).

- Low risk disease PSA < 10, T2a or less, Gleason 3+3, ISUP (Grade Group 1)
- Intermediate risk disease PSA 10–20, T2b, Gleason 3+4 (ISUP Grade Group 2) or 4+3 (ISUP Grade Group 3)
- High Risk disease PSA 20 or more, T2c, Gleason 4+4 or higher (ISUP Grade Group 4)

18.2 Prostate Cancer Management

18.2.1 Low Risk Disease

18.2.1.1 Active Surveillance in Low Risk Prostate Cancer

Hamdy compared active monitoring, radical prostatectomy, and external-beam radiotherapy for the treatment of clinically localized prostate cancer (Hamdy et al. 2016). The difference in mortality among the groups was not significant (P = 0.48 for the overall comparison) (Hamdy et al. 2016). In addition, no significant difference was seen among the groups in the number of deaths from any cause (169 deaths overall; P = 0.87 for the comparison among the three groups) (Hamdy et al. 2016). Metastases developed in more men in the active-monitoring group (33 men; 6.3 events per 1900 person-years; 95% CI, 4.5–8.8) than in the surgery group (13 men; 2.4 per 1000 person-years; 95% CI, 1.4–4.2) or the radiotherapy group (16 men; 3.0 per 1000 person-years; 95% CI, 1.9–4.9) (P = 0.004 for the overall comparison) (Hamdy et al. 2016). Higher rates of disease progression were seen in the active-monitoring group (112 men; 22.9 events per 1000 person-years; 95% CI, 19.0–27.5) than in the surgery group (46 men; 8.9 events per 1000 person-years; 95% CI, 6.7–11.9) or the radiotherapy group (46 men; 9.0 events per 1000 person-years; 95% CI, 6.7–12.0) (P < 0.001 for the overall comparison) (Hamdy et al. 2016). However, in young patients, with low risk prostate cancer, who are not keen on side effects of more radical therapies, this is seen as a suitable alternative, provided, it is as part of a structured active surveillance programme.

18.2.1.2 Brachytherapy in Low Risk Prostate Cancer

It was Anderson, that demonstrated the side effects of brachytherapy (Anderson et al. 2009). At 5 years, the cumulative risks of late urinary complications by grade were 8.6% for grade 1 complications, 6.5% for grade 2, 1.7% for grade 3%, and 0.5% for grade 4 (Anderson et al. 2009). The most common grade 2 late urinary complications were urethral stricture (four patients), incontinence requiring daily pads (three patients), and intermittent hematuria (three patients) (Anderson et al. 2009). Grade 3 complications were urinary retention requiring self-catheterization (two patients) and severe frequency with dysuria (two patients) (Anderson et al. 2009). The only grade 4 event was severe hemorrhagic cystitis (Anderson et al. 2009).

18.3   Intermediate Risk Disease

18.3.1   Robotic Radical Prostatectomy

Yaxley compared these two approaches in terms of functional and oncological outcomes and report the early postoperative outcomes at 12 weeks as part of a randomised controlled trial (Yaxley et al. 2016). Equivalence testing on the difference between the proportion of positive surgical margins between the two groups (15 [10%] in the radical retropubic prostatectomy group vs 23 [15%] in the robot-assisted laparoscopic prostatectomy group) showed that equality between the two techniques could not be established based on a 90% CI with a Δ of 10% (Yaxley et al. 2016). However, a superiority test showed that the two proportions were not significantly different ($P = 0.21$) (Yaxley et al. 2016).

Galfano assessed the effect of surgical experience on peri-operative, functional and oncological outcomes (Galfano 2021). Shorter console time (140 vs 120 min; $P = 0.001$) and a trend towards lower complications rates (13 vs 5.5%; $P = 0.038$) were observed in the expert group (Galfano 2021). Overall, console time, immediate urinary continence recovery and postoperative complications are optimal from the beginning and further quickly improve during the learning process, while PSM rates did not clearly improve over the first 50 cases (Galfano 2021).

18.3.2   EBRT vs. Robotic Radical Prostatectomy

Takizawa retrospectively analysed oncological outcomes between consecutive patients receiving RP (n = 86) and EBRT (n = 76) for localized prostate cancer (Takizawa et al. 2009). The 5-year biochemical progression-free survival did not differ between the RP and EBRT groups for low-risk and intermediate-risk patients (Takizawa et al. 2009). For high-risk patients, progression-free survival was lower in the RP group than in the EBRT group ($P = 0.002$) (Takizawa et al. 2009).

18.3.3   Radiotherapy and ADT

Bolla conducted a randomised phase 3 trial assessing long-term androgen suppression vs. external irradiation in patients with prostate cancer with high metastatic risk (Bolla 2010). The 10-year clinical disease-free survival was 22.7% in the radiotherapy-alone group and 47.7% in the combined treatment group (P < 0.0001) (Bolla 2010). The 10-year overall survival was 39.8% in patients receiving radiotherapy alone and 58.1% with combined treatment (HR 0.60, 95% CI 0.45–0.80, P = 0.0004), and 10-year prostate-cancer mortality was 30.4% and 10.3%, respectively (P < 0.0001) (Bolla 2010). This paper demonstrated combined therapy improves overall mortality, without increasing late cardiovascular toxicity (Bolla 2010).

18.3.4 *Radiotherapy and LDR Brachytherapy in Intermediate Risk Prostate Cancer*

Lee estimated the rate of acute and late Grade 3–5 genitourinary and gastrointestinal toxicity after treatment with external beam radiotherapy and permanent source brachytherapy in a multi-institutional, cooperative group setting (Lee et al. 2006). A total of 138 patients from 28 institutions were entered on this study (Lee et al. 2006). Acute toxicity information was available in 131 patients, and 127 patients were analyzable for late toxicity (Lee et al. 2006). Acute Grade 3 toxicity was documented in 10 of 131 patients (7.6%) (Lee et al. 2006). No Grade 4 or 5 acute toxicity has been observed (Lee et al. 2006). The 18-month month estimate of late Grade 3 genitourinary and gastrointestinal toxicity was 3.3% (Lee et al. 2006). No late Grade 4 or 5 toxicity has been observed (Lee et al. 2006).

18.4 High Risk Prostate Cancer

18.4.1 *Robotic Radical Prostatectomy Outcomes in High Risk Prostate Cancer*

Yee reported positive surgical margin rates PSM rates in 500 cases and particularly in clinically high-risk disease (Yee 2009). Yee reported PSM rates according to pathologic stage and D'Amico's risk stratification: low risk (prostate-specific antigen [PSA] <10, Gleason score [GS] 5–6, cT1-T2A), intermediate risk (PSA 10–20, GS 7, cT2B), and high risk (PSA >20, GS 8–10, cT3) (Yee et al. 2009). The overall PSM rate was 7.4%: pT2 = 3.1%, pT3 = 15.9%, and pT4 = 55.6% (Yee et al. 2009). PSMs occurred in 13 (4.9%) low, 10 (5.8%) intermediate, and 14 (22.6%) high D'Amico risk patients (Yee et al. 2009). Of the 62 high-risk patients, the median PSA was 6.9 (range 2.2–97.9); biopsy GS was 6–7 (26%) and 8–10 (74%) (Yee et al. 2009). For preoperatively palpable disease, the PSM rate was 9.9%: cT1 = 6.0%, cT2 = 7.7%, and cT3 = 26.3%. No PSMs occurred along the neurovascular bundle (Yee et al. 2009).

18.4.2   Multimodal Therapy in High Risk Disease

Guttilla conducted a prospective phase II trial including 35 patients with newly diagnosed high-risk localized or locally advanced prostate cancer treated with high-dose intensity-modulated radiation therapy preceded or not by radical prostatectomy, concurrent intensified-dose docetaxel-based chemotherapy and long-term androgen deprivation therapy (Guttilla et al. 2014). Acute gastro-intestinal and genito-urinary toxicity was grade 2 in 23% and 20% of patients, and grade 3 in 9% and 3% of patients, respectively (Guttilla et al. 2014). Acute blood/bone marrow toxicity was grade 2 in 20% of patients. No acute grade ≥ 4 toxicity was observed (Guttilla et al. 2014). Late gastro-intestinal and genito-urinary toxicity was grade 2 in 9% of patients each. No late grade ≥ 3 toxicity was observed (Guttilla et al. 2014). Actuarial 5-year biochemical and clinical recurrence-free survival rate was 55% (95% confidence interval, 35–75%) and 70% (95% confidence interval, 52–88%), respectively (Guttilla et al. 2014).

18.4.3   EBRT in High Risk Prostate Cancer

Blanchard analyzed the role of pelvic radiation therapy (RT) on the outcome in high-risk localized prostate cancer patients included in the Groupe d'Etude des TumeursUro-Genitales (GETUG) 12 trial (Blanchard et al. 2016). Patients with a non pre-treated high-risk localized prostate cancer and a staging lymphadenectomy were randomly assigned to receive either goserelin every 3 months for 3 years and four cycles of docetaxel plus estramustine or goserelin alone (Blanchard et al. 2016). In multivariate analysis, bPFS was had an inverse association with ascending stage. Gleason score 8 or higher and PSA higher than 20 ng/mL and positively impacted by the use of chemotherapy (Blanchard et al. 2016). There was no association between bPFS and use of pelvic elective nodal irradiation (ENI) in multivariate analysis even when analysis was restricted to pN0 patients (Blanchard et al. 2016). Pelvic ENI was not associated with increased acute or late patient reported toxicity (Blanchard et al. 2016).

18.4.4   Neoadjuvant Chemotherapy in High Risk Prostate Cancer

Febbo determined the clinical, pathologic, and molecular effects of neoadjuvant docetaxel chemotherapy in high-risk localized prostate cancer (Febbo et al. 2005). Patients received weekly docetaxel (36 mg/m^2) for 6 months, followed by radical prostatectomy, and were monitored with weekly visits, serum prostate-specific antigen measurements, and endorectal magnetic resonance imaging (MRI) (Febbo et al. 2005). The 19 patients enrolled received 82% of the planned chemotherapy (Febbo et al. 2005). Toxicity was mild to moderate; fatigue and taste disturbance were common (Febbo et al. 2005). Prostate-specific antigen declines of >50% were seen in 11 of 19 patients (58%; 95% confidence interval, 33–80%) and endorectal MRI showed maximum tumor volume reduction of at least 25% in 13 of 19 patients (68%; 95% confidence interval, 47–85%) and at least 50% in 4 patients (21%; 95% confidence interval, 6–46%) (Febbo et al. 2005). Sixteen patients completed chemotherapy and had radical prostatectomy; none achieved pathologic complete response (Febbo et al. 2005). However, the small size of this study has been noted.

Eastham hypothesized that chemohormonal therapy (CHT) with androgen-deprivation therapy plus docetaxel before RP would improve biochemical progression-free survival (BPFS) over RP alone (Eastham 2020). Seven hundred and eighty eight men were randomly assigned (Eastham 2020). The overall rates of grade 3 and 4 adverse events during chemotherapy were 26% and 19%, respectively (Eastham 2020). No difference was seen in 3-year BPFS between neoadjuvant CHT plus RP and RP alone (0.89 vs 0.84, respectively; 95% CI for the difference, −0.01 to 0.11; $P = .11$) (Eastham 2020). Neoadjuvant CHT was associated with improved overall BPFS (hazard ratio [HR], 0.69; 95% CI, 0.48–0.99), improved MFS (HR, 0.70; 95% CI, 0.51–0.95), and improved OS (HR, 0.61; 95% CI, 0.40–0.94) compared with RP alone (Eastham 2020).

18.4.5   Adjuvant Therapy in High Risk Prostate Cancer

Rosenthal assessed adjuvant combination chemotherapy (CT) with paclitaxel, estramustine, and oral etoposide plus LT Androgen suppression (AS) plus RT would improve overall survival (OS) in high risk prostate cancer (Rosenthal et al. 2015). Patients with high-risk PCa (prostate-specific antigen 20–100 ng/mL and Gleason score [GS] ≥ 7 or clinical stage $\geq$T2 and GS ≥ 8) were randomized to RT and AS (AS + RT) alone or with adjuvant CT (AS + RT + CT) (Rosenthal et al. 2015). CT was given as four 21-day cycles, delivered beginning 28 days after 70.2 Gy of RT (Rosenthal et al. 2015). AS was given as luteinizing hormone-releasing hormone for 24 months, beginning 2 months before RT plus an oral antiandrogen for 4 months before and during RT (Rosenthal et al. 2015). A total of 397 patients (380 eligible) were randomized (Rosenthal et al. 2015). The trial closed early because of excess thromboembolic toxicity in the CT arm (Rosenthal et al. 2015). The 10-year results for all randomized patients revealed no significant difference between the AS + RT and AS + RT + CT arms in OS (65% vs 63%; $P = .81$), biochemical failure (58% vs 54%; $P = .82$), local progression (11% vs 7%; $P = .09$), distant metastases (16% vs 14%; $P = .42$), or disease-free survival (22% vs 26%; $P = .61$) (Rosenthal et al. 2015).

References

Ahn J, Albanes D, Peters U, Schatzkin A, Lim U, Freedman M, Chatterjee N, Andriole GL, Leitzmann MF, Hayes RB; Prostate, Lung, Colorectal, and Ovarian Trial Project Team. Dairy products, calcium intake, and risk of prostate cancer in the prostate, lung, colorectal, and ovarian cancer screening trial. Cancer Epidemiol Biomarkers Prev. 2007;16(12):2623–30.

Anderson JF, Swanson DA, Levy LB, Kuban DA, Lee AK, Kudchadker R, Phan J, Bruno T, Frank SJ. Urinary side effects and complications after permanent prostate brachytherapy: the MD Anderson Cancer Center experience. Urology. 2009;74(3):601–5.

Antunes HP, Teixo R, Carvalho JA, Eliseu M, Marques I, Mamede A, Neves R, Oliveira R, Tavares-da-Silva E, Parada B, Abrantes AM, Figueiredo A, Botelho MF. Diabetes mellitus and prostate cancer metabolism: is there a relationship? Arch Ital Urol Androl. 2018;90(3):184–90.

Beilby J, Ambrosini GL, Rossi E, de Klerk NH, Musk AW. Serum levels of folate, lycopene, β-carotene, retinol and vitamin E and prostate cancer risk. Eur J Clin Nutr. 2010;64(10):1235–8.

Blanchard P, Faivre L, Lesaunier F, Salem N, Mesgouez-Nebout N, Deniau-Alexandre E, Rolland F, Ferrero JM, Houédé N, Mourey L, Théodore C, Krakowski I, Berdah JF, Baciuchka M, Laguerre B, Davin JL, Habibian M, Culine S, Laplanche A, Fizazi K. Outcome according to elective pelvic radiation therapy in patients with high-risk localized prostate cancer: a secondary analysis of the GETUG 12 Phase 3 randomized trial. Int J Radiat Oncol Biol Phys. 2016;94(1):85–92.

Bolla, M., G. Van Tienhoven, P. Warde, J. B. Dubois, R. O. Mirimanoff, G. Storme, J. Bernier, et al. "External Irradiation with or without Long-Term Androgen Suppression for Prostate Cancer with High Metastatic Risk: 10-Year Results of an Eortc Randomised Study." [In eng]. Lancet Oncol 11, no. 11 (Nov 2010):1066–73.

D'Amico AV, Whittington R, Malkowicz SB, Schultz D, Blank K, Broderick GA, Tomaszewski JE, Renshaw AA, Kaplan I, Beard CJ, Wein A. Biochemical outcome after radical prostatectomy, external beam radiation therapy, or interstitial radiation therapy for clinically localized prostate cancer. JAMA. 1998;280(11):969–74.

Downer MK, Allard CB, Preston MA, et al. Regular aspirin use and the risk of lethal prostate cancer in the Physicians' Health Study. Eur Urol. 2017;72(5):821–7.

Drost FH, Osses DF, Nieboer D, Steyerberg EW, Bangma CH, Roobol MJ, Schoots IG. Prostate MRI, with or without MRI-targeted biopsy, and systematic biopsy for detecting prostate cancer. Cochrane Database Syst Rev. 2019;4(4):CD012663.

Eastham JA, Heller G, Halabi S, Monk JP, 3rd, Beltran H, Gleave M, Evans CP, et al. "Cancer and Leukemia Group B 90203 (Alliance): Radical Prostatectomy with or without Neoadjuvant Chemohormonal Therapy in Localized, High-Risk Prostate Cancer." [In eng]. J Clin Oncol 38, no. 26 (Sep 10 2020):3042–50.

Epstein JI, Egevad L, Amin MB, Grading Committee. The 2014 International Society of Urological Pathology (ISUP) consensus conference on Gleason grading of prostatic carcinoma: definition of grading patterns and proposal for a new grading system. Am J SurgPathol. 2016;40: 244–52.

Febbo PG, Richie JP, George DJ, Loda M, Manola J, Shankar S, Barnes AS, Tempany C, Catalona W, Kantoff PW, Oh WK. Neoadjuvant docetaxel before radical prostatectomy in patients with high-risk localized prostate cancer. Clin Cancer Res. 2005;11(14):5233–40.

Fendler WP, Calais J, Eiber M, Flavell RR, Mishoe A, Feng FY, Nguyen HG, Reiter RE, Rettig MB, Okamoto S, Emmett L, Zacho HD, Ilhan H, Wetter A, Rischpler C, Schoder H, Burger IA, Gartmann J, Smith R, Small EJ, Slavik R, Carroll PR, Herrmann K, Czernin J, Hope TA. Assessment of 68Ga-PSMA-11 PET accuracy in localizing recurrent prostate cancer: a prospective single-arm clinical trial. JAMA Oncol. 2019;5(6):856–63.

Filippini T, Malavolti M, Borrelli F, Izzo AA, Fairweather-Tait SJ, Horneber M, Vinceti M. Green tea (Camellia sinensis) for the prevention of cancer. Cochrane Database Syst Rev. 2020;3(3):CD005004.

Galfano A, Secco S, Dell'Oglio P, Rha K, Eden C, Fransis K, Sooriakumaran P, et al. "Retzius-Sparing Robot-Assisted Radical Prostatectomy: Early Learning Curve Experience in Three Continents." [In eng]. BJU Int 127, no. 4 (Apr 2021): 412–17.

Guttilla A, Bortolus R, Giannarini G, Ghadjar P, Zattoni F, Gnech M, Palumbo V, Valent F, Garbeglio A, Zattoni F. Multimodal treatment for high-risk prostate cancer with high-dose intensity-modulated radiation therapy preceded or not by radical prostatectomy, concurrent intensified-dose docetaxel and long-term androgen deprivation therapy: results of a prospective phase II trial. Radiat Oncol. 2014;9:24.

Hamdy FC, Donovan JL, Lane JA, Mason M, Metcalfe C, Holding P, Davis M, Peters TJ, Turner EL, Martin RM, Oxley J, Robinson M, Staffurth J, Walsh E, Bollina P, Catto J, Doble A, Doherty A, Gillatt D, Kockelbergh R, Kynaston H, Paul A, Powell P, Prescott S, Rosario DJ, Rowe E, Neal DE, ProtecT Study Group. 10-Year outcomes after monitoring, surgery, or radiotherapy for localized prostate cancer. N Engl J Med. 2016;375(15):1415–24.

Karunasinghe N, Ng L, Wang A, Vaidyanathan V, Zhu S, Ferguson LR. Selenium Supplementation and Prostate Health in a New Zealand Cohort. Nutrients. 2019;12(1):2.

Kasivisvanathan V, Rannikko AS, Borghi M, Panebianco V, Mynderse LA, Vaarala MH, Briganti A, et al. "Mri-Targeted or Standard Biopsy for Prostate-Cancer Diagnosis." [In eng]. N Engl J Med 378, no. 19 (May 10 2018):1767–77.

Klein EA, Thompson IM Jr, Tangen CM, et al. Vitamin E and the risk of prostate cancer: the Selenium and Vitamin E Cancer Prevention Trial (SELECT). JAMA. 2011;306(14):1549–56.

Kristal AR, Till C, Platz EA, et al. Serum lycopene concentration and prostate cancer risk: results from the Prostate Cancer Prevention Trial. Cancer Epidemiol Biomarkers Prev. 2011;20(4):638–46.

Kristal AR, Till C, Song X, et al. Plasma vitamin D and prostate cancer risk: results from the Selenium and Vitamin E Cancer Prevention Trial. Cancer Epidemiol Biomarkers Prev. 2014;23(8):1494–504.

Lee WR, DeSilvio M, Lawton C, Gillin M, Morton G, Firat S, Baikadi M, Kuettel M, Greven K, Sandler H. A phase II study of external beam radiotherapy combined with permanent source brachytherapy for intermediate-risk, clinically localized adenocarcinoma of the prostate: preliminary results of RTOG P-0019. Int J Radiat Oncol Biol Phys. 2006;64(3):804–9.

López-Plaza B, Bermejo LM, Santurino C, Cavero-Redondo I, Álvarez-Bueno C, Gómez-Candela C. Milk and dairy product consumption and prostate cancer risk and mortality: an overview of systematic reviews and meta-analyses. Adv Nutr. 2019;10(Suppl 2):S212–23.

Meyer F, Galan P, Douville P, et al. Antioxidant vitamin and mineral supplementation and prostate cancer prevention in the SU.VI.MAX trial. Int J Cancer. 2005;116(2):182–6.

Micalli S, Territo A, Pirola GM, Ferrari N, Sighinolfi MC, Martorana E, Navarra M, Bianchi G. Effect of green tea catechins in patients with high-grade prostatic intraepithelial neoplasia: Results of a short-term double-blind placebo controlled phase II clinical trial. Arch Ital Urol Androl. 2017;89(3):197–202.

Mondul AM, Weinstein SJ, Moy KA, Männistö S, Albanes D. Circulating 25-Hydroxyvitamin D and Prostate Cancer Survival. Cancer Epidemiol Biomarkers Prev. 2016;25(4):665–9.

Rosenthal SA, Hunt D, Sartor AO, Pienta KJ, Gomella L, Grignon D, Rajan R, Kerlin KJ, Jones CU, Dobelbower M, Shipley WU, Zeitzer K, Hamstra DA, Donavanik V, Rotman M, Hartford AC, Michalski J, Seider M, Kim H, Kuban DA, Moughan J, Sandler H. A Phase 3 trial of 2 years of androgen suppression and radiation therapy with or without adjuvant chemotherapy for high-risk prostate cancer: final results of Radiation Therapy Oncology Group Phase 3 randomized trial NRG Oncology RTOG 9902. Int J Radiat Oncol Biol Phys. 2015;93(2):294–302.

Saini S. PSA and beyond: alternative prostate cancer biomarkers. Cell Oncol (Dordr). 2016;39(2):97–106.

El-Shater Bosaily A, Valerio M, Hu Y, Freeman A, Jameson C, Brown L, Kaplan R, et al. "The Concordance between the Volume Hotspot and the Grade Hotspot: A 3-D Reconstructive Model Using the Pathology Outputs from the Promis Trial." [In eng]. Prostate Cancer Prostatic Dis 19, no. 3 (Sep 2016):258–63.

Smith MR, Bae K, Efstathiou JA, et al. Diabetes and mortality in men with locally advanced prostate cancer: RTOG 92-02. J Clin Oncol. 2008;26(26):4333–9.

Sonni I, Eiber M, Fendler WP, Alano RM, Vangala SS, Kishan AU, Nickols N, Rettig MB, Reiter RE, Czernin J, Calais J. Impact of 68Ga-PSMA-11 PET/CT on Staging and Management of Prostate Cancer Patients in Various Clinical Settings: A Prospective Single-Center Study. J Nucl Med. 2020;61(8):1153–160.

Takizawa I, Hara N, Nishiyama T, Kaneko M, Hoshii T, Tsuchida E, Takahashi K. Oncological results, functional outcomes and health-related quality-of-life in men who received a radical prostatectomy or external beam radiation therapy for localized prostate cancer: a study on long-term patient outcome with risk stratification. Asian J Androl. 2009;11(3):283–90.

Wang P, Aronson WJ, Huang M, et al. Green tea polyphenols and metabolites in prostatectomy tissue: implications for cancer prevention. Cancer Prev Res (Phila). 2010;3(8):985–93.

Yaxley JW, Coughlin GD, Chambers SK, Occhipinti S, Samaratunga H, Zajdlewicz L, Dunglison N, et al. "Robot-Assisted Laparoscopic Prostatectomy Versus Open Radical Retropubic Prostatectomy: Early Outcomes from a Randomised Controlled Phase 3 Study." [In eng]. Lancet 388, no. 10049 (Sep 10 2016):1057–66.

Yee DS, Narula N, Amin MB, Skarecky DW, Ahlering TE. Robot-assisted radical prostatectomy: current evaluation of surgical margins in clinically low-, intermediate-, and high-risk prostate cancer. J Endourol. 2009;23(9):1461–5.

Zhang J, Dhakal IB, Lang NP, Kadlubar FF. Polymorphisms in inflammatory genes, plasma antioxidants, and prostate cancer risk. Cancer Causes Control. 2010;21(9):1437–44.

Zhao F, Vesprini D, Liu RSC, et al. Combining urinary DNA methylation and cell-free microRNA biomarkers for improved monitoring of prostate cancer patients on active surveillance. Urol Oncol. 2019;37(5):297.e9–297.e17.

Chapter 19
Physical Activity and Dietary Considerations for Men Diagnosed with Prostate Cancer

Ruth Ashston and John Saxton

19.1 Introduction

Prostate cancer is the most common malignancy in men aged 40–65 years. However, risk increases with age, abdominal obesity and family history (Gann 2002; Heidenreich et al. 2011) and the majority of cases are diagnosed in men over the age of 65 years (Smittenaar et al. 2016). Worldwide, more than a million men were estimated to have been diagnosed with prostate cancer in 2012, but with disease incidence varying across the world. Prostate cancer is a condition that in the early stages may not present any symptoms (Heidenreich et al. 2011; Ngollo et al. 2014), but its incidence is increasing with 299,923 (corresponding to 822 per day) new cases being registered in England in 2015; an increase of 3060 from the same time point in 2014 (Office Of National Statistics 2017). The prevalence of prostate cancer is estimated to show a 29% increase in the UK by 2035, but with this increase being at least partially attributable to further advances in screening and improved awareness of the disease (Smittenaar et al. 2016; Cancer Research UK 2015).

The majority of prostate cancers are classified as an adenocarcinoma, which are most common in the peripheral zone of the prostate. Over time, cancer cells multiply in an unregulated fashion to form a tumour that has the potential to invade nearby organs, including the seminal vesicles or rectum. Additionally, prostate tumour cells may develop the ability to travel in the blood stream or via the lymphatic system—a phenomenon known as metastasis. Most prostate cancers are

R. Ashston
School of Human Science, University of Derby, Derby, UK

J. Saxton (✉)
Department of Sport, Exercise and Rehabilitation, University of Northumbria, Newcastle upon Tyne, UK
e-mail: John.saxton@northumbria.ac.uk

© The Author(s), under exclusive license to Springer Nature Switzerland AG 2022
S. S. Goonewardene et al. (eds.), *Men's Health and Wellbeing*, https://doi.org/10.1007/978-3-030-84752-4_19

initially highly dependent on androgens for growth. Androgens are naturally occurring hormones that regulate the development and maintenance of male characteristics. The activation of androgen receptors involves several co-regulators (proteins that interact with transcription factors) that respond to a changing microenvironment to regulate specific gene targets involved in cell growth and survival. In a normal prostate epithelium, there is a balance between the rate of cell proliferation and the rate of apoptosis, however, in prostate cancer this process is unbalanced, leading to tumour growth.

19.2   Side-Effects of Treatment

Prostate cancer treatment, either through androgen deprivation therapy (ADT) in men with locally-advanced or metastatic disease or robot-assisted radical prostatectomy (minimally invasive surgery) in men whose cancer is confined to the prostate gland, carries a variety of side-effects, including urinary incontinence, negative mental state and erectile dysfunction (Lardas et al. 2017; Lehto et al. 2017). The side-effects of ADT are well documented and include an increase in body fat, unfavourable changes in blood-borne biomarkers of cardiometabolic health (e.g. high cholesterol levels, insulin resistance) and reduced bone mineral density (Keilani et al. 2017; Bourke et al. 2011). Robot-assisted radical prostatectomy, on the other hand, often results in erectile dysfunction and urinary incontinence, alongside increased body mass, reduced skeletal mass and impaired physical functioning, potentially leading to unfavourable blood-borne biomarkers of cardiometabolic health (Keilani et al. 2017; Adam et al. 2017; Santa Mina et al. 2013; Barocas et al. 2017; Joshu et al. 2011). Consequently, the side effects of both treatments can negatively impact a man's physical function and quality of life (Santa Mina et al. 2010; Keilani et al. 2017; Keogh and MacLeod 2012; Smith et al. 2001).

19.3   Physical Activity and Structured Exercise

Prostate cancer is linked to multiple non-modifiable risk factors such as family history, age and ethnicity (Gann 2002; Heidenreich et al. 2011). Additionally, a physically active lifestyle can help to prevent the accumulation of excess body fat and epidemiological evidence suggests that habitual physical activity can also reduce the risk of being diagnosed with high-grade disease (Cosimo et al. 2016; Friedenreich and Thune 2001; Moore et al. 2009). Physical activity is commonly defined as: "any bodily movement produced by skeletal muscles that results in energy expenditure and it can be categorized as occupational, sports, conditioning, household, or other activities" (Caspersen et al. 1985: 126). Walking accounts for most physical activity in many adults and has been shown to confer survival benefits after a prostate cancer diagnosis in observational studies. For example, men diagnosed with prostate cancer who engaged in ≥90 min of walking a week had a 61% (95% confidence interval: 18–84%) lower risk of prostate cancer mortality when compared to men who were relatively inactive (Kenfield et al. 2011; Richman et al. 2011).

Exercise is a subset of physical activity in that it is "planned, structured, and repetitive and has a final or an intermediate objective of the improvement or maintenance of physical fitness" (Caspersen et al. 1985). Few studies have investigated the relationship between structured exercise and prostate cancer risk. However, findings from one study suggest that a moderate amount of structured exercise (3–8.9 metabolic equivalent task hours (METs) a week e.g. playing golf, swimming) may lead to a reduced risk of prostate cancer and a lower risk of being diagnosed with high grade disease (Antonelli et al. 2009). Further to structured exercise reducing the risk of prostate cancer, after diagnosis it is plausible that as for physical activity, it could improve survival outcomes, as well as improving quality of life (Kenfield et al. 2011; Lynch and Leitzmann 2017).

19.4   Physical Activity Guidelines

Regular physical activity is recommended for men being treated for prostate cancer by both the National Institute for Health and Care Excellence and the American College of Sports Medicine (ACSM) (Meneses-Echavez et al. 2015; National Institute for Health and Care Excellence 2014b). This is due to the positive effects it can have on physical and mental health, as well as its potential role in reducing disease progression and reoccurrence after treatment (Chodzko-Zajko et al. 2009; Riebe et al. 2017; National Institute for Health and Care Excellence 2013). However, there are currently no published guidelines for prostate cancer patients specifically, and most advice has been aimed at those who have received ADT (National Institute for Health and Care Excellence 2014a). Commonly, doctors recommend that patients try to meet general public health guidelines on physical activity (Sutton et al. 2017). These guidelines state that healthy adults aged 19–64 years should complete at least 150 min of moderate intensity aerobic activity over the course of a week, alongside muscle strengthening activities on at least 2 days a week, as well as reducing sitting time (Health Do 2011). These are broadly consistent with physical activity guidelines recently published by the American College of Sports Medicine (Campbell et al. 2019) and World Cancer Research Fund/ACSM (Fund 2018) for cancer survivors.

The ACSM recommends that exercise, whether that be aerobic, resistance or a combination of both, during and after cancer treatment is generally safe and can bring about numerous health benefits (Campbell et al. 2019). They suggest moderate-intensity aerobic exercise at least three times per week, for at least 30 min, and for at least 8–12 weeks. Resistance exercise is recommended at least two times per week, using at least two sets of 8–15 repetitions of at least 60% of the one repetition maximum strength for a particular exercise. Men with prostate cancer can also refer to information produced by Prostate Cancer UK which offers brief advice, generally following Government guidelines regarding physical activity (Prostate Cancer UK 2015).

In summary, advice and support for physical activity and structured exercise is likely to become increasingly accepted as an important component of the treatment pathway for prostate cancer. A recent systematic review demonstrated that exercise is generally safe and well-tolerated in prostate cancer patients and should be considered an important component of prostate cancer care due its positive effects on fitness and physical function, and reduction of fatigue (Moe et al. 2017). A physically-active lifestyle could also help men to improve their survival chances and quality of life (Kenfield et al. 2011; Bourke et al. 2014). Finally, a physically-active lifestyle after a prostate cancer diagnosis could help men to reduce their risk of developing cardiometabolic disease (linked to ADT and unhealthy lifestyles), which is an important consideration as cardiovascular disease is the leading cause of death among men with prostate cancer (Allott et al. 2013).

19.5   Aerobic Exercise Training

Aerobic exercise includes exercises such as walking, running and cycling. The majority of studies investigating exercise in cancer patients have used aerobic exercise to improve patient outcomes with few exploring the benefits of resistance exercise (Jones et al. 2014; Windsor et al. 2004; Ashcraft et al. 2016; Zopf et al. 2017; Courneya et al. 2017; Tian et al. 2016). Aerobic exercise training is associated with multiple benefits including improvements in exercise tolerance (including aerobic exercise capacity), peripheral artery flow-mediated dilatation and body composition. In a study conducted by Santa Mina et al. (2013), aerobic exercise training was found to have more beneficial effects relating to reductions in body mass, waist circumference, and body mass index after 3 months compared with resistance exercise training in men being treated with ADT (Santa Mina et al. 2013). In addition, aerobic exercise has been reported to decrease fatigue, reduce depression and prevent deterioration in physical function during treatment for prostate cancer (Keogh and MacLeod 2012; Storer et al. 2012; Jones et al. 2014; Park et al. 2012). However, it has previously been suggested that resistance training could bring about more beneficial effects in this population than aerobic training (Champ et al. 2016; Teleni et al. 2016; Baumann et al. 2012). This could in part be due to prostate cancer patients already reporting high levels of aerobic activity (Ashton et al. 2019).

19.6   Resistance Exercise Training

Traditionally, resistance exercise has been used to improve muscular strength, mass and endurance, and extensive research has been conducted in this area. However, due to the likelihood that an increase in body fat and reduction in cardiopulmonary fitness will accompany muscular atrophy and a perception of poor body image during and after a prostate cancer diagnosis and throughout treatment, it is important that resistance exercise programmes are designed to address all these commonly experienced adverse health effects (Schmitz et al. 2010; Institute of Medicine National Research Council 2006). Previous studies in prostate cancer patients have focused on quality of life, upper and lower body strength, and fatigue as the primary outcomes (Santa Mina et al. 2013; Norris et al. 2015; Segal et al. 2009; Galvão et al. 2006; Winters-Stone et al. 2015) with few directly investigating cardiometabolic health, although several have reported cardiometabolic variables as secondary outcomes. Resistance exercise interventions are also effective for reducing fatigue and depression and may bring about more beneficial effects compared to aerobic training in relation to outcomes such as muscle function, quality of life and well-being, as well as reductions in blood pressure and waist circumference after prostate cancer treatments (Segal et al. 2009; Baumann et al. 2012; Mustian et al. 2009). Studies also show that resistance exercise can be manipulated to improve cardiopulmonary fitness and cardiometabolic health, whilst simultaneously achieving improvements in muscle size and strength. It has been suggested that resistance exercise performed in a circuit, with minimal rest between exercises, can maintain an elevated heart rate and improve aerobic exercise capacity, whilst improving strength, body composition and cardiometabolic health (Schmitz et al. 2010; Wilmore et al. 1978; Romero-Arenas et al. 2013; Paoli et al. 2013; Ashton et al. 2018).

19.7 Combined Training

It is likely that a combination of both aerobic and resistance exercise training would bring optimum health benefits to men with prostate cancer. A study by Galvao et al. (2010) was one of the first studies to demonstrate a reversal in muscle loss in prostate cancer patients receiving ADT after 12 weeks of combined exercise training (Galvão et al. 2010). The study required patients to complete eight resistance exercises for 2–4 sets per exercise and 15–20 min of cycling, walking, or jogging at 65–80% of maximum heart rate twice a week. In addition to a reversal in muscle loss (versus a control group), this study also demonstrated improvements in muscle strength, lean body mass, health-related quality of life and reduced fatigue.

Recent guidelines from the ACSM for cancer survivors suggest that combined exercise could improve common cancer-related side-effects, including anxiety, depressive symptoms, fatigue, impaired physical functioning, loss of bone mineral density and poor health-related quality of life (Campbell et al. 2019). However, the impact of combined exercise programmes for other outcomes, including peripheral neuropathy, pain and cognitive functioning, remain uncertain. Additionally, conflicting evidence has been reported for the effect of combined exercise on sexual function following prostate cancer treatment. For example, a systematic review by Bourke et al. (2016) reported inconclusive evidence regarding this health outcome. However Yunfeng et al. (2017) suggested that exercise is beneficial in this respect, with patients on ADT reporting less of a decline in sexual function versus a control group. It is evident that exercise has numerous beneficial effects for prostate cancer patients, yet, there are knowledge gaps which need addressing in future research.

19.8 Physical Activity and Exercise Advice in the Healthcare Setting

A prostate cancer diagnosis has been described as a 'teachable moment', in which patients become motivated to reduce unhealthy behaviours, especially if the advice is delivered by a trusted source such as a healthcare professional (Sutton et al. 2017; McBride et al. 2003; Horwood et al. 2014). Therefore, the diagnosis of prostate cancer presents an opportunity to implement and engage patients in physical activity or structured exercise. Men with prostate cancer have reported several barriers to taking part in structured exercise or becoming more physically active. Challenges to regular exercise include financial and environmental factors, insufficient knowledge about the health benefits of exercise and treatment-related side effects (i.e. urinary incontinence) (Hackshaw-McGeagh et al. 2017). In addition, fatigue and deconditioning as a result of cancer treatment can present physical barriers in cancer patients, which may lead to social isolation and lack of regular exercise (Fernandez et al. 2015). Other studies have reported lack of time, enjoyment and treatment side effects as barriers to exercise in cancer survivors (Fernandez et al. 2015; Ottenbacher et al. 2011; Craike et al. 2011; Hefferon et al. 2013; Spector et al. 2013). A study by Ottenbacher et al. (2011) showed that lack of willpower and the weather can be important barriers to exercise participation after prostate cancer treatment. All these barriers need to be addressed to ensure that programmes can be designed to meet the needs of men undergoing or recovering from prostate cancer treatment.

In the UK, the National Institute for Health and Care Excellence recommends structured exercise programmes, tailored to individual needs, to help self-manage the adverse health impacts of certain health conditions, such as myocardial infarction, stroke, chronic heart failure, chronic obstructive pulmonary disease, depression, lower back pain and chronic fatigue syndrome (National Institute for Health and Care Excellence 2014a). However, there are no such guidelines for patients who have been diagnosed with, or treated for, prostate cancer. This probably explains why fewer than half of cancer specialists in the United Kingdom regularly discuss exercise with men who have been diagnosed with prostate cancer (Sutton et al. 2017). Healthcare professionals have cited a variety of reasons for not providing such advice, including a lack of personal expertise, training and services to support structured exercise or physical activity, and with others simply reporting a lack of time (Sutton et al. 2017; Cowan 2016; Fortier et al. 2016). A recent study indicated when healthcare professionals do provide exercise advice for men with prostate cancer, the advice tends to be based on general Department of Health guidelines for adults (Sutton et al. 2017). This advice is likely to be inadequate as it relates to adults who are assumed to be healthy and is not tailored to the specific needs of men who have been treated for prostate cancer. There is, however, a sense that many health professionals acknowledge that exercise should be tailored to individual patient's pre-treatment and current level of fitness, any other relevant health considerations, mobility problems, treatment received, and their stage in the treatment process (Sutton et al. 2017). Finally, supported home-based exercise programmes may help to overcome some of the barriers associated with providing supervised exercise programmes within resource-constrained healthcare systems. Clinical populations already studied in home-based programmes include those with heart failure, prostate and breast cancer patients, type 2 diabetics, and those with osteoarthritis and good adherence rates have been demonstrated (Ashton et al. 2021; Hvid et al. 2016; Mustian et al. 2009; Safiyari-Hafizi et al. 2016; Plotnikoff et al. 2010; Bruce-Brand et al. 2012). Therefore, home-based training incorporating goal setting whilst maintaining or tapering a level of supervision may be more appropriate and cost-effective than fully supervised exercise for both the NHS and the patient.

19.9   Dietary Considerations

High amounts of animal fat in the diet and vitamin D deficiency may increase the risk of prostate cancer (Labbé et al. 2014; Farmer 2008; Hori et al. 2011; Mandair et al. 2014) and these factors, together with obesity, are thought to account for approximately 10% of the difference in incidence between black and white men (Farmer 2008). However, despite much research into the effects of dietary factors on prostate cancer risk, there is still much research to be done. After a prostate cancer diagnosis, a healthy diet is important for general health and can help maintain a healthy weight and assist in management of some of the side effects of prostate cancer treatment. For example, men on ADT often suffer a reduction in bone density and therefore need to ensure they are consuming enough calcium in their diet. Previous studies have explored the supplementation of calcium and vitamin D combined to help prevent bone loss in men undergoing ADT treatment, however, the results have been inconclusive (Datta and Schwartz 2012). Additionally, emerging evidence from studies of older adults suggest that protein supplementation, in conjunction with resistance training, could help to prevent the loss of skeletal muscle mass in men receiving ADT (Finger et al. 2015). However, few studies have examined this in men being treated with ADT and the potential disease-specific risks are unknown, although increasing protein intake is generally well tolerated (Daly et al. 2014; Chalé et al. 2013). In summary, further research is needed before the health benefits of specific food groups or dietary supplements in conjunction with exercise in men with prostate cancer can be fully understood.

19.10 Practical Tips

- Men who are not usually active should start gently for short periods of time (e.g. 10–15 min) and gradually exercise for longer as they become fitter.
- Moderate-intensity aerobic exercise is advised at least three times per week, for at least 30 min. Examples of aerobic exercise include walking, swimming, cycling and gardening. This can be easily achieved by doing simple things, such as getting off the bus one stop earlier, or using stairs rather than a lift or elevator.
- Moderate-intensity exercise means the heart should beat faster but maintaining a conversation should still be possible.
- Resistance exercise is recommended at least two times per week, using at least two sets of 8–15 repetitions of at least 60% of one repetition maximum. Gentle resistance exercise can be completed by lifting light weights, using elastic resistance bands or simply using body weight. A lot of exercises can be completed in the home with just a chair for equipment.
- Men should ensure adequate dietary intake of all essential macronutrients (protein, healthy fats, carbohydrates and water) and aim to maintain a healthy body composition (BMI: 18.5–24.9 kg/m^2; waist circumference: less than 94 cm) throughout their prostate cancer treatment and beyond.

References

Adam M, Tennstedt P, Lanwehr D, Tilki D, Steuber T, Beyer B, et al. Functional outcomes and quality of life after radical prostatectomy only versus a combination of prostatectomy with radiation and hormonal therapy. Eur Urol. 2017;71(3):330–6.

Allott EH, Masko EM, Freedland SJ. Obesity and prostate cancer: weighing the evidence. Eur Urol. 2013;63(5):800–9.

Antonelli JA, Jones LW, Bañez LL, Thomas J-A, Anderson K, Taylor LA, et al. Exercise and prostate cancer risk in a cohort of veterans undergoing prostate needle biopsy. J Urol. 2009;182(5):2226–31.

Ashcraft KA, Peace RM, Betof AS, Dewhirst MW, Jones LW. Efficacy and mechanisms of aerobic exercise on cancer initiation, progression, and metastasis: a critical systematic review of *in vivo* preclinical data. Cancer Res. 2016;76(14):4032–50.

Ashton RE, Tew GA, Aning JJ, Gilbert SE, Lewis L, Saxton JM. Effects of short-term, medium-term and long-term resistance exercise training on cardiometabolic health outcomes in adults: systematic review with meta-analysis. Br J Sports Med. 2018.

Ashton RE, Tew GA, Robson WA, Saxton JM, Aning JJ. Cross-sectional study of patient-reported fatigue, physical activity and cardiovascular status in men after robotic-assisted radical prostatectomy. Support Care Cancer. 2019;27(12):4763–70.

Ashton RE, Aning JJ, Tew GA, Robson WA, Saxton JM. Supported progressive resistance exercise training to counter the adverse side effects of robot-assisted radical prostatectomy: a randomised controlled trial. Support Care Cancer. 2021;29(8):4595–605.

Barocas DA, Alvarez J, Resnick MJ, Koyama T, Hoffman KE, Tyson MD, et al. Association between radiation therapy, surgery, or observation for localized prostate cancer and patient-reported outcomes after 3 years. JAMA. 2017;317(11):1126–40.

Baumann FT, Zopf EM, Bloch W. Clinical exercise interventions in prostate cancer patients–a systematic review of randomized controlled trials. Support Care Cancer. 2012;20(2):221–33.

Bourke L, Doll H, Crank H, Daley A, Rosario D, Saxton JM. Lifestyle intervention in men with advanced prostate cancer receiving androgen suppression therapy: a feasibility study. Cancer Epidemiol Biomarkers Prev. 2011;20(4):647–57.

Bourke L, Homer KE, Thaha MA, Steed L, Rosario DJ, Robb KA, et al. Interventions to improve exercise behaviour in sedentary people living with and beyond cancer: a systematic review. Br J Cancer. 2014;110(4):831–41.

Bourke L, Smith D, Steed L, Hooper R, Carter A, Catto J, et al. Exercise for men with prostate cancer: a systematic review and meta-analysis. Eur Urol. 2016;69(4):693–703.

Bruce-Brand RA, Walls RJ, Ong JC, Emerson BS, O'Byrne JM, Moyna NM. Effects of home-based resistance training and neuromuscular electrical stimulation in knee osteoarthritis: a randomized controlled trial. BMC Musculoskelet Disord. 2012;13:118.

Campbell KL, Winters-Stone KM, Wiskemann J, May AM, Schwartz AL, Courneya KS, et al. Exercise guidelines for cancer survivors: consensus statement from international multidisciplinary roundtable. Med Sci Sports Exerc. 2019;51(11):2375–90.

Cancer Research UK. Prostate Cancer Statistics 2015. 2015. https://www.cancerresearchuk.org/health-professional/cancer-statistics/statistics-by-cancer-type/prostate-cancer#heading-Zero

Caspersen CJ, Powell KE, Christenson GM. Physical activity, exercise, and physical fitness: definitions and distinctions for health-related research. Public Health Rep. 1985;100(2):126–31.

Chalé A, Cloutier GJ, Hau C, Phillips EM, Dallal GE, Fielding RA. Efficacy of whey protein supplementation on resistance exercise-induced changes in lean mass, muscle strength, and physical function in mobility-limited older adults. J Gerontol A Biol Sci Med Sci. 2013;68(6):682–90.

Champ CE, Francis L, Klement RJ, Dickerman R, Smith RP. Fortifying the treatment of prostate cancer with physical activity. Prostate Cancer. 2016;2016:9462975.

Chodzko-Zajko WJ, Proctor DN, Fiatarone Singh MA, Minson CT, Nigg CR, Salem GJ, Skinner JS. ACSM position stand: exercise and physical activity for older adults. Med Sci Sports Exerc. 2009;41(7):1510–30.

Cosimo DN, Fabrizio P, Riccardo L, Fabiana C, Stefano P, Alberto T, et al. Physical activity as a risk factor for prostate cancer diagnosis: a prospective biopsy cohort analysis. BJU Int. 2016;117(6B):E29–35.

Courneya KS, McNeil J, O'Reilly R, Morielli AR, Friedenreich CM. Dose-response effects of aerobic exercise on quality of life in postmenopausal women: results from the Breast Cancer and Exercise Trial in Alberta (BETA). Ann Behav Med. 2017;51(3):356–64.

Cowan RE. Exercise is medicine initiative: physical activity as a vital sign and prescription in adult rehabilitation practice. Arch Phys Med Rehabil. 2016;97(9 Suppl):S232–7.

Craike MJ, Livingston PM, Botti M. An exploratory study of the factors that influence physical activity for prostate cancer survivors. Support Care Cancer. 2011;19(7):1019–28.

Daly RM, O'Connell SL, Mundell NL, Grimes CA, Dunstan DW, Nowson CA. Protein-enriched diet, with the use of lean red meat, combined with progressive resistance training enhances lean tissue mass and muscle strength and reduces circulating IL-6 concentrations in elderly women: a cluster randomized controlled trial. Am J Clin Nutr. 2014;99(4):899–910.

Datta M, Schwartz GG. Calcium and vitamin D supplementation during androgen deprivation therapy for prostate cancer: a critical review. Oncologist. 2012;17(9):1171–9.

Farmer R. Prostate cancer: epidemiology and risk factors. Trends Urol Gynaecol Sex Health. 2008;13(3):32–4.

Fernandez S, Franklin J, Amlani N, DeMilleVille C, Lawson D, Smith J. Physical activity and cancer: a cross-sectional study on the barriers and facilitators to exercise during cancer treatment. Can Oncol Nurs J. 2015;25(1):37–48.

Finger D, Goltz FR, Umpierre D, Meyer E, Rosa LHT, Schneider CD. Effects of protein supplementation in older adults undergoing resistance training: a systematic review and meta-
analysis. Sports Med. 2015;45(2):245–55.

Fortier M, Guerin E, Segar ML. Words matter: reframing exercise is medicine for the general
population to optimize motivation and create sustainable behaviour change. Appl Physiol Nutr
Metab = Physiologie appliquee, nutrition et metabolisme. 2016;41(11):1212–5.

Friedenreich CM, Thune I. A review of physical activity and prostate cancer risk. Cancer Causes
Control. 2001;12(5):461–75.

Fund WCRFIWCR. Diet, nutrition, physical activity and cancer: a global perspective. 2018.
dietandcancerreport.org

Galvao DA, Nosaka K, Taaffe DR, Spry N, Kristjanson LJ, McGuigan MR, et al. Resistance training and reduction of treatment side effects in prostate cancer patients. Med Sci Sports Exerc.
2006;38(12):2045–52.

Galvão DA, Taaffe DR, Spry N, Joseph D, Newton RU. Combined resistance and aerobic exercise
program reverses muscle loss in men undergoing androgen suppression therapy for prostate
cancer without bone metastases: a randomized controlled trial. J Clin Oncol. 2010;28(2):340–7.

Gann PH. Risk factors for prostate cancer. Rev Urol. 2002;4(Suppl 5):S3–S10.

Hackshaw-McGeagh LE, Sutton E, Persad R, Aning J, Bahl A, Koupparis A, et al. Acceptability
of dietary and physical activity lifestyle modification for men following radiotherapy or radical
prostatectomy for localised prostate cancer: a qualitative investigation. BMC Urol. 2017;17:94.

Health Do. Factsheet 4: Physical activity guidelines for adults (19–64 years). 2011. https://www.
gov.uk/government/publications/uk-physical-activity-guidelines

Hefferon K, Murphy H, McLeod J, Mutrie N, Campbell A. Understanding barriers to exercise
implementation 5-year post-breast cancer diagnosis: a large-scale qualitative study. Health
Educ Res. 2013;28(5):843–56.

Heidenreich A, Bellmunt J, Bolla M, Joniau S, Mason M, Matveev V, et al. EAU guidelines on
prostate cancer. Part 1: Screening, diagnosis, and treatment of clinically localised disease. Eur
Urol. 2011;59(1):61–71.

Hori S, Butler E, McLoughlin J. Prostate cancer and diet: food for thought? BJU Int.
2011;107(9):1348–59.

Horwood JP, Avery KN, Metcalfe C, Donovan JL, Hamdy FC, Neal DE, et al. Men's knowledge
and attitudes towards dietary prevention of a prostate cancer diagnosis: a qualitative study.
BMC Cancer. 2014;14:812.

Hvid T, Lindegaard B, Winding K, Iversen P, Brasso K, Solomon TP, et al. Effect of a 2-year
home-based endurance training intervention on physiological function and PSA doubling time
in prostate cancer patients. Cancer Causes Control. 2016;27(2):165–74.

Institute of Medicine National Research Council. From cancer patient to cancer survivor: lost
in transition. In: Hewitt M, Greenfield S, Stovall E, editors. Washington, DC: The National
Academies Press; 2006. 534 p.

Jones LW, Hornsby WE, Freedland SJ, Lane A, West MJ, Moul JW, et al. Effects of nonlinear
aerobic training on erectile dysfunction and cardiovascular function following radical prostatectomy for clinically localized prostate cancer. Eur Urol. 2014;65(5):852–5.

Joshu CE, Mondul AM, Menke A, Meinhold C, Han M, Humphreys EB, et al. Weight gain is associated with an increased risk of prostate cancer recurrence after prostatectomy in the PSA era.
Cancer Prev Res (Phila). 2011;4(4):544–51.

Keilani M, Hasenoehrl T, Baumann L, Ristl R, Schwarz M, Marhold M, et al. Effects of resistance
exercise in prostate cancer patients: a meta-analysis. Support Care Cancer. 2017;25(9):2953–68.

Kenfield SA, Stampfer MJ, Giovannucci E, Chan JM. Physical activity and survival after prostate
cancer diagnosis in the health professionals follow-up study. J Clin Oncol. 2011;29(6):726–32.

Keogh JW, MacLeod RD. Body composition, physical fitness, functional performance, quality of
life, and fatigue benefits of exercise for prostate cancer patients: a systematic review. J Pain
Symptom Manage. 2012;43(1):96–110.

Labbé DP, Zadra G, Ebot EM, Mucci LA, Kantoff PW, Loda M, et al. Role of diet in prostate cancer: the epigenetic link. Oncogene. 2014;34:4683.

Lardas M, Liew M, van den Bergh RC, De Santis M, Bellmunt J, Van den Broeck T, et al. Quality of life outcomes after primary treatment for clinically localised prostate cancer: a systematic review. Eur Urol. 2017.

Lehto US, Tenhola H, Taari K, Aromaa A. Patients' perceptions of the negative effects following different prostate cancer treatments and the impact on psychological well-being: a nationwide survey. Br J Cancer. 2017;116(7):864–73.

Lynch BM, Leitzmann MF. An evaluation of the evidence relating to physical inactivity, sedentary behavior, and cancer incidence and mortality. Curr Epidemiol Rep. 2017;4(3):221–31.

Mandair D, Rossi RE, Pericleous M, Whyand T, Caplin ME. Prostate cancer and the influence of dietary factors and supplements: a systematic review. Nutr Metab (Lond). 2014;11:30.

McBride CM, Emmons KM, Lipkus IM. Understanding the potential of teachable moments: the case of smoking cessation. Health Educ Res. 2003;18(2):156–70.

Meneses-Echavez JF, Gonzalez-Jimenez E, Ramirez-Velez R. Supervised exercise reduces cancer-related fatigue: a systematic review. J Physiother. 2015;61(1):3–9.

Moe EL, Chadd J, McDonagh M, Valtonen M, Horner-Johnson W, Eden KB, et al. Exercise interventions for prostate cancer survivors receiving hormone therapy: systematic review. Transl J Am Coll Sports Med. 2017;2(1):1–9.

Moore SC, Peters TM, Ahn J, Park Y, Schatzkin A, Albanes D, et al. Age-specific physical activity and prostate cancer risk among white men and black men. Cancer. 2009;115(21):5060–70.

Mustian KM, Peppone L, Darling TV, Palesh O, Heckler CE, Morrow GR. A 4-week home-based aerobic and resistance exercise program during radiation therapy: a pilot randomized clinical trial. J Support Oncol. 2009;7(5):158–67.

National Institute for Health and Care Excellence. Integrating men being treated for prostate cancer into exercise referral schemes. NICE; 2013. www.nice.org.uk

National Institute for Health and Care Excellence. Prostate cancer: diagnosis and managment. Clinical guideline [CG175]. 2014a. https://www.nice.org.uk/guidance/cg175/ifp/chapter/Treatment-options

National Institute for Health and Care Excellence. Physical activity: exercise referral schemes – Public health guideline [PH54]. 2014b. https://www.nice.org.uk/guidance/ph54

Ngollo M, Dagdemir A, Karsli-Ceppioglu S, Judes G, Pajon A, Penault-Llorca F, et al. Epigenetic modifications in prostate cancer. Epigenomics. 2014;6(4):415–26.

Norris MK, Bell GJ, North S, Courneya KS. Effects of resistance training frequency on physical functioning and quality of life in prostate cancer survivors: a pilot randomized controlled trial. Prostate Cancer Prostatic Dis. 2015;18(3):281–7.

Office Of National Statistics. Cancer registration statistics. 2017. https://www.ons.gov.uk/peoplepopulationandcommunity/healthandsocialcare/conditionsanddiseases/bulletins/cancerregistrationstatisticsengland/firstrelease2015

Ottenbacher AJ, Day RS, Taylor WC, Sharma SV, Sloane R, Snyder DC, et al. Exercise among breast and prostate cancer survivors–what are their barriers? J Cancer Surviv. 2011;5(4):413–9.

Paoli A, Pacelli QF, Moro T, Marcolin G, Neri M, Battaglia G, et al. Effects of high-intensity circuit training, low-intensity circuit training and endurance training on blood pressure and lipoproteins in middle-aged overweight men. Lipids Health Dis. 2013;12(1):131.

Park SW, Kim TN, Nam JK, Ha HK, Shin DG, Lee W, et al. Recovery of overall exercise ability, quality of life, and continence after 12-week combined exercise intervention in elderly patients who underwent radical prostatectomy: a randomized controlled study. Urology. 2012;80(2):299–305.

Plotnikoff RC, Eves N, Jung M, Sigal RJ, Padwal R, Karunamuni N. Multicomponent, home-based resistance training for obese adults with type 2 diabetes: a randomized controlled trial. Int J Obes (Lond). 2010;34(12):1733–41.

Prostate Cancer UK. Living with cancer: your diet and physical activity. 2015. https://prostatecanceruk.org/prostate-information/living-with-prostate-cancer/your-diet-and-physical-activity#what-type-of-physical-activity-should-I-do

Richman EL, Kenfield SA, Stampfer MJ, Paciorek A, Carroll PR, Chan JM. Physical activity after diagnosis and risk of prostate cancer progression: data from the cancer of the prostate strategic urologic research endeavor. Cancer Res. 2011;71(11):3889–95.

Riebe D, Ehrman JK, Liguori G, Magal M, American College of Sports Medicine. ACSM's guidelines for exercise testing and prescription. 10th ed. Philadelphia: Wolters Kluwer; 2017.

Romero-Arenas S, Blazevich AJ, Martinez-Pascual M, Perez-Gomez J, Luque AJ, Lopez-Roman FJ, et al. Effects of high-resistance circuit training in an elderly population. Exp Gerontol. 2013;48(3):334–40.

Safiyari-Hafizi H, Taunton J, Ignaszewski A, Warburton DE. The health benefits of a 12-week home-based interval training cardiac rehabilitation program in patients with heart failure. Can J Cardiol. 2016;32(4):561–7.

Santa Mina D, Matthew AG, Trachtenberg J, Tomlinson G, Guglietti CL, Alibhai SM, et al. Physical activity and quality of life after radical prostatectomy. Can Urol Assoc J. 2010;4(3):180–6.

Santa Mina D, Alibhai SM, Matthew AG, Guglietti CL, Pirbaglou M, Trachtenberg J, et al. A randomized trial of aerobic versus resistance exercise in prostate cancer survivors. J Aging Phys Act. 2013;21(4):455–78.

Schmitz KH, Courneya KS, Matthews C, Demark-Wahnefried W, Galvao DA, Pinto BM, Irwin ML, Wolin KL, Segal RJ, Lucia A, Schneider CM, von Grenigen VE, Schwartz AL. American College of Sports Medicine Roundtable on Exercise guidelines for cancer survivors. Official Journal of the American College of Sports Medicine. 2010.

Segal RJ, Reid RD, Courneya KS, Sigal RJ, Kenny GP, Prud'Homme DG, et al. Randomized controlled trial of resistance or aerobic exercise in men receiving radiation therapy for prostate cancer. J Clin Oncol. 2009;27(3):344–51. http://onlinelibrary.wiley.com/o/cochrane/clcentral/articles/192/CN-00665192/frame.html

Smith JC, Bennett S, Evans LM, Kynaston HG, Parmar M, Mason MD, et al. The effects of induced hypogonadism on arterial stiffness, body composition, and metabolic parameters in males with prostate cancer. J Clin Endocrinol Metab. 2001;86(9):4261–7.

Smittenaar CR, Petersen KA, Stewart K, Moitt N. Cancer incidence and mortality projections in the UK until 2035. Br J Cancer. 2016;115(9):1147–55.

Spector D, Battaglini C, Groff D. Perceived exercise barriers and facilitators among ethnically diverse breast cancer survivors. Oncol Nurs Forum. 2013;40(5):472–80.

Storer TW, Miciek R, Travison TG. Muscle function, physical performance and body composition changes in men with prostate cancer undergoing androgen deprivation therapy. Asian J Androl. 2012;14(2):204–21.

Sutton E, Hackshaw-McGeagh LE, Aning J, Bahl A, Koupparis A, Persad R, et al. The provision of dietary and physical activity advice for men diagnosed with prostate cancer: a qualitative study of the experiences and views of health care professionals, patients and partners. Cancer Causes Control. 2017;28(4):319–29.

Teleni L, Chan RJ, Chan A, Isenring EA, Vela I, Inder WJ, et al. Exercise improves quality of life in androgen deprivation therapy-treated prostate cancer: systematic review of randomised controlled trials. Endocr Relat Cancer. 2016;23(2):101–12.

Tian L, Lu HJ, Lin L, Hu Y. Effects of aerobic exercise on cancer-related fatigue: a meta-analysis of randomized controlled trials. Support Care Cancer. 2016;24(2):969–83.

Wilmore JH, Parr RB, Girandola RN, Ward P, Vodak PA, Barstow TJ, et al. Physiological alterations consequent to circuit weight training. Med Sci Sports. 1978;10(2):79–84.

Windsor PM, Nicol KF, Potter J. A randomized, controlled trial of aerobic exercise for treatment-related fatigue in men receiving radical external beam radiotherapy for localized prostate carcinoma. Cancer. 2004;101(3):550–7.

Winters-Stone KM, Dobek JC, Bennett JA, Dieckmann NF, Maddalozzo GF, Ryan CW, et al. Resistance training reduces disability in prostate cancer survivors on androgen deprivation therapy: evidence from a randomized controlled trial. Arch Phys Med Rehabil. 2015;96(1):7–14.

Yunfeng G, Weiyang H, Xueyang H, Yilong H, Xin G. Exercise overcome adverse effects among prostate cancer patients receiving androgen deprivation therapy: an update meta-analysis. Medicine (Baltimore). 2017;96(27):e7368.

Zopf EM, Newton RU, Taaffe DR, Spry N, Cormie P, Joseph D, et al. Associations between aerobic exercise levels and physical and mental health outcomes in men with bone metastatic prostate cancer: a cross-sectional investigation. Eur J Cancer Care. 2017;26(6):e12575.

Chapter 20
Prostate Cancer Survivorship and Physical Health

Sanchia S. Goonewardene, Annie Young, Declan Cahill, and Raj Persad

20.1 Introduction

Over 2 million people in England have a diagnosis of cancer (National Cancer Survivorship Initiative 2008). Of this, over 250,000 have been diagnosed with prostate cancer (Maddams et al. 2009) and 130,000 people per year die. The Department of Health is spending £750 million on improving earlier diagnosis and prevention of cancer. During the next decade, a rapid increase in the number of new cancer diagnoses as well as a growing number of cancer survivors are predicted (Ganz 2009).

A cancer survivor is any person who has received a diagnosis of cancer from diagnosis until the end of life. Survivorship is defined by Macmillan Cancer Support, a leading UK cancer care and support charity, as someone who has completed initial cancer management with no evidence of apparent disease. According to the National Cancer Institute, cancer survivorship encompasses the "physical, psychosocial, and economic issues of cancer from diagnosis until the end of life." (National Cancer Institute).

Prostate cancer survivors require further investigation as there are concerns current follow-up methods are unsuitable (National Cancer Intelligence Network UK

S. S. Goonewardene (✉)
The Princess Alexandra Hospital, Harlow, UK

A. Young
The University of Warwick, Coventry, UK

D. Cahill
The Royal Marsden Hospital, London, UK

R. Persad
North Bristol NHS Foundation Trust, Bristol, UK
e-mail: rajpersad@bristolurology.com

© The Author(s), under exclusive license to Springer Nature Switzerland AG 2022
S. S. Goonewardene et al. (eds.), *Men's Health and Wellbeing*,
https://doi.org/10.1007/978-3-030-84752-4_20

2008). Due to the growing population of survivors of prostate cancer and the period of austerity for the NHS, patients are not getting the holistic care required during the survivorship phase. Concerns regarding permanent physical, psychosocial, and economic effects of cancer treatment were highlighted by the US Institute of Medicine Report (Hewitt et al. 2006). This defined landmarks for survivorship care: monitoring for recurrence, metastases or side effects and coordination between secondary and primary care.

The unmet needs of cancer survivors, the rising numbers, and pressures to utilise resources efficiently (National Cancer Survivorship Initiative, Department of Health 2008) are a significant burden on the health system. These issues have been raised by the National Cancer Survivorship Initiative (NCSI) (Giarelli 2004) which highlighted key shifts in attitude towards care.

20.2   Systematic Review Methods

A systematic review relating to physical activity and prostate cancer was conducted. This was to identify the impact of physical activity in prostate cancer. The search strategy aimed to identify all references related to prostate cancer AND physical activity. Search terms used were as follows: (Prostate cancer) AND (Physical activity). The following databases were screened from 2000 to June 2021:

- CINAHL
- MEDLINE (NHS Evidence)
- Cochrane
- AMed
- EMBASE
- PsychINFO
- SCOPUS
- Web of Science.

In addition, searches using Medical Subject Headings (MeSH) and keywords were conducted using Cochrane databases. Two UK-based experts in bladder cancer were consulted to identify any additional studies.

Studies were eligible for inclusion if they reported primary research. Papers were included if published after 2000 and had to be in English. Studies that did not conform to this were excluded. Only primary research was included (Fig. 20.1). The overall aim was to identify impact of physical activity on prostate cancer.

Abstracts were independently screened for eligibility by two reviewers and disagreements resolved through discussion or third-party opinion. Agreement level was calculated using Cohen's Kappa to test the intercoder reliability of this screening process (Cohen 1968). Cohens' Kappa allows comparison of inter-rater

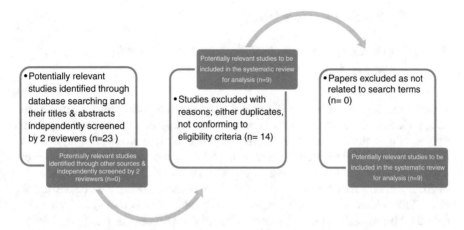

Fig. 20.1 Flow chart of studies identified through the systematic review (adapted from (Moher et al. 2009b) PRISMA)

reliability between papers using relative observed agreement. This also takes account of the comparison occurring by chance. The first reviewer agreed all nine papers to be included, the second, agreed on nine. Cohens kappa was therefore 1.0.

Data extraction was piloted by the researcher and amended in consultation with the research team (author and two academic supervisors). Data collected included authors, year and country of publication, study aims, setting, intervention aims, number of participants, study design, intervention components and delivery methods, comparison groups and outcome measures, notes and follow-up questions for the authors. Studies were quality assessed using the PRISMA criteria (Moher et al. 2009b) for randomised controlled trials, (Mays et al. 2005; Moher et al. 2009a) for the action research and qualitative studies and the Critical Skills Appraisal programme for cohort studies. This was also applied to randomised controlled trials and qualitative studies.

The search identified 23 papers (Fig. 20.1). All 9 out of 23 mapped to the search terms and eligibility criteria. The current systematic reviews were examined to gain further knowledge about the subject. 14 papers were excluded due to not conforming to eligibility criteria or adding to the evidence. Of the nine papers left, relevant abstracts were identified, and the full papers obtained (all of which were in English), to quality assure against search criteria. There was considerable heterogeneity of design among the included studies therefore a narrative review of the evidence was undertaken. There was significant heterogeneity within studies, including clinical topic, numbers, outcomes, as a result a narrative review was thought to be best. There were nine trials with a good level of evidence. These were from a range of countries with minimal loss to follow-up. All participants were identified appropriately.

20.3   Systematic Review Results

20.3.1   *Adherence to Physical Activity Guidelines in Prostate Cancer Survivorship*

Bruneau assessed adherence to PA guidelines in patients undergoing genetic counselling GC for prostate cancer (PCA) (Bruneau et al. 2020). A total of 158 men from the Genetic Evaluation of Men (GEM) study with a diagnosis or at risk for PCA completed a structured lifestyle survey, including questions about the number of days and intensity of PA over the past year (Bruneau et al. 2020). High proportions of GEM participants were overweight (44.9%) or obese (38.0%, p = 0.002) (Bruneau et al. 2020). Men with PCA engaged in less moderate (p = 0.019) and vigorous (p = 0.005) aerobic activity than men without PCA (Bruneau et al. 2020). Higher education was predictive of adherence to light (p = 0.008), moderate (p = 0.019), and vigorous (p = 0.002) intensity PA (Bruneau et al. 2020). Older age (p = 0.015) and higher education (p = 0.001) were predictive of adherence to strength-based recommendations (Bruneau et al. 2020).

20.4   Lifestyle Prescriptions and Physical Activity in Prostate Cancer

Lemanska assessed the feasibility and acceptability of a community pharmacy lifestyle intervention to improve physical activity and cardiovascular health of men with prostate cancer (Lemanska et al. 2019). Community pharmacy teams were trained to deliver a health assessment including fitness, strengths part of a phase 2 clinical trial (Lemanska et al. 2019). A computer algorithm generated a personalised lifestyle prescription for a home-based programme accompanied by supporting resources (Lemanska et al. 2019). Out of 403 invited men, 172 (43%) responded and 116 (29%) participated (Lemanska et al. 2019). Of these, 99 (85%) completed the intervention and 88 (76%) completed the 6-month follow-up (attrition 24%) (Lemanska et al. 2019). Certain components of the intervention were feasible and acceptable (e.g., community pharmacy delivery), while others were more challenging (e.g., fitness assessment) and will be refined for future studies (Lemanska et al. 2019). By 3 months, moderate to vigorous physical activity increased on average by 34 min (95% CI 6–62, p = 0.018), but this was not sustained over 6 months (Lemanska et al. 2019).

20.5   Spouse Input and Physical Activity for Prostate Cancer

Winters-Stone tested the feasibility and preliminary efficacy of a partnered strength training program on the physical and mental health of prostate cancer survivors (PCS) and spouse caregivers (Winters-Stone et al. 2016). Sixty-four couples were randomly assigned to 6 months of partnered strength training (Exercising Together, N = 32) or usual care (UC, N = 32) (Winters-Stone et al. 2016). Objective measures included body composition (lean, fat and trunk fat mass (kg), and % body fat) by DXA, upper and lower body muscle strength by one-repetition maximum, and physical function by the physical performance battery (PPB) (Winters-Stone et al. 2016). Couple retention rates were 100% for Exercising Together and 84% for UC (Winters-Stone et al. 2016). Median attendance of couples to Exercising Together sessions was 75%. Men in Exercising Together became stronger in the upper body (p < 0.01) and more physically active (p < 0.01) than UC (Winters-Stone et al. 2016). Women in Exercising Together increased muscle mass (p = 0.05) and improved upper (p < 0.01) and lower body (p < 0.01) strength and PPB scores (p = 0.01) more than UC (Winters-Stone et al. 2016).

20.6 Computer Based Programs to Improve Adherence to Physical Activity Guidelines

Golsteijn assessed the efficacy of a computer-tailored PA intervention in (four sub-groups of) prostate and colorectal cancer survivors (Golsteijn et al. 2018). Prostate and colorectal cancer patients and survivors were randomized to the OncoActive intervention group (N = 249), or a usual-care waiting-list control group (N = 229) (Golsteijn et al. 2018). OncoActive participants received a pedometer and computer-tailored PA advice, both Web-based via an interactive website and with printed materials (Golsteijn et al. 2018). Three months after baseline OncoActive participants significantly increased their self-reported PA (PA days: d = 0.46; MVPA: d = 0.23) (Golsteijn et al. 2018). Physical functioning (d = 0.23) and fatigue (d = −0.21) also improved significantly after 3 months (Golsteijn et al. 2018). Six months after baseline, self-reported PA (PA days: d = 0.51; MVPA: d = 0.37) and ActiGraph MVPA (d = 0.27) increased significantly, and ActiGraph days (d = 0.16) increased borderline significantly (p = .05; d = 0.16) (Golsteijn et al. 2018). Furthermore, OncoActive participants reported significantly improvements in physical functioning (d = 0.14), fatigue (d = −0.23) and depression (d = −0.32) (Golsteijn et al. 2018). Similar results were found for intention-to-treat analyses (Golsteijn et al. 2018). Higher increases in PA were found for colorectal cancer participants at 3 months, and for medium and highly educated participants' PA at 6 months (Golsteijn et al. 2018). Health outcomes at 6 months were more prominent in colorectal cancer participants and in women (Golsteijn et al. 2018).

Kenfield determined the feasibility and acceptability of a digital lifestyle intervention among men with prostate cancer (Kenfield et al. 2019). A 12-week pilot randomized controlled trial among 76 men with clinical stage T1-T3a prostate cancer (Kenfield et al. 2019). Eligibility included Internet access, no contraindications to aerobic exercise, and engaging in four or fewer of eight targeted behaviours at baseline (Kenfield et al. 2019). The primary outcomes were feasibility and acceptability based on recruitment and user data, and surveys, respectively (Kenfield et al. 2019). At baseline, men in both arms met a median of three targeted behaviours (Kenfield et al. 2019). Sixty-four men (n = 32 per arm) completed the study; 88%

completed 12-week assessments (intervention, 94%; control, 82%) (Kenfield et al. 2019). Intervention participants wore their Fitbits a median of 82d (interquartile range [IQR]: 72–83), replied to a median of 71% of text messages (IQR: 57–89%), and visited the website a median of 3d (IQR: 2–5) over 12 week. Median (IQR) absolute changes in the P8 score from baseline to 12 week were 2 (1, 3) for the intervention and 0 (−1, 1) for the control arm (Kenfield et al. 2019). The estimated mean score of the intervention arm was 1.5 (95% confidence interval: 0.7, 2.3) higher than that of the control arm at 12 week (ANCOVA p < 0.001) (Kenfield et al. 2019).

20.7   Group Activities for Physical Activity and Prostate Cancer

Bjerre conducted a pragmatic, multicentre, parallel randomised controlled trial in five Danish urological departments (Bjerre et al. 2019). In total, 214 men with prostate cancer were randomly assigned to either 6 months of free-of-charge football training twice weekly at a local club (football group [FG]) (n = 109) or usual care (usual care group [UG]) (n = 105), including brief information on physical activity recommendations at randomisation (Bjerre et al. 2019). Participants were on average 68.4 (SD 6.2) years old, 157 (73%) were retired, 87 (41%) were on castration-based treatment, 19 (9%) had received chemotherapy, and 41 (19%) had skeletal metastases at baseline (Bjerre et al. 2019). No statistically significant between-group difference was observed in change in prostate-cancer-specific quality of life (ITT: 1.9 points [95% CI −1.9 to 5.8], p = 0.325; PP: 3.6 points [95% CI −0.9 to 8.2], p = 0.119) (Bjerre et al. 2019). A statistically significant between-group difference was observed in change in total hip BMD, in favour of FG (0.007 g/cm^2 [95% CI 0.004–0.013], p = 0.037) (Bjerre et al. 2019). No differences were observed in change in lumbar spine BMD or lean body mass (Bjerre et al. 2019). Among patients allocated to football, 59% chose to continue playing football after the end of the 6-month intervention period (Bjerre et al. 2019). At 1-year follow-up in the PP population, FG participants had more improvement on the Mental Component Summary (2.9 [95% CI 0.0–5.7], p = 0.048 points higher) than UG participants, as well as a greater loss of fat mass (−0.9 kg [95% CI −1.7 to −0.1], p = 0.029) (Bjerre et al. 2019). There were no differences between groups in relation to fractures or falls (Bjerre et al. 2019). Hospital admissions were more frequent in UG compared to FG (33 vs 20; the odds ratio based on PP analysis was 0.34 for FG compared to UG) (Bjerre et al. 2019). There were three deaths in FG and four in UG (Bjerre et al. 2019). Main limitations of the study were the physically active control group and assessment of physical activity by means of self-report (Bjerre et al. 2019).

In the earlier stages of prostate cancer, effective treatments have created a need for research to focus on practices that may improve quality of life throughout survivorship (Ross Zahavich et al. 2013). This 14-week feasibility study involved a

7-week class-based yoga program (adherence phase), followed by 7 weeks of self-selected physical activity (maintenance phase) (Ross Zahavich et al. 2013). Demographic information, physical activity behaviour, quality of life, fatigue, stress, mood, and fitness variables were assessed at three time points (Ross Zahavich et al. 2013). Class attendance was 6.1 and 5.8 for prostate cancer survivors (n = 15) and their support persons (n = 10), respectively, for the seven classes (Ross Zahavich et al. 2013). Levels of perceived social support were higher for those who brought a support person (Ross Zahavich et al. 2013). Significant improvements with regard to stress, fatigue, and mood before and after yoga class (all Ps < .05) were reported by all participants (Ross Zahavich et al. 2013). No clinically significant changes were noted on prostate cancer survivor's quality of life or fatigue over the course of the 14-week study (Ross Zahavich et al. 2013).

20.8   Physical Activity and ADT

Dawson assessed 12 weeks of resistance training on body composition and MetS changes in prostate cancer patients on ADT (Dawson et al. 2018). Prostate cancer patients on ADT were randomized to resistance training and protein supplementation (TRAINPRO), resistance training (TRAIN), protein supplementation (PRO), or control stretching (STRETCH) (Dawson et al. 2018). Exercise groups (EXE = TRAINPRO, TRAIN) performed supervised exercise 3 days per week for 12 weeks, while non-exercise groups (NoEXE = PRO, STRETCH) performed a home-based stretching program (Dawson et al. 2018). TRAINPRO and PRO received 50 g·day^{-1} of whey protein (Dawson et al. 2018). A total of 37 participants were randomized; 32 participated in the intervention (EXE n = 13; NoEXE n = 19) (Dawson et al. 2018). At baseline, 43.8% of participants were sarcopenic and 40.6% met the criteria for MetS (Dawson et al. 2018). Post-intervention, EXE significantly improved lean mass (d = 0.9), sarcopenia prevalence (d = 0.8), body fat % (d = 1.1), strength (d = 0.8–3.0), and prostate cancer-specific quality of life (d = 0.9) compared to NoEXE (p < 0.05) (Dawson et al. 2018). No significant differences were observed between groups for physical function or MetS-related variables except waist circumference (d = 0.8) (Dawson et al. 2018).

20.9   Radical Prostatectomy: Pre-operative
       Physical Prehabilitation

Santa Mina et al. (2018) examined the feasibility and effects of prehabilitation on perioperative and postoperative outcomes in men undergoing radical prostatectomy (Santa Mina et al. 2018). This feasibility RCT compared prehabilitation (PREHAB) versus a control condition (CON) in 86 men undergoing radical prostatectomy

(Santa Mina et al. 2018). PREHAB consisted of home-based, moderate-intensity exercise prior to surgery (Santa Mina et al. 2018). Both groups received a preoperative pelvic floor training regimen (Santa Mina et al. 2018). Feasibility was assessed via rates of recruitment, attrition, intervention duration and adherence, and adverse events (Santa Mina et al. 2018). Clinical outcomes included surgical complications, and length of stay. The following outcomes were assessed at baseline, prior to surgery, and 4, 12, and 26-weeks postoperatively: 6-min walk test (6MWT), upper-extremity strength, quality of life, psychosocial wellbeing, urologic symptoms, and physical activity volume (Santa Mina et al. 2018). The recruitment rate was 47% and attrition rates were 25% and 33% for PREHAB and CON, respectively (Santa Mina et al. 2018). Adherence to PREHAB was 69% with no serious intervention-related adverse events (Santa Mina et al. 2018). After the intervention and prior to surgery, PREHAB participants demonstrated less anxiety (P = 0.035) and decreased body fat percentage (P = 0.001) compared to CON (Santa Mina et al. 2018). Four weeks postoperatively, PREHAB participants had greater 6MWT scores of clinical significance compared to CON (P = 0.006) (Santa Mina et al. 2018). Finally, compared to CON, grip strength and anxiety were also greater in the PREHAB at 26-weeks (P = 0.022) and (P = 0.025), respectively (Santa Mina et al. 2018).

20.10   Grey Literature Relating to Systematic Review

Shingler assessed a diet and physical activity intervention in men undergoing radical prostatectomy for localized prostate cancer as part of The feasibility of the Prostate cancer: Evidence of Exercise and Nutrition Trial (PrEvENT) dietary and physical activity modifications trial (Shingler et al. 2017). PrEvENT involved randomizing men to either a dietary and/or physical activity intervention (Shingler et al. 2017). Three overarching themes were identified: acceptability of the intervention, acceptability of the data collection methods and trial logistics (Shingler et al. 2017). Participants were predominantly positive about both the dietary and physical activity interventions and most men found the methods of data collection appropriate (Shingler et al. 2017). Recommendations for future trials include consideration of alternative physical activity options, such as cycling or gym sessions, increased information on portion sizes, the potential importance of including wives or partners in the dietary change process and the possibility of using the pedometer or other wearable technology as part of the physical activity intervention (Shingler et al. 2017).

Bourke conducted a multi-centre investigation was to examine what exercise referral is currently available for men on ADT as provided by the NHS and if a supervised, individually-tailored exercise training package (as per the national NICE guidelines CG175) is embedded within prostate cancer care (Bourke et al.

2018). A multi-centre investigation of current National Health Service (NHS) care involving a web-based survey of NHS prostate cancer care, five focus groups involving 26 men on ADT and 37 semi-structured interviews with healthcare professionals (HCPs) involved in the management of prostate cancer (Bourke et al. 2018). HCPs and men on ADT asserted that medical castration has a serious and debilitating impact on many features of men's quality of life (Bourke et al. 2018). There is support for exercise training programmes as part of cancer care and patients would support their initiation soon after diagnosis (Bourke et al. 2018). Involving the Multidisciplinary Team (MDT) is proposed as key to this. Critically, traditional values in oncology would need to be overcome for widespread acceptance (Bourke et al. 2018). Specialist further training for HCPs around behaviour change support could encourage this (Bourke et al. 2018). Given that these schemes are seen as a fundamental part of cancer care, it is felt the NHS should commission and support provision (Bourke et al. 2018). 79 representatives of 154 NHS trusts (51%) provided survey data on current delivery: only 17% could provide supervised exercise as per CG175 (Bourke et al. 2018).

20.11 Systematic Reviews Related to Physical Activity in Prostate Cancer

Davies conducted a review undertaken for the National Cancer Survivorship Initiative relating to physical activity (Davies et al. 2011). A total of 43 records were included in this review (Davies et al. 2011). Evidence from observational studies suggests that a low-fat, high-fibre diet might be protective against cancer recurrence and progression (Davies et al. 2011). There is more support for physical activity, with a dose response for better outcomes (Davies et al. 2011). When synthesized with findings from the World Cancer Research Fund review of RCTs investigating the effect of diet and physical activity interventions on cancer survival, evidence suggests that the mechanism of benefit from diet and physical activity pertains to body weight, with excess body weight being a risk factor, which is modifiable through lifestyle (Davies et al. 2011).

20.12 Conclusions

There is clearly a significant role physical activity has to play in prostate cancer survivorship, whether in radical therapy, on ADT or any part of the survivorship phase. There are key elements such as group activity and technological intervention that can help promote patient engagement.

References

Bjerre ED, Petersen TH, Jørgensen AB, Johansen C, Krustrup P, Langdahl B, Poulsen MH, et al. Community-based football in men with prostate cancer: 1-year follow-up on a pragmatic, multicentre randomised controlled trial. PLoS Med. 2019;16(10):e1002936. [In eng].

Bourke L, Turner R, Greasley R, Sutton E, Steed L, Smith D, Brown J, et al. A multi-centre investigation of delivering National Guidelines on Exercise Training for Men with Advanced Prostate Cancer Undergoing Androgen Deprivation Therapy in the UK NHS. PLoS One. 2018;13(7):e0197606. [In eng].

Bruneau M Jr, Milliron BJ, Sinclair E, Obeid E, Gross L, Bealin L, Smaltz C, Butryn M, Giri VN. Physical activity assessment among men undergoing genetic counseling for inherited prostate cancer: a teachable moment for improved survivorship. Support Care Cancer. 2020.

Cohen J. Weighted Kappa: nominal scale agreement with provision for scaled disagreement or partial credit. Psychological Bull. 1968;70(4):213–20. [In eng].

Davies NJ, Batehup L, Thomas R. The role of diet and physical activity in breast, colorectal, and prostate cancer survivorship: a review of the literature. Br J Cancer. 2011;105(Suppl 1):S52–73. [In eng].

Dawson JK, Dorff TB, Todd Schroeder E, Lane CJ, Gross ME, Dieli-Conwright CM. Impact of resistance training on body composition and metabolic syndrome variables during androgen deprivation therapy for prostate cancer: a pilot randomized controlled trial. BMC Cancer. 2018;18(1):368. [In eng].

Ganz PA. Survivorship: adult cancer survivors.Prim Care. 2009;36(4):721–41.

Giarelli E. A model of survivorship in cancer genetic care. Semin Oncol Nurs. 2004;20(3):196–202.

Golsteijn RHJ, Bolman C, Volders E, Peels DA, de Vries H, Lechner L. Short-term efficacy of a computer-tailored physical activity intervention for prostate and colorectal cancer patients and survivors: a randomized controlled trial. Int J Behav Nutr Phys Act. 2018;15(1):106. [In eng].

Hewitt M, Greenfield W, Stovall E. From Cancer Patient to Cancer Survivor: Lost in Transition. The National Academies Press: Washington; 2006.

Kenfield SA, Van Blarigan EL, Ameli N, Lavaki E, Cedars B, Paciorek AT, Monroy C, et al. Feasibility, acceptability, and behavioral outcomes from a technology-enhanced behavioral change intervention (prostate 8): a pilot randomized controlled trial in men with prostate cancer. Eur Urol. 2019;75(6):950–8. [In eng].

Lemanska A, Poole K, Griffin BA, Manders R, Saxton JM, Turner L, Wainwright J, Faithfull S. Community pharmacy lifestyle intervention to increase physical activity and improve cardiovascular health of men with prostate cancer: a phase II feasibility study. BMJ Open. 2019;9(6):e025114. [In eng].

Maddams J, Brewster D, Gavin A, et al. Cancer prevalence in the United Kingdom: estimates for 2008. Br J Cancer. 2009;101(3):541–47. https://doi.org/10.1038/sj.bjc.6605148.

Mays N, Pope C, Popay J. Systematically reviewing qualitative and quantitative evidence to inform management and policy-making in the health field. J Health Serv Res Policy. 2005;10(Suppl 1):6–20. [In eng].

Moher D, Liberati A, Tetzlaff J, Altman DG. Preferred reporting items for systematic reviews and meta-analyses: the prisma statement. Open Med. 2009a;3(3):e123–30. [In eng].

Moher D, Liberati A, Tetzlaff J, Altman DG. Preferred reporting items for systematic reviews and meta-analyses: the prisma statement. BMJ (Online). 2009b;339(7716):332–6. [In eng].

National Cancer Survivorship Initiative. Department of Health. 2008.

National Cancer Intelligence Network UK. Lifetime prevalence for all cancers and colorectal, lung, breast and prostate cancers 2008. http://library.ncin.org.uk/docs/080714-TCR-UK_prevalence.pdf.

Ross Zahavich AN, Robinson JA, Paskevich D, Culos-Reed SN. Examining a therapeutic yoga program for prostate cancer survivors. Integr Cancer Ther. 2013;12(2):113–25. [In eng].

Santa Mina D, Hilton WJ, Matthew AG, Awasthi R, Bousquet-Dion G, Alibhai SMH, Au D, et al. Prehabilitation for radical prostatectomy: a multicentre randomized controlled trial. Surg Oncol. 2018;27(2):289–98. [In eng].

Shingler E, Hackshaw-McGeagh L, Robles L, Persad R, Koupparis A, Rowe E, Shiridzinomwa C, et al. The feasibility of the prostate cancer: evidence of exercise and nutrition trial (prevent) dietary and physical activity modifications: a qualitative study. Trials. 2017;18(1):106. [In eng].

Winters-Stone KM, Lyons KS, Dobek J, Dieckmann NF, Bennett JA, Nail L, Beer TM. Benefits of partnered strength training for prostate cancer survivors and spouses: results from a randomized controlled trial of the exercising together project. J Cancer Surviv. 2016;10(4):633–44. [In eng].

Chapter 21
Mind and Body in Mens' Health

Oliver Brunckhorst, Robert Stewart, and Kamran Ahmed

21.1 Introduction

Men's health is a unique challenge to both the individual and to society. There is a changing shift in clinicians' perception of issues beyond the pure physical health of a patient. With newer treatments resulting in improving survival for many malignant conditions, there is an increasing realisation that living longer does not always equate to living well. Similarly, benign male conditions place distinctive challenges on the patient. In men's health, quality of life issues including urinary or sexual dysfunction are common, either as part of the diagnosis itself, or as a side-effect of the treatment undertaken. With their frequency, the important relationship between the mind and body is increasingly being recognised as they can have a profound impact on the mental wellbeing of patients. Mood disorders such as depression and anxiety are common, with other domains such as an individual's sense of

O. Brunckhorst (✉)
King's College London, Guys' Hospital Campus, MRC Centre for Transplantation,
London, UK
e-mail: Oliver.brunckhorst@kcl.ac.uk

R. Stewart
King's College London Institute of Psychiatry, Psychology and Neuroscience, London, UK

South London and Maudsley NHS Foundation Trust, London, UK
e-mail: robert.stewart@kcl.ac.uk

K. Ahmed
King's College London, London, UK

King's College Hospital, London, UK

Sheikh Khalifa Medical City, Abu Dhabi, UAE

Khalifa University, Abu Dhabi, UAE
e-mail: Kamran.ahmed@kcl.ac.uk

masculinity, perception of their body image and self-esteem also impacted. Furthermore, the effects these conditions have on social interactions and relationships are also severe. These domains of an individual's mental wellbeing are closely intertwined with each other, the physical health of an individual, and their social wellbeing.

The issues encountered are further compounded by the important fact that males are less likely to seek help (Moller-Leimkuhler 2002). This has been suggested to be one of the reasons men are at increased risk of completed suicide (Varnik 2012). With this in mind, the impact a condition can have on the mental wellbeing of a patient should not be overlooked. A knowledge of the components of mental wellbeing and how different male specific conditions can impact these is important. Subsequently, being familiar with the detection, prevention and treatment of these problems during follow-up will ensure that patients are managed holistically, to improve quality of life during treatment and beyond.

21.2 Defining Mental Wellbeing in Men's Health

The physical implications of disease and its treatment have long been a focus in men's health. However, the quality of life of individuals after treatment is being increasingly evaluated, with mental wellbeing an important aspect of this. Male specific conditions can have a large impact on this, and conversely can impact other physical aspects of the disease itself. There is no real set definition of what constitutes 'Wellbeing', being a broad and multi-dimensional concept consisting of several elements interplaying with each other (Prostate Cancer UK 2014). Important components previously highlighted include physical, mental and social wellbeing. These are similar to the core components of the biopsychosocial model of health, which considers the relationship between biological, psychological, and socio-environmental factors that contribute to an individual's wellbeing (Borrell-Carrio et al. 2004). These demonstrate the important link between the physical and psychological health, whereby ignoring one aspect can have negative consequences on the other. The strength of this link should not be underestimated. There is good evidence to demonstrate how individuals suffering from mental health symptoms and disorders such as depression and anxiety, have poorer functional outcomes (including urinary and sexual dysfunction) and even higher mortality in male cancers such as prostate, testicular and penile cancer (Jayadevappa et al. 2012; Tuinman et al. 2010; Yu et al. 2016). Therefore, the need to consider them in conjunction to ensure holistic management of diseases during patient care is clear.

There is no exact designation of what falls under 'Mental Wellbeing' in male conditions. This is largely due to the variation in experiences men encounter depending on the disease itself, the stage of their disease journey, treatments undergone and individual patient factors such as their views on masculinity, ethnicity, sexuality and relationships (UK 2014). There are however commonly experienced symptoms and issues in men's health which we will consider under the umbrella of mental

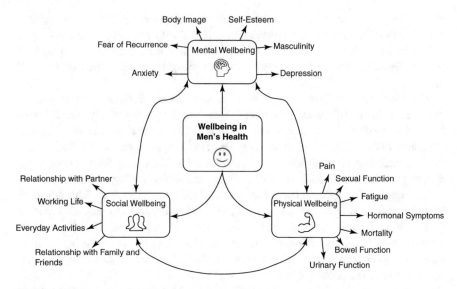

Fig. 21.1 Components of wellbeing in men's health

wellbeing, which should be considered during assessment and treatment in male conditions (Fig. 21.1).

The most commonly known and discussed of these are depressive and anxiety disorders. Whilst the term 'depression' is often used to describe low mood, for a formal diagnosis of a depressive disorder a number of core symptoms should be present on clinical evaluation. Whilst the exact terminology may differ between the International Classification of Diseases (ICD-10) and the Diagnostic and Statistical Manual of Mental Disorders (DSM-5), there is large overlap. Within the ICD-10 there is a requirement of at least two typical symptoms (depressed mood, anhedonia and fatigue) and two core symptoms (reduced concentration, reduced self-esteem, ideas of guilt, hopelessness, disturbed sleep, diminished appetite and suicidal thoughts) for at least 2 weeks (World Health Organization 2019). The severity of the episode is further classified into: i) mild, moderate and severe depending on the number of symptoms present; ii) with or without psychotic symptoms if severe; iii) recurrent if more than one discrete episode has been present. Similar to depression, 'anxiety' is often used to describe the symptom of 'excessive worry' and remains poorly defined in the literature relating physical and mental symptoms. However, it is when these symptoms are persistent and inappropriate that they can be regarded as a formal disorder. The most relevant of these—and on which many screening tools are based—are generalised anxiety disorder (GAD) and acute stress reactions. GAD is defined as anxiety that is generalized and persistent, not being restricted to particular circumstances. Symptoms can be variable but include persistent nervousness, autonomic symptoms (headache, sweating, trembling, palpitations) and excessive worrying (World Health Organization 2019). Acute stress reactions present with similar features, however, are associated to a physical or mental stress and are shorter in duration. Lastly, it is important to acknowledge that although depressive and anxiety disorders discussed are separated diagnostically, there is substantial evidence that they are highly overlapping and probably most common in community settings as mixed states. It is estimated that that three-quarters of individuals with depression have had a comorbid anxiety disorder in their lifetime (Lamers et al. 2011).

Outside formal disorders of depression and anxiety diagnosed using criteria mentioned above, an acknowledgement of the separate entity of significant symptoms of these disorders is important. These are often instead measured using psychometrics screening scales such as Patient Health Questionnaire-9 (PHQ-9) and the Generalised Anxiety Disorder-7 (GAD-7) scales. These vary from formal disorders in that they utilise lists of symptoms, often providing scores from which cut-offs are applied for defining people with a significant symptom load. This approach is commonly used in research settings or as a screening method instead due to its ease in administration as compared to the more time-consuming clinical interview required for a for formal diagnosis. However, whilst there is large overlap between formal disorders and significant symptoms, particularly within community settings, it is important to acknowledge they are separate entities. Disorders often focus on the more severe end of the spectrum than significant symptoms from psychometric scale. These scales are not substitutes for a clinical interview to diagnose disorders (Sharpley et al. 2020), but instead are useful to be aware of clinically as a screening tool for requiring further evaluation by a specialist.

Beyond the formal disorders and symptoms of depression and anxiety, there are several other constructs impacting men's mental wellbeing. Whilst linked with formal disorders, they are important to consider separately. One such concept is an individual's own view regarding their masculinity. As is expected, many male specific conditions have a profound impact on this, with men experiencing their conditions and treatment as a threat to their masculinity (Appleton et al. 2015). This is particularly common in individuals who hold onto more traditional hegemonic masculine ideals (Spendelow et al. 2018). Linked to these concepts, is the impact a condition can have on a man's view of his own body image. Treatments to male conditions can have significant implications to the shape and appearance of an individual, such as through scars or changes in body shape. Similarly, changes in bodily function—such as new urinary or sexual dysfunction—can impact a man's view of his own body. Linked to this is also the impact a condition can have on an individual's self-esteem. Many experience shame surrounding the sick role, vulnerability, loss of masculinity and sexuality (Ussher et al. 2017). Lastly, in malignant conditions, the component of fear of recurrence or progression is important. This is common across all cancer types, being strongly linked to anxiety, and is present in nearly a quarter of testicular cancer patients (Skaali et al. 2009).

There is therefore a great deal to consider when evaluating men's mental wellbeing broadly. Whilst, depression and anxiety symptoms or disorders are important components of this, there are other factors to consider during the disease process. These problems are common during follow up and are not only strongly interdependent within each other; but can also impact physical and social wellbeing through functional outcomes or social relationships of a man diagnosed with a male specific condition. A specific knowledge of how different male conditions impact these domains is therefore important and identifying these problems should form part of routine follow-up care.

21.3 Mind and Body in Benign Male Conditions

21.3.1 Lower Urinary Tract Symptoms

Lower urinary tract symptoms (LUTS) are common, with many elderly men having at least one symptom. LUTS include storage (urinary frequency, urgency and nocturia) voiding (hesitancy, incomplete voiding and terminal dribbling) and post-micturition symptoms (post-micturition dribble and the sensation of incomplete emptying). Whilst several causes exist this chapter focuses on the most common aetiology; individuals with symptoms secondary to bladder outlet obstruction caused by benign prostatic hyperplasia (Gravas et al. 2020). The symptoms, treatment received for them, resulting side-effects and impact of them on everyday life can have a major impact on quality of life in individuals affected.

Considering the impact LUTS and its treatment has on urinary and sexual function, there is surprisingly little evidence evaluating the impact on constructs of mental wellbeing such as masculinity and body image. Whilst it is likely these are affected; the majority of current evidence is focused around depression and anxiety. There is a demonstrated link between those who suffer from LUTS and subsequent depressive and anxiety symptoms. Around a third of men with LUTS have been found to suffer with both significant depressive and anxiety symptoms (Coyne et al. 2009). These mood symptoms further worsen as the burden of LUTS increases (Rom et al. 2012). The reverse relationship also appears true, with individuals who have pre-existing depressive symptoms commonly experiencing LUTS (Skalski et al. 2019). This likely bidirectional relationship between LUTS and depression is complex, with several contributing factors (Dunphy et al. 2015). Contributing factors include the symptoms themselves and the treatments received, including medical treatment with 5α-Reductase inhibitors which can have hormonal implications or operative management and their subsequent complications such as sexual dysfunction.

Regardless of the exact cause for these symptoms, they are important in the routine management of LUTS. The two are closely linked, with higher loads of depressive and anxiety symptoms leading to poorer quality of life outcomes secondary to urinary symptoms (Choi et al. 2017). Similarly, treating LUTS leads to an improvement in anxiety and depressive symptoms (Quek et al. 2004). Despite there being little evidence to demonstrate that treating depression or anxiety improves LUTS, in view of the current evidence to demonstrate their relationship, it is likely this could also play an important role in the management of patients. Therefore, screening patients with LUTS for depressive and anxiety symptoms should form part of a holistic care pathway for these men, with the subsequent multidisciplinary management of patients likely playing an important part in the management of LUTS to improve outcomes and quality of life experienced by these men (Dunphy et al. 2015).

21.3.2 Erectile Dysfunction

Erectile Dysfunction (ED) is a common symptom experienced by men, with over 50% of individuals affected in elderly age groups (Salonia et al. 2020). Erection is a complex process, involving the combination of neural, endocrine and vascular events meaning the aetiology can be complex. Whilst several organic causes exist, including vasculogenic or iatrogenic, psychogenic or mixed ED remains an important cause, particularly in younger individuals. That there is therefore a strong link between mind and body in these individuals is therefore not surprising. The presence of depressive and anxiety disorders are prevalent in individuals with ED, being present in 12.5% and 23% respectively (Rajkumar and Kumaran 2015). Having ED has also been found to increase your risk of developing a depressive disorder by nearly three times (Liu et al. 2018). Similarly, those with pre-existing depression are nearly 40% more likely to develop ED. The same is also seen when considering anxiety disorders where ED is prominent (Rosen 2001). This relationship is multi-faceted, reflecting the complexity of the physiology of erection. The autonomic element in erection is important when considering how these mood and stress disorders could be linked to ED. However, several other factors may be contributory, such as the concomitant use of anti-depressant medications such as selective serotonin reuptake inhibitors (SSRIs) (Rothschild 2000), or the presence of other associated co-morbidities such as alcohol abuse or dependence (Miller and Gold 1988).

Other components of mental and social wellbeing are also commonly related to ED. Self-esteem is commonly low in these men (Rosen 2001). Masculinity and how individuals perceive their body is also impacted (Chambers et al. 2017). With the role sexuality has on relationships and the current state of a relationship has on sexual satisfaction the social element of ED should also not be overlooked. With this strong interplay between the biological, psychological and social elements, as both causes for and complications of ED, these factors are important role in the management of patients. Whilst there is surprisingly little evidence to demonstrate that improving depression and anxiety improves erectile function, there is certainly a strong evidence for considering factors beyond the pure physical management of ED. Current European Association of Urology (EAU) guidelines highlight the importance of evaluating the current psychosocial status of a patient during initial work-up (Salonia et al. 2020). Similarly, the role of multidisciplinary management is emphasised with the suggestion of referring patients with psychiatric disorders to psychiatrists for assessment, particularly in younger patients with long term primary ED. Psychosexual therapy should also form a core component of the subsequent treatment of individuals found to contain a possible psychogenic element, with strong evidence for its efficacy (Fruhauf et al. 2013). Therefore, consideration of mental wellbeing elements during routine management of ED is vital at all stages of follow-up to ensure the best outcomes in men's health for this commonly experienced condition.

21.3.3   Male Factor Infertility

Globally, infertility is a large problem, affecting one in eight couples with a third of these being secondary to male factors alone (Agarwal et al. 2015). Experiencing difficulty in conceiving naturally, as well as undergoing reproductive therapies can be a difficult process for couples. Around one in five show signs of anxiety, depression or distress at some stage (Verkuijlen et al. 2016). Specifically, in men some estimates have put this even higher, with rates of up to 32% and 60% displaying signs of significant depressive or anxiety symptoms respectively (Pasch et al. 2016). These symptoms have been associated with several other aspects of mental and social wellbeing in men, including self-esteem and self-confidence due to feeling of shame and a loss of a male's own sense of masculinity. Additionally, it has been found that subsequent issues with a man's view of their own body image is impacted, becoming increasingly preoccupied with one's weight and ability to be satisfied with their own body (Akhondi et al. 2011).

The frequency of mental wellbeing issues is compounded by the evidence to demonstrate the impact these conditions can have on subsequent fertility outcomes themselves. Depression has been negatively correlated with decreased sperm motility, count and volume (Wdowiak et al. 2017). Subsequently, there is some evidence to suggest poorer clinical pregnancy rates in individuals undergoing assisted reproductive techniques (Matthiesen et al. 2011). However, it is important to note that the current evidence remains variable with a large focus on females specifically or couples jointly. Additionally, with the multifactorial factors at play in these individual's true causation is difficult to be established. Regardless, with the frequency and breadth of symptoms experienced, combined with the potential for poorer fertility outcomes, mental wellbeing in male factor infertility should be considered during care. Whilst there is little evidence for the role of improving mental wellbeing to better fertility outcomes, there is certainly a role for psychological or educational interventions in couples to improve depression and anxiety symptoms experienced (Verkuijlen et al. 2016). With the role these conditions have on general quality of life and mental wellbeing in these young men these should be considered during their management.

21.3.4 *Peyronie's Disease*

Peyronie's disease (PD) is an acquired connective tissue disorder characterised by fibrosis of the tunica albuginea resulting in the development of penile deformity, pain, shortening and sexual dysfunction. It is likely underdiagnosed, with prevalence estimates of probable PD at around 11% in the United States (Salonia et al. 2020). Whist the exact aetiology is not entirely clear, it is believed that repetitive microvascular injury leads to abnormal wound healing and formation of a fibrotic penile plaque resulting in curvature (Salonia et al. 2020). The physical sequelae of penile curvature can be such that it impacts satisfactory sexual intercourse. There can be difficulty in achieving penetrative intercourse or in certain sexual positions previously enjoyed by men and their partners. In sexually active males it is therefore not surprising that there can be subsequent psychological implications for patients (Terrier and Nelson 2016).

Depression is common in this cohort; with significant depressive symptoms demonstrated in nearly half of patients some studies (Nelson and Mulhall 2013). Similarly, distress experienced by these patient's by the disease itself is severe in a third of patients (Coyne et al. 2015). Other components of mental wellbeing are also impacted, with men's perception of their own body image c severely impacted along with their own self-esteem (Rosen et al. 2008). Subsequent self-image and masculinity are impacted, with increasing concerns about physical appearance, sexual self-image and sexual confidence, sometimes resulting in social isolation and stigmatisation. Unsurprisingly, with the sexual dysfunction that may result from PD, relationship strains are also common with over 50% reporting problems, particularly in those who have the complete inability to have sexual intercourse or those who experience penile shortening (Smith et al. 2008). Additionally, these psychological components are important in subsequent sexual satisfaction experienced, with individuals who display a higher sense of shame having subsequent poorer sexual satisfaction (Davis et al. 2017). It is therefore important to consider these aspects during follow, minimising the impact PD can have on mental wellbeing. Initial evaluation should include psychosocial assessments—including the use of validated screening tools—as well as the availability of interdisciplinary expertise to improve the quality of life of PD patients (Nelson and Mulhall 2013). However, whilst integrating this into normal care is important, there is unfortunately limited real evidence as to what the exact best methods to detect and treat this cohort of men is, as well as the impact this has on sexual function subsequently. More evidence is certainly required through future research in view of the prevalence and impact on quality of life PD appears to have on men.

21.4   Mind and Body in Male Cancers

21.4.1   *Prostate Cancer*

Prostate cancer represents a large proportion of the global male cancer incidence, accounting for 13.5% of cases (Bray et al. 2018). Five year survival rates are high at 83%, meaning aspects beyond the pure survival of patients are becoming increasingly appreciated (De Angelis et al. 2014). The physical wellbeing during prostate cancer survivorship is well understood, with complications including urinary, bowel and sexual dysfunction most commonly investigated. However, the relationship between these and the mental well-being of individuals is now increasingly appreciated. Prostate cancer places unique strains on men's health, which go beyond the pure psychological impact of the diagnosis of cancer itself (Fervaha et al. 2019). Treatment options such as androgen deprivation therapy (ADT) can have implications for mental wellbeing through its hormonal interactions and profound changes in body shape. Similarly, the physical complications of the disease and radical treatment options such as surgery—including urinary and sexual dysfunction—also place a large strain on a man's mental wellbeing. Lastly, the impact these can have on other relationships, such as those with the patients' partner, are also important. With this complex impact prostate cancer has on mental wellbeing, it is not surprising that this has become a well investigated area.

There is increasing evidence to demonstrate the link between prostate cancer and poorer mental health. Formal depressive disorders have been estimated to be present in around 5% of patients, with high rates of significant depressive and anxiety symptoms in around 17% for both (Brunckhorst et al. 2021). However, with varying treatments received in prostate cancer these figures can vary. ADT for example, is a well-established risk factor for the development of depression. However, the limited evidence available so far suggests there is little difference between those receiving radical as compared to those on active surveillance for low risk disease (Sharpley et al. 2020). Regardless of treatment received, this high psychological morbidity appears to translates into significant suicidality, with recent suicidal ideation seen in nearly 10% of patients (Brunckhorst et al. 2021). Importantly, this appears to translate into an increased risk of suicide mortality when compared to the general population (Guo et al. 2018). These figures demonstrate the high psychological impact of prostate cancer on men.

Other domains of mental wellbeing are also impacted. With the impact prostate cancer has on the male body and impacting sexual function, men's sense of masculinity can be severely altered in this group. This is largely secondary to threats to an individual's own masculinity with the loss of the 'male' prostate organ after surgery (Appleton et al. 2015), their loss of sexual functioning (Araujo et al. 2019) and some feminisation of physical features secondary to hormonal therapy (Farrington et al. 2020). These features similarly impact the body image views of individuals. There are changes in perception of their own appearance (Chambers et al. 2018), and the loss of a perceived vital part of their body meaning men have described feeling 'deficient' (Gentili et al. 2019). These two domains are strongly linked with men's self-esteem. Feelings of shame are common secondary to the individuals diagnosis itself, repeated intimate medical examinations and their inability to perform sexually (Evans et al. 2005). All of these are strongly interlinked and are also related to other psychological issues such as low mood and anxiety. Therefore, the importance of considering them within the context of an individual's mental wellbeing should not be overlooked.

Not only are problems common during in patients with prostate cancer, but they also have important implications for the treatment outcomes of the cancer itself. Co-existing depression and anxiety disorders impact treatment choices with regards to active surveillance patients and its subsequent adherence (Sciarra et al. 2018). Additionally, functional outcomes have been demonstrated to be impacted with poorer sexual function (Tavlarides et al. 2013) and urinary bother post treatment (Sciarra et al. 2018). Importantly, mortality is also impacted, with studies demonstrating higher mortality in those with depressive disorders, even when adjusting for other confounders and being present regardless of prostate cancer risk level (Lin et al. 2018; Prasad et al. 2014). The causes for the poorer functional outcomes are not surprising, considering the endocrine and physiological disruption conditions such as depression and anxiety can have. These can subsequently impact sexual and urinary function. The causes for increased mortality may however be more complex. Whilst confounding social factors such as sociodemographic status, other com-morbidities and delayed diagnosis may also play a role, biological factors such as impaired immune system responses secondary to depression have also previously been hypothesised (Sharpley et al. 2020).

Regardless of causation, this relationship between mind and body in prostate cancer warrants attention during clinical care. These should be considered in detail during routine follow-up through improved detection and specialist input for their treatment, which is which is demonstrated to improve the quality of life of cancer patients (Galway et al. 2012). Adding these to routine care could potentially offer a further adjunct to improving the quality of life and outcomes in this complex group of male patients.

21.4.2   Testicular Cancer

As the most common non-haematological malignancy affecting men between the ages of 14 and 44, representing 5% of urological neoplasms testicular cancer remains an important condition in men's health (Cheng et al. 2018). This predominantly younger demographic combined with the fact that earlier detection and improved chemotherapy regimens result in 5-year survival rates in excess of 90% is important when considering the mental wellbeing of these patients. These sometimes young and usually sexually active men have to live with the side-effects of treatments and the implications of disease for longer. Problems such as infertility, fatigue, erectile dysfunction or organ dysfunction are frequent post-chemotherapy or surgery meaning quality of life issues are important in this cohort (Fung et al. 2019).

Numerous studies have demonstrated the psychosocial impact and the subsequent compromise in quality of life testicular cancer has. Whilst depressive symptoms and depressive disorders have not been found to be higher in this cohort, anxiety symptoms and disorders are prevalent at around 19% and more frequent than in the general population (Dahl et al. 2005; Smith et al. 2018). Similarly, issues surrounding perception around own body image and attractiveness are commonly impacted. As such problems surrounding masculinity (Gurevich et al. 2004) and self-esteem are commonly seen (Tuinman et al. 2006). This is particularly the case in post-orchidectomy patients with the loss of an organ that is highly associated with perceptions of masculinity, attractiveness and body image. Problems appear to be particularly common in younger men (Carpentier et al. 2011) and are important as they have implications for sexual functioning and relationships in these patients. Those with more negative views on their own body image appear to experience more sexual dysfunction (Rossen et al. 2012). Additionally, individuals with higher anxiety symptom loads have also been demonstrated to be at increased risk of having ED (Dahl et al. 2005).

Social wellbeing is also impacted, particularly surrounding the relationship with the patients sexual partner which can become increasingly strained (Carpentier et al. 2011) and re-entry into working life after treatment (Schepisi et al. 2019). Therefore, considering the wide consequences of a testicular cancer diagnosis, extending to all aspects of a man's wellbeing, a holistic management pathway is required. However, young men can be difficult to engage within some aspects of care, meaning a proactive approach is required. As with many men's health conditions identifying problems using validated methods is important. There are however surprisingly few evidence-based interventions available within the literature once problems are identified and a multidisciplinary approach is therefore vital. It is therefore clear that this unique male cancer poses challenges to patients mental, physical and social wellbeing, requiring to be acknowledged and addressed throughout the care pathway.

21.4.3 Penile Cancer

Penile cancer is a relatively rare cancer with incidences of around 1 in 100,000 individuals, predominantly affecting those over 60 (Hernandez et al. 2008). Similar to other male specific cancers, 5-year survival rates are high at over 90%, meaning quality of life issues in survivorship are of importance. Additionally, treatment options often include surgery which can be invasive, including partial or total penectomy. Considering the physical implications of penile cancer, patients can experience a profound impact on their psychological wellbeing with around half of patients experiencing some psychiatric symptoms (Maddineni et al. 2009). This is particularly the case with more aggressive surgical treatments where around a third of patients experience significant anxiety or depressive symptoms, with rates higher for those undergoing total penectomy. With the penis being strongly linked to many individuals sense of masculinity it also not unexpected that men can subsequently struggle with their self-image after treatment (Bullen et al. 2010). Masculinity is particularly impacted in those undergoing more radical surgical treatments (Sosnowski et al. 2019). Previous research also highlights how relationships are affected postoperatively, with the linking factor between many of the issues encountered being subsequent alterations in sexual functioning.

Unfortunately, due to the relatively rare nature of penile cancer, and due to the management often being undertaken in specialist centres, there remains little evidence on the impact of disease on mental wellbeing beyond the obvious sexual implications. Similarly, the impact of mental wellbeing problems on cancer outcomes is not known. Additionally, validated psychometric tools used in penile cancer for the assessment of problems encountered are sparse with few evidence-based prevention or intervention strategies available once these are encountered. Considering the prevalence and psychosexual implications of penile cancer and its treatments it is likely these issues are important for ensuring the best quality of life outcomes. Whilst ensuring detection and suitable treatment options, including psychosexual counselling are available within the specialist managing team, further work is certainly required to ensure this is conducted in the most beneficial manner for penile cancer patients.

21.4.4 Improving Mental Wellbeing in Men's Health

Across all male conditions explored in this chapter the mental wellbeing of men appears to be regularly impacted. Whilst all conditions display some unique reasonings behind the trends identified, common features exist. These include the close relationship mental wellbeing domains display between each other and with other important physical complications of disease; particularly sexual function. With this in consideration, general principles of improving mental wellbeing in men's health include a large deal of overlap between conditions. Several concepts have been discussed previously already through individual conditions; these include the importance of detecting, preventing and treating mental wellbeing problems at all stages during a man's health journey (Fig. 21.2). Although it is important to understand all of these concepts, the availability of appropriate expertise within the team through interdisciplinary collaboration at each stage cannot be underestimated. It is through this that the best holistic outcomes of individuals can be achieved.

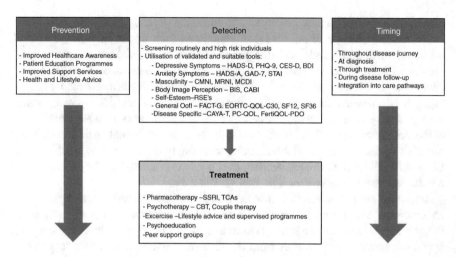

Fig. 21.2 Improving mental wellbeing in men's health. Index: *BDI* Beck's depression Inventory, *BIS* Body Image Scale, *CBT* Cognitive Behavioural Therapy, *CABI* Concern About Body Image, *CES-D* Center for Epidemiologic Studies Depression Scale, *CMNI* Conformity to Masculine Norms Inventory, *EORTC-QOL-C30* European Organisation for Research and Treatment of Cancer Quality of Life of Cancer Patients, *FACT-G* Functional Assessment of Cancer Therapy-General, *GAD-7* Generalised Anxiety Disorder 7, *HADS* Hospital Anxiety And Depression Scale, *MCDI* Masculinity in Chronic Disease Inventory, *MRN* Male Role Norms Inventory, *PHQ-9* Patient Health Questionnaire-9, *RSES* Rosenberg's Self-Esteem Inventory, *SF* Short Form, *SSRI* Selective Serotonin Reuptake Inhibitor, *STAI* State-Trait Anxiety Inventory, *TCA* TricyclicAntidepressant

21.5 Prevention

With mental wellbeing problems being so common during and after treatment, a key feature of any programme to improve these should include the prevention of them arising in the first place. An important first step is the increasing awareness amongst healthcare professionals caring for individuals suffering from male specific conditions (Kiffel and Sher 2015). An understanding of the prevalence and unique mental wellbeing problems encountered allows problems to be addressed in the first place through incorporating measures within the current care pathway. Without this, issues will continue to be overlooked, particularly in men, a group where individuals can be less forthcoming about issues, often suffering in silence.

In line with improving knowledge, patient education about the condition itself has also been demonstrated to be a key aspect. Informing patients of the impact their condition can have on their own mental wellbeing allows patients to remain vigilant, realise what they may experience is not unusual, and can make them more comfortable sharing their experiences with healthcare professionals or with family (Kiffel and Sher 2015). Furthermore, research has shown that educating the patient and relative about their condition itself, treatment options, side-effects that can occur and what can be done to improve general health (e.g. diet, exercise, smoking cessation) improves general quality of life and some psychological symptoms such as depression (Badger et al. 2011). In conditions such as prostate cancer where some treatment options such as ADT pose unique challenges to patients these have previously been tailored specifically to these subgroups (Wibowo et al. 2020). These education programmes have previously utilised various modalities including literature provision, online material, short courses and enhanced multidisciplinary counselling during care pathways (Bourke et al. 2015). However, whilst extensive evidence demonstrates the effectiveness of these in malignant male conditions, there is surprisingly less available for men suffering from benign conditions such as ED and PD. With their unique challenges it is however likely that they would have a positive impact for all patients in men's health.

Improved support during the disease process has additionally been highlighted as important. Support arising from relationships is important for mental wellbeing in testicular cancer for example (Tuinman et al. 2006). If available, encouraging active engagement and support from this source, provides one source to prevent deterioration in mental wellbeing. Additionally, peer support groups are common amongst cancer patients and are useful for addressing the unmet needs experienced by many male patients (King et al. 2015). Lastly, increased multidisciplinary support during the care pathway, such as from psychosexual therapists and specialist nurses is vital for improving information deliverance, addressing problems early and providing coping strategies (King et al. 2015; Bourke et al. 2015).

Improving overall patient health through lifestyle changes may be also important to enhance mental wellbeing in men's health. Whilst dietary changes are commonly recommended across the scope of male conditions, there is limited high quality evidence for their effectiveness in preventing mental wellbeing problems to arise (Zuniga et al. 2020). However, with the role they have on other physical outcomes such as with erectile function, ongoing practice remains to encourage a habit of wellness through healthy eating. When considering exercise, the evidence is much more convincing. Aerobic, resistance or high intensity interval training or formal training programmes have been consistently associated with improved quality of life, fatigue and mental wellbeing in cancer patients in multiple randomised trials (Mishra et al. 2014; Bourke et al. 2015). As such supervised programmes are recommended as part of follow up care in patients with ADT in prostate cancer by the EAU. Again, there is less investigation of this within benign male conditions; however, as with dietary recommendations, in view of its impact on other aspects of health of patients, this commonly remains part of advice given to patients.

21.6   Detecting Mental Wellbeing Problems

Accurate identification of problems arising during the disease journey is important for improving mental wellbeing. However, challenges for screening initiatives include the broad range of domains which encompass mental wellbeing, the large number of self-report tools available to measure them, and their varying suitability

Table 21.1 Available assessment inventories for evaluation of depressive and anxiety symptoms in men's health

Assessment tool	Number of items	Properties assessed	Response options	Scoring	Men's health condition utilisation
Depressive symptoms					
Beck's Depressive Inventory (BDI)	21	Affective, cognitive, somatic and vegetative symptoms, reflecting DSM criteria	4-point scale from 0 (symptom absent) to 3 (severe symptoms)	<14: Minimal depression 14–19: Mild depression 20–28: Moderate depression 29–63: Severe depression	ED, Infertility, LUTS, Prostate Ca
Center for Epidemiologic Studies Depression Scale (CES-D)	20	Depressed affect, anhedonia, somatic activity and interpersonal challenges	4-point scale from 0 (rarely or never) to 3 (most or all of the time)	0–16: No depressive symptoms 16–23: Moderate symptoms 24–60: Severe symptoms	Infertility, LUTS, Peyronie's Disease, Prostate Ca, Testicular Ca
Hospital Anxiety and Depression Scale- Depression Subscale (HADS-D)	7	Designed for people with physical health problems, focusing on cognitive symptoms of depression rather than somatic symptoms that might reflect the physical condition.	4-point scale from 0 (not present) to 3 (considerable)	<8: No depressive symptoms 8–10: Mild symptoms 11–14: Moderate symptoms 15–21: Severe symptoms	ED, Infertility, LUTS, Penile Ca, Prostate Ca, Testicular Ca
Patient Health Questionnaire-9 (PHQ-9)	9	Based on nine symptoms criteria for major depressive disorder from DSM including low mood, anhedonia, sleep disturbance, somatic symptoms and suicidal ideation	4-point scale from 0 (not at all) to 3 (nearly every day)	<5: No significant symptoms 5–9: Mild symptoms 10–14: Moderate symptoms 15–19 Moderately severe 20–27: Severe symptoms	ED, Infertility, LUTS, Penile Ca, Peyronie's Disease, Prostate Ca, Testicular Cancer

(continued)

Table 21.1 (continued)

Assessment tool	Number of items	Properties assessed	Response options	Scoring	Men's health condition utilisation
Anxiety symptoms					
Generalised Anxiety Disorder-7 (GAD-7)	7	Based on seven symptom criteria for GAD from DSM including worrying, feeling anxious, restless and irritable	4-point scale from 0 (not at all) to 3 (nearly every day)	<5: No significant symptoms 5–9: Mild anxiety symptoms 10–14: Moderate symptoms 15–21: Severe symptoms	ED, Infertility, LUTS, Prostate Ca, Testicular Ca
Hospital Anxiety and Depression Scale-Anxiety Subscale (HADS-A)	7	Developed for people with physical health problems so focused around cognitive symptoms of GAD rather than purely physical symptoms which may reflect condition.	4-point scale from 0 (not present) to 3 (considerable)	<8: No anxiety symptoms 8–10: Mild symptoms 11–14: Moderate symptoms 15–21: Severe symptoms	Infertility, LUTS, Prostate Ca, Penile Ca, Testicular Ca
State-Trait Anxiety Inventory (STAI)	40	Evaluates state anxiety (current state of anxiety right now) and trait anxiety (stable aspects of anxiety proneness including calmness, confidence, and security)	4-point scale from 1 to 4 with response depending on state or trait subscale	Each subscale scored out of 80 For each ≥30 suggest moderate anxiety and scores ≥45 suggest severe anxiety	ED, Infertility, LUTS, Prostate Cancer, Testicular Cancer

Index: *Ca* cancer, *DSM* diagnostic and statistical manual of mental disorders, *ED* erectile dysfunction, *LUTS* lower urinary tract symptoms

within the targeted population. When considering the core domains of depression and anxiety, the gold standard for diagnosing a formal disorder is a structured clinical interview by a trained mental health professional. However, in routine practice clinicians looking after patients with male specific conditions may not have the time or importantly the expertise to administer this interview. This makes the use of patient self-report psychometric tools the most appropriate method to identify significant symptoms in this setting (Sharpley et al. 2020).

A large number of inventories are available for measuring depressive and anxiety symptoms. These often draw their items from standard symptom lists set out in the DSM or ICD diagnostic criteria (Table 21.1). Many are extensively validated, although not often within the setting of a male specific condition specifically where no gold standard tools exist at present. Despite this, numerous have been utilised within research and clinical settings previously or are recommended by current guidelines. When considering depression, the Hospital Anxiety and Depression–Depression Subscale (HADS-D), PHQ-9 and the Center for Epidemiologic Studies Depression Scale (CES-D) have been extensively previously used for both malignant and benign conditions. Additionally, the Beck's Depression Inventory (BDI) is one tool recommended for erectile dysfunction specifically by the EAU (Salonia et al. 2020). Similarly, for evaluating anxiety symptoms in male specific conditions the HADS—anxiety subscale (HADS-A), GAD-7 and the State-Trait Anxiety Inventory (STAI) have been utilised extensively being predominantly geared towards screening for detecting generalized anxiety disorder. It is however important to realise these tools are not diagnostic for their formal disorders when using them for screening (Sharpley et al. 2020), but instead meeting thresholds for significant symptoms should warrant further evaluation by specialist within the multidisciplinary team.

Other mental wellbeing domains can be more difficult to measure in clinical practice (Table 21.2). This is largely due to the varying definitions and people's own views surrounding masculinity, body image and self-esteem and what constitutes a deficiency in these constructs. However, tools which have been developed for evaluating masculinity, and have previously been utilised in malignant male conditions, include the Conformity to Masculine Norms Inventory, the Male Role Norms Inventory and the Masculinity in Chronic Disease Inventory (which is developed and evaluated for prostate cancer) (Bowie et al. 2020). Similarly, for evaluating self-body image perception, the Body Image Scale and the Concern About Body Image have been used within a men's health context. Lastly, for self-esteem the Rosenberg Self-Esteem Scale offers the most utilised scale, particularly within malignant conditions. Whilst offering a method to evaluate these domains and thereby highlighting areas which can be addressed further through increased support, it is important to bear in mind their limitations. As the issues encountered can vary depending on age and condition itself there is variability in the constitution of these various tools, and they may not reflect the issues encountered by an individual patient. As an example, many do not address loss of sexuality within their evaluations, a key issue in many conditions within this chapter. Additionally, many of these, particularly those evaluating masculinity and body image were developed around more "traditional" male perceptions of masculinity and are not always reflective of all individuals view of what constitutes a man in modern society. Therefore, whilst available, further work is certainly required to develop tools which reflect the individual conditions and patients experience within the unique setting of men's health.

Table 21.2 Available assessment inventories for evaluation of masculinity, body image and self-esteem in men's health

Assessment tool	Number of items	Properties assessed	Response options	Scoring	Men's health condition utilisation
Masculinity					
Conformity to Masculine Norms Inventory—Abbreviated	22	Winning, emotional control, risk-taking, violence, dominance, self-reliance, primacy of work, power over women, pursuit of status	4-point scale from 0 (strongly disagree) to 3 (strongly agree)	Summed up to 66 maximum with higher scores reflecting higher conformity to traditional masculinity norms	Infertility, Penile Ca, Prostate Ca, Testicular Ca
Male Role Norms Inventory—Revised	7	Measuring extent to which men feel they can display emotion	6-point scale from 1 (strongly disagree) to 6 (strongly agree)	Mean score, no clear cut-offs utilised	Infertility, Prostate Ca
Masculinity in Chronic Disease Inventory	28	Strength, sexual importance, family responsibility, emotional self-reliance, optimistic capacity, action approach	5-point scale from 1 (not at all true) to 6 (very true)	Summed to 110. Higher scores correlate to higher endorsement of masculine ideals. No cut-off	Prostate Ca
Body image perception					
Body Image Scale (BIS)	10	Self-consciousness, physical attractiveness, satisfaction with appearance, difficulty seeing self-naked, wholeness of body, dissatisfaction with scars	4-point scale from 0 (not at all) to 3 (very much)	Summative up to 30, no clearly defined cut-offs for significance	Infertility, Prostate Ca, Testicular Ca
Concerns About Body Image	3	Loss of muscle, negative feelings about boy look, feelings body is getting soft/flabby	4-point scale from not at all to very much	Recalibrated to scores range from 0 to 100, no clear cut-offs	Prostate Cancer
Self-esteem					
Rosenberg Self-Esteem Scale	10	Positive and negative statements related to own feelings of self-esteem	4-point scale from 0 (strongly disagree) to 3 (strongly agree)	Summative with half reverse scored, 30 indicating highest self-esteem, <15 indicates low self-esteem	ED, Infertility, Penile Ca, Prostate Ca, Testicular Ca

Index: Ca Cancer, *ED* erectile dysfunction

Table 21.3 Available assessment inventories for evaluation of health-related quality of life and disease specific inventories

Assessment tool	Number of items	Mental wellbeing properties assessed	Response options	Scoring	Men's health condition utilisation
Health related quality of life					
European Organisation for Research and Treatment of Cancer Quality of Life of Cancer Patients (EORTC-QOL-C30)	30	Focus around physical and social wellbeing, but items on low mood and anxiety	4-point scale from 1 (not at all) to 4 (very much)	Scores scaled to 0–100 with higher scale scores indicating better functioning. No cut-offs	Penile Ca, Prostate Ca, Testicular Ca
Functional Assessment of Cancer Therapy-General (FACT-G)	27	Four subscales: emotional well-being amongst physical well-being, social well-being, and functional well-being	5-point scale from 1 (not at all) to 5 (very true)	Summed, negative items reverse scored, total score up to 108, each subscale scored separately	Prostate Ca, Testicular Ca
Short Form 12 and 36 (SF12 and SF36)	12 or 36	Mental Component Summary with additional Physical Component Summary	Variable on question, binary yes/ no or 5–6-point scales	Adjusted scores calculated for each component to 100. Higher scores for higher functioning	ED, Infertility, LUTS, Penile Ca, Peyronie's Disease, Prostate Ca, Testicular Ca
Disease specific tools					
Cancer Assessment for Young Adults-Testicular (CAYA-T)	90	Body image strength, masculine self-image, adult self-image cognitive emotional regulation, amongst other physical, sexual and relationship domains	3-point scale from 0 (none of the time) to 2 (much or most of the time)	Summative for each domain with lower scores indicating areas of concern. No cut-offs	Testicular Ca

(continued)

Table 21.3 (continued)

Assessment tool	Number of items	Mental wellbeing properties assessed	Response options	Scoring	Men's health condition utilisation
Fertility Quality of Life Questionnaire (FertiQOL)	36	Emotional, mind/body amongst other relationship, social, tolerability to treatment and environment domains	Variable on domain/question, 4–5 point scales	Each domain scores adjusted to scale from 0 to 100 with higher scores reflecting higher QofL	Fertility
Memorial Anxiety Scale for Prostate Cancer (MAX-PC)	18	Prostate ca related anxiety, looking at general anxiety related to prostate ca and treatment, fear of recurrence, and anxiety related to PSA testing	4-point scale, 0 to 3, wording dependent on domain	Summative of 0–54, scores ≥27 commonly used as significant Prostate ca related anxiety	Prostate Ca
Peyronie's Disease Questionnaire (PDQ)	15	Peyronie's psychological and physical symptoms domain. However, predominant focus on physical and sexual symptoms.	Variable according to item from binary to 10-point scale	Summative, psychological and physical domain 0–24 with higher scores showing increased bother	Peyronie's Disease
Prostate Cancer-Related Quality of Life Scales (PC-QOL)	44	Masculine self-esteem, health worry, PSA concern, regret and cancer control amongst urinary, sexual and social impact of disease	5-point scale from 1 (not at all) to 5 (very much)	Standardised to 0–100, usually higher scores mean higher QofL but reverse for health worry and regret	Prostate Ca

Index: *Ca* cancer, *ED* erectile dysfunction, *LUTS* lower urinary tract symptoms, *QofL* quality of life

There are several disease specific tools currently available in the context of men's health (Table 21.3). They measure varying domains of mental wellbeing, to vastly differing lengths and some are specific to problems encountered for individual conditions. Several examples include sub domains of broad quality of life patient reported outcome measure tools available in specific cancer types. Within prostate cancer the Prostate Cancer-Related Quality of Life Scale (PC-QOL) contains sub domains to evaluate anxiety and masculine self-esteem. In testicular cancer the Cancer Assessment for Young Adults-Testicular (CAYA-T) contains questions addressing positive self-image, body image and masculine self-image. Additionally, several cancer or generic quality of life tools such as the Functional Assessment of Cancer Therapy-General (FACT-G), European Organisation for Research and Treatment of Cancer Quality of Life of Cancer Patients (EORTC-QOL-C30) and the Short Form 12 and 36 (SF12 and SF36) contain subdomains evaluating 'Emotional Wellbeing' or 'Psychological Wellbeing'. These tools are however not based on diagnostic criteria for many of the psychological conditions discussed and are often heavily focused on the physical wellbeing of individuals and may be less suited for measuring specific mental wellbeing domains.

21.7 Treatment

The initiation and delivery of pharmacological treatment or psychotherapy for many formal mental health disorders are likely beyond the everyday clinical scope of healthcare practitioners primarily responsible for looking after the conditions discussed in this chapter. This highlights the importance of multidisciplinary expertise within the caring team of these patients which will aid in the initiation of treatment once problems have been highlighted. However, having a basic understanding of available modalities is important. Pharmacotherapy for anxiety and depressive disorder and symptoms in men's health can include the use of antidepressants such as tricyclic antidepressants (TCAs) or selective serotonin reuptake inhibitors (SSRIs) (Sharpley et al. 2020). These are important to be aware of in view of their side effect profile and effect they can have on the condition being treated. For example, TCAs through their anticholinergic effects have important urinary side effects such as increased risk of retention; of important in prostate cancer and LUTS patients. Additionally, SSRIs can also impact sexual function with erectile dysfunction, ejaculatory disorders and reduced libido (Salonia et al. 2020). Therefore, being aware of these and their impact on individual patients is useful when considering the management of male patients.

There are additionally several psychotherapeutic interventions commonly utilised including cognitive behaviour therapy, psychotherapy or couples therapy which have demonstrated evidence to improve depressive symptoms in groups such as prostate cancer (Sharpley et al. 2020). Similarly, these interventions can also help with other domains of mental wellbeing such as masculinity and self-esteem (Wootten et al. 2017). Other psychosocial measures previously discussed as prevention measures are also helpful for improving symptoms once they arise. Peer support groups, psychoeducation and exercise are important and have demonstrated improvement on depressive or anxiety symptoms in prostate or testicular malignancy (Newby et al. 2015; Sharpley et al. 2020). These are also important in improving the more peripheral domains including such as masculinity, self-esteem and body image perceptions (Adams et al. 2018; Langelier et al. 2018). Therefore, treating mental wellbeing problems once they arise is likely to require a multimodal and multidisciplinary approach to ensure the best quality of life outcomes for patients in men's health.

21.8   Conclusions

Male specific conditions have been demonstrated to have a profound impact on the mental wellbeing of an individual across this chapter. Repeatedly mental health conditions such as depression and anxiety, as well as more peripheral entities such as masculinity, self-esteem and body image are implicated, regardless of the benign or malignant conditions experienced. The components have been shown to be inter-linked across the spectrum of conditions and are important in the physical and social wellbeing of a patient. An increased awareness of the problems encountered by patients is an important first step in addressing and preventing them. Subsequently, looking for them regularly through screening using well validated tools should be routine for many conditions. This ensures the early identification and subsequent referral for addressing them within the multidisciplinary team. The majority of conditions are addressable once detected, having a subsequent large resultant improvement in men's overall quality of life and perhaps even improved outcomes for the condition itself.

Acknowledgements Oliver Brunckhorst and Kamran Ahmed acknowledge research support from the Medical Research Council (MRC) Centre for Transplantation and funding from the King's Medical Research Trust (KMRT). Kamran Ahmed further acknowledges funding from the Royal College of Surgeons of England, The Urology Foundation, CoptcoatCharity and the Pelican Foundation. Robert Stewart is part-funded by: i) the National Institute for Health Research (NIHR) Biomedical Research Centre at the South London and Maudsley NHS Foundation Trust and King's College London; ii) a MRC Mental Health Data Pathfinder Award to King's College London; iii) an NIHR Senior Investigator Award; iv) the NIHR Applied Research Collaboration South London (NIHR ARC South London) at King's College Hospital NHS Foundation Trust. The views expressed are those of the authors and not necessarily those of the NIHR or the Department of Health and Social Care.

References

Adams SC, DeLorey DS, Davenport MH, Fairey AS, North S, Courneya KS. Effects of high-intensity interval training on fatigue and quality of life in testicular cancer survivors. Br J Cancer. 2018;118(10):1313–21.

Agarwal A, Mulgund A, Hamada A, Chyatte MR. A unique view on male infertility around the globe. Reprod Biol Endocrinol. 2015;13:37.

Akhondi MM, Dadkhah A, Bagherpour A, Ardakani ZB, Kamali K, Binaafar S, et al. Study of body image in fertile and infertile men. J Reprod Infertil. 2011;12(4):295–8.

Appleton L, Wyatt D, Perkins E, Parker C, Crane J, Jones A, et al. The impact of prostate cancer on men's everyday life. Eur J Cancer Care (Engl). 2015;24(1):71–84.

Araujo JS, Nascimento LC, Zago MMF. Embodied hegemonies: moral dilemmas in the onset of prostate cancer. Rev Esc Enferm USP. 2019;53:e03494.

Badger TA, Segrin C, Figueredo AJ, Harrington J, Sheppard K, Passalacqua S, et al. Psychosocial interventions to improve quality of life in prostate cancer survivors and their intimate or family partners. Qual Life Res. 2011;20(6):833–44.

Borrell-Carrio F, Suchman AL, Epstein RM. The biopsychosocial model 25 years later: principles, practice, and scientific inquiry. Ann Fam Med. 2004;2(6):576–82.

Bourke L, Boorjian SA, Briganti A, Klotz L, Mucci L, Resnick MJ, et al. Survivorship and improving quality of life in men with prostate cancer. Eur Urol. 2015;68(3):374–83.

Bowie J, Brunckhorst O, Stewart R, Dasgupa P, Ahmed K. A systematic review of tools used to assess body image, masculinity and self-esteem in men with prostate cancer. Psycho-Oncology. 2020;29(11):1761–71.

Bray F, Ferlay J, Soerjomataram I, Siegel RL, Torre LA, Jemal A. Global cancer statistics 2018: GLOBOCAN estimates of incidence and mortality worldwide for 36 cancers in 185 countries. CA Cancer J Clin. 2018;68(6):394–424.

Brunckhorst O, Hashemi S, Martin A, George G, Van Hemelrijck M, Dasgupa P, et al. Depression, anxiety and suicidality in patients with prostate cancer: a systematic review and meta-analysis of observational studies. Prostate Cancer Prostatic Dis. 2021;24(2):281–9.

Bullen K, Edwards S, Marke V, Matthews S. Looking past the obvious: experiences of altered masculinity in penile cancer. Psychooncology. 2010;19(9):933–40.

Carpentier MY, Fortenberry JD, Ott MA, Brames MJ, Einhorn LH. Perceptions of masculinity and self-image in adolescent and young adult testicular cancer survivors: implications for romantic and sexual relationships. Psychooncology. 2011;20(7):738–45.

Chambers SK, Chung E, Wittert G, Hyde MK. Erectile dysfunction, masculinity, and psychosocial outcomes: a review of the experiences of men after prostate cancer treatment. Transl Androl Urol. 2017;6(1):60–8.

Chambers SK, Hyde MK, Laurie K, Legg M, Frydenberg M, Davis ID, et al. Experiences of Australian men diagnosed with advanced prostate cancer: a qualitative study. BMJ Open. 2018;8(2):e019917.

Cheng L, Albers P, Berney DM, Feldman DR, Daugaard G, Gilligan T, et al. Testicular cancer. Nat Rev Dis Primers. 2018;4(1):29.

Choi WS, Heo NJ, Lee YJ, Son H. Factors that influence lower urinary tract symptom (LUTS)-related quality of life (QoL) in a healthy population. World J Urol. 2017;35(11):1783–9.

Coyne KS, Wein AJ, Tubaro A, Sexton CC, Thompson CL, Kopp ZS, et al. The burden of lower urinary tract symptoms: evaluating the effect of LUTS on health-related quality of life, anxiety and depression: EpiLUTS. BJU Int. 2009;103(Suppl 3):4–11.

Coyne KS, Currie BM, Thompson CL, Smith TM. Responsiveness of the Peyronie's Disease Questionnaire (PDQ). J Sex Med. 2015;12(4):1072–9.

Dahl AA, Haaland CF, Mykletun A, Bremnes R, Dahl O, Klepp O, et al. Study of anxiety disorder and depression in long-term survivors of testicular cancer. J Clin Oncol. 2005;23(10):2389–95.

Davis S, Ferrar S, Sadikaj G, Binik Y, Carrier S. Shame, catastrophizing, and negative partner responses are associated with lower sexual and relationship satisfaction and more negative affect in men with Peyronie's disease. J Sex Marital Ther. 2017;43(3):264–76.

De Angelis R, Sant M, Coleman MP, Francisci S, Baili P, Pierannunzio D, et al. Cancer survival in Europe 1999–2007 by country and age: results of EUROCARE-5–a population-based study. Lancet Oncol. 2014;15(1):23–34.

Dunphy C, Laor L, Te A, Kaplan S, Chughtai B. Relationship between depression and lower urinary tract symptoms secondary to benign prostatic hyperplasia. Rev Urol. 2015;17(2):51–7.

Evans J, Butler L, Etowa J, Crawley I, Rayson D, Bell DG. Gendered and cultured relations: exploring African Nova Scotians' perceptions and experiences of breast and prostate cancer. Res Theory Nurs Pract. 2005;19(3):257–73.

Farrington AP, Wilson G, Limbrick H, Swainston K. The lived experience of adjustment to prostate cancer. Psychol Men Masc. 2020;21(3):369–79.

Fervaha G, Izard JP, Tripp DA, Rajan S, Leong DP, Siemens DR. Depression and prostate cancer: a focused review for the clinician. Urol Oncol. 2019;37(4):282–8.

Fruhauf S, Gerger H, Schmidt HM, Munder T, Barth J. Efficacy of psychological interventions for sexual dysfunction: a systematic review and meta-analysis. Arch Sex Behav. 2013;42(6):915–33.

Fung C, Dinh PC, Fossa SD, Travis LB. Testicular cancer survivorship. J Natl Compr Canc Netw. 2019;17(12):1557–68.

Galway K, Black A, Cantwell M, Cardwell CR, Mills M, Donnelly M. Psychosocial interventions to improve quality of life and emotional wellbeing for recently diagnosed cancer patients. Cochrane Database Syst Rev. 2012;11:CD007064.

Gentili C, McClean S, Hackshaw-McGeagh L, Bahl A, Persad R, Harcourt D. Body image issues and attitudes toward exercise amongst men undergoing androgen deprivation therapy (ADT) following diagnosis of prostate cancer. Psychooncology. 2019;28(8):1647–53.

Gravas S, Cornu JN, Gacci M, Gratzke C, Herrmann TRW, Mamoulakis C, et al. European Association of Urology Guidelines on Management of Non-neurogenic Male LUTS. 2020. https://uroweb.org/guideline/treatment-of-non-neurogenic-male-luts/

Guo Z, Gan S, Li Y, Gu C, Xiang S, Zhou J, et al. Incidence and risk factors of suicide after a prostate cancer diagnosis: a meta-analysis of observational studies. Prostate Cancer Prostatic Dis. 2018;21(4):499–508.

Gurevich M, Bishop S, Bower J, Malka M, Nyhof-Young J. (Dis)embodying gender and sexuality in testicular cancer. Soc Sci Med. 2004;58(9):1597–607.

Hernandez BY, Barnholtz-Sloan J, German RR, Giuliano A, Goodman MT, King JB, et al. Burden of invasive squamous cell carcinoma of the penis in the United States, 1998-2003. Cancer. 2008;113(10 Suppl):2883–91.

Jayadevappa R, Malkowicz SB, Chhatre S, Johnson JC, Gallo JJ. The burden of depression in prostate cancer. Psychooncology. 2012;21(12):1338–45.

Kiffel J, Sher L. Prevention and management of depression and suicidal behavior in men with prostate cancer. Front Public Health. 2015;3:28.

King AJ, Evans M, Moore TH, Paterson C, Sharp D, Persad R, et al. Prostate cancer and supportive care: a systematic review and qualitative synthesis of men's experiences and unmet needs. Eur J Cancer Care (Engl). 2015;24(5):618–34.

Lamers F, van Oppen P, Comijs HC, Smit JH, Spinhoven P, van Balkom AJ, et al. Comorbidity patterns of anxiety and depressive disorders in a large cohort study: the Netherlands Study of Depression and Anxiety (NESDA). J Clin Psychiatry. 2011;72(3):341–8.

Langelier DM, Cormie P, Bridel W, Grant C, Albinati N, Shank J, et al. Perceptions of masculinity and body image in men with prostate cancer: the role of exercise. Support Care Cancer. 2018;26(10):3379–88.

Lin PH, Liu JM, Hsu RJ, Chuang HC, Chang SW, Pang ST, et al. Depression negatively impacts survival of patients with metastatic prostate cancer. Int J Environ Res Public Health. 2018;15(10).

Liu Q, Zhang Y, Wang J, Li S, Cheng Y, Guo J, et al. Erectile dysfunction and depression: a systematic review and meta-analysis. J Sex Med. 2018;15(8):1073–82.

Maddineni SB, Lau MM, Sangar VK. Identifying the needs of penile cancer sufferers: a systematic review of the quality of life, psychosexual and psychosocial literature in penile cancer. BMC Urol. 2009;9:8.

Matthiesen SM, Frederiksen Y, Ingerslev HJ, Zachariae R. Stress, distress and outcome of assisted reproductive technology (ART): a meta-analysis. Hum Reprod. 2011;26(10):2763–76.

Miller NS, Gold MS. The human sexual response and alcohol and drugs. J Subst Abuse Treat. 1988;5(3):171–7.

Mishra SI, Scherer RW, Snyder C, Geigle P, Gotay C. Are exercise programs effective for improving health-related quality of life among cancer survivors? A systematic review and meta-analysis. Oncol Nurs Forum. 2014;41(6):E326–42.

Moller-Leimkuhler AM. Barriers to help-seeking by men: a review of sociocultural and clinical literature with particular reference to depression. J Affect Disord. 2002;71(1–3):1–9.

Nelson CJ, Mulhall JP. Psychological impact of Peyronie's disease: a review. J Sex Med. 2013;10(3):653–60.

Newby TA, Graff JN, Ganzini LK, McDonagh MS. Interventions that may reduce depressive symptoms among prostate cancer patients: a systematic review and meta-analysis. Psychooncology. 2015;24(12):1686–93.

Pasch LA, Holley SR, Bleil ME, Shehab D, Katz PP, Adler NE. Addressing the needs of fertility treatment patients and their partners: are they informed of and do they receive mental health services? Fertil Steril. 2016;106(1):209–15e2.

Prasad SM, Eggener SE, Lipsitz SR, Irwin MR, Ganz PA, Hu JC. Effect of depression on diagnosis, treatment, and mortality of men with clinically localized prostate cancer. J Clin Oncol. 2014;32(23):2471–8.

Prostate Cancer UK. Research into wellbeing services for men with prostate cancer – final report. 2014. https://prostatecanceruk.org/media/2460122/Report-Wellbeing-services-for-men-with-prostate-cancer.pdf

Quek KF, Razack AH, Chua CB, Low WY, Loh CS. Effect of treating lower urinary tract symptoms on anxiety, depression and psychiatric morbidity: a one-year study. Int J Urol. 2004;11(10):848–55.

Rajkumar RP, Kumaran AK. Depression and anxiety in men with sexual dysfunction: a retrospective study. Compr Psychiatry. 2015;60:114–8.

Rom M, Schatzl G, Swietek N, Rucklinger E, Kratzik C. Lower urinary tract symptoms and depression. BJU Int. 2012;110(11 Pt C):E918–21.

Rosen RC. Psychogenic erectile dysfunction. Classification and management. Urol Clin North Am. 2001;28(2):269–78.

Rosen R, Catania J, Lue T, Althof S, Henne J, Hellstrom W, et al. Impact of Peyronie's disease on sexual and psychosocial functioning: qualitative findings in patients and controls. J Sex Med. 2008;5(8):1977–84.

Rossen P, Pedersen AF, Zachariae R, von der Maase H. Sexuality and body image in long-term survivors of testicular cancer. Eur J Cancer. 2012;48(4):571–8.

Rothschild AJ. Sexual side effects of antidepressants. J Clin Psychiatry. 2000;61(Suppl 11):28–36.

Salonia A, Bettocchi C, Carvalho J, Corona G, Jones TH, Kadioglu A, et al. European Association of Urology Guidelines on Sexual and Reproductive Health. 2020. https://uroweb.org/guideline/sexual-and-reproductive-health/

Schepisi G, De Padova S, De Lisi D, Casadei C, Meggiolaro E, Ruffilli F, et al. Psychosocial issues in long-term survivors of testicular cancer. Front Endocrinol (Lausanne). 2019;10:113.

Sciarra A, Gentilucci A, Salciccia S, Von Heland M, Ricciuti GP, Marzio V, et al. Psychological and functional effect of different primary treatments for prostate cancer: a comparative prospective analysis. Urol Oncol. 2018;36(7):340e7–e21.

Sharpley CF, Christie DRH, Bitsika V. Depression and prostate cancer: implications for urologists and oncologists. Nat Rev Urol. 2020;17(10):571–85.

Skaali T, Fossa SD, Bremnes R, Dahl O, Haaland CF, Hauge ER, et al. Fear of recurrence in long-term testicular cancer survivors. Psychooncology. 2009;18(6):580–8.

Skalski M, Przydacz M, Sobanski JA, Cyranka K, Klasa K, Datka W, et al. Coexistence of lower urinary tract symptoms (LUTS) with depressive symptoms in patients suffering from depressive disorders. Psychiatr Pol. 2019;53(4):939–53.

Smith JF, Walsh TJ, Conti SL, Turek P, Lue T. Risk factors for emotional and relationship problems in Peyronie's disease. J Sex Med. 2008;5(9):2179–84.

Smith AB, Rutherford C, Butow P, Olver I, Luckett T, Grimison P, et al. A systematic review of quantitative observational studies investigating psychological distress in testicular cancer survivors. Psychooncology. 2018;27(4):1129–37.

Sosnowski R, Wolski JK, Zi Talewicz U, Szyma Ski M, Baku AR, Demkow T. Assessment of selected quality of life domains in patients who have undergone conservative or radical surgical treatment for penile cancer: an observational study. Sex Health. 2019;16(1):32–8.

Spendelow JS, Eli Joubert H, Lee H, Fairhurst BR. Coping and adjustment in men with prostate cancer: a systematic review of qualitative studies. J Cancer Surviv. 2018;12(2):155–68.

Tavlarides AM, Ames SC, Diehl NN, Joseph RW, Castle EP, Thiel DD, et al. Evaluation of the association of prostate cancer-specific anxiety with sexual function, depression and cancer aggressiveness in men 1 year following surgical treatment for localized prostate cancer. Psychooncology. 2013;22(6):1328–35.

Terrier JE, Nelson CJ. Psychological aspects of Peyronie's disease. Transl Androl Urol. 2016;5(3):290–5.

Tuinman MA, Hoekstra HJ, Fleer J, Sleijfer DT, Hoekstra-Weebers JE. Self-esteem, social support, and mental health in survivors of testicular cancer: a comparison based on relationship status. Urol Oncol. 2006;24(4):279–86.

Tuinman MA, Hoekstra HJ, Vidrine DJ, Gritz ER, Sleijfer DT, Fleer J, et al. Sexual function, depressive symptoms and marital status in nonseminoma testicular cancer patients: a longitudinal study. Psychooncology. 2010;19(3):238–47.

Ussher JM, Perz J, Rose D, Dowsett GW, Chambers S, Williams S, et al. Threat of sexual disqualification: the consequences of erectile dysfunction and other sexual changes for gay and bisexual men with prostate cancer. Arch Sex Behav. 2017;46(7):2043–57.

Varnik P. Suicide in the world. Int J Environ Res Public Health. 2012;9(3):760–71.

Verkuijlen J, Verhaak C, Nelen WL, Wilkinson J, Farquhar C. Psychological and educational interventions for subfertile men and women. Cochrane Database Syst Rev. 2016;3:CD011034.

Wdowiak A, Bien A, Iwanowicz-Palus G, Makara-Studzinska M, Bojar I. Impact of emotional disorders on semen quality in men treated for infertility. Neuro Endocrinol Lett. 2017;38(1):50–8.

Wibowo E, Wassersug RJ, Robinson JW, Santos-Iglesias P, Matthew A, McLeod DL, et al. An educational program to help patients manage androgen deprivation therapy side effects: feasibility, acceptability, and preliminary outcomes. Am J Mens Health. 2020;14(1):1557988319898991.

Wootten AC, Meyer D, Abbott JM, Chisholm K, Austin DW, Klein B, et al. An online psychological intervention can improve the sexual satisfaction of men following treatment for localized prostate cancer: outcomes of a Randomised Controlled Trial evaluating My Road Ahead. Psychooncology. 2017;26(7):975–81.

World Health Organization. International Classifications of Diseases-10 (ICD-10). 2019. https://icd.who.int/browse10/2019/en#/F32

Yu C, Hequn C, Longfei L, Minfeng C, Zhi C, Feng Z, et al. Sexual function after partial penectomy: a prospectively study from China. Sci Rep. 2016;6:21862.

Zuniga KB, Chan JM, Ryan CJ, Kenfield SA. Diet and lifestyle considerations for patients with prostate cancer. Urol Oncol. 2020;38(3):105–17.

Chapter 22
Engaging Men with Men's Health

Sachin Perera, Nathan Lawrentschuk, and Ray Swann

22.1 Introduction

Many aspects of men's health are often not adequately addressed in today's society. This leads to worse outcomes and can impact overall care. It is also common for men to present with more advanced disease or display more severe symptoms of a condition, as a result of a lack of engagement with their health.

Low help seeking can lead to poor course of life outcomes. This has certainly not been improved by the COVID-19 pandemic. The pandemic has seen an increased burden on our healthcare system, from both primary and secondary means. Secondary effects of the pandemic have meant disease presentations that would normally be prioritised have been triaged to be addressed at later stages. Additionally, the rates of men presenting to their healthcare practitioner for the first time, have markedly reduced. Whether this is directly linked to the pandemic or an exacerbation of an underlying issue with help seeking behaviours is unclear. However, the resultant effects in men's health are still reflected in the statistics.

S. Perera
Peter MacCallum Cancer Centre, Melbourne, Australia

Department of Urology, Royal Melbourne Hospital, Parkville, Australia
e-mail: Sachin.Perera@mh.org.au

N. Lawrentschuk (✉)
Department of Urology, Royal Melbourne Hospital, Parkville, Australia

Department of Surgery, University of Melbourne, Melbourne, Australia

EJ Whitten Prostate Cancer Research Centre at Epworth Healthcare, Melbourne, Australia

R. Swann
Department of Urology, Royal Melbourne Hospital, Parkville, Australia

Department of Surgery, University of Melbourne, Melbourne, Australia
e-mail: ray.swann1@unimelb.edu.au

Current literature surrounding men's health identifies some key barriers to engagement. They include, but are not limited to:

- Men as a cohort are less likely to disclose mental health issues to family, friends or colleagues. Rather, it is more common for men to adopt other unhelpful coping mechanisms, such as alcohol or drugs.
- Societal constructs for men have built an image of stoicism. This has led to ongoing reluctance to seek help in order to conform to this image.
- The focus on men's health is too narrow. It needs to be broadened and diversified to be more accessible to vulnerable minorities.
- There is a lack of validated high-quality resources available to the public. This continues to be a significant issue, especially when it pertains to malignancy.

However, despite such barriers being identified, these areas continue to be neglected and poorly addressed.

The approaches that have seen the most gains in garnering health engagement recently, have been multifactorial and focus on traditional values. The World Health Organisation (WHO) has outlined an approach to increase male engagement in their healthcare through their "*The Solid Facts*" campaign. Local adaptations of this campaign have shown that social determinants of health specific to men include: addressing lifestyle factors (work, political and spiritual choices), family values (including protective factors such as children or healthy relationships with partners) and societal values (such as redefining masculinity). It has proven that a wholistic approach, with emphasis on previously underacknowledged areas, enhances male engagement with healthcare.

22.2 Mental Health and Recreational Drug Use

Mental health has been broadly identified as one of the key areas in men's health that lacks appropriate action. The current focus on men's mental health centres on suicide statistics in the community (Ellis et al. 2012). Shockingly, these numbers make up 80% of total suicides per year in some countries. Though tragic, such high numbers of suicide have as a result, overshadowed the underlying message around the health conditions affecting male morbidity (such as major depressive disorder) (Rice et al. 2018; River 2014). Further, despite easy identification of these statistics, there has been very little research into preventative measures that should be taken at a local and government level to bring these incidences down (Shand et al. 2015).

Looking further then into the current understanding of mental health conditions affecting males, we can identify the following broad 'themes' as possible barriers to help-seeking actions. These include drug use, high-risk behaviours and controversially, toxic masculinity (Seidler et al. 2016). In Australia, a topic closely married to the issue around mental health, is the use or misuse of alcohol and other recreational drugs. This issue is pervasive amongst the population and most prominent amongst men. As a nation, Australia has conceded to an unhealthy relationship with alcohol, though little has been done at a local level to address the issue. This is evidenced by prevailing attitudes amongst some men to sideline unhealthy drinking habits until a negative outcome precipitates from this habit (Ferguson et al. 2019). The same can be said, though perhaps less commonly, to illicit substance abuse.

22.3   Social Health

Social health is an emerging topic of discussion in relation to men's health. It involves introspection of a man's identity at home and in society. A key component of this is the idea of supportive networking, such as the "Men's Shed" movement (Taylor et al. 2018; Nurmi et al. 2018; Wilson et al. 2019). This initiative creates an accessible symbol (the backyard shed) as a talking point for networking, policy change and mental health advocacy.

The concept of rebranding health messages is not, however, isolated to the Men's Shed. There has also been some conjecture around expanding it on a broader scale. For example, it has been identified that men's health at a policy level, has failed to recognise appropriate screening tools for chronic disease. This compared to the steps taken for women's health with cervical cancer vaccination and pap smears (Holden et al. 2010).

Masculinity and the identity of the man in society, have also been raised in literature as possible barriers to seeking help (Liddle et al. 2019; Seidler et al. 2020). Longitudinal analysis of men's attitudes to health have found that social determinants of accessing health are independent to previously identified help-seeking barriers (Rice et al. 2017). Social health is a thus far under investigated, but an increasingly important part of raising male engagement with healthcare. It is thought therefore, that this should be a key focus for future research.

22.4   Sexual Health and Cultural Awareness

As with a many of the identified themes in the current literature, sexuality and culture are woven into the discussion with shared connections across categories. Male sexuality and sexual health care issues that strike directly at the heart of the male ego and may pre-date current understandings of the term "sexual health".

The social stigma that exists with discussing matters related sexual health is still prevalent and has been identified as a significant barrier to accessing healthcare (Litras et al. 2015; Wilkinson et al. 2016; Su et al. 2016; Stroud and Lockwood 2012). Analysis of this data further identifies minority groups as being particularly vulnerable. For example, men who sleep with men (MSM) and indigenous populations, have been flagged as two demographics who lack the specific resources and culturally appropriate access to obtain care (Brown et al. 2017). These issues are compounded by the cultural barriers that inherently exist for all aboriginal people in general (Tsey et al. 2014).

Pre-existing biases or assumptions regarding minorities in healthcare have unfortunately made a prominent feature in this area of investigation (Su et al. 2016). Encouragingly though, there are steps being taken to trial novel methods of delivering culturally appropriate care to these groups (Tsey et al. 2014; Isaacs et al. 2013).

Overall, sexual health remains a stigmatised topic of discussion in both majority and minority circles. It is an area in need of particular attention in order to prevent a significant disease burden on our healthcare system.

22.5   Malignancy and Literature Inconsistency

Chronic disease prevention is the ultimate goal of engaging men with healthcare. An area in need of critical advocacy, that is currently neglected in the literature, is that of penile and prostate malignancies. Whilst there is an abundance of literature designed for peer review at a scientific level, there is little easily-accessible, high-quality information available for patients or the public (Kannan et al. 2019; Ilic 2013; Teh et al. 2019). This lack of consistent, reliable information on male malignancies gives rise to the biggest barrier for better male healthcare outcomes—engagement. The result of this has led to many men educating themselves via methods which are unwholesome and potentially detrimental to the overall surveillance and treatment of their disease or symptoms (Litras et al. 2015).

From birth, men are marked for earlier graves than their female counterparts. According to the Australian Institute of Health and Wellbeing, the current life expectancy for an Australian man at birth is 81 years compared to 85 years for females. Though there is undoubtedly a multitude of physiological, social and lifestyle reasons for this gap, it also highlights a potential burden of disease for men in the latter stages of life.

As it stands genitourinary (GU) cancers are the most common form of cancer in Australian men. This group includes renal, prostate and penile cancers, with prostate being the most common. Of this number, 3500 males die each year from prostate cancer (Moolla et al. 2020), with many more suffering from the chronic disease. Penile cancer, whilst more scarce than prostate, has a higher morbidity and mortality rate, accounting for 1–2% of all male malignancies (Teh et al. 2018).

Using penile cancer as an example, we can highlight the distinct lack of validated internet resources available to the public and clinicians. Such disparities in resource availability across language, socio-economic and geographic barriers, ultimately affect health outcomes (Teh et al. 2018; Corboy et al. 2014). The lack of resource consistency compounds existing issues with regard to male cancers. For instance, prostate cancer is already associated with a strong stigma, whereby newly diagnosed males often enter a cycle of self-isolation and self-blame for their disease (Ettridge et al. 2018; Hyde et al. 2017). This again, is fed by a lack of succinct and consistent information regarding the disease that is readily available (Hodyl et al. 2020).

In recent times there has been a shift in focus to incorporate prostate cancer awareness. Movements such as *'Movember',* have increased an emphasis on men reporting their symptoms to general practitioners. Unfortunately, however, this attention pales in comparison to other prominent public health campaigns targeted at women's health advocacy and engagement. A large part of the reasoning for this is believed to sadly just be due to the cost of screening outweighing the benefits when it comes to penile and prostate cancers.

The disparity in engagement for men is again compounded by the quality of resources available to investigate their symptoms. As a result, these diseases often present later at more advance stages, adding to the burden of chronic disease on the healthcare system (Chang et al. 2018). Penile cancers in particular, lack consistent validation of source material on the internet, which could be contributing to the trend of late presentations seen in the community. This theme is also reflected in prostate cancer resources, with unvalidated commercial resources making up 28% of all available information on the internet (Lawrentschuk et al. 2009).

Following from the concept of "Men's Shed" in a mental health setting, some work has been completed with regard to cancer support groups. These groups seek to address the isolation felt by a cancer diagnosis, and advocate for ongoing surveillance and management. Interestingly, these span across other malignancies affecting men too, including colorectal cancer (Martopullo et al. 2020; Oberoi et al. 2016).

Ultimately, successful engagement of men in men's health will involve a multifactorial approach. The need to address key areas previously overlooked, is laid bare in this section and examined in depth.

22.6   Best Practice

Programs and initiatives that have shown promise in enhancing male engagement with healthcare have shared traits. These include focus on protective family and relationship factors, encouragement of networking and de-stigmatising disease.

Going forward, research into men's health should build upon the lessons learned from successful initiatives such as the Men's shed movement. Specifically, providing a discussion platform for men who suffer the same disease or are concerned about their health, has been a powerful tool in encouraging engagement. The next steps for this kind of initiative would be to implement it into policy.

As discussed, the lack of large-scale screening for male specific disease has put the onus of health awareness on the individual. Whilst empowering, it is incumbent on the healthcare system and policymakers to offer the resources necessary for men to investigate and discuss symptoms in order to make informed choices.

Discussions in relation to male health, also need to occur more consistently from an early age. In Australia, the current early childhood management of chronic disease is focused on tangible interventions, such as the cervical cancer vaccination for high school aged girls. We believe that there is an opportunity here to educate boys. Educating young boys about their bodies and what to expect both mentally and physically as they grow, will lead to a paralleled success as has been seen with the women's health initiatives.

With this in mind, it is surmised that resource availability needs to be multifaceted, in order to cater to a diverse population and cross platforms. For example, in the digital age it is insufficient to simply rely on physical brochures at the GP's office or handouts from a clinic. The leading source of information now comes from a few finger movements and clicks of a mouse. Academia therefore needs to adjust to meet this demand, in order to prevent information being sourced from less reputable or potentially harmful sources.

Malignancy specific to men causes a significant burden of disease on the healthcare system. As with any chronic disease, prevention is the key and as discussed, the road to this is an early diagnosis. Help-seeking behaviours must be encouraged and made the norm. Barriers to help-seeking are well-documented, but unfortunately still lack tangible countermeasures. Healthcare can only work in parallel to a disease process once presented. It is therefore reliant upon the man to ask for help.

The stigma of weakness when it comes to male disease, may be of the greatest barrier to engaging men with healthcare. While there is no one answer to this issue, it will involve addressing the areas of masculinity and modern values. The effort to do so cannot be solely led by politicians, healthcare professionals or hospitals. Rather, to successfully remove this stigma, there must be a societal effort. Long established programs like *Movember* have shown great promise in raising awareness and generating platforms for engagement. The conversation now needs to move past mere awareness, and step into action.

Finally, a word on the current coronavirus pandemic. As a disease whose only effective treatment option is self-isolation, COVID-19 has exacerbated issues surrounding engaging men with healthcare. As a whole the pandemic has led to worsening of diseases that were causing considerable strain on the healthcare system prior to 2020 for both men and women. For example, the incidence of mental health presentations has dramatically increased since March of 2020. What the pandemic means for men's health specifically is possibly further preventing help-seeking behaviour. There is very little data on how the pandemic has affected presentations for chronic disease, but we can expect these issues to declare themselves later than the system has already seen as we emerge from the pandemic.

The vast body of literature and discussion on men's health clearly identifies the issues and provides a clear starting point for the way forward. Palatable, accessible and validated information is the first step in achieving this goal.

References

Brown G, Leonard W, Lyons A, Power J, Sander D, McColl W, et al. Stigma, gay men and biomedical prevention: the challenges and opportunities of a rapidly changing HIV prevention landscape. Sex Health. 2017;14(1):111–8.

Chang DTS, Abouassaly R, Lawrentschuk N. Quality of health information on the internet for prostate cancer. Adv Urol. 2018;2018:6705152.

Corboy D, McLaren S, Jenkins M, McDonald J. The relationship between geographic remoteness and intentions to use a telephone support service among Australian men following radical prostatectomy. Psycho-Oncology. 2014;23(11):1259–66.

Ellis LA, Collin P, Davenport TA, Hurley PJ, Burns JM, Hickie IB. Young men, mental health, and technology: implications for service design and delivery in the digital age. J Med Internet Res. 2012;14(6):e160.

Ettridge KA, Bowden JA, Chambers SK, Smith DP, Murphy M, Evans SM, et al. "Prostate cancer is far more hidden…": Perceptions of stigma, social isolation and help-seeking among men with prostate cancer. Eur J Cancer Care. 2018;27(2):e12790.

Ferguson N, Savic M, McCann TV, Emond K, Sandral E, Smith K, et al. "I was worried if I don't have a broken leg they might not take it seriously": experiences of men accessing ambulance services for mental health and/or alcohol and other drug problems. Health Expect. 2019;22(3):565–74.

Hodyl NA, Hogg K, Renton D, von Saldern S, McLachlan R. Understanding the preferences of Australian men for accessing health information. Aust J Primary Health. 2020;26(2):153–60.

Holden CA, Allan CA, McLachlan RI. 'Male friendly' services. Aust Fam Physician. 2010;39(1-2):9–10.

Hyde MK, Newton RU, Galvão DA, Gardiner RA, Occhipinti S, Lowe A, et al. Men's help-seeking in the first year after diagnosis of localised prostate cancer. Eur J Cancer Care. 2017;26(2).

Ilic D. Educating men about prostate cancer in the workplace. Am J Mens Health. 2013;7(4):285–94.

Isaacs AN, Maybery D, Gruis H. Help seeking by Aboriginal men who are mentally unwell: a pilot study. Early Interv Psychiatry. 2013;7(4):407–13.

Kannan A, Kirkman M, Ruseckaite R, Evans SM. Prostate care and prostate cancer from the perspectives of undiagnosed men: a systematic review of qualitative research. BMJ Open. 2019;9(1):e022842.

Lawrentschuk N, Abouassaly R, Hackett N, Groll R, Fleshner NE. Health information quality on the internet in urological oncology: a multilingual longitudinal evaluation. Urology. 2009;74(5):1058–63.

Liddle JLM, Lovarini M, Clemson LM, Jang H, Lord SR, Sherrington C, et al. Masculinity and preventing falls: insights from the fall experiences of men aged 70 years and over. Disabil Rehabil. 2019;41(9):1055–62.

Litras A, Latreille S, Temple-Smith M. Dr Google, porn and friend-of-a-friend: where are young men really getting their sexual health information? Sex Health. 2015;12(6):488–94.

Martopullo C, Oberoi D, Levin G, Qureshi M, Morgan-Maver E, Korzeniewski O, et al. "In the same boat"-a mixed-methods exploration of reasons why male gastrointestinal cancer patients joined a professionally led men-only cancer support group. J Cancer Survivorship Res Pract. 2020;14(3):261–72.

Moolla Y, Adam A, Perera M, Lawrentschuk N. 'Prostate cancer' information on the internet: fact or fiction? Curr Urol. 2020;13(4):200–8.

Nurmi MA, Mackenzie CS, Roger K, Reynolds K, Urquhart J. Older men's perceptions of the need for and access to male-focused community programmes such as Men's Sheds. Ageing Soc. 2018;38(4):794–816.

Oberoi DV, Jiwa M, McManus A, Hodder R, de Nooijer J. Help-seeking experiences of men diagnosed with colorectal cancer: a qualitative study. Eur J Cancer Care. 2016;25(1):27–37.

Rice SM, Aucote HM, Parker AG, Alvarez-Jimenez M, Filia KM, Amminger GP. Men's perceived barriers to help seeking for depression: longitudinal findings relative to symptom onset and duration. J Health Psychol. 2017;22(5):529–36.

Rice SM, Telford NR, Rickwood DJ, Parker AG. Young men's access to community-based mental health care: qualitative analysis of barriers and facilitators. J Mental Health (Abingdon, England). 2018;27(1):59–65.

River J. Expanding our understanding of suicidal men's help-seeking practices. Austral Nurs Midwifery J. 2014;22(5):34.

Seidler ZE, Dawes AJ, Rice SM, Oliffe JL, Dhillon HM. The role of masculinity in men's help-seeking for depression: a systematic review. Clin Psychol Rev. 2016;49:106–18.

Seidler ZE, Rice SM, Kealy D, Oliffe JL, Ogrodniczuk JS. What gets in the way? Men's perspectives of barriers to mental health services. Int J Social Psychiatry. 2020;66(2):105–10.

Shand FL, Proudfoot J, Player MJ, Fogarty A, Whittle E, Wilhelm K, et al. What might interrupt men's suicide? Results from an online survey of men. BMJ Open. 2015;5(10):e008172.

Stroud P, Lockwood C. Obstacles to the take-up of mental health-care provision by adult males in rural and remote areas of Australia: a systematic review protocol. JBI Libr Systemat Rev. 2012;10(42 Suppl):1–12.

Su JY, Belton S, Ryder N. Why are men less tested for sexually transmitted infections in remote Australian Indigenous communities? A mixed-methods study. Cult Health Sex. 2016;18(10):1150–64.

Taylor J, Cole R, Kynn M, Lowe J. Home away from home: health and wellbeing benefits of men's sheds. Health Promot J Aust. 2018;29(3):236–42.

Teh J, Op't Hoog S, Nzenza T, Duncan C, Wang J, Radojcic M, et al. Penile cancer information on the internet: a needle in a haystack. BJU Int. 2018;122(Suppl 5):22–6.

Teh J, Wei J, Chiang G, Nzenza TC, Bolton D, Lawrentschuk N. Men's health on the web: an analysis of current resources. World J Urol. 2019;37(6):1043–7.

Tsey K, Chigeza P, Holden CA, Bulman J, Gruis H, Wenitong M. Evaluation of the pilot phase of an Aboriginal and Torres Strait Islander Male Health Module. Aust J Primary Health. 2014;20(1):56–61.

Wilkinson AL, Pedrana AE, El-Hayek C, Vella AM, Asselin J, Batrouney C, et al. The impact of a social marketing campaign on HIV and sexually transmissible infection testing among men who have sex with men in Australia. Sex Transm Dis. 2016;43(1):49–56.

Wilson NJ, Cordier R, Parsons R, Vaz S, Ciccarelli M. An examination of health promotion and social inclusion activities: a cross-sectional survey of Australian community Men's Sheds. Health Promot J Aust. 2019;30(3):371–80.

Chapter 23
Cardiovascular Health in Men

ThetSuSu and Husain Shabeeh

23.1 Cardiovascular Health and Cardiovascular Disease

Cardiovascular disease (CVD) is related to the circulatory system which composes of the heart and blood vessels. Its primary role is to carry oxygen and nutrients to the tissues of the body and removes the carbon dioxide and toxins from them. Cardiovascular disease is the generic term for the conditions affecting the heart or the blood vessels.

CVD is one of the leading causes of death worldwide, accountable for over 17 million annual global deaths (Rafael Lozano et al. 2012; Mackay and Mensah 2004). It is usually associated with the build-up of fatty deposits inside the arteries (atherosclerosis) causing the narrowing of the arteries and increase risk of blood clots.

There are many different types of CVD, but the most common diseases are coronary heart disease (CHD), stroke and transient ischaemic attack, peripheral arterial disease and aortic disease.

23.1.1 Coronary Heart Disease

Coronary heart disease (CHD) is the disease which occurs when the oxygen rich blood supply to the heart muscle is reduced or blocked. It is associated with atherosclerosis and formation of atheroma (build-up of fatty deposits) in the coronary

ThetSuSu (✉)
Croydon University Hospital, London, UK
e-mail: thetsu.su@nhs.net

H. Shabeeh
Croydon University Hospital and Kings College Hospital, London, UK

© The Author(s), under exclusive license to Springer Nature
Switzerland AG 2022
S. S. Goonewardene et al. (eds.), *Men's Health and Wellbeing*,
https://doi.org/10.1007/978-3-030-84752-4_23

arteries resulting the narrowing of the arteries. When the atheroma ruptures, it damages the endothelium of the artery which is the lining of the interior surface of blood vessels. The endothelium has multiple functions such as maintaining haemostatic balance, regulating vascular tone and angiogenesis. Damage to the endothelium leads to the formation of blood clot and subsequent narrowing and blockage of coronary arteries. This can then cause angina pectoris, heart attack, heart failure, abnormal heart rhythm (arrhythmia) and sudden cardiac death.

The most common presenting symptom is angina. It is usually a constricting discomfort that typically occurs in the front of the chest, but may radiate to the neck, shoulders, jaw or arms. It is commonly brought on by physical exertion or emotional stress and relieved by rest or medications which dilate the blood vessels (vasodilatation). The cause is limitation of blood flow in the coronary arteries due to atherosclerosis. Some may have atypical symptoms such as gastrointestinal discomfort, breathlessness or nausea.

There are different types of investigations for coronary artery disease, ranging from non-invasive tests such as stress tests, cardiac commuted tomography (CT), magnetic resonance imaging (MRI) perfusion scans; to invasive tests such as coronary angiography. The treatment options also vary from medical therapy, percutaneous coronary intervention including angioplasty and stents, coronary artery bypass graft surgery. The treatment depends on many factors including severity of disease, co-morbidity, anatomical consideration and others.

23.1.2 Stroke and Transient Ischemic Attack

Stroke is either ischaemic where the blood supply is stopped due to a blood clot or haemorrhagic where a weakened blood vessel bursts. Both events can cause brain damage and possible death. A transient ischaemic attack (TIA) occurs when the blood flow to the brain is only temporarily disrupted.

The common symptoms are facial drooping on one side, weakness of arm or leg (one arm/leg or one side of the body) and slurred speech.

Treatment depends on types of stroke and usually treated with medication which includes medicines to prevent and dissolve blood clots, reduce blood pressure and cholesterol. In some cases, procedures may be required to remove blood clots.

23.1.3 Peripheral Arterial Disease

Peripheral arterial disease (PAD) occurs when there is a blockage in the arteries of the limbs usually the legs.

This can cause dull or cramping leg pain which is worse on walking and relieved by rest, hair loss on the legs and feet, numbness or weakness in the legs, skin discolouration such as turning pale or blue and persistent open sores on the feet and legs.

Like coronary disease, it is usually caused by a build-up of fatty deposits which are made up of cholesterol and other waste substances in the walls of leg arteries. This results in the arteries becoming narrower and restricts blood flow to the legs.

PAD is largely treated through lifestyle changes and medication. An angioplasty or bypass graft surgery is usually recommended for critical limb ischaemia which is the condition in which blood flow to the legs becomes severely restricted.

23.1.4 Aortic Disease

Aortic diseases are a group of conditions affecting the aorta. This is the largest blood vessel in the body which carries blood from the heart to the rest of the body. One of most common aortic diseases is an aortic aneurysm in which the aorta becomes weakened and bulges outwards.

Abdominal aortic aneurysm (AAA) does not usually cause any obvious symptoms and is often incidentally found during screening or tests done for other reasons.

People at a higher risk of getting an AAA include all men aged 66 or over and women aged 70 or over with one or more of the following risk factors: hypertension, chronic obstructive pulmonary disease (COPD), high blood cholesterol, family history of AAA, cardiovascular disease such as heart disease or stroke and tobacco smoking.

Treatment depends on how large the AAA is. Surgery is usually recommended for large AAA and otherwise this requires regular surveillance with interval scanning.

23.2 Risk Factors for Cardiovascular Disease

Given the fact that they are commonly caused by fatty deposition (atherosclerosis), the risk factors are same for all CVD. Men generally develop CVD at a younger age and have a higher risk of developing coronary heart disease (CHD) than women (George et al. 2015; Leening et al. 2014). CHD mortality rate in middle-aged men is higher than in middle-aged women and a difference that may persist throughout most of their lifetime (Ali et al. 2015; Mozaffarian et al. 2015). In men, cardiovascular risk profile linearly increases over time, as well as atherosclerotic process continuously develops.

Various risk factors increase a person's risk of developing CVD (NHS UK and NICE CVD risk assessment and management 2020) (Fig. 23.1).

In terms of family history of CVD, you are considered to have if your first-degree relatives were diagnosed with CVD before 55 years of age in men and 65 years of age in women.

Being overweight or obese increases your risk of developing diabetes and high blood pressure which are the risk factors for CVD. You are at an increased risk of

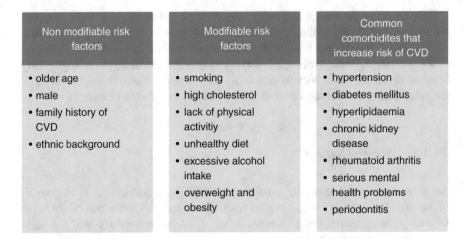

Non modifiable risk factors	Modifiable risk factors	Common comorbidites that increase risk of CVD
• older age • male • family history of CVD • ethnic background	• smoking • high cholesterol • lack of physical activitiy • unhealthy diet • excessive alcohol intake • overweight and obesity	• hypertension • diabetes mellitus • hyperlipidaemia • chronic kidney disease • rheumatoid arthritis • serious mental health problems • periodontitis

Fig. 23.1 Risk factors in Cardiovascular disease

CVD if body mass index (BMI) is 25 or above or a man with a waist measurement of 94 cm (about 37 in.) or more (UK NHS cardiovascular disease web) and for a woman with waist measurement of 80 cm (about 31.5 in.) or more (UK NHS cardiovascular disease web). This highlights the vast importance of lifestyle modification in reducing cardiovascular risk.

Low socio-economic status, lack of social support, stress at work and in family life, depression, anxiety and other mental health disorders contribute to the risk of developing CVD and a worse prognosis of CVD, with the absence of these conditions being associated with a lower risk of developing CVD and a better prognosis (ESC 2016 CVD prevention guideline).

23.3   Recommended Risk Assessment Scores for CVD

There are different recommended risk assessment scores in different countries. In UK, it is recommended that a person's 10-year CVD risk should be assessed at least every 5 years from the age of 40 years and over. This is of course not needed for those who already have CVD or are at high risk of developing it as they require management of their risk (e.g., people with type 1 diabetes mellitus or chronic kidney disease stage 3 and above) or aged 85 years or over.

High-risk persons qualify for intensive lifestyle advice and may need drug treatment. Commonly used risk assessment scores include for example the Framingham Risk Score which is recommended in US.

23.4   Erectile Dysfunction and Cardiovascular Disease

An association between erectile dysfunction (ED) and CVD has been clearly recognized. Numerous studies have shown that erectile dysfunction as an independent risk marker for cardiovascular disease. ED is felt to precede CAD, stroke and PAD by a period that usually ranges from 2 to 5 years (average 3 years) (Vlachopoulos et al. 2013). A meta-analysis showed that patients with ED compared with patients without ED have a 44% higher risk for total CV events, 62% for AMI, 39% for stroke and 25% for all-cause mortality (Miner et al. 2014; Vlachopoulos et al. 2013).

It is important to distinguish between predominantly vasculogenic erectile dysfunction and psychogenic or non-vasculogenic organic erectile dysfunction such as hormonal and neurogenic.

The common pathophysiologic bases for erectile dysfunction and CVD are endothelial dysfunction, inflammation either due to atherosclerotic blockage or factors affecting endothelial function that prevent adequate blood vessel dilatation (vasodilation) during sexual stimulation and low testosterone. Besides having common pathophysiology, ED and CVD also share risk factors including age, hypercholesterolaemia, hypertension, DM, smoking, obesity, metabolic syndrome, sedentary lifestyle and depression.

For all men with erectile dysfunction, particularly with vasculogenic erectile dysfunction, studies and current evidence suggest that initial risk stratification be based on the risk scores and those who have intermediate risk and high risk should have further cardiovascular work up and investigations. An example of this has been proposed by Miner and colleagues in 2014 as see in Fig. 23.2 below.

23.5   Lifestyle Modifications

An adage is 'Prevention is better than cure'—a healthy lifestyle has been clearly shown to significantly lower the risk of cardiovascular disease (Cooney et al. 2009; Liu et al. 2012). Lifestyle modification is important in cardiovascular risk reduction for both primary prevention i.e., in those who do not have CVD and secondary prevention i.e., those who have CVD. Some of the ways that CVD risk can be reduced through lifestyle modification include:

1. Stop Smoking.

Smoking and other tobacco use is a significant risk factor for CVD. Smoking cessation is crucial in reducing risk of CVD. Some studies suggest that smoking cessation in heavy cigarette smokers can reduce the risk of cardiovascular disease by 39% within 5 years (Duncan et al. 2014).

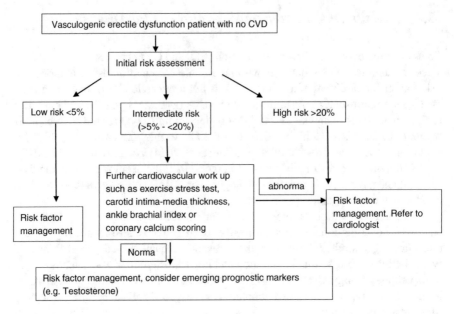

Fig. 23.2 Prevention of cardiovascular disease

2. Healthy diet

A healthy, balanced diet is recommended for a healthy cardiovascular system and reduction of the risk of developing CVD. A healthy diet plays a major role in multiple factors including weight management and lipid control.

A balanced diet has been suggested to include low levels of saturated fat (found in foods such as fatty cuts of meat, lard, cheese, cream, cake and biscuits, sausage, bacon, pastries etc), low salt, low sugar, plenty of fibre and wholegrain foods (such as brown rice, wholewheat pasta and brown, wholemeal bread), plenty of fruit and vegetables.

In UK for example, the National Institute for Heath and Care Excellence (NICE) advises that people at high risk of or with cardiovascular disease are to eat a diet in which total fat intake is 30% or less, saturated fats are 7% or less, dietary cholesterol is less than 300 mg/day of total energy intake (NICE CG181). The average man aged 19–64 years should eat no more than 30 g of saturated fat a day. Tips to eat less fat are choosing foods that are lower in fat and to grill, bake or steam rather than frying or roasting, use monounsaturated fats such as olive oil, rapeseed oil.

NICE and National Health Service (NHS) UK suggest eating at least five portions of fruits and vegetable per day and at least two portions of fish per week including a portion of oily fish. One portion of fruit and vegetable is 80 g of fresh, canned or frozen fruit and vegetable, 30 g of dried fruit, 150 mL of glass of fruit juice or smoothie. Starchy carbohydrates should make up just over a third of the food you eat. They include potatoes, bread, rice, pasta and cereals.

European Society of Cardiology (ESC) have also produced guidelines, as seen in table below (Fig. 23.3).

Saturated fatty acids to account for <10% of total energy intake, through replacement by polyunsaturated fatty acids.
Trans unsaturated fatty acids: as little as possible, preferably no intake from processed food and <1% of total energy intake from natural origin.
<5 g of salt per day
30-45 g of fibre per day, preferably from wholegrain products.
≥ 200g of fruit per day
≥ 200g of vegetables per day
Fish 1-2 times per week, one of which to be oily fish.
30 grams unsalted nuts per day
Consumption of alcoholic beverages should be limited to 2 glasses per day (20g/day of alcohol) for men and 1 glass per day (10g/day of alcohol) for women.
Sugar-sweetened soft drinks and alcoholic beverages consumption must be discouraged.

Fig. 23.3 Lifestyle modifications in cardiovascular disease

23.6   Regular Exercise

Physical activity is associated with reduction in cardiovascular risk (ESC 2016 CVD prevention). It is important to aim to be physically active every day and reduce time spent sitting or lying down and break up long periods of not moving with some activity. The European Society of Cardiology (ESC) guidelines from 2016 suggest doing at least 150 min of moderate intensity aerobic activity or 75 min of vigorous intensity aerobic activity or a mix of moderate and vigorous aerobic activity every week (ESC 2016 CVD prevention; NICE CG181).

Examples of moderate intensity activities are walking briskly (4.8–6.5 km/h), slow cycling (15 km/h), painting/decorating, vacuuming, gardening (mowing lawn), golf (pulling clubs in trolley), tennis (doubles), ballroom dancing, water aerobics and hiking (ESC 2016 CVD prevention).

Vigorous physical activities include race-walking, jogging or running, bicycling >15 km/h, heavy gardening (continuous digging or hoeing), swimming laps, tennis (single) (ESC 2016 CVD prevention).

23.7   Reduce Alcohol Consumption

At the general public level, alcohol consumption is associated with multiple health risks. For example, the current UK alcohol guidelines suggest that to reduce the health risk from alcohol, it is safest not drink more than 14 units a week on a regular basis (UK CMO's alcohol guidance). For regular drinker who drinks as much as 14 units per week, it is best to spread your drinking evenly over 3 or more days. A unit of alcohol is roughly equivalent to half a pint of normal-strength lager or a single measure (25 mL) of spirits. A small glass of wine (125 mL) is about 1.5 units.

23.8   Weight Management

A combination of regular exercise, healthy diet and behaviour modification are mainstay therapy for obesity. UK NICE guidance for obesity management and ESC 2016 CVD prevention suggested to aim to get BMI below 25 but not to aim below 20 because all-cause mortality appears to increase at BMI levels <20 (ESC 2016; NICE obesity 2014; Berrington de Gonzalez et al.). The waist circumference ≥102 cm in men and ≥88 cm in women represents the threshold at which weight reduction should be initiated (2016 European Guidelines on cardiovascular disease prevention in clinical practice). Medical therapy with drugs and/or bariatric surgery are additional options. BMI can be simply calculated by using weight and height. The formula is BMI = kg/m^2 where kg is body weight in kilogram and m^2 is height in metres squared.

23.9   Optimize the Control of Concomitant Cardiovascular Disease Risks!

It is particularly important to optimize the treatment and control of associated CVD risks especially hypertension, diabetes and high cholesterol.

23.9.1   Hyperlipidaemia (High Cholesterol)

According to the 2019 ESC Dyslipidaemia guidelines, treatment targets and goals for cholesterol levels such LDL, non-HDL cholesterol levels are depending on risk of CVD. For examples, in low-risk group, goal of LDL cholesterol is <3 mmol/L (<116 mg/dL), in moderate risk group, target level of LDL is <2.6 mmol/L and non-HDL is <3.4 mmol/L. There is no goal for triglycerides, but it is good to maintain <1.7 mmol/L (<150 mg/dL) as it indicates lower risk (ESC Dyslipidaemia 2019). Lifestyle modifications are crucial steps in lipid control, and there may be use of medications such as statin depend on your risk and lipid level.

23.9.2   Hypertension

Hypertension is one of the most important risk factors for CVD and has an independent and continuous relationship with the incidence of several cardiovascular events (ESC 2018 HTN; Forouzanfar et al. 2017; Lewington et al. 2002). According to

NICE and 2018 ESC Hypertension Guidelines, hypertension is defined as clinic blood pressure above 140/90 mmHg and subsequent ambulatory blood pressure monitoring (ABPM) daytime average or home blood pressure (HBPM) ranging from 135/85 mmHg or higher.

2018 ESC Hypertension guideline stated that several studies have shown that a 10-mmHg reduction in systolic BP (upper reading) or a 5-mmHg reduction in diastolic BP (lower reading) is associated with significant reductions in all major cardiovascular events by 20%, all-cause mortality by 10–15%, stroke by 35%, coronary events by 28% and heart failure by 40% (ESC HTN 2018).

Reduce and maintain blood pressure below 135/85 mmHg for adults aged under 80 and below 145/85 for adults aged 80 and above. Lifestyle modifications (as mention above) are the first step in the management of hypertension. Antihypertensive medications are needed in addition to lifestyle advice to adults with persistent hypertension.

23.9.3 Diabetes Mellitus

Diabetes is a chronic metabolic condition characterised by insulin resistance and insufficient insulin production resulting in high blood glucose level. It is commonly associated with obesity, physical inactivity, high blood pressure, disturbed blood lipid level and therefore is recognised to significantly increase cardiovascular risk.

Integrated dietary advice with other aspects of lifestyle modifications such as increasing physical activity and losing weight are the fundamental part in management of diabetes in addition to medical therapy. Adults with diabetes mellitus are encouraged to have meals which include high-fibre, low carbohydrate diets such as fruit, vegetable, wholegrains and pluses; include low-fat dairy products and oily fish. For those with diabetes who are overweight, larger degrees of weight loss in the longer term will have advantageous metabolic impact (NICE NG 28).

European Society of Cardiology (ESC) guideline in 2016 on CVD prevention have suggested risk factors goal and target levels for important CVD risk factors as below (Fig. 23.4).

23.10 Conclusion

Cardiovascular disease is a significant issue for men. As men age, cardiovascular health becomes a higher priority. Cardiovascular disease remains one of the leading major health problems and causes of death in men.

Smoking	No exposure to tobacco in any form.
Diet	Low in saturated fat with a focus on wholegrain products, vegetables, fruit and fish.
Physical activity	At least 150 minutes a week of moderate aerobic physical activities (30 minutes for 5 days/week) or 75 minutes a week of vigorous aerobic physical activities(15 minutes for 5 days/week) or a combination.
Body weight	BMI 20-25 kg/m2. Waist circumference <94cm (men) or <80cm (women).
Blood pressure	<140/90 mmHg
Lipids LDL is the primary target	Very high risk:<1.8 mmol/L (<70 mg/dL), or a reduction of at least 50% if the baseline is between 1.8 and 3.5 mmol/L (70 and 135 mg/dL) High risk :<2.6 mmol/L (<100mg/dL) or a reduction of at least 50% if the baseline is between 2.6 and 5.1 mmol/L. Low to moderate risk: <3.0 mmol/L (<115 mg/dL)
Triglycerides	No target but <1.7 mmol/L (<150,g/dL) indicates lower risk and higher levels indicate a need to look for other risk factors.
Diabetes	HbA1c <7% (<53 mmol/mol)

Fig. 23.4 ESC guideline on cardiovascular disease prevention

References

2016 European Guidelines on cardiovascular disease prevention in clinical practice.

2018 ESC guidelines for Management of HTN.

2019 ESC guidelines for the Management of Dyslipidaemia.

Ali MK, Jaacks LM, Kowalski AJ, et al. Noncommunicable diseases: three decades of global data show a mixture of increases and decreases in mortality rates. 2015.

Berrington de Gonzalez A, Hartge P, Cerhan JR, Flint AJ, Hannan L, MacInnis RJ, Moore SC, Tobias GS, Anton-Culver H, Freeman LB, Beeson WL, Clipp SL, English DR, Folsom AR, Freedman DM, Giles G, Hakansson N, Henderson KD, Hoffman-Bolton J, Hoppin JA, Koenig KL, Lee IM, Linet MS, Park Y, Pocobelli G, Schatzkin A, Sesso HD, Weiderpass E, Willcox BJ, Wolk A, Zeleniuch-Jacquotte A, Willett WC, Thun MJ. Body-mass index and mortality among 1.46 million white adults. N Engl J Med. 2010 Dec 2;363(23):2211–9. https://doi.org/10.1056/NEJMoa1000367. Erratum in: N Engl J Med. 2011;365(9):869. PMID: 21121834; PMCID: PMC3066051.

Cardiovascular disease NHS.

Cardiovascular disease risk assessment and prevention NICE.

Cooney MT, Dudina A, Whincup P, Capewell S, Menotti A, Jousilahti P, Njolstad I, Oganov R, Thomsen T, Tverdal A, Wedel H, Wilhelmsen L, Graham I. Reevaluating the Rose approach: comparative benefits of the population and highrisk preventive strategies. Eur J Cardiovasc Prev Rehabil. 2009.

Coronary heart disease NHS, Stroke TIA NHS, Peripheral arterial disease NHS.

Duncan MS, Freiburg MS, Greevy RA Jr, et al. Association of Smoking Cessation with Subsequent Risk of Cardiovascular Disease. [Published online August 20, 2019]. JAMA. 2014. https://doi.org/10.1001/jama.2019.10298.

Forouzanfar MH, Liu P, Roth GA, Ng M, Biryukov S, Marczak L, Alexander L, Estep K, Hassen Abate K, Akinyemiju TF, Ali R, Alvis-Guzman N, Azzopardi P, Banerjee A, Barnighausen T,

Basu A, Bekele T, Bennett DA, Biadgilign S. Global burden of hypertension and systolic blood pressure of at least 110 to 115 mm Hg, 1990-2015. JAMA. 2017.

George J, Rapsomaniki E, Pujades-Rodtriguez M, et al. How does cardiovascular disease first present in women and men? Incidence of 12 cardiovascular diseases in a contemporary cohort of 1,937,36 people. Circulation. 2015;132:1320–8.

Leening MJ, Ferket BS, Steyerberg EW, et al. Sex differences in lifetime risk and first manifestation of cardiovascular disease: prospective population-based cohort study. BMJ. 2014.

Lewington S, Clarke R, Qizilbash N, Peto R, Collins R. Age-specific relevance of usual blood pressure to vascular mortality: a meta-analysis of individual data for one million adults in 61 prospective studies. Lancet. 2002.

Liu K, Daviglus ML, Loria CM, Colangelo LA, Spring B, Moller AC, Lloyd-Jones DM. Healthy lifestyle through young adulthood and the presence of low cardiovascular disease risk profile in middle age: The Coronary Artery Risk Development in (Young) Adults (CARDIA) study. 2012.

Lozano R, et al. Global and regional mortality from 235 causes of death for 20 age groups in 1990 and 2010: a systematic analysis for the Global Burden of Disease Study 2010. Lancet. 2012;380(9859):2095–128.

Mackay J, Mensah G, editors. The atlas of heart disease and stroke. Geneva: World Health Organization; 2004.

Miner M, et al. Evaluation of erectile dysfunction without established cardiovascular disease. 2014.

Mozaffarian D, Benjanin EJ, Go AS, et al. Heart disease and stroke statistics – 2015 update: a report from the American Heart Association. 2015.

NICE. Cardiovascular disease: risk assessment and reduction, including lipid modification. Clinical guideline [CG181]. 2014. Last updated: 27 September 2016.

NICE guideline for hypertension.

NICE obesity. Identification, assessment and management. Clinical guideline. 2014.

NICE pathway on lifestyle advice on diet and physical activity, NHS Eat well Web page.

NICE pathway type 2 DM in adults – patient education and lifestyle advice.

UK alcohol unit guidance: CMOs' low risk drinking guidelines.

Vlachopoulos CV, Terentes-Printzios DG, Ioakeimidis NK, Aznaouridis KA, Stefanadis CI. Prediction of cardiovascular events and all-cause mortality with erectile dysfunction: a systematic review and meta-analysis of cohort studies. Circ Cardiovasc Qual Outcomes. 2013;6:99–109.

Printed in the United States
by Baker & Taylor Publisher Services